Macroeconometrics and Time Series Analysis

The New Palgrave Economics Collection
Editors: Steven N. Durlauf, University of Wisconsin-Madison, USA & Lawrence E. Blume, Cornell University, USA

Also in the series: Behavioural and Experimetal Economics
Economic Growth
Game Theory
Microeconometrics
Monetary Economics

Series Standing Order ISBN 978-0-230-24014-8 hardcover
Series Standing Order ISBN 978-0-230-24013-1 paperback

To receive future titles in this series as they are published quote one of the ISBNs listed above to set up a standing order: contact your bookseller; write to Customer Services Department, Macmillan Distribution Ltd, Houndmills, Basingstoke, Hampshire, RG21 6XS; or email orders@palgrave.com.

Macroeconometrics and Time Series Analysis

Edited by
Steven N. Durlauf
University of Wisconsin-Madison, USA

Lawrence E. Blume
Cornell University, USA

palgrave
macmillan

All articles first published in *The New Palgrave Dictionary of Economics*, 2nd Edition
Edited by Steven N. Durlauf and Lawrence E. Blume
in eight volumes, 2008

First published 2010 by
PALGRAVE MACMILLAN

Palgrave Macmillan in the UK is an imprint of Macmillan Publishers Limited, registered in England, company number 785998, of Houndmills, Basingstoke, Hampshire RG21 6XS.

Palgrave Macmillan in the US is a division of St Martin's Press LLC, 175 Fifth Avenue, New York, NY 10010.

Palgrave Macmillan is the global academic imprint of the above companies and has companies and representatives throughout the world.

ISBN 978-0-230-23884-8 hardback
ISBN 978-0-230-23885-5 paperback

This book is printed on paper suitable for recycling and made from fully managed and sustained forest sources. Logging, pulping and manufacturing processes are expected to conform to the environmental regulations of the country of origin.

A catalogue record for this book is available from the British Library.

A catalog record for this book is available from the Library of Congress.

10 9 8 7 6 5 4 3 2 1
19 18 17 16 15 14 13 12 11 10

Printed and bound in the United States of America

Contents

List of Contributors

LAURENT E. CALVET
HEC School of Management, France

TIMOTHY COGLEY
New York University, USA

FRANCIS X. DIEBOLD
University of Pennsylvania, USA

JESÚS FERNÁNDEZ-VILLAVERDE
University of Pennsylvania, USA

A. RONALD GALLANT
Duke University, USA

CLIVE W. J. GRANGER
University of California San Diego, USA

JAMES D. HAMILTON
University of California San Diego, USA

LARS PETER HANSEN
University of Chicago, USA

ANDREW HARVEY
University of Cambridge, UK

DAVID F. HENDRY
University of Oxford, UK

YONGMIAO HONG
Cornell University, USA

SVEND HYLLEBERG
Aarhus University, Denmark

LUTZ KILIAN
University of Michigan, USA

ANDROS KOURTELLOS
University of Cyprus, Cyprus

G. M. KUERSTEINER
University of California Davis, USA

KURT F. LEWIS
University of Iowa, USA

OLIVER B. LINTON
London School of Economics, UK

HELMUT LÜTKEPOHL
European University Institute, Italy

BRUCE MIZRACH
Rutgers University, USA

CHARLES R. NELSON
University of Washington, USA

MARC NERLOVE
University of Maryland, USA

PIERRE PERRON
Boston University, USA

PETER C. B. PHILLIPS
Yale University, USA

WERNER PLOBERGER
Washington University in St Louis, USA

DALE J. POIRIER
University of California Irvine, USA

JAMES B. RAMSEY
New York University, USA

P. M. ROBINSON
London School of Economics, UK

JUAN F. RUBIO-RAMÍREZ
Duke University, USA

THOMAS J. SARGENT
New York University, USA

FRANK SCHORFHEIDE
University of Pennsylvania, USA

NEIL SHEPHARD
University of Oxford, UK

CHRISTOPHER A. SIMS
Princeton University, USA

MARK F. J. STEEL
University of Warwick, UK

THANASIS STENGOS
University of Guelph, Canada

THOMAS M. STOKER
Massachusetts Institute of Technology, USA

TIMO TERÄSVIRTA
Aarhus University, Denmark

TIMOTHY J. VOGELSANG
Michigan State University, USA

MARK W. WATSON
Princeton University, USA

KENNETH D. WEST
University of Wisconsin-Madison, USA

CHARLES H. WHITEMAN
University of Iowa, USA

TAO ZHA
Federal Reserve Bank of Atlanta, USA

General Preface

All economists of a certain age remember the "little green books". Many own a few. These are the offspring of *The New Palgrave: A Dictionary of Economics*; collections of reprints from *The New Palgrave* that were meant to deliver at least a sense of the *Dictionary* into the hands of those for whom access to the entire four volume, four million word set was inconvenient or difficult. *The New Palgrave Dictionary of Economics, Second Edition* largely resolves the accessibility problem through its online presence. But while the online search facility provides convenient access to specific topics in the now eight volume, six million word *Dictionary of Economics*, no interface has yet been devised that makes browsing from a large online source a pleasurable activity for a rainy afternoon. To our delight, *The New Palgrave*'s publisher shares our view of the joys of dictionary-surfing, and we are thus pleased to present a new series, the "little blue books", to make some part of the *Dictionary* accessible in the hand or lap for teachers, students, and those who want to browse. While the volumes in this series contain only articles that appeared in the 2008 print edition, readers can, of course, refer to the online *Dictionary* and its expanding list of entries.

The selections in these volumes were chosen with several desiderata in mind: to touch on important problems, to emphasize material that may be of more general interest to economics beginners and yet still touch on the analytical core of modern economics, and to balance important theoretical concerns with key empirical debates. The 1987 Eatwell, Milgate and Newman *The New Palgrave: A Dictionary of Economics* was chiefly concerned with economic theory, both the history of its evolution and its contemporary state. The second edition has taken a different approach. While much progress has been made across the board in the 21 years between the first and second editions, it is particularly the flowering of empirical economics which distinguishes the present interval from the 61 year interval between Henry Higgs' *Palgrave's Dictionary of Political Economy* and *The New Palgrave*. It is fair to say that, in the long run, doctrine evolves more slowly than the database of facts, and so some of the selections in these volumes will age more quickly than others. This problem will be solved in the online *Dictionary* through an ongoing process of revisions and updates. While no such solution is available for these volumes, we have tried to choose topics which will give these books utility for some time to come.

Steven N. Durlauf
Lawrence E. Blume

Introduction

This collection of entries covers one of the most important changes in economic methodology between the 1987 and 2008 editions of the *New Palgrave*, namely the nature and role of time series analysis. This explosion has two sources. First, the 1987 edition did not reflect the status of rational expectations as a central feature of modern macroeconomic analysis. As a result, the edition had little on the work of Lars Hansen, Thomas Sargent, and Christopher Sims which has come to dominate current empirical macroeconomics. In contrast, the 2008 edition has much material on topics ranging from the cross-equations restrictions generated by rational expectations, generalized methods of moments approaches to estimation which reflect the interpretation of data interrelationships as first order conditions, and vector autoregression methods that summarize the dymamic properties of data.

Second, time series econometrics experienced a quantum leap in technical sophistication, as exemplified in the work by Robert Engle, John Geweke, Clive Granger, Peter Philips and Peter Robinson. This new work represents a substantial relaxation of the statistical assumptions that had previously been imposed on economic data, as exemplified in the work on ARCH models, unit roots and cointegration and long memory. These new approaches not only have examined the domain of empirical processes which may be subjected to formal analysis but have extended the sorts of questions macroeconomists ask. Examples of this include cointegration, which provides a way of formalizing long run restrictions on comovements across data series that may or may not be consistent with economic theory and unit roots, which changes the way that trend/cycle distinctions are formulated. So, while these entries are among the most mathematically sophisticated of those that appear in the new edition, they are of great relevance to empirical work.

Steven N. Durlauf
Lawrence E. Blume

aggregation (econometrics)

Aggregation refers to the connection between economic interactions at the micro and the macro levels. The micro level refers to the behaviour of individual economic agents. The macro level refers to the relationships that exist between economy-wide totals, averages or other economic aggregates. For instance, in a study of savings behaviour refers to the process that an individual or household uses to decide how much to save out of current income, whereas the aggregates are total or per-capita savings and income for a national economy or other large group. The econometrics of aggregation refers to modelling with the individual–aggregate connection in mind, creating a framework where information on individual behaviour together with co-movements of aggregates can be used to estimate a consistent econometric model.

In economic applications one encounters many types and levels of aggregation: across goods, across individuals within households, and so on. We focus on micro to macro as outlined above, and our 'individual' will be a single individual or a household, depending on the context. We hope that this ambiguity does not cause confusion.

At a fundamental level, aggregation is about handling detail. No matter what the topic, the microeconomic level involves purposeful individuals who are dramatically different from one another in terms of their needs and opportunities. Aggregation is about how all this detail distils in relationships among economic aggregates. Understanding economic aggregates is essential for understanding economic policy. There is just too much individual detail to conceive of tuning policies to the idiosyncrasies of many individuals.

This detail is referred to as individual heterogeneity, and it is pervasive. This is a fact of empirical evidence and has strong econometric implications. If you ignore or neglect individual heterogeneity, then you can't get an interpretable relationship between economic aggregates. Aggregates reflect a smear of individual responses and shifts in the composition of individuals in the population; without careful attention, the smear is unpredictable and uninterpretable.

Suppose that you observe an increase in aggregate savings, together with an increase in aggregate income and in interest rates. Is the savings increase primarily arising from wealthy people or from those with moderate income? Is the impact of interest rates different between the wealthy and others? Is the response different for the elderly than for the young? Has future income for most people become more risky?

How could we answer these questions? The change in aggregate savings is a mixture of the responses of all the individuals in the population. Can we disentangle it to understand the change at a lower level of detail, like rich versus poor, or young versus old? Can we count on the mixture of responses underlying aggregate savings to be stable? These are questions addressed by aggregation.

Recent progress on aggregation and econometrics has centred on explicit models of individual heterogeneity. It is useful to think of heterogeneity as arising from three broad categories of differences. First, individuals differ in tastes and incomes. Second, individuals differ in the extent to which they participate in markets. Third, individuals differ in the situations of wealth and income risk that they encounter depending on the market environment that exists. Our discussion of recent solutions is organized around these three categories of heterogeneity. For deeper study and detailed citations, see the surveys by Blundell and Stoker (2005), Stoker (1993) and Browning, Hansen and Heckman (1999).

The classical aggregation problem provides a useful backdrop for understanding current solutions. We now review its basic features, as originally established by Gorman (1953) and Theil (1954). Suppose we are studying the consumption of some product by households in a large population over a given time period t. Suppose that the quantity purchased q_{it} is determined by household resources m_{it}, or 'income' for short, as in the formula:

$$q_{it} = \alpha_i + \beta_i m_{it}$$

Here α_i represents a base level consumption, and β_i represents household i's marginal propensity to spend on the product.

For aggregation, we are interested in what, if any, relationship there is between average quantity and average income:

$$q_t = \frac{1}{n_t}\sum_{i=1}^{n_t} q_{it} \quad \text{and} \quad m_t = \frac{1}{n_t}\sum_{i=1}^{n_t} m_{it}$$

where all households have been listed as $i = 1, \ldots, n_t$. Let's focus on one version of this issue, namely, what happens if some new income becomes available to households, either through economic growth or a policy. How will the change in average quantity purchased $\Delta\bar{q}$ be related to the change in average income $\Delta\bar{m}$?

Suppose that household i gets Δm_i in new income. Their change in quantity purchased is the difference between purchases at income $m_{it} + \Delta m_i$ and at income m_{it}, or

$$\Delta q_i = \beta_i \cdot \Delta m_i$$

Now, the average quantity change is $\Delta\bar{q} = \sum_i \Delta q_i / n_t$, so that

$$\Delta\bar{q} = \frac{1}{n_t}\sum_{i=1}^{n_t} \beta_i \cdot \Delta m_i \tag{1}$$

In general, it seems we need to know a lot about who gets the added income – which i's get large values of Δm_i and which i's get small values of Δm_i. With a transfer policy, any group of households could be targeted for the new income, and their specific set of values of β_i would determine $\Delta\bar{q}$. A full schedule of how much new income goes to each household i as well as how they spend it (that is, Δm_i and β_i), seems like a lot of detail to keep track of, especially if the population is large. Can we ever get by knowing just the change in average income $\Delta\bar{m} = \sum_i \Delta m_i / n_t$?

There are two situations where we can, where a full schedule is not needed:

1. Each household spends in exactly the same way, namely, $\beta_i = \beta$ for all i, so that who gets the new income doesn't affect $\Delta\bar{q}$.
2. The distribution of income transfers is restricted in a convenient way.

Situation 1 is (common) micro linearity, which is termed *exact aggregation*. Another way to understand the structure is to write (1) in the covariance formulation:

$$\Delta\bar{q} = \bar{\beta} \cdot \Delta\bar{m} + \frac{1}{n_t}\sum_{i=1}^{n_t}(\beta_i - \bar{\beta}) \cdot (\Delta m_i - \Delta\bar{m}) \tag{2}$$

where we denote the average spending propensity as $\bar{\beta} = \sum_i \beta_i / n_t$. With exact aggregation there is no variation in β_i, so that $\beta_i = \beta = \bar{\beta}$ and the latter term always vanishes. That is, it doesn't matter who gets the added income because everyone spends the same way. When there is variation in β_i, matters are more complicated unless it can be assured that the new income were always given to households in a way that is uncorrelated with the propensities β_i. 'Uncorrelated transfers' provide an example of a Situation 2, but that is a distribution restriction that is hard to verify with empirical data.

Under uncorrelated transfers, we can also interpret the relationship between $\Delta\bar{q}$ and $\Delta\bar{m}$, that is, the macro propensity is the average propensity $\bar{\beta}$. There are other distributional restrictions that give a constant macro propensity, but a different one from the parameter produced by uncorrelatedness. For instance, suppose that transfers of new income always involved fixed shares of the total amount. That is, household i gets

$$\Delta m_i = s_i \Delta\bar{m} \tag{3}$$

In this case, average purchases are

$$\Delta\bar{q} = \frac{1}{n_t}\sum_{i=1}^{n_t}\beta_i \cdot (s_i \Delta\bar{m}) = \tilde{\beta}_{wtd} \cdot \Delta\bar{m} \tag{4}$$

where $\tilde{\beta}_{wtd}$ is the weighted average $\tilde{\beta}_{wtd} \equiv \sum_i \beta_i s_i / n_t$. This is a simple aggregate relationship, but the coefficient $\tilde{\beta}_{wtd}$ applies only for the distributional scheme (3); it matters who gets what share of the added income. Aside from being a weighted average of $\{\beta_i\}$, there is no reason for $\tilde{\beta}_{wtd}$ to be easily interpretable – for instance, if households with low β_i's have high s_i's, then $\tilde{\beta}_{wtd}$ will be low. If your aim was to estimate the average propensity $\bar{\beta}$, there is no reason to believe that the bias $\tilde{\beta}_{wtd} - \bar{\beta}$ will be small.

Empirical models that take aggregation into account apply structure to individual responses and to allowable distributional shifts. Large populations are modelled, so that compositional changes are represented via probability distributions, and expectations are used instead of averages (for example, mean quantity $E_t(q)$ is modelled instead of the sample average q_t). Individual heterogeneity is the catch-all term for individual differences, and they must be characterized. Distribution restrictions must be applied where heterogeneity is important. For instance, in our example structure on the distribution of new income is required for dealing with the heterogeneity in β_i, but not for the heterogeneity in α_i.

Progress in empirical modelling has come about because of the enhanced availability of micro data over time. The forms of behavioural models in different research areas have been tightly characterized, which is necessary for understanding how to account for aggregation. That is, when individual heterogeneity is characterized empirically, the way is clear to understanding what distributional influences are relevant and must be taken into account. We discuss recent examples of this below.

Some solutions to aggregation problems

Demand models and exact aggregation

It is well known that demand patterns of individual households vary substantially with whether households are rich or poor, and vary with many observable demographic characteristics, such as household (family) size, age of head and ages of children, and so on. As surveyed in Blundell (1988), traditional household demand models relate household commodity expenditures to price levels, total household budget (income) and observable household characteristics. Aggregate demand models relate (economy-wide) aggregate commodity expenditures to price levels and the distribution of income and characteristics in the population. Demand models illustrate exact aggregation, a practical approach for accommodating heterogeneity at the micro and macro levels. These models assume that demand parameter values are the same for all individuals, but explicitly account for observed differences in tastes and income.

For instance, suppose we are studying the demand for food and we are concerned with the difference in demands for households of small size versus large size. We model food purchases for household i as part of static allocation of the budget m_{it} to $j=1,\ldots,J$ expenditure categories, where food is given by $j=1$, and price levels at time t are given by $P_t=(p_{1t},\ldots,p_{Jt})$. Small families are indicated by $z_{it}=0$ and large families by $z_{it}=1$.

Expenditure patterns are typically best fit in budget share form. For instance, a translog model of the food share takes the form

$$w_{1it}=\frac{p_{1t}q_{1it}}{m_{it}}=\frac{1}{D(p_t)}\left[\alpha_1+\sum_{i=1}^{J}\beta_{1j}\ln p_{jt}+\beta_m\ln m_{it}+\beta_z z_{it}\right] \tag{5}$$

where $D(p_t)=1+\sum_{i=1}^{J}\beta_j\ln p_{jt}$. The parameters ($\alpha_1$ and all β's) are the same across households, and the price levels (p_{jt}'s) are the same for all households but vary with t. Individual heterogeneity is represented by the budget m_{it} and the family size indicator z_{it}. We have omitted an additive disturbance for simplicity, which would represent another source of heterogeneity. The important thing for aggregation is that model (5) is intrinsically linear in the individual heterogeneity. That is, we can write

$$w_{1it} = b_1(p_t) + b_m(p_t) \cdot \ln m_{it} + b_z(p_t) \cdot z_{it} \tag{6}$$

The aggregate share of food in the population is the mean of food expenditures divided by mean budget, or

$$W_{1t} = \frac{E_t(m_{it}\, w_{1it})}{E_t(m_{it})} = b_1(p_t) + b_m(p_t) \cdot \frac{E_t(m_{it}\ \ln\ m_{it})}{E_t(m_{it})} + b_z(p_t) \cdot \frac{E_t(m_{it}z_{it})}{E_t(m_{it})} \tag{7}$$

The aggregate share depends on prices, the parameters (α_1 and all β's) and two statistics of the joint distribution of m_{it} and z_{it}. The first,

$$S_{mt} = \frac{E_t(m_{it} \ln m_{it})}{E_t(m_{it})} \tag{8}$$

is an entropy term that captures the size distribution of budgets, and the second

$$S_{zt} = \frac{E_t(m_{it}z_{it})}{E_t(m_{it})} \tag{9}$$

is the percentage of total expenditure accounted for by households with $z_{it} = 1$, that is, large families.

The expressions (6) and (7) illustrate *exact aggregation* models. Heterogeneity in tastes and budgets (incomes) are represented in an intrinsically linear way. For aggregate demand, all one needs to know about the joint distribution of budgets m_{ti} and household types z_{it} is a few statistics; here S_{mt} and S_{zt}.

The obvious similarity between the individual model (6) and the aggregate model (7) raises a further question. How much bias is introduced by just fitting the individual model with aggregate data, that is, putting $E_t(m_{it})$ and $E_t(z_{it})$ in place of m_{it} and z_{it}, respectively? This can be judged by the use of *aggregation factors*. Define the factors π_{mt} and π_{zt} as

$$\pi_{mt} = \frac{S_{mt}}{\ln E_t(m_{it})} \quad \text{and} \quad \pi_{zt} = \frac{S_{zt}}{E_t(z_{it})}$$

so that the aggregate share is

$$W_{1t} = \frac{E_t(m_{it}\, w_{1it})}{E_t(m_{it})} = b_1(p_t) + b_m(p_t) \cdot \pi_{mt} \cdot \ln E_t(m_{it}) + b_z(p_t) \cdot \pi_{zt} \cdot E_t(z_{it})$$

One can learn about the nature of aggregation bias by studying the factors π_{mt} and π_{zt}. If they are both roughly equal to 1 over time, then no bias would be introduced by fitting the individual model with aggregate data. If they are roughly constant but not equal to 1, then constant biases are introduced. If the factors are time varying, more complicated bias would result. In this way, with exact aggregation models, aggregation factors can depict the extent of aggregation bias.

The current state of the art in demand analysis uses models in exact aggregation form. The income (budget) structure of shares is adequately represented as quadratic in ln m_{it}, as long as many demographic differences are included in the analysis. This means that aggregate demand depends explicitly on many statistics of the income-demographic distribution, and it is possible to gauge the nature and sources of aggregation bias using factors as we have outlined. See Banks, Blundell and Lewbel (1997) for an example of demand modelling of British expenditure data, including the computation of various aggregation factors.

Exact aggregation modelling arises naturally in situations where linear models have been found to provide adequate explanations of empirical data patterns. This is not always the case, as many applications require models that are intrinsically nonlinear. We now discuss an example of this kind where economic decisions are discrete.

Market participation and wages

Market participation is often a discrete decision. Labourers decide whether to work or not, firms decide whether to enter a market or exit a market. There is no 'partial' participation in many circumstances, and changes are along the extensive margin. This raises a number of interesting issues for aggregation.

We discuss these issues using a simple model of labour participation and wages. We consider two basic questions. First, how is the fraction of working (participating) individuals affected by the distribution of factors that determine whether each individual chooses to work? Second, what is the structure of average wages, given that wages are observed only for individuals who choose to work? The latter question is of interest for interpreting wage movements: if average wages go up, is that because (*a*) most individual wages went up or (*b*) low-wage individuals become unemployed, or leave work? These two reasons give rise to quite different views of the change in economic welfare associated with an increase in average wages.

The standard empirical model for individual wages expresses log wage as a linear function of time effects, schooling and demographic (cohort) effects. Here we begin with

$$\ln w_{it} = r(t) + \beta \cdot S_{it} + \varepsilon_{it} \tag{10}$$

where $r(t)$ represents a linear trend or other time effects, S_{it} is the level of training or schooling attained by individual i at time t, and ε_{it} are all other idiosyncratic factors. This setting is consistent with a simple skill price model, where $w_{ti} = R_t H_{it}$ with skill price $R_t = e^{r(t)}$ and skill (human capital) level $H_{it} = e^{\beta S_{it} + \varepsilon_{it}}$. We take eq. (10) to apply to all individuals, with the wage representing the available or offered wage, and β the return to schooling. However, we observe that wage only for individuals who choose to work.

We assume that individuals decide whether to work by first forming a reservation wage

$$\ln w_{it}^{*} = s^{*}(t) + \alpha \ln B_{it} + \beta^{*} \cdot S_{it} + \zeta_{it}$$

where $s(t)$ represents time effects, B_{it} is the income or benefits available when individual i is out of work at time t, S_{it} is schooling as before, and ζ_{it} are all other individual factors. Individual i will work at time t if their offered wage is as big as their reservation wage, or $w_{it} \geq w_{it}^{*}$. We denote this by the participation indicator I_{it}, where $I_{it}=1$ if i works and $I_{it}=0$ if i doesn't work. This model of participation can be summarized as

$$\begin{aligned} I_{it} &= 1\left[w_{it} \geq w_{it}^{*}\right] = 1[\ln w_{it} - \ln w_{it}^{*} \geq 0] \\ &= 1\left[s(t) - \alpha \ln B_{it} + \gamma \cdot S_{it} + v_{it} \geq 0\right] \end{aligned} \tag{11}$$

where $s(t) \equiv r(t) - s^{*}(t)$, $\gamma \equiv \beta - \beta^{*}$ and $v_{It} \equiv \varepsilon_{it} - \zeta_{it}$.

If the idiosyncratic terms ε_{it}, v_{it} are stochastic errors with zero means (conditional on B_{it},S_{it}) and constant variances, then (10) and (11) is a standard selection model. That is, if we observe a sample of wages from working individuals, they will follow (10) subject to the proviso that $I_{it}=1$. This can be accommodated in estimation by assuming that ε_{it}, v_{it} have a joint normal distribution. That implies that the log wage regression of the form (10) can be corrected by adding a standard selection term as

$$\ln w_{it} = r(t) + \beta \cdot S_{it} + \frac{\sigma_{\varepsilon v}}{\sigma_{v}} \lambda \left[\frac{s(t) - \alpha \ln B_{it} + \gamma S_{it}}{\sigma_{v}}\right] + \eta_{t}. \tag{12}$$

Here, σ_v is the standard deviation of v and $\sigma_{\varepsilon v}$ is the covariance between ε and v. $\lambda(\cdot)=\phi(\cdot)/\Phi(\cdot)$ is the 'Mills ratio', where ϕ and Φ are the standard normal p.d.f. and c.d.f respectively. This equation is properly specified for a sample of working individuals – that is, we have $E(\eta_t|S_{it}, B_{it}, I_{it}=1)=0$. For a given levels of benefits and schooling, eq. (11) gives the probability of participating in work as

$$E_t[I_{it}|B_{it}, S_{it}] = \Phi\left[\frac{s(t) - \alpha \ln B_{it} + \gamma \cdot S_{it}}{\sigma_{v}}\right] \tag{13}$$

where $\Phi[\cdot]$ is the normal c.d.f.

For studying average wages, the working population is all individuals with $I_{it}=1$. The fraction of workers participating is therefore the (unconditional) probability that $\alpha \ln B_{it} - \gamma \cdot S_{it} - v_{it} \leq s(t)$. This probability is the expectation of I_{it} in (11), an intrinsically nonlinear function in observed heterogeneity B_{it} and S_{it} and unobserved heterogeneity v_{it}, so we need some explicit distribution assumptions. In particular, assume that the participation index $\alpha \ln B_{it} - \gamma \cdot S_{it} - v_{it}$ is normally distributed with mean $\mu_t = \alpha E_t(\ln B_{it}) - \gamma E_t(S_{it})$ and variance

$$\sigma_t^2 = \alpha^2 Var_t(\ln B_{it}) + \beta^2 Var_t(S_{it}) - 2a\beta \cdot Cov_t(\ln B_{it}, S_{it}) + \sigma_v^2. \tag{14}$$

Now we can derive the labour participation rate (or one minus the unemployment rate) as

$$E_t[I_{it}] = \Phi\left[\frac{s(t) - \alpha E_t(\ln B_{it}) + \gamma E_t(S_{it})}{\sigma_t}\right] \tag{15}$$

where again $\Phi[\cdot]$ is the normal c.d.f. This formula relates the participation rate to average out-of-work benefits $E_t(\ln B_{it})$ and average training $E_t(S_{it})$, as well as their variances and covariances through σ_t. The specific relation depends on the distributional assumption adopted; (15) relies on normality of the participation index in the population.

For wages, a similar analysis applies. Log wages are a linear function (10) applicable to the full population. However, for participating individuals, the intrinsically nonlinear selection term is introduced, so that we need explicit distributional assumptions. Now suppose that log wage $\ln w_{it}$ and the participation index $\alpha \ln B_{it} - \gamma \cdot S_{it} - v_{it}$ are joint normally distribution. It is not hard to derive the expression for average log wages of working individuals

$$\begin{aligned} E_t[\ln w_{it}|I_{it} = 1] &= r(t) + \beta \cdot E_t(S_{it}|I = 1) \\ &\quad + \frac{\sigma_{\varepsilon v}}{\sigma_t}\lambda\left[\frac{s(t) - \alpha E_t(\ln B_{it}) + \gamma E_t(S_{it})}{\sigma_t}\right] \cdot \end{aligned} \tag{16}$$

This is an interesting expression, which relates average log wage to average training of the workers as well as to the factors that determine participation.

However, we are not interested in average log wages, but rather average wages $E_t(w_{it})$. The normality structure we have assumed is enough to derive a formulation of average wages, although it is a little complex to reproduce in full here. In brief, Blundell, Reed and Stoker (2003) show that the average wages of working individuals $E\ [w_{it}|I_{it} = 1]$ can be written as

$$\ln\ E[w_{it}|I_{it} = 1] = r(t) + \beta \cdot E_t(S_{it}) + \Omega_t + \Psi_t \tag{17}$$

where Ω_t, Ψ_t are correction terms that arise as follows. Ω_t corrects for the difference between the log of an average and the average of a log, as

$$\Omega_t \equiv \ln E_t(w_{it}) - E_t(\ln w_{it}) + \Omega_t.$$

Ψ_t corrects for participation, as

$$\Psi_t \equiv \ln E[w_{it}|I_{it} = 1] - \ln E_t\ (w_{it}).$$

Recall our original question, about whether an increase in average wages is due to an increase in individual wages or to increased unemployment of low-wage workers. That is captured in (17). That is, Ψ_t gives the participation effect, and the other terms capture changes in average wage $E_t(w_{it})$ when all are participating. As such, this analysis provides a vehicle for separating overall wage growth from compositional effects due to participation.

Blundell, Reed and Stoker (2003) analyse British employment using a framework similar to this, but also allowing for heterogeneity in hours worked. Using out-of-work benefits as an instrument for participation, they find that over 40 per cent of observed aggregate wage growth from 1978 to 1996 arises from selection and other compositional effects.

We have now discussed aggregation and heterogeneity with regard to tastes and incomes, and market participation. We now turn to heterogeneity with regard to risks and market environments.

Consumption and risk environments

Consumption and savings decisions are clearly affected by preference heterogeneity, as we discussed earlier. The present spending needs of a large family clearly differ from those of a small family or a single individual, the needs of teenage children differ from those of preschoolers, the needs of young adults differ from those of retirees, and on and on. These aspects are very important, and need to be addressed as they were in demand models above. Browning and Lusardi (1996) survey the extensive evidence on heterogeneity in consumption, and Attanasio (1999) is an excellent comprehensive survey of work on consumption.

We use consumption and savings to illustrate another type of heterogeneity, namely, that of wealth and income risks. That is, with forward planning under uncertainty, the risk environment of individuals or households becomes relevant. There can be individual shocks to income, such as a work layoff or a health problem, or aggregate shocks, such as an extended recession or stock market boom. Each of these shocks can differ in its duration – a temporary layoff can be usefully viewed as transitory, whereas a debilitating injury may affect income for many years. In planning consumption, it is important to understand the role of income risks and wealth risks. When there is no precautionary planning, such as when consumers have quadratic preferences, income risks do not become intertwined with other heterogeneous elements. However, when there is risk aversion, then the precise situation of individual income risks and insurance markets is relevant.

A commonly used model for income is to assume multiplicative permanent and transitory components, with aggregate and individual shocks, as in

$$\Delta \ln y_{it} = (\eta_t + \Delta u_t) + (\varepsilon_{it} + \Delta v_{it}).$$

Here $\eta_t + \Delta u_t$ is the common aggregate shock, with η_t a permanent component and Δu_t transitory. The idiosyncratic shock is $\varepsilon_{it} + \Delta v_{it}$, where ε_{it} is permanent and Δv_{it} transitory.

For studying individual level consumption with precautionary planning, it is standard practice to assume constant relative risk aversion (CRRA) preferences and assume that the interest rate r_t is small. This, together with the income process above, gives a log-linear approximation to individual consumption growth

$$\Delta \ln c_{it} = \rho r_t + (\beta + \varphi r_t)' z_{it} + k_1 \sigma_{At} + k_2 \sigma_{it} + \kappa_1 \eta_t + \kappa_2 \varepsilon_{it}. \tag{18}$$

Here, z_{it} reflects heterogeneity in preferences, such as differences in demographic characteristics. σ_{At} is the variance of aggregate risk and σ_{it} s the variance of idiosyncratic risk (with each conditional on what is known at time $t-1$), so that these terms reflect precautionary planning. Finally, η_t and ε_{it} arise because of adjustments that are made as permanent shocks are revealed. At time $t-1$ these shocks are not possible to forecast, but then they are incorporated in the consumption plan once they are revealed. In terms of the level of consumption c_{it}, eq. (18) is written as

$$c_{it} = exp(\ln c_{it-1} + \rho r_t + (\beta + \varphi r_t)' z_{it} + k_1 \sigma_{At} + k_2 \sigma_{it} + \kappa_1 \eta_t + \kappa_2 \varepsilon_{it}).$$

This is an intrinsically nonlinear model in the following heterogeneous elements: $\ln c_{it-1}$, z_{it}, σ_{it} and ε_{it}. For aggregation, it seems we would need a great deal of distributional structure.

Here is where we can see the role of the risk environment, or markets for insurance for income risks. That is, if there were complete markets with insurance for all risks, then all risk terms vanish from consumption growth. When complete insurance exists for idiosyncratic risks only, then the idiosyncratic terms σ_{it} and ε_{it} vanish from consumption growth, since less precautionary saving is needed. Otherwise, the idiosyncratic risk terms σ_{it} and ε_{it} represent heterogeneity that must be accommodated just like preference differences (and in other settings, participation differences).

In the realistic situation where risks are not perfectly insurable, we require distributional assumptions in order to formulate aggregate consumption. For instance, suppose that we assume that $(\ln c_{it-1}, (\beta + \varphi r_t)' z_{it}, \varepsilon_{it})$ is joint normally distributed with $E_t(\varepsilon_{it}) = 0$, and that idiosyncratic risks are drawn from the same distribution for each consumer (so $\sigma_{it} = \sigma_{It}$ for each i), and that a stability assumption applies to the distribution of lagged consumption. Blundell and Stoker (2005) show that aggregate consumption growth is

$$\Delta \ln E_t(c_{it}) = \rho r_t + (\beta + \varphi r_t)' E_t(z_{it}) + k_1 \sigma_{At} + k_2 \sigma_{It} + \kappa_1 \eta_t + \Lambda_t.$$

This model explains aggregate consumption growth in terms of the mean of preference heterogeneity, risk terms, and an aggregation factor Λ_t. The factor Λ_t is comprised of variances and covariances of the heterogeneous elements $\ln c_{it-1}$, z_{it} and ε_{it}. Thus, this model reflects how aggregate consumption will vary as the individual incomes become more or less risky, and captures how the income risk interplays with previous consumption values.

In overview, as micro consumption models are nonlinear, distributional restrictions are essential. On this point, an empirical fact is that the distribution of household consumption is often observed to be well approximated by a lognormal distribution, and so such lognormal restrictions may have empirical validity. Also relevant here is the empirical study of income and wealth risks, which has focused on earnings processes; see Meghir and Pistaferri (2004) for a recent contribution.

Micro to macro and vice versa

We now turn to two related uses of aggregation structure that have emerged in the literature.

Aggregation as a solution to microeconometric estimation

Consider a situation where the estimation of a model at the micro level is the primary goal of empirical work. Some recent work uses aggregation structure to enhance or permit micro-level parameter identification and estimation. Since aggregation structure provides a bridge between models at the micro level and the aggregate level, it permits all data sources – individual-level data and aggregate-level data – to be used for identification and estimation of economic parameters. Sometimes it is necessary to combine all data sources to identify economic effects (for example, Jorgenson, Lau and Stoker, 1982), and sometimes one can study (micro) economic effects with aggregate data alone (for example, Stoker, 1986). Recent work has developed more systematic methods of using aggregate data to improve micro-level estimates. In particular, one can match aggregate data with simulated moments from the individual data as part of the estimation process.

To see how this can work, suppose we have data on labour participation over several time periods (or groups). We assume that the participation decision is given by the model (11) with normal unobserved heterogeneity, as discussed above. We normalize $\sigma_v = 1$ and take $s(t) = \psi$, a constant, so that the unknown parameters of the participation model are α, γ and ψ. The data situation is as follows; for each group $t = 1, \ldots, T$, we observe the proportion of labour participants P_t and a random sample of benefits and schooling values, $\{B_{it}, S_{it}, i = 1, \ldots, n_t\}$. Given the (probit) expression (13), estimation can be based on matching the observed proportion P_t to the simulated moment

$$P_t(\alpha, \gamma, \psi) = \frac{1}{n_t} \sum_{i=1}^{n_t} \Phi\left[\psi - \alpha \ln B_{it} + \gamma \cdot S_{it}\right].$$

For instance, we could estimate by least squares over groups, by choosing $\hat{\alpha}, \hat{\gamma}, \hat{\psi}$ to minimize

$$\sum_{t=1}^{T} (P_t - P_t(\alpha, \gamma, \psi))^2.$$

Note that this approach does not require a specific assumption on the joint distribution of B_{it} and S_{it} for each t, as the random sample provides the distributional information needed to link the parameters to the observed proportion P_t.

It turns out that this approach for estimation is extremely rich, and was essentially mapped out by Imbens and Lancaster (1994). It has become a principal method of estimating demands for differentiated products, for use in structural models of industrial organization. See Berry, Levinsohn and Pakes (2004) for good coverage of this development.

Can macroeconomic interaction solve aggregation problems?

The basic heuristic that underlies much macroeconomic modelling is that, because of markets, individuals are very coordinated in their actions, so that individual heterogeneity likely has a secondary impact. In simplest terms, the notion is that common reactions across individuals will swamp any behavioural differences. This idea is either just wrong or, at best, very misleading for economic analysis. But that is not to deny that in real world economies there are many elements of commonality in reactions across individuals. Households face similar prices, interest rates and opportunities for employment. Extensive insurance markets effectively remove some individual differences in risk profiles. Optimal portfolio investment can have individuals choosing the same (efficient) basket of securities.

The question whether market interactions can minimize the impact of individual heterogeneity is a classic one, and by and large the answers are negative. However, there has been some recent work with calibrated stochastic growth models that raises some possibilities. A principal example of this is Krusell and Smith (1998), which we now discuss briefly. The Krusell–Smith set-up has infinitely lived consumers, with the same preferences within each period, but with different discount rates and wealth holdings. Each consumer has a chance of being unemployed each period, so there are transitory individual income shocks. Production arises from labour and capital, and there are transitory aggregate productivity shocks. Consumers can insure for the future by investing in capital only. Thus, insurance markets are incomplete, and consumers cannot hold negative capital amounts.

To make savings and portfolio decisions, consumers must predict future prices. To do this, each consumer must keep track of the evolution of the entire distribution of wealth holdings, in principle. This is a lot of information to know, just like what is needed for standard aggregation solutions as discussed earlier. Krusell–Smith's simulations show, however, that this forecasting problem is much easier than one would suspect. That is, for consumer planning and for computing equilibrium, consumers get very close to optimal solutions by keeping track of only two things: mean wealth in the economy and the aggregate productivity shock. This is approximate aggregation, a substantial simplification of the information requirements that one would expect.

The source of this simplification, as well as its robustness, is a topic of active current study. One aspect is that most consumers, especially those with lowest discount rates, save enough to insure their risk so that their propensity to save out of wealth is essentially constant. Those consumers also hold a large fraction of the wealth, so that saving is essentially linear in wealth. This means that there is (approximate) exact aggregation structure, with the mean of wealth determining how much aggregate saving is undertaken. That is, the nature of savings and wealth accumulation approximately solves the aggregation problem for individual forecasting. Aggregate consumption, however, does not exhibit the same simplification. Many low-wealth consumers become unemployed and encounter liquidity constraints. Their consumption is much more sensitive to current output than that of wealthier consumers.

These results depend on the specific formulation of the growth model. Krusell and Smith (2006) survey work that suggests that their type of approximate aggregation can be obtained under a variety of variations of the basic model assumptions. As such, this work raises a number of fascinating issues on the interplay between economic interaction, aggregation and individual heterogeneity. However, it remains to be seen whether the structure of such calibrated models is empirically relevant to actual economies, or whether forecasting can be simplified even with observed variation in saving propensities of wealthy households.

Future progress

Aggregation problems are among the most difficult in empirical economics. The progress that has been made recently is arguably due to two complementary developments. First is the enormous expansion in the availability of data on the behaviour of individual agents, including consumers, households, firms, and so on, in both repeated cross-section and panel data form. Second is the enormous expansion in computing power that facilitates the study of large data sources. These two trends can be reasonably expected to continue, which makes the prospects for further progress quite bright.

There is sufficient variety and complexity in the issues posed by aggregation that progress may arise from many approaches. For instance, we have noted how the possibility of approximate aggregation has arisen in computable stochastic growth models. For another instance, it is sometimes possible to derive properties of aggregate relationships with very weak assumptions on individual behaviour, as in Hildenbrand's (1994) work of the law of demand.

But is seems clear to me that the best prospects for progress lie with careful microeconomic modelling and empirical work. Such work is designed to ferret out economic effects in the presence of individual heterogeneity, and can also establish what are 'typical' patterns of heterogeneity in different applied contexts. Knowledge of typical patterns of heterogeneity is necessary for characterizing the distributional structure that will facilitates aggregation, and such distributional restrictions can then be refuted or validated with actual data. That is, enhanced understanding of the standard structure in the main application areas of empirical economics, such as with commodity demand, consumption and saving and labour supply, will lead naturally to an enhanced understanding of aggregation problems and accurate interpretation of aggregate relationships. There has been great progress of this kind in the past few decades, and there is no reason to think that such progress won't continue or accelerate.

THOMAS M. STOKER

Bibliography

Attanasio, O. 1999. Consumption. In *Handbook of Macroeconomics*, vol. 1B, ed. J. Taylor and M. Woodford. Amsterdam: North-Holland.

Banks, J., Blundell, R. and Lewbel, A. 1997. Quadratic Engel curves, indirect tax reform and welfare measurement. *Review of Economics and Statistics* 79, 527–39.
Berry, S., Levinsohn, J. and Pakes, A. 2004. Differentiated products demand systems from a combination of micro and macro data: the new car market. *Journal of Political Economy* 112, 68–105.
Blundell, R. 1988. Consumer behaviour: theory and empirical evidence. *Economic Journal* 98, 16–65.
Blundell, R., Reed, H. and Stoker, T. 2003. Interpreting aggregate wage growth. *American Economic Review* 93, 1114–31.
Blundell, R. and Stoker, T. 2005. Aggregation and heterogeneity. *Journal of Economic Literature* 43, 347–91.
Browning, M., Hansen, L. and Heckman, J. 1999. Micro data and general equilibrium. In *Handbook of Macroeconomics*, vol. 1A, ed. J. Taylor and M. Woodford. Amsterdam: North-Holland.
Browning, M. and Lusardi, A. 1996. Household saving: micro theories and micro facts. *Journal of Economic Literature* 34, 1797–855.
Gorman, W.M. (Terence). 1953. Community preference fields. *Econometrica* 21, 63–80.
Hildenbrand, W. 1994. *Market Demand: Theory and Empirical Evidence.* Princeton: Princeton University Press.
Imbens, G. and Lancaster, T. 1994. Combining micro and macro data in microeconometric models. *Review of Economic Studies* 61, 655–80.
Jorgenson, D., Lau, L. and Stoker, T. 1982. The transcendental logarithmic model of aggregate consumer behavior. In *Advances in Econometrics*, vol. 1, ed. R. Basmann, and G. Rhodes. Greenwich, CT: JAI Press.
Krusell, P. and Smith, A. 1998. Income and wealth heterogeneity in the macroeconomy. *Journal of Political Economy* 106, 867–96.
Krusell, P. and Smith, A. 2006. Quantitative macroeconomic models with heterogeneous agents. In *Advances in Economics and Econometrics, Proceedings of the Ninth World Congress of the Econometric Society*, ed. R. Blundell, W. Newey and T. Persson. Cambridge: Cambridge University Press.
Meghir, C. and Pistaferri, L. 2004. Income variance dynamics and heterogeneity. *Econometrica* 72, 1–32.
Stoker, T. 1986. Simple tests of distributional effects on macroeconomic equations. *Journal of Political Economy* 94, 763–95.
Stoker, T. 1993. Empirical approaches to the problem of aggregation over individuals. *Journal of Economic Literature* 31, 1827–74.
Theil, H. 1954. *Linear Aggregation of Economic Relations.* Amsterdam: North-Holland.

ARCH models

Introduction of model and basic properties

The key properties of financial time series appear to be that: (*a*) marginal distributions have heavy tails and thin centres (leptokurtosis); (*b*) the scale appears to change over time; (*c*) return series appear to be almost uncorrelated over time but to be dependent through higher moments (see Mandelbrot, 1963; Fama, 1965). Linear models like the autoregressive moving average (ARMA) class cannot capture well all these phenomena, since they only really address the conditional mean $\mu_t = E(y_t|y_{t-1},\ldots)$ and in a rather limited way. This motivates the consideration of nonlinear models. For a discrete time stochastic process y_t, the conditional variance $\sigma_t^2 = \mathrm{var}(y_t|y_{t-1},\ldots)$ of the process is a natural measure of risk for an investor at time $t-1$. Empirically it appears to change over time and so it is important to have a model for it. Engle (1982) introduced the autoregressive conditional heteroskedasticity (ARCH) model

$$\sigma_t^2 = \omega + \gamma y_{t-1}^2, \quad t = 0, \pm 1, \ldots,$$

where for simplicity we rewrite $y_t \mapsto y_t - \mu_t$ and suppose that the process started in the infinite past. This model makes σ_t^2 vary over time depending on the realization of past squared returns. For σ_t^2 to be a valid conditional variance it is necessary that $\omega > 0$ and $\gamma \geq 0$, in which case $\sigma_t^2 > 0$ for all t. Suppose also that $y_t = \varepsilon_t \sigma_t$ with ε_t i.i.d. mean zero and variance one. Provided $\gamma < 1$, the process y_t is weakly (covariance) stationary and has finite unconditional variance $\sigma^2 = E(\sigma_t^2) = E(y_t^2) = \omega/(1-\gamma)$. This can be proven rigorously under a variety of assumptions on the initialization of the process (see Nelson, 1990). The meaning of this is that the process fluctuates about the long-run value σ^2 and forecasts converge to this value as the forecast horizon lengthens.

The ARCH process is dynamic like ARMA models and indeed we can write the process as an AR(1) in y_t^2, that is,

$$y_t^2 = \omega + \gamma y_{t-1}^2 + \eta_t,$$

where $\eta_t = y_t^2 - \sigma_t^2 = \sigma_t^2(\varepsilon_t^2 - 1)$ is a mean zero, uncorrelated sequence, that is heteroskedastic. Therefore, we generally have dependence in σ_t^2, y_t^2, and because of the parameter restrictions, positive dependence that is, $\mathrm{cov}(\sigma_t^2, \sigma_{t-j}^2) > 0$ and $\mathrm{cov}(y_t^2, y_{t-j}^2) > 0$. As far as the second order properties (that is, the covariance function) of the process y_t^2, this is identical to that of an AR(1) process. However, it should be remembered that y_t^2 is heteroskedastic itself and that the form of the heteroskedasticity has to be particularly extreme since y_t^2 is kept non-negative.

One feature of linear models like the ARMA class is that the marginal distribution of the variable is normally distributed whenever the shocks are i.i.d. normally distributed. This is not the case for the ARCH class of processes. Specifically, the marginal distribution of y_t will be heavy tailed even if $\varepsilon_t = (y_t - \mu_t)/\sigma_t$ is standard normal. Suppose ε_t is standard normal (and the process is weakly stationary), then the excess kurtosis of y_t is $\kappa_4 = 6\gamma^2/(1 - 3\gamma^2) \geq 0$ provided $\gamma^2 < 1/3$. If $\gamma \geq 1/3^{1/2}$, then $E(y_t^4) = \infty$. For leptokurtic ε_t, the restriction on γ for finite fourth moment is even more severe. Although the ARCH(1) model implies heavy tails and volatility clustering, it does not in practice generate enough of either. The constraint on γ for finite fourth moment severely restricts the amount of persistence; it is an undesirable feature that the same parameter controls both persistence and heavy tailedness, although if one allows non-normal distributions for ε_t, this link is broken on one side at least.

The extension to the ARCH(p) process with p lags, while more flexible, becomes very complicated to estimate without restrictions on the coefficients. Bollerslev (1986) introduced the GARCH(p, q) process

$$\sigma_t^2 = \omega + \sum_{k=1}^{p} \beta_k \sigma_{t-k}^2 + \sum_{j=1}^{q} \gamma_j (y_{t-j} - \mu_{t-j})^2,$$

whose $p = 1$, $q = 1$ GARCH(1, 1) special case contains only three parameters and usually does a better job than an unrestricted ARCH(12), say, according to a variety of statistical criteria. The GARCH(1, 1) process is probably still the most widely used model. As with the ARCH process one needs restrictions on the parameters to make sure that σ_t^2 is positive with probability one. For the GARCH(1, 1) it is necessary that $\gamma, \beta \geq 0$ and $\omega > 0$. Interestingly, for higher order processes it is not necessary that $\omega, \gamma_j, \beta_j \geq 0$ for all j: see Nelson and Cao (1992). For example, in GARCH(1, 2) the conditions are that $\beta, \gamma_1 \geq 0$ and $\beta\gamma_1 + \gamma_2 \geq 0$. Provided $\sum_{k=1}^{p} \beta_k + \sum_{j=1}^{q} \gamma_j < 1$, the process y_t is weakly stationary and has finite unconditional variance

$$\sigma^2 = E(\sigma_t^2) = \frac{\omega}{1 - \sum_{k=1}^{p} \beta_k - \sum_{j=1}^{q} \gamma_j}.$$

As for the ARCH process, the series y_t has higher kurtosis than ε_t.

Drost and Nijman (1993) provide an important classification of ARCH models according to the precise properties required of the error terms. The strong GARCH process is where

$$\varepsilon_t = \frac{y_t - \mu_t}{\sigma_t} \text{ i.i.d. } E(\varepsilon_t) = 0 \text{ and } E(\varepsilon_t^2) = 1.$$

It is generally this case that has been investigated in the literature. It is a very strong assumption by the standards of most modern econometrics, where usually only conditional moment restrictions are imposed, but is a complete specification that is useful for deriving properties like stationarity. The strong Gaussian case is where ε_t is additionally normally distributed. The semi-strong GARCH process is where

$$E[\varepsilon_t | y_{t-1}, y_{t-2}, \ldots] = 0 \text{ and } E[\varepsilon_t^2 | y_{t-1}, y_{t-2}, \ldots] = 1.$$

These assumptions are weaker and turn out to be sufficient in many cases for consistent estimation. They are quite weak assumptions and restrict only the conditional mean and conditional variance of the process, allowing a variety of behaviour in the potentially time varying distribution of ε_t. Drost and Nijman (1993) show that conventional strong and semi-strong GARCH processes are not closed under temporal aggregation, meaning that if a process is GARCH at the daily frequency that the weakly or monthly data may not be GARCH, either weak or strong.

Strong stationarity and mixing

Consider the GARCH(1, 1) process

$$y_t = \sigma_t \varepsilon_t, \quad \sigma_t^2 = \omega + \beta\sigma_{t-1}^2 + \gamma y_{t-1}^2$$

with ε_t i.i.d. and $\omega > 0$ and $\beta, \gamma \geq 0$. A sufficient condition for strong stationarity is that $E[\ln(\beta + \gamma\varepsilon_t^2)] < 0$ (see Nelson, 1990). If additionally, $E(\varepsilon_t)=0$ and $\mathrm{var}(\varepsilon_t)=1$, then the necessary and sufficient condition for weak stationarity is that $\beta + \gamma < 1$. By Jensen's inequality $E[\ln(\beta + \gamma\varepsilon_t^2)] < \ln E[(\beta + \gamma\varepsilon_t^2)] = \ln(\beta + \gamma)$, so it can be that $E[\ln(\beta + \gamma\varepsilon_t^2)] < 0$ even when $\beta + \gamma \geq 1$, that is, there are strongly stationary processes that are not weakly stationary.

There are many measures of dependence in time series. Mixingness is the property that dependence dies out with horizon. It can be measured in different ways: covariance mixing, strong mixing, and beta mixing are the main concepts. A stationary sequence $\{X_t, \ t=0, \ \pm 1, \dots\}$ is said to be covariance mixing if $\mathrm{cov}(X_t, X_{t+k}) \to 0$ as $k \to \infty$. A stationary sequence $\{X_t, \ t=0, \ \pm 1, \dots\}$ is said to be strong mixing (α-mixing) if

$$\alpha(k) = \sup_{A \in \mathscr{F}^n_{-\infty}, B \in \mathscr{F}^\infty_{n+k}} |P(AB) - P(A)P(B)| \to 0$$

as $k \to \infty$, where $\mathscr{F}^n_{-\infty}$ and $\mathscr{F}^\infty_{n+k}$ are two σ-fields generated by $\{X_t, t \leq n\}$ and $\{X_t, t \geq n+k\}$, respectively. We call $\alpha(\cdot)$ the mixing coefficient. A stationary sequence $\{X_t, \ t=0, \ \pm 1, \ \dots\}$ is said to be β-mixing if

$$\beta(k) = \sup_{A \in \mathscr{F}^n_{-\infty}, B \in \mathscr{F}^\infty_{n+k}} |P(A|B) - P(A)| \to 0$$

as $k \to \infty$. We call $\beta(\cdot)$ the mixing coefficient. We have $2\alpha(k) \leq \beta(k)$. The covariance mixing property is only well defined for weakly stationary processes, so it is natural here to work with the more general notions of α and β mixing. A sufficient condition that a GARCH(1, 1) process is β-mixing with exponential decay is that it is weakly stationary, Carrasco and Chen (2002), but this is not necessary. More recently it has been shown that IGARCH is strong mixing under some conditions (see Meitz and Saikkonen, 2004). One problem is that when you combine a GARCH process with other processes for the mean, the mixingness is not preserved and has still to be established. The weaker concept of near epoch dependence can be established, though in quite a general class of models (Hansen, 1991). Why does mixing matter? It is a key property that allows one to learn from the data through the law of large numbers and central limit theorems.

IGARCH models

In practice, estimated GARCH parameters lie close to the boundary of the weakly stationary region. This prompts consideration of the process where $\sum_{k=1}^{p}\beta_k + \sum_{j=1}^{q}\gamma_j = 1$, which is called the integrated GARCH or IGARCH. In this case, the process y_t with i.i.d. Gaussian innovations is strongly stationary but not covariance stationary, since the unconditional variance is infinite (although the conditional variance is finite with probability 1). This is in contrast to linear unit root processes in which the process is neither weakly nor strongly stationary and these two notions coincide. Also, in contrast to the linear case, differencing does not induce weak stationarity, that is, $y_t^2 - y_{t-1}^2$ is not weakly stationary (although its mean is constant over time).

The exponentially weighted moving average model (sometimes called the J.P. Morgan model) is a variant on the IGARCH model in which there is no intercept ω and a unit root:

$$y_t = \sigma_t \varepsilon_t, \quad \sigma_t^2 = \beta\sigma_{t-1}^2 + (1-\beta)y_{t-1}^2.$$

It is a very simple process with only one parameter and is widely used by practitioners, with particular values of the parameter β. Write $\sigma_t^2 = \sigma_{t-1}^2[\beta + (1-\beta)\varepsilon_{t-1}^2]$, so that $\ln\sigma_t^2$ is a random walk, that is,

$$\ln\sigma_t^2 = \ln\sigma_{t-1}^2 + \eta_{t-1}, \; \eta_{t-1} = \ln(\beta + (1-\beta)\varepsilon_{t-1}^2),$$

and hence is not strongly stationary. On the other hand, the process y_t is informally weakly stationary since $E[y_t^2|\mathscr{F}_{-\infty}^0] = E[\sigma_t^2|\mathscr{F}_{-\infty}^0] = \prod_{s=1}^{t} E[\beta + (1-\beta)\varepsilon_s^2]\sigma_0^2 = \sigma_0^2$ for all t. The properties of this process depend on the moments of η_{t-1}. If $E[\eta_{t-1}] > 0$, then $\sigma_t^2 \to \infty$ with probability 1. If $E[\eta_{t-1}] < 0$, then $\ln\sigma_t^2 \to -\infty$ with probability 1 as $t \to \infty$ and so $\sigma_t^2 \to 0$ with probability 1. If $E[\eta_{t-1}] = 0$, then $\ln\sigma_t^2$ is a driftless random walk and the process just wanders everywhere. If we assume $E[\varepsilon_t^2] = 1$, then by Jensen's inequality $E[\eta_{t-1}] < 0$, and the process $\sigma_t^2 \to 0$ with probability 1 as $t \to \infty$ whatever the initialization. Thus the process is essentially degenerate and is not plausible, despite being widely used.

Functional form

The news impact curve is the relationship between σ_t^2 and $y_{t-1} = y$ holding past values σ_{t-1}^2 constant at some level σ^2. This is an important relationship that describes how new information affects volatility. For the GARCH process, the news impact curve is

$$m(y, \sigma^2) = \omega + \gamma y^2 + \beta\sigma^2.$$

It is separable in σ^2, it is an even function of news y, $m(y, \sigma^2)=m(-y, \sigma^2)$, and it is a quadratic function of y. The symmetry property implies that cov $(y_t^2, y_{t-j}) = 0$ for symmetric about zero ε_t.

The GARCH process does not allow 'leverage effects' or asymmetric news impact curves. Because of limited liability, we might expect that negative and positive shocks have different effects on volatility. Nelson (1991) introduced the exponential GARCH model. Let $h_t = \log \sigma_t^2$ and let

$$h_t = \omega + \sum_{j=1}^{p} \gamma_j[\theta\varepsilon_{t-j} + \delta|\varepsilon_{t-j}|] + \sum_{k=1}^{q} \beta_k h_{t-k},$$

where $\varepsilon_t = (y_t - \mu_t)/\sigma_t$ is i.i.d. with mean zero and variance one. Nelson's paper contains four innovations. First, it models the log, not the level. Therefore there are no parameter restrictions to ensure that $\sigma_t^2 \geq 0$. Second, it allows asymmetric effect of past shocks ε_{t-j} on current volatility, that is, the news impact curve is allowed to be asymmetric. For example, $\text{cov}(y_t^2, y_{t-j}) \neq 0$ even when ε_t is symmetric about zero. Third, it makes the innovations ε_t i.i.d. It follows that h_t is a linear process so that strong and weak stationarity coincide where they ought to (for h_t anyway). On the other hand estimation and forecasting is quite tricky because of the repeated exponential/logarithmic transformations involved. The final innovation was to allow heavy tailed innovations based on the so-called generalized error distribution (GED) that nests the Gaussian as a special case.

An alternative approach to allowing asymmetric news impact curve is the Glosten, Jagannathan and Runkle (1993) model

$$\sigma_t^2 = \omega + \beta\sigma_{t-1}^2 + \gamma y_{t-1}^2 + \delta y_{t-1}^2 1(y_{t-1} < 0).$$

In this case, the news impact curve is asymmetric but still has quadratic tails. It is a simple enough modification, that it has similar probabilistic properties to the GARCH(1, 1) process. There are many other variations on the basic GARCH model, too many to list here, but the interested reader can find a fuller description in the survey paper of Bollerslev, Engle and Nelson (1994).

One might expect that risk and return should be related: see Merton (1973) for an example. The GARCH-in-Mean process captures this idea. This process is

$$y_t = g(\sigma_t^2; b) + \varepsilon_t\sigma_t,$$

for various functional forms of g, for example, linear and log-linear and for some given GARCH specification of σ_t^2. Engle, Lilien and Robins (1987) used this model on interest rate data (see also Pagan and Hong, 1991). Here, b are parameters to be estimated along with the parameters of the error variance. Some authors find small but significant effects.

Estimation

The standard approach to estimation of these models has been through estimation of the (conditional) Gaussian quasi-likelihood criterion

$$L_T(\theta) = \sum_{t=1}^{T} \ell_t(\theta) = -\sum_{t=1}^{T} \log \sigma_t^2(\theta) - \frac{1}{2}\sum_{t=1}^{T} \left(\frac{y_t - \mu_t(\theta)}{\sigma_t(\theta)}\right)^2,$$

where $\sigma_t^2(\theta)$ and perhaps $\mu_t(\theta)$ are built up by recursions from some starting values. There are several possibilities regarding starting values: (a) $\sigma_0^2(\theta) = \omega/1 - \beta - \gamma$, (b) $\sigma_0^2(\theta) = T^{-1}\sum_{t=1}^{T} y_t^2$, and (c) $\sigma_0^2(\theta) = y_1^2$. Approach (a) imposes weak stationarity and would not be appropriate were IGARCH to be thought plausible, while value (b) sort of requires weak stationarity for the asymptotic properties to follow through. The likelihood function is maximized with respect to the parameter values usually using some derivative-based algorithm like BHHH and sometimes imposing inequality restrictions (like those required for $\sigma_t^2 \geq 0$ with probability 1 or for σ_t^2 to be weakly stationary) and sometimes not.

The (quasi) MLE (QMLE) can be expected to be consistent provided only the conditional mean and the conditional variance are correctly specified (Bollerslev and Wooldridge, 1992), that is, semi-strong not strong GARCH is required and conditional normality is certainly not required. This is true because the score function $\partial\ell_t(\theta_0)/\partial\theta$ is a martingale difference sequence. Robust standard errors can be constructed in the usual way

$$\left[\frac{\partial \ell_T(\hat{\theta})}{\partial\theta\,\partial\theta^\top}\right]^{-1} \left[\sum_{t=1}^{T} \frac{\partial \ell_t}{\partial\theta} \frac{\partial \ell_t}{\partial\theta^\top}(\hat{\theta})\right] \left[\frac{\partial \ell_T(\hat{\theta})}{\partial\theta\,\partial\theta^\top}\right]^{-1}, \qquad (1)$$

although the default option in many software packages is to compute standard errors as if Gaussianity held.

The distribution theory is difficult to establish from primitive conditions even for simple models. There is one important point about these asymptotics – that one does not need moments on y_t (for example, one does not need weak stationarity). Lumsdaine (1996) established consistency and asymptotic normality allowing the IGARCH case but under strong stationarity and symmetric unimodal i.i.d. ε_t with $E[\varepsilon_t^{32}] < \infty$. Lee and Hansen (1994) proved the same result under weaker conditional moment conditions and allows for semi-strong processes with some higher-level assumptions. Jensen and Rahbek (2004) established consistency and asymptotic normality of the QMLE in strong GARCH model without strict stationarity. Hall and Yao (2003) assume weak stationarity and show that if $E(\varepsilon_t^4) < \infty$ the asymptotic normality holds, but also establish limiting behaviour (non-normal) under weaker moment conditions. No results have yet been published for consistent and asymptotically normality of EGARCH from primitive conditions, although simulation evidence does suggest normality is a good approximation in large samples.

Typically, one finds small intercepts and a large parameter on the lagged dependent volatility; see Lumsdaine (1995) and Brooks, Burke and Persand (2001) for simulation evidence. These two parameter estimates are often highly correlated. Engle and Sheppard (2001) suggested a method they called target variance to obviate the computational difficulties sometimes encountered in estimating GARCH models. For a weakly stationary GARCH(1, 1) process we have $E(y_t^2) = \omega/(1-\beta-\gamma)$ so that $\omega = E(y_t^2)(1-\beta-\gamma)$. They suggest replacing $E(y_t^2)$ by $\sum_{t=1}^{T} y_t^2/T$ in the likelihood so that one only has two parameters to chose. This results in a much more stable performance of most algorithms. The downside with this approach is that distribution theory is much more complicated due to the lack of martingale property, and in particular one needs to use Newey–West standard errors.

It is quite common now to estimate GARCH models using different objective functions suggested by alternative specifications of the error distribution like the t or the GED distribution that Nelson (1991) favoured. These objective functions often have additional parameters such as the degrees of freedom that have to be computed. They lead to greater efficiency when the chosen specification is correct, but otherwise can lead to inconsistency, as was shown by Newey and Steigerwald (1997).

Long memory

The GARCH(1, 1) process $\sigma_t^2 = \omega + \beta\sigma_{t-1}^2 + \gamma y_{t-1}^2$ is of the form

$$\sigma_t^2 = c_0 + \sum_{j=1}^{\infty} c_j y_{t-j}^2 \tag{2}$$

for constants c_j satisfying $c_j = \gamma\beta^{j-1}$, provided the process is weakly stationary, which requires $\gamma + \beta < 1$. These coefficients decay very rapidly so the actual amount of memory is quite limited. There is some empirical evidence on the autocorrelation function of y_t^2 for high frequency data that suggests a slower decay rate than would be implied by these coefficients. Long memory models essentially are of the form (2) but with slower decay rates. For example, suppose that $c_j = j^{-\theta}$ for some $\theta > 0$. The coefficients satisfy $\sum_{j=1}^{\infty} c_j^2 < \infty$ provided $\theta > 1/2$. Fractional integration (FIGARCH) leads to such an expansion. There is a single parameter called d that determines the memory properties of the series, and

$$(1-L)^d \sigma_t^2 = \omega + \gamma\sigma_{t-1}^2(\varepsilon_{t-1}^2 - 1),$$

where $(1-L)^d$ denotes the fractional differencing operator. When $d = 1$ we have the standard IGARCH model. For $d \neq 1$ we can define the binomial expansion of $(1-L)^{-d}$ in the form given above. See Robinson (1991) and Bollerslev and Mikkelson (1996) for models and evidence of long memory. The evidence for long memory is often based on sample autocovariances of y_t^2, and this may be questionable due to a paper of Mikosch and Stărică (2000).

Multivariate models

In practice we observe many closely related series, and so it may be important to model their behaviour jointly. Define the conditional covariance matrix

$$\Sigma_t = E(y_t y_t^\top | \mathscr{F}_{-\infty}^{t-1})$$

for some $n \times 1$ vector of mean zero series y_t. Bollerslev, Engle and Wooldridge (1988) introduced the most general generalization of the univariate GARCH(1, 1) process

$$h_t = \text{vech}(\Sigma_t) = A + Bh_{t-1} + C\text{vech}(y_{t-1} y_{t-1}^\top),$$

where A is an $n(n+1)/2 \times 1$ vector, while B, C are $n(n+1)/2 \times n(n+1)/2$ matrices. In practice, there are too many parameters. Also, the restrictions on the parameters to ensure that Σ_t is positive definite are very complicated in this formulation. For weak stationarity one requires that the matrix $I{-}B{-}C$ is nonsingular and positive definite in which case the unconditional variance matrix is $\text{unvech}((I{-}B{-}C)^{-1}A)$. The conditions for strong stationarity are rather complicated to state.

The so-called BEKK model is a special case that addresses these issues. It is of the form

$$\Sigma_t = AA^\top + B\Sigma_{t-1}B^\top + Cy_{t-1}y_{t-1}^\top C^\top$$

for $n \times n$ matrices A, B, C. This gives a big reduction in number of parameters and imposes symmetry and positive definiteness automatically. There are still many parameters that have to be estimated simultaneously, of the order n^2, and this limits the applicability and interpretability of this model.

Bollerslev (1990) introduced the constant conditional covariance (CCC) model, which greatly reduces the parameter explosion issue. This involves standard univariate dynamic models for each of the conditional variances and a constant correlation assumption, that is,

$$\Sigma_t = D_t R D_t, \quad D_t = \text{diag}\{\sigma_{it}\} \tag{3}$$

$$\sigma_{it}^2 = \omega_i + \beta_i \sigma_{i,t-1}^2 + \gamma_i y_{i,t-1}^2 \tag{4}$$

and $R = (R_{ij})$ is a time invariant matrix

$$R_{ij} = \frac{E[\varepsilon_{it}\varepsilon_{jt}]}{(E[\varepsilon_{it}^2]E[\varepsilon_{jt}^2])^{1/2}} = E[\varepsilon_{it}\varepsilon_{jt}],$$

where $\varepsilon_{it} = y_{it}/\sigma_{it}$. The values R_{ij} are restricted to lie in $[-1, 1]$ and the matrix R is symmetric and positive definite but otherwise unrestricted. This model generates time varying conditional covariances, but the dynamics are all driven by the conditional variances as the correlations are constant. The estimation of R is quite straightforward: use the sample correlation matrix of the standardized residuals

$\varepsilon_{it} = y_{it}/\hat{\sigma}_{it}$. The estimated matrix R is guaranteed to be symmetric and positive definite because it is a correlation matrix and consequently the estimated Σ_t shares these properties.

Engle and Sheppard (2001) introduced the dynamic conditional covariance (DCC) model where we replace in (3) and (4)

$$R_{ij,t} = \frac{q_{ij,t}}{(q_{ii,t}q_{jj,t})^{1/2}}$$

$$q_{ij,t} = c_{ij} + b_{ij}q_{ij,t-1} + a_{ij}\varepsilon_{i,t-1}\varepsilon_{j,t-1}.$$

If we assume also that $a_{ij} = a$, $b_{ij} = b$, and $c_{ij} = c$ for all $i \neq j$ one can show that the resulting covariance matrix Σ_t is guaranteed to be symmetric and positive definite. This model allows slightly more flexibility in allowing the correlations to vary over time, but because of the need to impose positive definiteness it still imposes common dynamics on the correlations, which may be too restrictive.

The approach that brings the most flexible dimensionality reduction is based on the ideas of factor analysis. Suppose that for $y_t \in \mathbb{R}^n$, $f_t \in \mathbb{R}^k$:

$$y_t = Cf_t + u_t \tag{5}$$

$$\begin{pmatrix} f_t \\ u_t \end{pmatrix} | I_{t-1} \sim \begin{pmatrix} 0 \\ 0 \end{pmatrix}, \begin{pmatrix} \Lambda_t & 0 \\ 0 & \Gamma \end{pmatrix}, \tag{6}$$

where $Y_{t-1} = \{y_{t-1}, \dots\}$ is the observed information and $I_t = \{y_t, f_t, y_{t-1}, f_{t-1}, \dots\}$ contains both observed series and the latent factors $F_{t-1} = \{f_t, f_{t-1}, \dots\}$. Suppose that rank$(C) = k$ and that Λ_t is a $k \times k$ positive definite time varying matrix. It follows that $y_t|I_{t-1} \sim 0, C\Lambda_t C^\top + \Gamma$ (Sentana, 1998). The implied Σ_t is of reduced rank and depends on only order nK (time-varying associated) parameters so there is a big reduction in dimensionality. This model includes as a special case the Diebold and Nerlove (1989) model where Γ, Λ_t are diagonal and $\lambda_{jjt} = \text{var}[f_{jt}|I_{t-1}] = \omega_j + \beta_j\lambda_{jj,t-1} + \gamma_j f_{j,t-1}^2$, in which case $\lambda_{jjt} \notin Y_{t-1}$. This process is closed under block marginalization – that is, subsets of y_t do not have the same structure. Estimation is complicated by the latent variables. This framework also includes the Engle, Ng and Rothschild (1990) factor GARCH model $\Sigma_t = \Sigma_0 + \sum_{k=1}^{K} \delta_k \delta_k^\top \sigma_{kt}^2$, where $K < n$, and σ_{kt}^2 is the conditional variance of a certain portfolio k, with time invariant weights α_k, that is, $y_{kt}^p = \alpha_k^\top y_t$ with $\alpha_k^\top i = 1$. They assume also that σ_{kt}^2 are standard univariate GARCH(1, 1) processes, that is, for some parameters $(\omega_k, \beta_k, \gamma_k)$, $\sigma_{kt}^2 = \omega_k + \beta_k\sigma_{k,t-1}^2 + \gamma_k(\gamma_k^\top y_{t-1})^2$. This model is written in terms of observables and consequently its estimation is somewhat easier, but it suffers from the fact that it is not closed under block marginalization – that is, subsets of y_t do not have the same structure. Sentana (1998) shows how it is nested in the general model (5) and (6).

Nonparametric and semiparametric models

There have been a number of contributions to ARCH modelling from the nonparametric or semiparametric point of view; see Hafner (1998) for an overview. Engle and González-Rivera (1991) suggested treating the error distribution in a GARCH process nonparametrically, that is,

$$y_t = \mu_t + \varepsilon_t \sigma_t \qquad \sigma_t^2 = \omega + \beta \sigma_{t-1}^2 + \gamma (y_{t-1} - \mu_{t-1})^2,$$

where μ_t depends on observed covariates and parameters, while ε_t is i.i.d. with density f that is not restricted in shape. This is motivated by the great deal of evidence that the density of the standardized residuals $\varepsilon_t = (y_t - \mu_t)/\sigma_t$ is non-Gaussian. They proposed an estimation algorithm that involved estimating f from the data. Linton (1993) and Drost and Klaassen (1997) have shown that one can achieve significant efficiency improvements depending on the shape of the error density.

An alternative line of research has been to treat the functional form of $\sigma_t^2(y_{t-1}, y_{t-2}, \ldots)$ nonparametrically. In particular, suppose that

$$\sigma_t^2 = g(y_{t-1}, \ldots, y_{t-p})$$

for some unknown function g and fixed lag length p. This allows for a general shape to the news impact curve and nests all the usual parametric ARCH processes. See Pagan and Hong (1991) and Härdle and Tsybakov (1997) for some applications. This model is somewhat limited in the dependence it allows in comparison with the GARCH(1, 1) process, which is a function of all past y's. Also, the curse of dimensionality means that the usual estimation methods do not work well in practice for large p, that is, $p > 4$.

One compromise approach to avoiding the curse of dimensionality is to use additive models, whence

$$\sigma_t^2 = \sum_{j=1}^{p} g_j(y_{t-j}) \tag{7}$$

for some unknown functions g_j. The functions g_j are allowed to be of general functional form but only depend on y_{t-j}. This class of processes nests many parametric ARCH models. The functions g_j can be estimated by kernel regression techniques (see Masry and Tjøstheim, 1995). Yang, Härdle and Nielsen (1999) proposed an alternative nonlinear ARCH model in which the conditional mean is again additive, but the volatility is multiplicative $\sigma_t^2 = c_v \prod_{j=1}^{d} \sigma_j^2(y_{t-j})$. Kim and Linton (2004) generalize this model to allow for arbitrary, but known, transformations, that is, $G(\sigma_t^2) = c_v + \sum_{j=1}^{d} \sigma_j^2(y_{t-j})$, where $G(.)$ is known function like log or level. Linton and Mammen (2005) considered the case where $\sigma_t^2 = \sum_{j=0}^{\infty} \beta^{j-1} g(y_{t-j})$, which nests the GARCH(1, 1) process when $g(y) = \omega + \gamma y^2$.

One final semiparametric approach has been to model the coefficients of a GARCH process as changing over time, thus

$$\sigma_t^2 = \omega(x_{tT}) + \beta(x_{tT}) \sigma_{t-1}^2 + \gamma(x_{tT})(y_{t-1} - \mu_{t-1})^2,$$

where ω, β, and γ are smooth functions of a variable x_{tT}, for example, $x_{tT} = t/T$. This class of processes is non-stationary but can be viewed as locally stationary along the lines of Dahlhaus (1997).

OLIVER B. LINTON

See also **continuous and discrete time models; time series analysis.**

The author would like to thank the Economic and Social Science Research Council of the United Kingdom for financial support through a research fellowship.

Bibliography

Bollerslev, T. 1986. Generalized autoregressive conditional heteroskedasticity. *Journal of Econometrics* 31, 307–27.

Bollerslev, T. 1990. Modelling the coherence in short-run nominal exchange rates: a multivariate generalized autoregressive conditional heteroskedasticity. *Review of Economics and Statistics* 72, 498–505.

Bollerslev, T., Engle, R.F. and Wooldridge, J.M. 1988. A capital asset pricing model with time varying covariances. *Journal of Political Economy* 96, 116–31.

Bollerslev, T. and Mikkelson, H.O. 1996. Modelling and pricing long memory in stock market volatility. *Journal of Econometrics* 73, 151–184, 498–505.

Bollerslev, T., Chou, R.Y. and Kroner, K. 1992. ARCH modelling in finance. *Journal of Econometrics* 52, 5–59.

Bollerslev, T., Engle, R.F. and Nelson, D. 1994. ARCH models. In *The Handbook of Econometrics*, vol. 4, ed. D.F. McFadden and R.F. Engle III. Amsterdam: North-Holland.

Bollerslev, T. and Wooldridge, J.M. 1992. Quasi maximum likelihood estimation and inference in dynamic models with time varying covariances. *Econometric Reviews* 11, 143–72.

Brooks, C., Burke, S.P. and Persand, G. 2001. Benchmarks and the accuracy of GARCH model estimation. *International Journal of Forecasting* 17, 45–56.

Carrasco, M. and Chen, X. 2002. Mixing and moment properties of various GARCH and stochastic volatility models. *Econometric Theory* 18, 17–39.

Dahlhaus, R. 1997. Fitting time series models to nonstationary processes. *Annals of Statistics* 25, 1–37.

Diebold, F.S. and Nerlove, M. 1989. The dynamics of exchange-rate volatility: a multivairate latent-factor ARCH model. *Journal of Applied Econometrics* 4, 1–22.

Drost, F.C. and Klaassen, C.A.J. 1997. Efficient estimation in semiparametric GARCH models. *Journal of Econometrics* 81, 193–221.

Drost, F.C. and Nijman, T.E. 1993. Temporal aggregation of GARCH processes. *Econometrica* 61, 909–27.

Engle, R.F. 1982. Autoregressive conditional heteroscedasticity with estimates of the variance of U.K. inflation. *Econometrica* 50, 987–1008.

Engle, R.F. and Bollerslev, T. 1986. Modeling the persistence of conditional variances. *Econometric Reviews* 5, 1–50.

Engle, R.F. and González-Rivera, G. 1991. Semiparametric ARCH models. *Journal of Business and Economic Statistics* 9, 345–59.

Engle, R.F., Lilien, D.M. and Robins, R.P. 1987. Estimating time varying risk premia in the term structure: the ARCH-M model. *Econometrica* 19, 3–29.

Engle, R.F. and Ng, V.K. 1993. Measuring and testing the impact of news on volatility. *Journal of Finance* 48, 1749–78.

Engle, R.F. and Ng, V.K. and Rothschild, M. 1990. Asset pricing with a FACTOR-ARCH covariance structure: empirical estimates for treasury bills. *Journal of Econometrics* 45, 213–37.

Engle, R.F. and Sheppard, K. 2001. Theoretical and empirical properties of dynamic conditional correlation multivariate GARCH. Working Paper No. 8554. Cambridge, MA: NBER.

Fama, E.F. 1965. The behavior of stock market prices. *Journal of Business* 38, 34–105.

Glosten, L.R., Jagannathan, R. and Runkle, D.E. 1993. On the relation between the expected value and the volatility of the nominal excess returns on stocks. *Journal of Finance* 48, 1779–801.

Hafner, C. 1998. *Nonlinear Time Series Analysis with Applications to Foreign Exchange Rate Volatility.* Heidelberg: Physica.

Hall, P. and Yao, Q. 2003. Inference in ARCH and GARCH models with heavy tailed errors. *Econometrica* 71, 285–317.

Hansen, B.A. 1991. GARCH(1,1) processes are near epoch dependent. *Economics Letters* 36, 181–6.

Hansen, P.R. and Lunde, A. 2005. A forecast comparison of volatility models: does anything beat a GARCH(1,1). *Journal of Applied Econometrics* 20, 873–89.

Härdle, W., Tsybakov, A.B. and Yang, L. 1996. Nonparametric vector autoregression. Discussion Paper, SFB 373. Berlin: Humbodt-Universität.

Härdle, W. and Tsybakov, A.B. 1997. Locally polynomial estimators of the volatility function. *Journal of Econometrics* 81, 223–42.

Jensen, S.T. and Rahbek, A. 2004. Asymptotic normality of the QMLE of ARCH in the nonstationary case. *Econometrica* 72, 641–6.

Kim, W. and Linton, O. 2004. A local instrumental variable estimation method for generalized additive volatility models. *Econometric Theory* 20, 1094–139.

Lee, S.-W. and Hansen, B.E. 1994. Asymptotic theory for the GARCH(1,1) quasi-maximum likelihood estimator. *Econometric Theory* 10, 29–52.

Linton, O.B. 1993. Adaptive estimation in ARCH models. *Econometric Theory* 9, 539–69.

Linton, O.B. and Mammen, E. 2005. Estimating semiparametric ARCH(∞) models by kernel smoothing methods. *Econometrica* 73, 771–836.

Lumsdaine, R. 1995. Finite-sample properties of the maximum likelihood estimator in GARCH(1,1) and IGARCH(1,1) models: a Monte Carlo investigation. *Journal of Business and Economic Statistics* 13, 1–10.

Lumsdaine, R.L. 1996. Consistency and asymptotic normality of the quasi-maximum likelihood estimator in IGARCH(1,1) and covariance stationary GARCH(1,1) models. *Econometrica* 64, 575–96.

Mandelbrot, B. 1963. The variation of certain speculative prices. *Journal of Business* 36, 394–419.

Masry, E. and Tjøstheim, D. 1995. Nonparametric estimation and identification of nonlinear ARCH time series: strong convergence and asymptotic normality. *Econometric Theory* 11, 258–89.

Meitz, M. and Saikkonen, P. 2004. Ergodicity, mixing, and existence of moments of a class of Markov models with applications to GARCH and ACD models. Working Paper Series in Economics and Finance No. 573. Stockholm School of Economics.

Merton, R.C. 1973. An intertemporal capital asset pricing model. *Econometrica* 41, 867–87.

Mikosch, T. and Stărică, C. 2000. Limit theory for the sample autocorrelations and extremes of a GARCH(1,1) process. *Annals of Statistics* 28, 1427–51.

Nelson, D.B. 1990. Stationarity and persistence in the GARCH(1,1) model. *Econometric Theory* 6, 318–34.

Nelson, D.B. 1991. Conditional heteroskedasticity in asset returns: a new approach. *Econometrica* 59, 347–70.

Nelson, D.B. and Cao, C.Q. 1992. Inequality constraints in the univariate GARCH model. *Journal of Business and Economic Statistics* 10, 229–35.

Newey, W.K. and Steigerwald, D.G. 1997. Asymptotic bias for quasi-maximum-likelihood estimators in conditional heteroskedasticity models. *Econometrica* 65, 587–99.

Pagan, A.R. and Hong, Y.S. 1991. Nonparametric estimation and the risk premium. In *Nonparametric and Semiparametric Methods in Econometrics and Statistics*, ed. W. Barnett, J. Powell and G.E. Tauchen. Cambridge: Cambridge University Press.

Pagan, A.R. and Schwert, G.W. 1990. Alternative models for conditional stock volatility. *Journal of Econometrics* 45, 267–90.

Robinson, P.M. 1991. Testing for strong serial correlation and dynamic conditional heteroskedasticity in multiple regression. *Journal of Econometrics* 47, 67–84.

Robinson, P.M. and Zaffaroni, P. 2006. Pseudo-maximum likelihood estimation of ARCH(∞) models. *Annals of Statistics* 34, 1049–1074.

Sentana, E. 1998. The relation between conditionally heteroskedastic factor models and factor GARCH models. *Econometrics Journal* 1, 1–9.

Yang, L., Härdle, W. and Nielsen, J.P. 1999. Nonparametric autoregression with multiplicative volatility and additive mean. *Journal of Time Series Analysis* 20, 579–604.

Bayesian methods in macroeconometrics

Macroeconometrics encompasses a large variety of probability models for macroeconomic time series as well as estimation and inference procedures to study the determinants of economic growth, to examine the sources of business cycle fluctuations, to understand the propagation of shocks, to generate forecasts, and to predict the effects of economic policy changes. Bayesian methods are a collection of inference procedures that permit researchers to combine initial information about models and their parameters with sample information in a logically coherent manner by use of Bayes' theorem. Both prior and post-data information is represented by probability distributions.

Unfortunately, the term 'macroeconometrics' is often narrowly associated with large-scale system-of-equations models in the Cowles Commission tradition that were developed from the 1950s to the 1970s. These models came under attack on academic grounds in the mid 1970s. Lucas (1976) argued that the models are unreliable tools for policy analysis because they are unable to predict the effects of policy regime changes on the expectation formation of economic agents in a coherent manner. Sims (1980) criticized the fact that many of the restrictions that are used to identify behavioural equations in these models are inconsistent with dynamic macroeconomic theories and proposed the use of vector autoregressions (VARs) as an alternative. Academic research on econometric models in the Cowles tradition reached a trough in the early 1980s and never recovered. The state-of-the-art is summarized in a monograph by Fair (1994).

I am adopting a modern view of macroeconometrics in this article and will portray an active research area that is tied to modern dynamic macroeconomic theory. Reviewing Bayesian methods in macroeconometrics in a short essay is a difficult task. My review is selective and not representative of Bayesian time-series analysis in general. I have chosen some topics that I believe are important, but the list is by no means exhaustive. I focus on the question how Bayesian methods are used to address some of the challenges that arise in the econometric analysis of dynamic stochastic general equilibrium (DSGE) models and VARs. A more extensive treatment can be found in the survey article by An and Schorfheide (2007).

DSGE models

The term 'DSGE model' is often used to refer to a broad class of dynamic macroeconomic models that spans the standard neoclassical growth model discussed in King, Plosser and Rebelo (1988) as well as the monetary model with numerous real and nominal frictions developed by Christiano, Eichenbaum and Evans (2005).

A common feature of these models is that decision rules of economic agents are derived from assumptions about preferences and technologies by solving intertemporal optimization problems. Moreover, agents potentially face uncertainty with respect to, for instance, total factor productivity or the nominal interest rate set by a central bank. This uncertainty is generated by exogenous stochastic processes or shocks that shift technology or generate unanticipated deviations from a central bank's interest-rate feedback rule. Conditional on distributional assumptions for the exogenous shocks, the DSGE model generates a joint probability distribution for the endogenous model variables such as output, consumption, investment, and inflation.

What are the goals?

While macroeconometric methods are used to address many different questions, several issues stand out. Business cycle analysts are interested in identifying the sources of fluctuations; for instance, how important are monetary policy shocks for movements in aggregate output? We would like to understand the propagation of shocks; for example, what happens to aggregate hours worked in response to a technology shock? Moreover, researchers ask questions about structural changes in the economy: has monetary policy changed in the early 1980s? Why did the volatility of many macroeconomic time series drop in the mid 1980s? Macroeconometricians are also interested in forecasting the future: how will inflation and output growth rates evolve over the next eight quarters? Finally, an important aspect of macroeconometrics is to predict the effect of policy changes: how will output and inflation respond to an unanticipated change in the nominal interest rate? Is it desirable to adopt an inflation targeting regime?

What are the challenges?

In principle one could proceed as follows: specify a DSGE model that is sufficiently rich to address the substantive economic question of interest; derive its likelihood function and fit the model to historical data; answer the questions based on the estimated DSGE model. Unfortunately, this is easier said than done. A trade-off between theoretical coherence and empirical fit poses the first challenge to macroeconometric analysis.

Under certain regularity conditions DSGE models can be well approximated by VARs that satisfy particular cross-coefficient restrictions. The DSGE model is misspecified if these restrictions are at odds with the data and the model has difficulties in tracking and forecasting historical time series. Misspecification was quite apparent for the first generation of DSGE models and has led Kydland, Prescott, and their followers since the early 1980s to abandon formal econometric procedures and advocate a calibration approach, outlined for instance in Kydland and Prescott (1996). Recent Bayesian and non-Bayesian research, however, has resulted in formal econometric tools that are general enough to explicitly account for misspecification problems that arise in the context of DSGE models. Examples of Bayesian approaches are Canova (1994), Dejong, Ingram, and Whiteman (1996), Geweke (1999), Schorfheide (2000), Del Negro and Schorfheide (2004), and Del Negro et al. (2006).

The presence of misspecification might suggest that we should simply ignore the cross-coefficient restrictions implied by dynamic economic theories in the empirical work and try to answer the questions posed above directly by VARs. Unfortunately, there is no free lunch. VARs have many free parameters, and without restrictions on their coefficients tend to generate poor forecasts. VARs do not provide a tight economic interpretation of economic dynamics in terms of the behaviour of rational, optimizing agents. Moreover, it is difficult to predict the effects of rare policy regime changes on the expectation formation and the behaviour of economic agents since these are not explicitly modelled. While the most recent generation of DSGE models comes much closer to matching the empirical fit of VARs, as documented in Smets and Wouters (2003), a trade-off between theoretical coherence and empirical fit remains.

A second challenge is identification. The parameters of a model are identifiable if no two parameterizations of that model generate the same probability distribution for the observables. In VARs the mapping between the one-step-ahead forecast errors of the endogenous variables and the underlying structural shocks is not unique, and additional restrictions are necessary to identify, say, a monetary policy or a technology shock. Many of the popular identification schemes and the controversies surrounding them are surveyed in Cochrane (1994), Christiano and Eichenbaum (1999) and Stock and Watson (2001).

DSGE models can be locally approximated by linear rational expectations (LRE) models. While tightly parameterized compared to VARs, LRE models can generate delicate identification problems. Suppose a model implies that $y_t = \theta \mathbb{E}_t[y_{t-1}] + u_t$, where u_t is an independently distributed random variable with mean zero. If $0 \leq \theta < 1$, then the only stable law of motion for y_t that satisfies the rational expectations restrictions is $y_t = u_t$, which means that θ is not identifiable. More elaborate examples are discussed in Beyer and Farmer (2004), Lubik and Schorfheide (2004; 2006), and Canova and Sala (2006). Unfortunately, it is in many cases difficult to detect identification problems in DSGE models, since the mapping from the structural parameters into the autoregressive law of motion for y_t is highly nonlinear and typically can be evaluated only numerically.

Many regularities of macroeconomic time series are indicative of nonlinearities, for instance, the rise and fall of inflation in the 1970s and early 1980s and time-varying volatility of many macroeconomic time series; see, for example, Cogley and Sargent (2005), Sargent, Williams, and Zha (2006), and Sims and Zha (2006). In VARs nonlinear dynamics are typically generated with time-varying coefficients, whereas most DSGE models are nonlinear and only for convenience approximated by linear rational expectations models. Conceptually the analysis of nonlinear models is very similar to the analysis of linear models, but the implementation of the computations is often more cumbersome and poses a third challenge.

How can Bayesian analysis help?

Bayesian analysis is conceptually straightforward. Pre-sample information about parameters is summarized by a prior distribution $p(\theta)$. We can also assign discrete probabilities to distinct models although the distinction between models and parameters is somewhat artificial. The prior is combined with the conditional distribution of the data given the parameters (likelihood function) $p(Y|\theta)$. The application of Bayes' theorem yields the posterior model probabilities and parameter distributions $p(\theta|Y)$. Markov chain Monte Carlo methods can be used to generate θ draws from the posterior. Based on these draws one can numerically approximate the relevant moments of the posterior and make inference about taste and technology parameters as well as the relative importance and the propagation of the various shocks.

The literature on Bayesian estimation of DSGE models began with work by Landon-Lane (1998), DeJong, Ingram and Whiteman (2000), Schorfheide (2000), and Otrok (2001). DeJong, Ingram and Whiteman (2000) estimate a stochastic growth model and examine its forecasting performance, Otrok (2001) fits a real business cycle with habit formation and time-to-build to the data to assess the welfare costs of business cycles, and Schorfheide (2000) considers cash-in-advance monetary DSGE models. The Bayesian analysis of VAR dates at least back to Doan, Litterman and Sims (1984).

Since DSGE models are to some extent micro-founded, macroeconomists require their parameterization to be consistent with microeconometric evidence on, for instance, labour supply elasticities and the frequency with which firms adjust their prices. If information in the estimation sample were abundant and model misspecification were not a concern, then there would be little need for a prior distribution that summarizes information contained in other data-sets. However, in the estimation of DSGE model this additional information plays an important role.

The prior is used to down-weigh the likelihood function in regions of the parameter space that are inconsistent with out-of-sample information and in which the structural model becomes uninterpretable. The shift from prior to posterior can be an indicator of tensions between different sources of information. If the likelihood function peaks at a value that is at odds with, say, the micro-level information that has been used to construct the prior distribution then marginal data density $\int p(Y|\theta)p(\theta)d\theta$ will be low. If two models have equal prior probabilities, then the ratio of their marginal data densities determine the posterior model odds. Hence, in a posterior odds comparison a DSGE model will automatically be penalized for not being able to reconcile two sources of information with a single set of parameters.

Identification problems manifest themselves through ridges and multiple peaks of equal height in the likelihood function. While Bayesian inference is based on the same likelihood function as classical maximum likelihood estimation, it can bring to bear additional information that may help to discriminate between different parameterizations of a model. If, for instance, the likelihood function is invariant to a subvector θ_1 of θ then the posterior distribution of θ_1 conditional on the remaining parameters will simply equal to the prior distribution. Hence, a comparison of priors and posteriors can provide important insights about the extent to which the data provide information about the parameters of interest. Regardless, the posterior provides a coherent summary of pre-sample and sample information and can be used for inference and decision making. This insight has been used, for instance, by Lubik and Schorfheide (2004) to assess whether monetary policy in the 1970s was conducted in a way that would allow expectations to be self-fulfilling and cause business cycle fluctuations unrelated to fundamental shocks.

Bayesian inference is well suited for model comparisons. Under a loss function that is zero if the correct model is chosen and 1 otherwise, it is optimal to select the model that has the highest posterior probability. However, in many applications, in particular related to the comparison of two possibly misspecified DSGE models, this zero–1 loss function is not very attractive because it does provide little insight into the dimensions along which the structural models should be improved. Schorfheide (2000) provides a framework for the comparison of two or more potentially misspecified DSGE models. A VAR plays the role of a reference model. If the DSGE models are indeed misspecified the VAR will attain the highest posterior probability and the model comparison is based on the question: given a particular loss function, which DSGE model best mimics the dynamics captured by the VAR?

VARs typically have many more parameters than DSGE models and the role of prior distributions is mainly to reduce the effective dimensionality of this parameter space to avoid over-fitting. More interestingly, if one interprets the DSGE model as a set of restrictions on the VAR, then the DSGE model induces a degenerate prior for the VAR coefficients. If the researcher is concerned about potential misspecification of the DSGE model, a natural approach is to relax the DSGE model restrictions and construct a non-degenerate prior distribution that concentrates most of its mass near the restrictions. This approach was originally proposed by Ingram and Whiteman (1994) and has been further developed by Del Negro and Schorfheide (2004), who provide a framework for the joint estimation of VAR and DSGE model parameters. The framework generates a continuum of intermediate specifications that differ according to the degree by which the restrictions are relaxed. This degree is measured by a hyperparameter and the posterior distribution of the hyperparameter can be interpreted as a measure of fit.

Incorporating model and parameter uncertainty into a decision is straightforward in a Bayesian set-up. Levin et al. (2006), for instance, study the effect of optimal monetary policy under parameter uncertainty in the context of an estimated DSGE model. Let δ denote a decision, such as the choice of a monetary policy rule or a tax rate, and $L(\delta, \theta)$ be a loss function that is used to evaluate the decision. The optimal choice minimizes the posterior risk $\int L(\delta, \theta)p(\theta|Y)d\theta$. The calculation of the risk is facilitated by Markov chain Monte Carlo methods that enable a numerical evaluation of expected losses. If the parameter θ in the loss function is replaced by a future observation y' and $p(\theta|Y)$ is replaced by the predictive distribution $p(y'|Y)$, the decision-theoretic framework can also be used to generate forecasts from the Bayes model.

Finally, with respect to the analysis of nonlinear models, Bayesian methods are in some instances very helpful. Data-augmentation techniques let researchers efficiently deal with numerical complications that arise in models with latent state variables, such as regime-switching models or VARs with time-varying coefficients as in Cogley and Sargent (2005) and Sims and Zha (2006). On the other hand, the need to compute a likelihood function can create serious obstacles. For instance, the computation of the likelihood function for a DSGE model solved with a nonlinear solution method requires a computational-intensive particle filter as in Fernández-Villaverde and Rubio-Ramírez (2006).

Conclusion

The Bayesian paradigm provides a rich framework for inference and decision making with modern macroeconometric models such as DSGE models and VARs. The econometric methods can be tailored to cope with the challenges in this literature: potential model misspecification and a trade-off between theoretical coherence and empirical fit, identification problems, and estimation of models with many parameters based on relatively few observations. Advances in Bayesian computations let the researcher efficiently deal with numerical complications that arise in models with latent state variables, such as regime-switching models, or nonlinear state-space models.

FRANK SCHORFHEIDE

See also **vector autoregressions.**

Bibliography

An, S. and Schorfheide, F. 2007. Bayesian analysis of DSGE models. *Econometric Reviews*, 26, pp. 113–72.

Beyer, A. and Farmer, R. 2004. On the indeterminacy of New-Keynesian economics. Working Paper No. 323, European Central Bank.

Canova, F. 1994. Statistical inference in calibrated models. *Journal of Applied Econometrics* 9, S123–S144.

Canova, F. and Sala, L. 2006. Back to square one: identification issues in DSGE models. Working Paper No. 583, European Central Bank.

Christiano, L. and Eichenbaum, M. 1999. Monetary policy shocks: what have we learned and to what end? *Handbook of Macroeconomics*, vol. 1A, ed. J. Taylor and M. Woodford. Amsterdam: North-Holland.

Christiano, L., Eichenbaum, M. and Evans, C. 2005. Nominal rigidities and the dynamic effects of a shock to monetary policy. *Journal of Political Economy* 113, 1–45.

Cochrane, J. 1994. Shocks. *Carnegie Rochester Conference Series* 41, 295–364.

Cogley, T. and Sargent, T. 2005. Drifts and volatilities: monetary policies and outcomes in the post WWII U.S. *Review of Economic Dynamics* 8, 262–302.

DeJong, D., Ingram, B. and Whiteman, C. 1996. A Bayesian approach to calibration. *Journal of Business Economics and Statistics* 14, 1–9.

DeJong, D., Ingram, B. and Whiteman, C. 2000. A Bayesian approach to dynamic macroeconomics. *Journal of Econometrics* 98, 201–23.

Del Negro, M. and Schorfheide, F. 2004. Priors from equilibrium models for VARs. *International Economic Review* 45, 643–73.

Del Negro, M., Schorfheide, F., Smets, F. and Wouters, R. 2006. On the fit of New Keynesian models. *Journal of Business and Economic Statistics* (forthcoming).

Doan, T., Litterman, R. and Sims, C. 1984. Forecasting and conditional projections using realistic prior distributions. *Econometric Reviews* 3, 1–100.

Fair, R. 1994. *Testing Macroeconomic Models.* Cambridge, MA: Harvard University Press.

Fernández-Villaverde, J. and Rubio-Ramírez, J. 2006. Estimating macroeconomic models: a likelihood approach. *Review of Economic Studies* (forthcoming).

Geweke, J. 1999. Computational experiments and reality. *Computing in Economics and Finance* 1999, No. 401. Society for Computational Economics, Department of Economics, Boston College.

Ingram, B. and Whiteman, C. 1994. Supplanting the Minnesota prior – forecasting macroeconomic time series using real business cycle model priors. *Journal of Monetary Economics* 34, 497–510.

Levin, A., Onatski, A., Williams, J. and Williams, N. 2006. Monetary policy under uncertainty in micro-founded macroeconometric models. In *NBER Macroeconomics Annual 2005*, ed. M. Gertler and K. Rogoff. Cambridge, MA: MIT Press.

King, R., Plosser, C. and Rebelo, S. 1988. Production, growth, and business cycles: I. Neoclassical model. *Journal of Monetary Economics* 81, 819–40.

Kydland, F. and Prescott, E. 1996. The computational experiment: An econometric tool. *Journal of Economic Perspectives* 10(1), 69–85.

Landon-Lane, J. 1998. Bayesian comparison of dynamic macroeconomic models. Ph.D. thesis, University of Minnesota.

Lubik, T. and Schorfheide, F. 2004. Testing for indeterminacy: an application to U.S. monetary policy. *American Economic Review* 94, 190–217.

Lubik, T. and Schorfheide, F. 2006. A Bayesian look at New Open Economy Macroeconomics. In *NBER Macroeconomics Annual 2005*, ed. M. Gertler and K. Rogoff. Cambridge, MA: MIT Press.

Lucas, R., Jr. 1976. Econometric policy evaluation: a critique. In *The Phillips Curve and Labor Markets*, ed. K. Brunner and A. Meltzer. Amsterdam: North-Holland.

Otrok, C. 2001. On measuring the welfare cost of business cycles. *Journal of Monetary Economics* 47, 61–92.

Sargent, T., Williams, N. and Zha, T. 2006. Shocks and government beliefs: the rise and fall of American inflation. *American Economic Review* 96, 1193–224.

Schorfheide, F. 2000. Loss function-based evaluation of DSGE models. *Journal of Applied Econometrics* 15, 645–70.

Sims, C. 1980. Macroeconomics and reality. *Econometrica* 48, 1–48.

Sims, C. and Zha, T. 2006. Were there regime switches in U.S. monetary policy? *American Economic Review* 96, 54–81.

Smets, F. and Wouters, R. 2003. An estimated stochastic dynamic general equilibrium model of the Euro area. *Journal of the European Economic Association* 1, 1123–75.

Stock, J. and Watson, M. 2001. Vector autoregressions. *Journal of Economic Perspectives* 15(4), 101–16.

Bayesian time series analysis

Bayesian methods

The importance of Bayesian methods in econometrics has increased rapidly since the early 1990s. This has, no doubt, been fuelled by an increasing appreciation of the advantages that Bayesian inference entails. In particular, it provides us with a formal way to incorporate the prior information we often possess before seeing the data, it fits perfectly with sequential learning and decision making, and it directly leads to exact small sample results. In addition, the Bayesian paradigm is particularly natural for prediction, since we take into account all parameter or even model uncertainty. The predictive distribution is the sampling distribution where the parameters are integrated out with the posterior distribution and provides exactly what we need for forecasting, often a key goal of time-series analysis.

Usually, the choice of a particular econometric model is not pre-specified by theory, and many competing models can be entertained. Comparing models can be done formally in a Bayesian framework through so-called posterior odds, which is the product of the prior odds and the Bayes factor. The Bayes factor between any two models is the ratio of the likelihoods integrated out with the corresponding prior and summarizes how the data favour one model over another. Given a set of possible models, this immediately leads to posterior model probabilities. Rather than choosing a single model, a natural way to deal with model uncertainty is to use the posterior model probabilities to average out the inference (on observables or parameters) corresponding to each of the separate models. This is called Bayesian model averaging. The latter was already mentioned in Leamer (1978) and recently applied to economic problems in, for example, Fernández, Ley and Steel (2001) (for growth regressions) and in Garratt et al. (2003) and Jacobson and Karlsson (2004) (for macroeconomic forecasting).

An inevitable prerequisite for using the Bayesian paradigm is the specification of prior distributions for all quantities in the model that are treated as unknown. This has been the source of some debate, a prime example of which is given by the controversy over the choice of prior on the coefficients of simple autoregressive models. The issue of testing for a unit root (deciding whether to difference the series before modelling it through a stationary model) is subject to many difficulties from a sampling-theoretical perspective. Comparing models in terms of posterior odds provides a very natural Bayesian approach to testing, which does not rely on asymptotics or approximations. It is, of course, sensitive to how the competing models are defined (for example, do we contrast the stationary model with a pure unit root model or a model with a root larger than or equal to 1?) and to the choice of prior. The latter issues have lead to some controversy in the literature, and prompted a special issue of the *Journal of Applied Econometrics* with animated discussion around the paper by Phillips (1991). The latter paper advocated the use of Jeffreys' principles to represent prior ignorance about the parameters (see also the discussion in Bauwens, Lubrano and Richard, 1999, Ch. 6).

Like the choice between competing models, forecasting can also be critically influenced by the prior. In fact, prediction is often much more sensitive than parameter inference to the choice of priors (especially on autoregressive coefficients) and Koop, Osiewalski and Steel (1995) show that imposing stationarity through the prior on the autoregressive coefficient in a simple AR(1) model need not lead to stabilization of the predictive variance as the forecast horizon increases.

Computational algorithms

Partly, the increased use of Bayesian methods in econometrics is a consequence of the availability of very efficient and flexible algorithms for conducting inference through simulation in combination with ever more powerful computing facilities, which have made the Bayesian analysis of non-standard problems an almost routine activity. Particularly, Markov chain Monte Carlo (MCMC) methods have opened up a very useful class of computational algorithms and have created a veritable revolution in the implementation of Bayesian methods. Whereas Bayesian inference before 1990 was at best a difficult undertaking in practice, reserved for a small number of specialized researchers and limited to a rather restricted set of models, it has now become a very accessible procedure which can fairly easily be applied to almost any model. The main idea of MCMC methods is that inference about an analytically intractable posterior (often in high dimensions) is conducted through generating a Markov chain which converges to a chain of drawings from the posterior distribution. Of course, predictive inference is also immediately available once one has such a chain of drawings. Various ways of constructing such a Markov chain exist, depending on the structure of the problem. The most commonly used are the Gibbs sampler and the Metropolis Hastings sampler. The use of data augmentation (that is, adding auxiliary variables to the sampler) can facilitate implementation of the MCMC sampler, so that often the analysis is conducted on an augmented space including not only the model parameters but also things like latent variables and missing observations. An accessible reference to MCMC methods is, for example, Gamerman (1997).

As a consequence, we are now able to conduct Bayesian analysis of time series models that have been around for a long time (such as ARMA models) but also of more recent additions to our catalogue of models, such as Markov switching and nonparametric models, and the literature is vast. Therefore, I will have to be selective and will try to highlight a few areas which I think are of particular interest. I hope this can give an idea of the role that Bayesian methods can play in modern time series analysis.

ARIMA and ARFIMA models

Many models used in practice are of the simple ARIMA type, which have a long history and were formalized in Box and Jenkins (1970). ARIMA stands for 'autoregressive integrated moving average' and an ARIMA(p, d, q) model for an observed series $\{y_t\}$, $t = 1, \ldots, T$ is a model where the dth difference $z_t = y_t - y_{t-d}$ is taken to induce stationarity of the series. The process $\{z_t\}$ is then modelled as $z_t = \mu + \varepsilon_t$ with

$$\varepsilon_t = \phi_1 \varepsilon_{t-1} + \cdots + \phi_p \varepsilon_{t-p} + u_t - \theta_1 u_{t-1} - \cdots - \theta_q u_{t-q}$$

or in terms of polynomials in the lag operator L (defined through $L^s x_t = x_{t-s}$):

$$\phi(L)\varepsilon_t = \theta(L)u_t$$

where $\{u_t\}$ is white noise and usually distributed as $u_t \sim N(0, \sigma^2)$. The stationarity and invertibility conditions are simply that the roots of $\phi(L)$ and $\theta(L)$, respectively, are outside the unit circle. An accessible and extensive treatment of the use of Bayesian methods for ARIMA models can be found in Bauwens, Lubrano and Richard (1999). The latter book also has a useful discussion of multivariate modelling using vector autoregressive (VAR) models and cointegration.

The MCMC samplers used for inference in these models typically use data augmentation. Marriott et al. (1996) use a direct conditional likelihood evaluation and augment with unobserved data and errors to conduct inference on the parameters (and the augmented vectors $\varepsilon_a = (\varepsilon_0, \varepsilon_{-1}, \ldots, \varepsilon_{1-p})'$ and $u_a = (u_0, u_{-1}, \ldots, u_{1-q})$). A slightly different approach is followed by Chib and Greenberg (1994), who consider a state space representation and use MCMC on the parameters augmented with the initial state vector.

ARIMA models will either display perfect memory (if there are any unit roots) or quite short memory with geometrically decaying autocorrelations (in the case of a stationary ARMA model). ARFIMA ('autoregressive fractionally integrated moving average') models (see Granger and Joyeux, 1980) have more flexible memory properties, due to fractional integration which allows for hyperbolic decay.

Consider $z_t = \Delta y_t - \mu$, which is modelled by an ARFIMA(p, δ, q) model as:

$$\phi(L)(1 - L)^{\delta} z_t = \theta(L)u_t,$$

where $\{u_t\}$ is white noise with $u_t \sim N(0, \sigma^2)$, and $\delta \in (-1, 0.5)$. The fractional differencing operator $(1 - L)^{\delta}$ is defined as

$$(1 - L)^{\delta} = \sum_{j=0}^{\infty} c_j(\delta) L^j,$$

where $c_0(\cdot) = 1$ and for $j > 0$:

$$c_j(a) = \prod_{k=1}^{j} \left(1 - \frac{1 + a}{k}\right).$$

This model takes the entire past of z_t into account, and has as a special case the ARIMA(p, 1, q) for y_t (for $\delta = 0$). If $\delta > -1$, z_t is invertible (Odaki, 1993) and for $\delta < 0.5$ we have stationarity of z_t. Thus, we have three regimes:

$\delta \in (-1, -0.5)$: y_t trend-stationary with long memory

$\delta \in (-0.5, 0)$: z_t stationary with intermediate memory
$\delta \in (0, 0.5)$: z_t stationary with long memory.

Of particular interest is the impulse response function $I(n)$, which captures the effect of a shock of size one at time t on y_{t+n}, and is given by

$$I(n) = \sum_{i=0}^{n} c_i(-\delta - 1)J(n - i),$$

with $J(i)$ the standard ARMA(p, q) impulse responses (that is, the coefficients of $\phi^{-1}(L)\theta(L)$). Thus, $I(\infty)$ is 0 for $\delta<0$, $\theta(1)/\phi(1)$ for $\delta=0$ and ∞ for $\delta>0$. Koop et al. (1997) analyse the behaviour of the impulse response function for real US GNP data using a set of 32 possible models containing both ARMA and ARFIMA models for z_t. They use Bayesian model averaging to conduct predictive inference and inference on the impulse responses, finding about one-third of the posterior model probability concentrated on the ARFIMA models. Koop et al. (1997) use importance sampling to conduct inference on the parameters, while MCMC methods are used in Pai and Ravishanker (1996) and Hsu and Breidt (2003).

State space models

The basic idea of such models is that an observable y_t is generated by an observation or measurement equation

$$y_t = F_t' \, \theta_t + v_t,$$

where $v_t \sim N(0, V_t)$, and is expressed in terms of an unobservable state vector θ_t (capturing, for example, levels, trends or seasonal effects) which is itself dynamically modelled through a system or transition equation

$$\theta_t = G_t\theta_{t-1} + w_t,$$

with $w_t \sim N(0, W_t)$ and all error terms $\{v_t\}$ and $\{w_t\}$ are mutually independent. Normality is typically assumed, but is not necessary and a prior distribution is required to describe the initial state vector θ_0. Models are defined by the (potentially time-varying) quadruplets $\{F_t, G_t, V_t, W_t\}$ and the time-varying states θ_t make them naturally adaptive to changing circumstances. This feature also fits very naturally with Bayesian methods, which easily allow for sequential updating. These models are quite general and include as special cases, for example, ARMA models, as well as stochastic volatility models, used in finance (see below).

There is a relatively long tradition of state space models in econometrics and a textbook treatment can already be found in Harvey (1981). Bayesian methods for such models were discussed in, for example, Harrison and Stevens (1976), and a very extensive treatment is provided in West and Harrison (1997), using the terminology 'dynamic linear models'. An accessible introduction to Bayesian analysis with these models can be found in Koop (2003, Ch. 8).

Online sequential estimation and forecasting with the simple Normal state space model above can be achieved with Kalman filter recursions, but more sophisticated models (or estimation of some aspects of the model besides the states) usually require numerical methods for inference. In that case, the main challenge is typically the simulation of the sequence of unknown state vectors. Single-state samplers (updating one state vector at a time) are generally less efficient than multi-state samplers, where all the states are updated jointly in one step. Efficient algorithms for multi-state MCMC sampling schemes have been proposed by Carter and Kohn (1994) and de Jong and Shephard (1995). For fundamentally non-Gaussian models, the methods in Shephard and Pitt (1997) can be used. A recent contribution of Harvey, Trimbur and van Dijk (2006) uses Bayesian methods for state space models with trend and cyclical components, exploiting informative prior notions regarding the length of economic cycles.

Markov switching and mixture models

Markov switching models were introduced by Hamilton (1989) and essentially rely on an unobserved regime indicator s_t, which is assumed to behave as a discrete Markov chain with, say, K different levels. Given $s_t = i$ the observable y_t will be generated by a time series model which corresponds to regime i, where $i = 1, \ldots, K$. These models are often stationary ARMA models, and the switching between regimes will allow for some non-stationarity, given the regime allocations. Such models are generally known as hidden Markov models in the statistical literature.

Bayesian analysis of these models is very natural, as that methodology provides an immediate framework for dealing with the latent states, $\{s_t\}$, and a simple MCMC framework for inference on both the model parameters and the states was proposed in Albert and Chib (1993). A bivariate version of the Hamilton model is analysed in Paap and van Dijk (2003), who also examine the cointegration relations between the series modelled and find evidence for cointegration between US per capita income and consumption. Using a similar model, Smith and Summers (2005) examine the synchronization of business cycles across countries and find strong evidence in favour of the multivariate Markov switching model over a linear VAR model.

When panel data are available, another relevant question is whether one can find clusters of entities (such as countries or regions) which behave similarly, while allowing for differences between the clusters. This issue is addressed from a fully Bayesian perspective in Frühwirth-Schnatter and Kaufmann (2006), where model-based clustering (across countries) is integrated with a Markov switching framework (over time). This is achieved by a finite mixture of Markov switching autoregressive models, where the number of elements in the mixture corresponds to the number of clusters and is treated as an unknown parameter. Frühwirth-Schnatter and Kaufmann (2006) analyse a panel of growth rates of industrial production in 21 countries and distinguish two clusters with different business cycles. This also feeds into the important debate on the existence of so-called convergence clubs in terms of income per capita as discussed in Durlauf and Johnson (1995) and Canova (2004).

Another popular way of inducing nonlinearities in time series models is through so-called threshold autoregressive models, where the choice of regimes is not governed by an underlying Markov chain but depends on previous values of the observables. Bayesian analyses of such models can be found in, for example, Geweke and Terui (1993) and are extensively reviewed in Bauwens, Lubrano and Richard (1999, Ch. 8). The use of Bayes factors to choose between various nonlinear models, such as threshold autoregressive and Markov switching models is discussed in Koop and Potter (1999).

Geweke and Keane (2006) present a general framework for Bayesian mixture models where the state probabilities can depend on observed covariates. They investigate increasing the number of components in the mixture, as well as the flexibility of the components and the specification of the mechanism for the state probabilities, and find their mixture model approach compares well with ARCH-type models (as described in the next section) in the context of stock return data.

Models for time-varying volatility

The use of conditional heteroskedasticity initially introduced in the ARCH (autoregressive conditional heteroskedasticity) model of Engle (1982) has been extremely successful in modelling financial time series, such as stock prices, interest rates and exchange rates. The ARCH model was generalized to GARCH (generalized ARCH) by Bollerslev (1986). A simple version of the GARCH model for an observable series $\{y_t\}$, given its past which is denoted by I_{t-1}, is the following:

$$y_t = u_t\sqrt{h_t} \tag{1}$$

where $\{u_t\}$ is white noise with mean zero and variance one. The conditional variance of y_t given I_{t-1} is then h_t, which is modelled as

$$h_t = \omega + \sum_{i=1}^{p} \alpha_i y_{t-i}^2 + \sum_{j=1}^{q} \beta_j h_{t-i} \tag{2}$$

where all parameters are positive and usually $p = q = 1$ is sufficient in practical applications. Bayesian inference for such models was conducted through importance sampling in Kleibergen and van Dijk (1993) and, with MCMC methods, in Bauwens and Lubrano (1998).

An increasingly popular alternative model allows for the variance h_t to be determined by its own stochastic process. This is the so-called stochastic volatility model, which in its basic form replaces (2) by the assumption that the logarithm of the conditional volatility is driven by its own AR(1) process

$$\ln(h_t) = \alpha + \delta \ln(h_{t-1}) + v_t,$$

where $\{v_t\}$ is a white noise process independent of $\{u_t\}$ in (1). Inference in such models requires dealing with the latent volatilities, which are incidental parameters and have to be integrated out in order to evaluate the likelihood. MCMC sampling of the model parameters and the volatilities jointly is a natural way of handling this. An MCMC sampler where each volatility was treated in a separate step was introduced in Jacquier, Polson and Rossi (1994), and efficient algorithms for multi-state MCMC sampling schemes were suggested by Carter and Kohn (1994) and de Jong and Shephard (1995). Many extensions of the simple stochastic volatility model above have been proposed in the literature, such as correlations between the $\{u_t\}$ and $\{v_t\}$ processes, capturing leverage effects, or fat-tailed distributions for u_t. Inference with these more general models and ways of choosing between them are discussed in Jacquier, Polson and Rossi (2004).

Recently, the focus in finance has shifted more towards continuous-time models, and continuous-time versions of stochastic volatility models have been proposed. In particular, Barndorff-Nielsen and Shephard (2001) introduce a class of models where the volatility behaves according to an Ornstein–Uhlenbeck process, driven by a positive Lévy process without Gaussian component (a pure jump process). These models introduce discontinuities (jumps) into the volatility process. Barndorff-Nielsen and Shephard (2001) also consider superpositions of such processes. Bayesian inference in such models through MCMC methods is complicated by the fact that the model parameters and the latent volatility process are often highly correlated in the posterior, leading to the problem of over-conditioning. Griffin and Steel (2006b) propose MCMC methods based on a series representation of Lévy processes, and avoid over-conditioning by dependent thinning methods. In addition, they extend the model by including a jump component in the returns, leverage effects and separate risk pricing for the various volatility components in the superposition. An application to stock price data shows substantial empirical support for a superposition of processes with different risk premiums and a leverage effect. A different approach to inference in such models is proposed in Roberts, Papaspiliopoulos and Dellaportas (2004), who suggest a re-parameterization to reduce the correlation between the data and the process. The re-parameterized process is then proposed only in accordance with the parameters.

Semi- and nonparametric models

The development and use of Bayesian nonparametric methods has been a rapidly growing topic in the statistics literature, some of which is reviewed in Müller and Quintana (2004). However, the latter review does not include applications to time series, which have been perhaps less prevalent than applications in other areas, such as regression, survival analysis and spatial statistics.

Bayesian nonparametrics is sometimes considered an oxymoron, since Bayesian methods are inherently likelihood-based, and thus require a complete probabilistic specification of the model. However, what is usually called Bayesian nonparametrics corresponds to models with priors defined over infinitely dimensional parameter spaces (functional spaces) and this allows for very flexible procedures, where the data are allowed to influence virtually all features of the model.

Defining priors over collections of distribution functions requires the use of random probability measures. The most popular of these is the so-called Dirichlet process prior introduced by Ferguson (1973). This is defined for a space Θ and a σ-field B of subsets of Θ. The process is parameterized in terms of a probability measure H on (Θ, B) and a positive scalar M. A random probability measure, F, on (Θ, B) follows a Dirichlet process $DP(MH)$ if, for any finite measurable partition, $B_1, \ldots, B_k$, the vector $(F(B_1), \ldots, F(B_k))$ follows a Dirichlet distribution with parameters $(MH(B_1), \ldots, MH(B_k))$. The distribution H centres the process and M can be interpreted as a precision parameter.

The Dirichlet process is (almost surely) discrete and, thus, not always suitable for modelling observables directly. It is, however, often incorporated into semiparametric models using the hierarchical framework

$$y_i \sim g(y_i|u_i) \quad \text{with} \quad u_i \sim F \quad \text{and} \quad F \sim \mathrm{DP}(MH), \tag{3}$$

where $g(\cdot)$ is a probability density function. This model is usually referred to as a 'mixture of Dirichlet processes'. The marginal distribution for y_i is a mixture of the distribution characterized by $g(\cdot)$. This basic model can be extended: the density $g(\cdot)$ or the centring distribution H can be (further) parameterized, and inference can be made about these parameters. In addition, inference can be made about the mass parameter M. Inference in these models with the use of MCMC algorithms has become quite feasible, with methods based on MacEachern (1994) and Escobar and West (1995).

However, the model in (3) assumes independent and identically distributed observations and is, thus, not directly of interest for time series modelling. A simple approach followed by Hirano (2002) is to use (3) for modelling the errors of an autoregressive model specification. However, this does not allow for the distribution to change over time. Making the random probability measure F itself depend on lagged values of the variable under consideration y_t (or, generally, any covariates) is not a straightforward extension. Müller, West and MacEachern (1997) propose a solution by modelling y_t and y_{t-1} jointly, using a mixture of Dirichlet processes. The main problem with this approach is that the resulting model is not really a conditional model for y_t given y_{t-1}, but incorporates a contribution from the marginal model for y_{t-1}. Starting from the stick-breaking representation of a Dirichlet process, Griffin and Steel (2006a) introduce the class of order-based dependent Dirichlet processes, where the weights in the stick-breaking representation induce dependence between distributions that correspond to similar values of the covariates (such as time). This class induces a Dirichlet process at each covariate value, but allows for dependence. Similar weights are associated with similar orderings of the elements in the representation and these orderings are derived from a point process in such a way that distributions that are close in covariate space will tend to be highly correlated. One proposed construction (the arrivals ordering) is particularly suitable for time series and is applied to stock index returns, where the volatility is modelled through an order-based dependent Dirichlet process. Results illustrate the flexibility and the feasibility of this approach. Jensen (2004) uses a Dirichlet process prior on the wavelet representation of the observables to conduct Bayesian inference in a stochastic volatility model with long memory.

Conclusion: where are we heading?

In conclusion, Bayesian analysis of time series models is alive and well. In fact, it is an ever growing field, and we are now starting to explore the advantages that can be gained from using Bayesian methods on time series data. Bayesian counterparts to the classical analysis of existing models, such as AR(F)IMA models, are by now well-developed and a lot of work has already been done there to make Bayesian inference in these models a fairly routine activity. The main challenge ahead for methodological research in this field is perhaps to further develop really novel models that not merely constitute a change of inferential paradigm but are inspired by the new and exciting modelling possibilities that are available through the combination of Bayesian methods and MCMC computational algorithms. In particular, nonparametric Bayesian time-series modelling falls in that category and I expect that more research in this area will be especially helpful in increasing our understanding of time series data.

MARK F.J. STEEL

See also **ARCH models; Bayesian methods in macroeconometrics; long memory models; state space models; stochastic volatility models; time series analysis.**

Bibliography

Albert, J. and Chib, S. 1993. Bayes inference via Gibbs sampling of autoregressive time series subject to Markov mean and variance shifts. *Journal of Business and Economic Statistics* 11, 1–15.

Barndorff-Nielsen, O. and Shephard, N. 2001. Non-Gaussian OU based models and some of their uses in financial economics. *Journal of the Royal Statistical Society Series B* 63, 167–241 (with discussion).

Bauwens, L. and Lubrano, M. 1998. Bayesian inference on GARCH models using the Gibbs sampler. *Econometrics Journal* 1, C23–C46.

Bauwens, L., Lubrano, M. and Richard, J.F. 1999. *Bayesian Inference in Dynamic Econometric Models*. Oxford: Oxford University Press.

Bollerslev, T. 1986. Generalized autoregressive conditional heteroskedasticity. *Journal of Econometrics* 31, 307–27.

Box, G. and Jenkins, G. 1970. *Time Series Analysis: Forecasting and Control*. San Francisco: Holden Day.

Canova, F. 2004. Testing for convergence clubs in income per capita: a predictive density approach. *International Economic Review* 45, 49–77.

Carter, C. and Kohn, R. 1994. On Gibbs sampling for state space models. *Biometrika* 81, 541–53.

Chib, S. and Greenberg, E. 1994. Bayes inference in regression models with ARMA(p,q) errors. *Journal of Econometrics* 64, 183–206.

de Jong, P. and Shephard, N. 1995. The simulation smoother for time series models. *Biometrika* 82, 339–50.

Durlauf, S. and Johnson, P. 1995. Multiple regimes and cross-country growth behaviour. *Journal of Applied Econometrics* 10, 365–84.
Engle, R. 1982. Autoregressive conditional heteroskedasticity with estimates of the variance of United Kingdom inflation. *Econometrica* 50, 987–1008.
Escobar, M. and West, M. 1995. Bayesian density-estimation and inference using mixtures. *Journal of the American Statistical Association* 90, 577–88.
Ferguson, T.S. 1973. A Bayesian analysis of some nonparametric problems. *Annals of Statistics* 1, 209–230.
Fernández, C., Ley, E. and Steel, M. 2001. Model uncertainty in cross-country growth regressions. *Journal of Applied Econometrics* 16, 563–76.
Frühwirth-Schnatter, S. and Kaufmann, S. 2006. Model-based clustering of multiple time series. *Journal of Business and Economic Statistics* (forthcoming).
Gamerman, D. 1997. *Markov Chain Monte Carlo: Stochastic Simulation for Bayesian Inference*. Boca Raton, FL: Chapman and Hall/CRC.
Garratt, A., Lee, K., Pesaran, H. and Shin, Y. 2003. Forecast uncertainties in macroeconometric modelling: an application to the UK economy. *Journal of the American Statistical Association* 98, 829–38.
Geweke, J. and Terui, N. 1993. Bayesian threshold autoregressive models for nonlinear time series. *Journal of Time Series Analysis* 14, 441–54.
Geweke, J. and Keane, M. 2006. Smoothly mixing regressions. *Journal of Econometrics* (forthcoming).
Granger, C. and Joyeux, R. 1980. An introduction to long-memory time series models and fractional differencing. *Journal of Time Series Analysis* 1, 15–39.
Griffin, J. and Steel, M. 2006a. Order-based dependent Dirichlet processes. *Journal of the American Statistical Association* 101, 179–94.
Griffin, J. and Steel, M. 2006b. Inference with non-Gaussian Ornstein-Uhlenbeck processes for stochastic volatility. *Journal of Econometrics* 134, 605–44.
Hamilton, J. 1989. A new approach to the economic analysis of nonstationary time series and the business cycle. *Econometrica* 57, 357–84.
Harrison, P. and Stevens, C. 1976. Bayesian forecasting. *Journal of the Royal Statistical Society Series B* 38, 205–47 (with discussion).
Harvey, A. 1981. *Time Series Models*. Oxford: Philip Allen.
Harvey, A., Trimbur, T. and van Dijk, H. 2006. Trends and cycles in economic time series: a Bayesian approach. *Journal of Econometrics* (forthcoming).
Hirano, K. 2002. Semiparametric Bayesian inference in autoregressive panel data models. *Econometrica* 70, 781–99.
Hsu, N. and Breidt, F. 2003. Bayesian analysis of fractionally integrated ARMA with additive noise. *Journal of Forecasting* 22, 491–514.
Jacobson, T. and Karlsson, S. 2004. Finding good predictors for inflation: a Bayesian model averaging approach. *Journal of Forecasting* 23, 479–96.
Jacquier, E., Polson, N. and Rossi, P. 1994. Bayesian analysis of stochastic volatility models. *Journal of Business and Economic Statistics* 12, 371–417 (with discussion).
Jacquier, E., Polson, N. and Rossi, P. 2004. Bayesian analysis of stochastic volatility models with fat-tails and correlated errors. *Journal of Econometrics* 122, 185–212.
Jensen, M.J. 2004. Semiparametric Bayesian inference of long-memory stochastic volatility models. *Journal of Time Series Analysis* 25, 895–922.
Kleibergen, F. and van Dijk, H. 1993. Non-stationarity in GARCH models: a Bayesian analysis. *Journal of Applied Econometrics* 8, S41–S61.
Koop, G. 2003. *Bayesian Econometrics*. Chichester: Wiley.

Koop, G., Ley, E., Osiewalski, J. and Steel, M. 1997. Bayesian analysis of long memory and persistence using ARFIMA models. *Journal of Econometrics* 76, 149–69.

Koop, G., Osiewalski, J. and Steel, M. 1995. Bayesian long-run prediction in time series models. *Journal of Econometrics* 69, 61–80.

Koop, G. and Potter, S. 1999. Bayes factors and nonlinearity: evidence from economic time series. *Journal of Econometrics* 88, 251–81.

Leamer, E. 1978. *Specification Searches: Ad Hoc Inference with Nonexperimental Data*. New York: Wiley.

MacEachern, S. 1994. Estimating Normal means with a conjugate style Dirichlet process prior. *Communications in Statistics, B* 23, 727–41.

Marriott, J., Ravishanker, N., Gelfand, A. and Pai, J. 1996. Bayesian analysis of ARMA processes: complete sampling-based inference under exact likelihoods. In *Bayesian Analysis in Statistics and Econometrics*, ed. D. Berry, K. Chaloner and J. Geweke. New York: Wiley.

Müller, P. and Quintana, F. 2004. Nonparametric Bayesian data analysis. *Statistical Science* 19, 95–110.

Müller, P., West, M. and MacEachern, S. 1997. Bayesian models for nonlinear autoregressions. *Journal of Time Series Analysis* 18, 593–614.

Odaki, M. 1993. On the invertibility of fractionally differenced ARIMA processes. *Biometrika* 80, 703–09.

Paap, R. and van Dijk, H. 2003. Bayes estimation of Markov trends in possibly cointegrated series: an application to U.S. consumption and income. *Journal of Business and Economic Statistics* 21, 547–63.

Pai, J. and Ravishanker, N. 1996. Bayesian modeling of ARFIMA processes by Markov chain Monte Carlo methods. *Journal of Forecasting* 16, 63–82.

Phillips, P. 1991. To criticize the critics: an objective Bayesian analysis of stochastic trends. *Journal of Applied Econometrics* 6, 333–473 (with discussion).

Roberts, G., Papaspiliopoulos, O. and Dellaportas, P. 2004. Bayesian inference for non-Gaussian Ornstein-Uhlenbeck stochastic volatility processes. *Journal of the Royal Statistical Society Series B* 66, 369–93.

Shephard, N. and Pitt, M. 1997. Likelihood analysis of non-Gaussian measurement time series. *Biometrika* 84, 653–67.

Smith, P.A. and Summers, P.M. 2005. How well do Markov switching models describe actual business cycles? The case of synchronization. *Journal of Applied Econometrics* 20, 253–74.

West, M. and Harrison, P. 1997. *Bayesian Forecasting and Dynamic Models*, 2nd edn. New York: Springer Verlag.

central limit theorems

At the end of the 17th century, the mathematician Abraham de Moivre first used the normal distribution as an approximation for the percentage of successes in a large number of experiments. Later on, Laplace generalized his results, but it took 20th century mathematics to give an exact and complete description of this subject. So let me now describe the modern approach. We assume that for each n we have given a sequence $X_{1,n}, \ldots, X_{n,n}$ of random variables, which we assume to be independent. Then we want to 'approximate' the distribution of

$$S_n = \sum_{i=1}^{n} X_{i,n}$$

by a standard normal distribution, whose density equals

$$\frac{1}{\sqrt{2\pi}} \int_A \exp\left(-\frac{x^2}{2} \right) dx.$$

Let us denote by $P(B)$ the probability of an event B. If X is a random variable, than let us denote by $E(X)$ its expectation. For $A \subseteq \mathbf{R}$ let $[X \in A]$ be the event that X takes a value into A. Written in formal terms, we want to establish that

$$\lim_{n \to \infty} P([S_n \in A]) = \frac{1}{\sqrt{2\pi}} \int_A \exp\left(-\frac{x^2}{2} \right) dx \tag{1}$$

or

$$\lim_{n \to \infty} Ef(S_n) = \frac{1}{\sqrt{2\pi}} \int f(x) \exp\left(-\frac{x^2}{2} \right) dx. \tag{2}$$

The first question we have to ask ourselves is the nature of the approximation. Clearly it is impossible to approximate the distribution of S_n for *all* sets. Consider the binomial distribution discussed above. In this case, each S_n can only take a finite number of values. Therefore the possible values for all S_n lie for all n in a countable set, which has zero probability under the normal distribution.

So we have to aim at a compromise: the smaller the class of sets A or functions f, the more 'convergent' sequences S_n we have. The most successful compromise is the convergence in distribution of the random variables (or the weak convergence of the probability distributions). We postulate that (2) holds for all bounded, continuous functions f. This requirement can be shown to be equivalent to postulating that (1) holds for all sets A so that the boundary of A (that is, the difference between closure of A and inner points of A) has zero probability under the limiting measure. So in our case, where the limiting distribution is normal, (1) holds if A is an interval (a, b): the boundary consists of two points, namely a and b. Equation (1) does not hold if, for example, A is the set of all rational numbers in (0, 1): then the boundary equals [0, 1], which obviously has non-zero probability under the normal distribution (see Billingsley, 1999).

It is noteworthy that there are many more equivalent ways to define convergence in distribution for unidimensional random variables; for example, convergence in distribution is equivalent to the convergence of the cumulative distribution functions to the cumulative distribution function of the limiting distribution in all points where the latter is continuous. Another well-known criterion is the convergence of the characteristic functions.

Now we are in a position to formulate our first main theorem, the central limit theorem (CLT) of Lindeberg and Feller (see Billingsley, 1995).

Suppose we have given a triangle array of random variables $X_{i,n}$, so that for each n the $X_{i,n}$ are independent, not necessarily identically distributed. We furthermore have

$$EX_{i,n} = 0,$$

$$\sum_{i=1}^{n} \mathrm{Var}(X_{i,n}) = 1.$$

Then the following two propositions are equivalent:

- The 'Lindeberg' condition: For all $\delta > 0$

$$\sum_{i=1}^{n} E\big(X_{i,n}^2 I[|X_{i,n}| > \delta]\big) \tag{L}$$

  converges to zero.
- Our sums

$$S_n = \sum_{i=1}^{n} X_{i,n}$$

  converge in distribution to a standard normal and the 'Feller' condition is satisfied:

$$\max_{1 \le i \le n} \mathrm{Var}(X_{i,n}) \to 0. \tag{F}$$

  It seems plausible to assume the Feller condition (F). It simply states that the maximal contribution of an individual $X_{i,n}$ to the variance of the sum gets arbitrarily small. This seems reasonable. The Lindeberg condition (L) which is necessary for our theorem is a little stronger. Not only the maximum, but the *total* contribution of the $X_{i,n}$ taking 'large' values to the variance of the sum, must vanish asymptotically!

It is quite easy to establish that (L) is fulfilled if

$$X_{i,n} = \frac{1}{\sqrt{n}} X_i, \tag{3}$$

where the X_i are independent and identically distributed. In the general case, a sufficient condition is the 'Lyapunov condition': for some fixed $\varepsilon > 0$ we have

$$\sum_{i=1}^{n} E(|X_{i,n}|^{2+\varepsilon}) \to 0.$$

So we need a little more than second moments to establish convergence to a standard normal. Practitioners often assume that the requirements of the theorems are fulfilled automatically. This assumption is quite dangerous. We need a little more than lack of outliers; the contribution to the variance of the largest values must be negligible.

This relation between higher moments and goodness of the approximation with a standard normal is extensive. Under the assumption of at least three absolute moments, the theorem of Berry–Esseen shows that in the case (3) of independent, identically distributed X_i the maximal difference between the cumulative distribution functions of S_n and the standard normal is $1/\sqrt{n}$. Related are 'coupling' results. One can show that – possibly on a richer probability space – there exist *exactly* normally distributed random variables U_n. In particular, if the X_i have a Laplace transform, then the 'Hungarian construction' allows one to construct U_n so that the difference to S_n is $O(\log(n)/\sqrt{n})$. If the X_i 'only' have fourth moments, then it is easy (for the insider: use Skorohod embedding) to construct U_n so that the difference to S_n is of the order of $1/\sqrt[4]{n}$.

All these bounds are very interesting from the theoretical point of view. Playing around with numbers for n with realistic sample sizes, one can easily see that the bounds found that way are unrealistic. Although these bounds cannot be improved, they are a little pessimistic. Nevertheless, they indicate when we venture into dangerous territory: a lack of fourth moments indicates a 'slow' convergence.

So the normal approximation is a useful first-order approximation of the distributions of sums of random variables. To improve this approximation, various techniques are used. Since the 19th century, Edgeworth expansions have proved useful. Nowadays, however, cheap computing makes direct calculation of distributions by Monte Carlo simulation possible.

Independent, non-normal limit theorems

Let us define $X_{i,n}$ to be independent, identically distributed and taking the value of zero with probability $1-\lambda/n$ and one with probability λ/n with some $\lambda > 0$. Now one has an easy example where the Lindeberg condition is not fulfilled. (For $\delta < 1$, $\Sigma_{i=1}^{n} E(X_{i,n}^2 I(|X_{i,n}| > \delta) = \lambda$, since $X_{i,n}$ can take only the values 0 and 1). Nevertheless, it is well known that $\sum_{i=1}^{n} X_{i,n}$ converges in distribution to a Poisson distribution with intensity λ. So the normal distribution is not the only limiting distribution of sums of independent random variables. One can, however, show that the normal and the Poisson distribution and mixtures (with possibly an infinite number of components) of these distributions are the only possible limits of sums S_n of independent, identically distributed random variables $X_{i,n}$. These limiting distributions are called 'infinitely divisible'. A precise formula for the logarithm of the characteristic function is given by the formula of Levy–Khinchin.

We even have some analogon, some generalization of the normal distribution. The properly normalized sum of normally distributed random variables is normal again. Can we generalize this property? Let us assume that

$$X_{i,n} = a_n(X_i - b_n), \tag{4}$$

where the X_i are independent and identically distributed, and the a_n are scale factors, and let us assume that the distribution of the S_n is identical to the distribution of the X_i. These distributions are called the 'stable' distributions. Their density is determined essentially by two parameters, traditionally called α and β. α determines the 'tail behaviour' and varies between 0 and 2, and β determines the symmetry. For $\alpha = 2$, we have the normal distribution, for $\alpha < 2$ the distributions are more heavily tailed: in general, one has only moments of order smaller than α. There is no closed form for their densities in the general case, only the characteristic functions can be expressed by elementary functions. One special case ($\alpha = 1$) is the Cauchy distribution with density

$$\frac{1}{\pi(1 + x^2)}.$$

The index α determines the scale factors a_n: in general, one has $a_n = n^{\frac{1}{\alpha}}$.

Convergence of sums to stable distributions can be achieved in more general circumstances. In general, under certain conditions on the 'tail' of the X_i (the probabilities exceeding 'large' values have to obey certain regularity conditions) the sums of the $X_{i,n}$ defined by (4) one can ensure convergence (see Ibragimov and Linnik, 1971).

Central limit theorems for dependent random variables

Many econometric applications involve sums of dependent random variables. Hence it is important to remove the requirement of independence. Traditionally, one tried to replace independence by some form of 'mixing'.

Independence of two σ-algebras $\mathfrak{A}$ and $\mathfrak{B}$ can be defined in various ways. Usually one defines $\mathfrak{A}$ and $\mathfrak{B}$ to be independent if for all $A \in \mathfrak{A}$ and $B \in \mathfrak{B}$

$$P(A \cap B) = P(A)P(B).$$

Another usual definition is that for all $A \in \mathfrak{A}$

$$P(A/\mathfrak{B}) = P(A),$$

where $P(\cdot/\cdot)$ should denote the conditional probability. Consequently, one can measure the 'degree of dependence' of σ-algebras $\mathfrak{A}$ and $\mathfrak{B}$ by

$$\alpha(\mathfrak{A}, \mathfrak{B}) = \sup_{A \in \mathfrak{A}, B \in \mathfrak{B}} |P(A \cap B) - P(A)P(B)|$$

or

$$\psi(\mathfrak{A}, \mathfrak{B}) = \sup_{A \in \mathfrak{A}} |P(A/\mathfrak{B}) - P(A)|.$$

Suppose one has give a process X_t. Then one defines the 'mixing coefficients'

$$\alpha_k = \sup_t \alpha(\mathfrak{A}_\sigma(X_t, X_{t+1,\dots}), \mathfrak{A}_\sigma(X_{t-k}, X_{t-1-k,\dots}))$$

or

$$\psi_k = \sup_t \psi(\mathfrak{A}_\sigma(X_t, X_{t+1,\dots}), \mathfrak{A}_\sigma(X_{t-k}, X_{t-1-k,\dots})).$$

Typically, conditions like

$$\sum \sqrt{\alpha_k} < \infty$$

or

$$\psi_k \to 0$$

are sufficient conditions for a CLT. So the CLT remains valid for stationary processes if the random variables in questions get less and less dependent if the time difference gets larger and larger (Ibragimov and Linnik, 1971; Davidson, 1994).

CLT for martingale differences

One of the most important applications is the CLT for martingale differences. A process X_t is a 'martingale difference' if for all t

$$E(X_t/\mathfrak{F}_{t-1}) = 0,$$

where $\mathfrak{F}_{t-1}$ is an increasing sequence of σ-algebras which contain at least X_{t-1}, $X_{t-2}, \dots$. Then we have a result perfectly analogous to the case of independent random variables.

Suppose we have given a triangle array $X_{t,T}$, $t = 1, \dots, T$, of martingale differences with σ-algebras $\mathfrak{F}_{t-1,T}$ and the following two conditions are satisfied:

- the conditional Lindeberg condition

$$\sum_{t=1}^{T} E(X_{t,T}^2 I_{[|X_{t,T}| \geq \varepsilon]}/\mathfrak{F}_{t-1,T}) \to 0,$$

- the norming condition

$$\sum_{t=1}^{T} E(X_{t,T}^2/\mathfrak{F}_{t-1,T}) \to 1,$$

  where the convergence should be understood to be in probability. Then

$$S_n = \sum_{i=1}^{n} X_{i,n}$$

  converges in distribution to a standard normal distribution (Davidson, 1994; Hall and Heyde, 1980).

This limit theorem is one of the most important ones for applications in econometrics. It is relatively easily seen that derivatives of log-likelihood functions are martingale differences. Hence this theorem is instrumental in establishing the limit theorems for maximum likelihood estimators.

An easy consequence of the theorem is that for every (strictly) stationary, ergodic martingale difference X with $\sigma^2 = E(X_t^2) < \infty$ we have an almost classical CLT:

$$\frac{1}{\sigma\sqrt{n}}\sum_{i=1}^{n} X_i$$

which converges in distribution to a standard normal.

Gordin's theorem

Martingale differences form a large class of processes. Unfortunately, however, this class is not sufficiently large for many important applications (martingale differences must be, for example, uncorrelated). As an alternative, one might use mixing conditions. These conditions are, however, hard to verify. They usually involve inequalities involving *all* events from the σ-algebras involved. Hence a theorem allowing for general, autocorrelated processes with conditions which are easy to verify is an important tool in theoretical econometrics. Such a result was found by Gordin in 1969. Hayashi (2000) demonstrates the versatility of the theorem.

Suppose we have a stationary, ergodic process X_i, $i \in Z$ so that $EX_i^2 < \infty$. Assume that $\mathfrak{F}_i$ are adapted σ-algebras (that is, X_i are $\mathfrak{F}_i$-measurable), and let

$$\varepsilon_i = E(X_i/\mathfrak{F}_1) - E(X_i/\mathfrak{F}_0).$$

Then let us assume that

$$\sum_{i=1}^{\infty} \sqrt{E\varepsilon_i^2} < \infty.$$

Then

$$\frac{1}{\sqrt{n}}\sum_{i=1}^{n} X_i$$

converges in distribution to a normal distribution with zero mean and variance σ_{LT}^2, where

$$\sigma_{LT}^2 = E\left(\sum_{i=1}^{\infty} \varepsilon_i\right)^{2'}.$$

σ_{LT}^2 is usually called the 'long-term variance'.

Conclusion

Almost all theorems about limit distributions of estimators and test statistics depend on central limit theorems. So it should not be surprising that central limit theorems and their generalizations are an active field of research. Especially, generalizations of the concept of convergence in distribution to more general spaces generate theorems, which are important from the theoretical as well as the practical point of view. Billingsley (1999) and Davidson (1994) give an introduction to these 'functional limit theorems'.

WERNER PLOBERGER

See also **functional central limit theorems.**

Bibliography

Billingsley, P. 1995. *Probability and Measure*, 3rd edn. New York: Wiley.

Billingsley, P. 1999. *Convergence of Probability Measures*, 2nd edn. New York: Wiley-Interscience.

Davidson, J. 1994. *Stochastic Limit Theory: An Introduction for Econometricians*. Oxford: Oxford University Press.

Hall, P. and Heyde, C.C. 1980. *Martingale Limit Theory and its Application*. New York: Academic.

Hayashi, F. 2000. *Econometrics*. Princeton, NJ: Princeton University Press.

Ibragimov, I.A. and Linnik, Yu.V. 1971. *Independent and Stationary Sequences of Random Variables*. Groningen: Wolters-Noordhoff.

cointegration

Cointegration means that two or more time series share common stochastic trends. Thus, while each series exhibits smooth or trending behaviour, a linear combination of the series exhibits no trend. For example, short-term and long-term interest rates are highly serially correlated (so they are smooth and in this sense exhibit a stochastic trend), but the difference between long rates and short rates – the 'term spread' – is far less persistent and shows no evidence of a stochastic trend. Long rates and short rates are cointegrated.

The concept of cointegration was formalized by Clive W.J. Granger in a series of papers in the 1980s (Granger, 1981; Granger and Weiss, 1983; Granger, 1986; Engle and Granger, 1987), and in 2003 Granger received the Nobel Prize in Economics for this work. A flurry of research activity followed Granger's original contributions in this area and produced a practical set of econometric procedures for analysing cointegrated time series.

Mathematical structure of I(1) cointegrated models

Let X_t denote a scalar I(1) stochastic process, with moving average representation $X_t = c(\mathrm{L})\varepsilon_t$, where ε_t is a scalar white noise process, and $c(\mathrm{L}) = \sum_{i=0}^{\infty} c_i \mathrm{L}^i$ is a polynomial in the lag operator L, and where the moving average coefficients, c_i decay sufficiently rapidly so that $\sum_{i=1}^{\infty} i|c_i| < \infty$. The Beveridge–Nelson decomposition (see TREND/CYCLE DECOMPOSITION) implies that X_t can be represented as $X_t = \tau_t + a_t$, where τ_t is a random walk, so that $\tau_t = \tau_{t-1} + e_t$, where e_t is white noise and a_t has a moving average representation $a_t = d(\mathrm{L})\varepsilon_t$, where $\sum_{i=1}^{\infty} |d_i| < \infty$. Thus, X_t can be expressed as the sum of a stochastic trend, τ_t, and an I(0) process, a_t.

When X_t is an $n \times 1$ vector of I(1) processes, a similar result implies that $X_t = A\tau_t + a_t$, where A is a matrix of constants, τ_t is a vector of random-walk stochastic trends, and a_t is a vector of I(0) processes. Because X_t contains n elements, the vector τ_t will generally contain n stochastic trends. However, when τ_t contains only $k < n$ stochastic trends, A is $n \times k$, so that $\alpha' A = 0$, for any vector α in the null space of the column space of A. This means that $\alpha' X_t = \alpha' a_t$, so that the linear combination $\alpha' X_t$ does not depend on the stochastic trends. In this case, the time series making up X_t are said to be cointegrated. Any non-zero vector α that satisfies $\alpha' A = 0$ will annihilate the stochastic trend in $\alpha' X_t$, and vectors with this property are called cointegrating vectors. When A has full column rank, the number of linearly independent cointegrating vectors is $r = n - k$, which is called the cointegrating rank of the process.

For example, suppose that X_t contains $n = 3$ series representing interest rates on one-month, three-month and six-month US treasury bills. Suppose that $X_{it} = \tau_t + a_{it}$, for $i = 1, 2, 3$, where τ_t is a common stochastic trend shared by the three interest rates.

Then $X_t = A\tau_t + a_t$, where $k = 1$ (there is a single stochastic trend), $A = (1\ 1\ 1)'$ (the trend has an equal effect on each of the interest rates) and $\alpha_1 = (1\ 0\ -1)'$ and $\alpha_2 = (1\ 0\ -1)'$ are two linearly independent cointegrating vectors, so that $r = 2$ and $\alpha_1{}'X_t$ and $\alpha_2{}'X_t$ denote the interest rate term spreads.

Vector moving average models (VMAs) and vector autoregressions (VARs) are often used to represent the linear properties of vector stochastic processes. The Granger representation theorem (see Engle and Granger, 1987) shows that VMAs and VARs for cointegrated processes have special structures. In general, the VMA for an I(1) vector process is $X_t = D(L)\varepsilon_t$, where ε_t is white noise with full rank covariance matrix. When X_t is not cointegrated, the $n \times n$ matrix $D(1)$, which contains the sum of the moving average coefficients, has rank n. But, when X_t is cointegrated, $D(1)$ has rank $k<n$, where k denotes the number of stochastic trends. When X_t is not cointegrated, the VAR for X_t can be written in terms of ΔX_t and has the form $\Phi(L)X_t = \varepsilon_t$, where $\Phi(L)$ is a stable lag polynomial (so its roots are outside the unit circle) and ε_t is white noise. When X_t is cointegrated, the VAR has the form $\Phi(L)X_t = \beta\alpha'X_{t-1} + \varepsilon_t$, where α is an $n \times r$ matrix with columns that are the linearly independent cointegrating vectors. Thus, the cointegrated VAR expresses the elements of ΔX_t as functions of its own lags, but also includes the r regressors $\alpha'X_{t-1}$ in each of the VAR's n equations. The variables $\alpha'X_{t-1}$ are called 'error-correction terms' and the cointegrated VAR is called a 'vector error correction model' (VECM).

Watson (1994) provides a summary of the algebra linking these various representations of the cointegrated model.

Testing for cointegration

The time series making up X_t are cointegrated if the linear combinations $\alpha'X_t$ are I(0) random variables. If X_t is not cointegrated, then $\alpha'X_t$ will be I(1) for any non-zero vector α. Tests of cointegration ask whether $\alpha'X_t$ is I(1) or I(0).

Consider the simple case in which there is only one potential cointegrating vector, so that $\alpha'X_t$ is a scalar. Cointegration can then be tested using a unit root test applied to $\alpha'X_t$. The straightforward application of a unit root test requires that α is known, so that the scalar variable $\alpha'X_t$ can be calculated directly from the data. This is possible in many empirical applications (such as the interest rate example described above) where the value of α can be pre-specified.

Thus, suppose that α is known, and consider the competing hypotheses $H_{I(1)}$: $\alpha'X_t$ is I(1) and $H_{I(0)}$: $\alpha'X_t$ is I(0). The hypothesis $H_{I(1)}$ means that the elements of X_t are not cointegrated and the hypothesis $H_{I(0)}$ means that the elements are cointegrated. Under $H_{I(1)}$ the autoregressive model for $\alpha'X_t$ contains a unit root, while under $H_{I(0)}$, the autoregressive model for $\alpha'X_t$ is stable.

The null $H_{I(1)}$ can be tested against the alternative $H_{I(0)}$ using an augmented Dickey–Fuller (ADF) unit root test or the modified ADF test developed in Elliott, Rothenberg and Stock (1996). The null $H_{I(0)}$ can be tested against $H_{I(1)}$ using the best local test proposed by Nyblom (1989), modified for serial correlation as described in

Kwiatkowski et al. (1992), or a point-optimal test as discussed in Jansson (2004). (There are important practical considerations associated with the choice of the long-run-variance estimator (see HETEROSKEDASTICITY AND AUTOCORRELATION CORRECTIONS) used in tests for the $H_{I(0)}$ null hypothesis because of the high degree of serial correlation under the alternative. See Müller (2005) for discussion.)

When α is not known, the unit root tests described in the last paragraph use $\hat{\alpha}'X_t$ in place of $\alpha'X_t$, where $\hat{\alpha}$ is an estimator of α. For example, Engle and Granger (1987) suggest estimating α by regressing the first element of X_t onto the other elements of X_t using OLS, and carrying out an ADF test using the residuals from this regression. Estimation of α changes the distribution of the ADF test statistic from what it is when α is known, so that critical values for the Engle–Granger test are different than the standard ADF critical values. As described in Phillips and Ouliaris (1990) and Hansen (1992) the correct critical values depend on the number of elements in X and on the properties of the deterministic trends in the model. Stock and Watson (2007) tabulate choices of critical values from the Phillips and Ouliaris (1990) and Hansen (1992) papers that are appropriate for data that follow I(1) processes that may or may not contain drift, and thus serve as conservative critical values. Modifications for tests of the $H_{I(0)}$ null versus the $H_{I(1)}$ alternative are discussed in Shin (1994) and Jansson (2005).

The tests outlined above are useful for testing whether a single series $\alpha'X_t$ is I(0) or I(1), but in many applications there may be more than one potential cointegrating relation ($r>1$) so that it is useful to have tests for hypothesis that postulate different values of r. That is, it is useful to entertain hypotheses of the form $H_j : r = j$, for $j=0$, 1, ..., n. The hypothesis $r = 0$ means that there is no cointegration, $r = 1$ means that there is a single cointegrating vector, and so forth. As discussed in Johansen (1988), these tests are easily formulated and carried out using the VECM model. Recall that the VECM model has the form $\Phi(L)X_t = \beta\alpha'X_{t-1} + \varepsilon_t$. Consider the null and alternative hypotheses $H_o : r = r_o$ *vs.* $H_a : r = r_a$ where $r_a > r_o$, and write the VECM as $\Phi(L)X_t = \beta_o\alpha_o'X_{t-1} + \tilde{\beta}\tilde{\alpha}'X_{t-1} + \varepsilon_t$, where α_o contains the r_o cointegrating vectors under the null and $\tilde{\alpha}$ contains the additional cointegrating vectors under the alternative. Under the null hypothesis, the variables $\tilde{\alpha}'X_{t-1}$ do not enter the VECM, while under the alternative these variables enter the VECM. Thus, the null and alternative can be written as $H_o : \tilde{\beta} = 0$ versus $H_a : \tilde{\beta} \neq 0$. As in the case of $r = 1$, the tests depend on whether the cointegrating vectors are known or unknown. When the cointegrating vectors are known, the regressors $\alpha_o'X_{t-1}$ and $\tilde{\alpha}'X_{t-1}$ can be constructed from the data, and the Wald test for $\tilde{\beta} = 0$ can be constructed using the usual regression formula. When the cointegrating vectors are unknown, the testing problem is more difficult, but Johansen (1988) provides an simple formula for the likelihood ratio test statistic. In either case, the critical values for the test are 'non-standard', that is, they are not based on the χ^2 or F distributions. Critical values for the tests depend on the values of $r_a - r_o$, the number of cointegrating vectors that are known and unknown, and the presence or absence of constants and time trends in the model. The various critical values are tabulated in Horvath and Watson (1995).

Estimating unknown cointegrating coefficients

Unknown coefficients in cointegrating vectors are typically estimated using least squares and Gaussian maximum likelihood estimators (MLEs). The properties of these estimators can be understood by considering a simple bivariate model

$$X_{1t} = \theta X_{2t} + \eta_{1t}$$

$$X_{2t} = X_{2t-1} + \eta_{2t}$$

where $\eta_t = [\eta_{1t}\ \ \eta_{2t}]' \sim$ iid N $(0,\ \Sigma)$. In this example, there is one common trend that coincides with X_{2t}, the cointegrating vector is $\alpha = (1 - \theta)'$ where θ is an unknown parameter, the error correction term is $\alpha' X_t = n\eta_{1t}$ which is potentially correlated with the innovation in the common trend, η_{2t}, and the assumption of normality is used to motivate the Gaussian MLE of θ.

The OLS estimator of θ has several interesting properties (Stock, 1987). Even though X_{2t} and η_{1t} are correlated, the OLS estimator is consistent; indeed it is 'super-consistent' in the sense that $\hat{\theta}^{OLS} - \theta \sim O_p(T^{-1})\hat{\theta}^{OLS}$, so that $\hat{\theta}^{OLS}$ converges to θ faster than the usual $\sqrt{T}$ rate familiar from regressions involving I(0) variables. These results follow because, in the cointegrated model, the regressor X_{2t} is I(1) and therefore is much more variable than an I(0) regressor ($\sum_{t=1}^{T} X_{2t}^2 \sim O_p(T^{-2})$ in this I(1) regression instead of $O_p(T^{-1})$ in the usual I(0) regression), and the correlation between X_{2t} and η_{1t} is non-zero, but vanishes as the sample size becomes large. (The covariance is constant, but the variance of X_{2t} increases linearly with t, so the correlation vanishes as t increases.)

Despite these intriguing and powerful features, the OLS estimator has two properties that make it unsatisfactory for many uses. First, while OLS is consistent, the correlation between the regressor and error term induces a bias in the large sample distribution of the estimator, and this bias can be severe in sample sizes typically encountered in applied work (Stock, 1987). Second, the large-sample distribution of the OLS estimator is non-normal, and this complicates statistical inference. For example, the standard interval $\hat{\theta}^{OLS} \pm 1.96SE(\hat{\theta}^{OLS})$ does not provide a 95 per cent confidence set even in large samples. Interestingly, Gaussian maximum likelihood estimators share the super consistency properties of OLS, but do not suffer from these unsatisfactory properties (Johansen, 1988; Phillips, 1991).

To construct the Gaussian MLE, factor the joint density of $\{X_t\}_{t=1}^{T}$ into the density of $\{X_{1t}|(X_{2t})_{t=1}^{T}\}_{t=1}^{T}$ and the density of $\{X_{2t}\}_{t=1}^{T}$. The density of $\{X_{2t}\}_{t=1}^{T}$ does not depend on θ, and the density of $X_{1t}|(X_{2t})_{t=1}^{T}$ is characterized by the Gaussian linear regression $X_{1t} = \theta X_{2t} + \beta X_{2t} + v_t$, where β is the regression coefficient from the regression of η_{1t} onto η_{2t} $(=\Delta X_{2t})$, v_t is the error in this regression, and $v_t|(X_{2t})_{t=1}^{T} \sim$ iid $\text{N}(0,\ \sigma_2)$. Simple calculations (Phillips, 1991) can then be used to show that $\hat{\theta}^{MLE} - \theta \sim O_p(T^{-1})$ and that $\hat{\theta}^{MLE}|(X_{2t})_{t=1}^{T} \sim \text{N}(\theta, V)$, where V depends on $(X_{2t})_{t=1}^{T}$. Thus, $\hat{\theta}^{MLE}$ is consistent, is conditionally normally distributed and unbiased, and $(\hat{\theta}^{MLE} - \theta)/V^{1/2} \sim$ N(0, 1), so that inference about θ can be carried out using standard methods associated with the Gaussian linear regression model. Thus, for example, $\hat{\theta}^{MLE} \pm 1.96SE(\hat{\theta}^{MLE})$ provides a valid 95 per cent confidence set for θ, where $SE\ (\hat{\theta}^{MLE})$ is computed using the usual regression formula.

While these results may appear quite special (X_t is bivariate and η_t is normally distributed and serially uncorrelated) they carry over to more general models with minor modifications. For example, X_{1t} and X_{2t} may each be vectors and the regression $X_{1t} = \theta X_{2t} + \beta X_{2t} + \nu_t$ becomes a multivariate regression. Under weak assumptions on the distribution of η_t, there is sufficient averaging so that $V^{-1/2}(\hat{\theta}^{MLE} - \theta) \xrightarrow{d} N(0, I)$, meaning that the assumption of normality for η is not critical (although $\hat{\theta}^{MLE}$ still refers to the MLE computed by maximizing the Gaussian likelihood). Serial correlation in η_t can be handled in a variety of ways. For example, Saikkonen (1991) and Stock and Watson (1993) consider the 'dynamic OLS' (DOLS) regression $X_{1t} = \theta X_{2t} + \Sigma_{i=-k}^{k}\beta_i X_{2t-i} + \nu_t$, which includes enough leads and lags of ΔX_{2t} to insure that ν_t is (linearly) independent of $(X_{2t})_{t=1}^{T}$. Phillips and Hansen (1990) and Park (1992) develop adjustments based on long-run covariance matrix estimators, and Johansen (1988) derives the exact Gaussian MLE based on the VECM. Under general assumptions, all of the estimators are asymptotically equivalent.

Alternative models for the common trends

The concept of cointegration involves variables that share common persistent 'trend' components. The statistical analysis outlined above utilized a particular model of the trend component, namely, the driftless unit root process $\tau_t = \tau_{t-1} + e_t$. Analysis of this model highlights many of the key features of cointegrated processes, but more general models are often needed for empirical analysis. For example, constant terms are often added to the model to capture non-zero means of error correction terms or drifts in the trend process. These constant terms change the distribution of test statistics for cointegration in ways familiar from the effect of constants and time trends in Dickey–Fuller unit root tests (see Hamilton, 1994). Hansen (1992) and Johansen (1994) contain useful discussion of the key issues. Higher-order integrated processes (for example, I(2) processes) are discussed in Johansen (1995), Granger and Lee (1990), and Stock and Watson (1993). Hylleberg et al. (1990) discuss cointegration at seasonal frequencies. Robinson and Hualde (2003) and the references cited therein discuss cointegration in fractionally integrated models.

Elliott (1998) discusses cointegrated models in which the trend follows a 'near-unit-root' process – an AR process with largest autoregressive root very close to 1.0. (Formally, the asymptotics use a local-to-unity nesting with largest root AR root equal to $1-c/T$, where c is a constant.) Elliott shows that, while the basic cointegrated model remains unchanged in this case, the properties of Gaussian maximum likelihood estimators of unknown cointegrating coefficients change in important ways. In particular, the Gaussian MLEs are no longer conditionally unbiased, and confidence intervals constructed using Gaussian approximations (for example, $\hat{\theta}^{MLE} \pm 1.96SE(\hat{\theta}^{MLE})$) can be very misleading.

Elliott's critique is important because small deviations from exact unit roots cannot be detected with high probability, and yet small deviations may undermine the validity of statistical inferences constructed using large-sample normal approximation applied to Gaussian MLEs.

Several papers have sought to address the Elliott critique by developing methods with good performance for a range of autoregressive roots close to, but not exactly equal to 1.0. For example, Wright (2000) argues that if θ_0 is the true value of a cointegrating coefficient, then $X_{1t} - \theta_0 X_{2t}$ will be I(0), but if θ_0 is not the true value then $X_{1t} - \theta_0 X_{2t}$ will be highly persistent. He suggests testing that $\theta = \theta_0$ by testing the $H_{I(0)}$ null for the series $X_{1t} - \theta_0 X_{2t}$. Alternative testing procedures in this context are proposed in Stock and Watson (1996) and Jansson and Moreira (2006).

MARK W. WATSON

See also **heteroskedasticity and autocorrelation corrections; trend/cycle decomposition; unit roots.**

Bibliography

Elliott, G. 1998. The robustness of cointegration methods when regressors almost have unit roots. *Econometrica* 66, 149–58.

Elliott, G., Rothenberg, T.J. and Stock, J.H. 1996. Efficient tests for an autoregressive unit root. *Econometrica* 64, 813–36.

Engle, R.F. and Granger, C.W.J. 1987. Co-integration and error correction: representation, estimation, and testing. *Econometrica* 55, 251–76.

Granger, C.W.J. 1981. Some properties of time series data and their use in econometric specification. *Journal of Econometrics* 16, 121–30.

Granger, C.W.J. 1986. Developments in the study of co-integrated economic variables. *Oxford Bulletin of Economics and Statistics* 48, 213–28.

Granger, C.W.J. and Lee, T.H. 1990. Multicointegration. *Advances in Econometrics* 8, 71–84.

Granger, C.W.J. and Weiss, A.A. 1983. Time series analysis of error-correction models. In *Studies in Econometrics, Time Series, and Multivariate Statistics, in Honor of T. W. Anderson*, ed. S. Karlin, T. Amemiya and L.A. Goodman. San Diego: Academic.

Hamilton, J.D. 1994. *Time Series Analysis.* Princeton, NJ: Princeton University Press.

Hansen, B. 1992. Efficient estimation and testing of cointegrating vectors in the presence of deterministic trends. *Journal of Econometrics* 53, 86–121.

Horvath, M.T.K. and Watson, M.W. 1995. Testing for cointegration when some of the cointegrating vectors are prespecified. *Econometric Theory* 11, 952–84.

Hylleberg, S., Engle, R.F., Granger, C.W.J. and Yoo, B.S. 1990. Seasonal integration and cointegration. *Journal of Econometrics* 44, 215–38.

Jansson, M. 2004. Stationarity testing with covariates. *Econometric Theory* 20, 56–94.

Jansson, M. 2005. Point optimal tests of the null of hypothesis of cointegration. *Journal of Econometrics* 124, 187–201.

Jansson, M. and Moreira, M. 2006. Optimal inference in regression models with integrated regressors. *Econometrica* 74, 681–714.

Johansen, S. 1988. Statistical analysis of cointegrating vectors. *Journal of Economic Dynamics and Control* 12, 231–54.

Johansen, S. 1994. The role of the constant and linear terms in cointegration analysis of non-stationary variables. *Econometric Reviews* 13, 205–29.

Johansen, S. 1995. A statistical analysis of cointegration for I(2) variables. *Econometric Theory* 11, 25–59.

Kwiatkowski, D., Phillips, P.C.B., Schmidt, P. and Shin, Y. 1992. Testing the null hypothesis of stationarity against the alternative of a unit root. *Journal of Econometrics* 54, 159–78.

Müller, U.K. 2005. Size and power of tests for stationarity in highly autocorrelated time series. *Journal of Econometrics* 128, 195–213.

Nyblom, J. 1989. Testing for the constancy of parameters over time. *Journal of the American Statistical Association* 84, 223–30.

Park, J.Y. 1992. Canonical cointergrating regressions. *Econometrica* 60, 119–43.

Phillips, P.C.B. 1991. Optimal inference in cointegrated systems. *Econometrica* 59, 283–306.

Phillips, P.C.B. and Hansen, B.E. 1990. Statistical inference on instrumental variables regression with I(1) processes. *Review of Economic Studies* 57, 99–124.

Phillips, P.C.B. and Ouliaris, S. 1990. Asymptotic properties of residual based test for cointegration. *Econometrica* 58, 165–93.

Robinson, P.M. and Hualde, J. 2003. Cointegration in fractional systems of unknown orders. *Econometrica* 71, 1727–66.

Saikkonen, P. 1991. Asymptotically efficient estimation of cointegrating regressions. *Econometric Theory* 7, 1–21.

Shin, Y. 1994. A residual-based test of the null of cointegration against the alternative of no cointegration. *Econometric Theory* 10, 91–115.

Stock, J.H. 1987. Asymptotic properties of least squares estimates of cointegrating vectors. *Econometrica* 55, 1035–56.

Stock, J.H. and Watson, M.W. 1993. A simple estimator of cointegrated vectors in higher-order integrated systems. *Econometrica* 61, 783–820.

Stock, J.H. and Watson, M.W. 1996. Confidence sets in regression with highly serially correlated regressors. Manuscript, Department of Economics, Princeton University.

Stock, J.H. and Watson, M.W. 2007. *Introduction to Econometrics*, 2nd edn. Boston: Pearson-Addison Wesley.

Watson, M.W. 1994. Vector autoregression and cointegration, vol. 4. *Handbook of Economics*, ed. R.F. Engle and D.L. McFadden. Amsterdam: North-Holland.

Wright, J.H. 2000. Confidence sets for cointegrating coefficients based on stationarity tests. *Journal of Business and Economic Statistics* 18, 211–22.

continuous and discrete time models

Discrete time models are generally only an approximation, and the error induced by this approximation can under some conditions be important.

Most economists recognize that the use of discrete time is only as an approximation, but assume (usually implicitly) that the error of approximation involved is trivially small relative to the other sorts of simplification and approximation inherent in economic theorizing. We consider below first the conditions under which this convenient assumption may be seriously misleading. We discuss briefly how to proceed when the assumption fails and the state of continuous time economic theory.

Approximation theory

Some economic behaviour does involve discrete delays, and most calculated adjustments in individual patterns of behaviour seem to occur following isolated periods of reflection, rather than continually. These notions are sometimes invoked to justify economic theories built on a discrete time scale. But to say that there are elements of discrete delay or time discontinuity in behaviour does not imply that discrete time models are appropriate. A model built in continuous time can include discrete delays and discontinuities. Only if all delays were discrete multiples of a single underlying time unit, and synchronized across agents in the economy, would modelling with a discrete time unit be appropriate.

Nonetheless, sometimes discrete models can avoid extraneous mathematical complexity at little cost in approximation error. It is easy enough to argue that time is in fact continuous and to show that there are in principle cases where use of discrete time models can lead to error. But it is also true in practice that more often than not discrete time models, translated intuitively and informally to give implications for the real continuous time world, are not seriously misleading. The analytical task, still not fully executed in the literature, is to understand why discrete modelling usually is adequate and thereby to understand the special circumstances under which it can be misleading.

The basis for the usual presumption is that, when the time unit is small relative to the rate at which variables in a model vary, discrete time models can ordinarily provide good approximations to continuous time models. Consider the case, examined in detail in Geweke (1978), of a dynamic multivariate distributed leg regression model, in discrete time.

$$Y(t) = A^{*}X(t) + U(t), \tag{1}$$

where * stands for convolution, so that

$$A^{*}X(t) = \sum_{s=-\infty}^{\infty} A(s)X(t-s). \tag{2}$$

We specify that the disturbances are uncorrelated with the independent variable vector X, that is, $\text{cov}[X(t), U(s)] = 0$, all t, s. The natural assumption is that, if approximation error from use of discrete time is to be small, $A(s)$ must be smooth as a function of s, and that in this case (1) is a good approximation to a model of the form

$$y(t) = a^*x(t) + u(t) \tag{3}$$

where

$$a^*x(t) = \int_{-\infty}^{\infty} a(s)x(t-s)ds \tag{4}$$

and y, a and x are functions of a continuous time parameter and satisfy $y(t) = Y(t)$, $x(t) = X(t)$ and $a(t) = A(t)$ at integer t. In this continuous time model we specify, paralleling the stochastic identifying assumption in discrete time, $\text{cov}[x(t), u(s)] = 0$, all t, s. If the discrete model (2) corresponds in this way to a continuous time model, the distributed lag coefficient matrices $A(s)$ are uniquely determined by a and the serial correlation properties of x.

We should note here that, though this framework seems to apply only to the case where X is a simple discrete sampling of x, not to the time-averaged case where $X(t)$ is the integral of $x(s)$ from $t-1$ to t, in fact both cases are covered. We can simply redefine the x process to be the continuously unit-averaged version of the original x process. This redefinition does have some effect on the nature of limiting results as the time unit goes to zero (since the unit-averaging transformation is different at each time unit) but turns out to be qualitatively of minor importance. Roughly speaking, sampling a unit-averaged process is like sampling a process whose paths have derivatives of one higher order than the unaveraged process.

Geweke shows that under rather general conditions

$$\sum_{s=-\infty}^{\infty} ||A(s) - \tau a(s\tau)||^2 \to 0 \tag{5}$$

as the time unit τ goes to zero, where $|| \; ||$ is the usual root-sum-of-squared-elements norm. In this result, the continuous time process x and lag distribution a are held fixed while the time interval corresponding to the unit in the discrete time model shrinks.

This is the precise sense in which the intuition that discrete approximation does not matter much is correct. But there are important limitations on the result. Most obviously, the result depends on a in (3) being an ordinary function. In continuous time, well-behaved distributed lag relations like (3) are not the only possible dynamic relation between two series. For example, if one replaces (3) by

$$y(t) = \alpha(\mathrm{d}/\mathrm{d}t)x(t) + u(t), \tag{6}$$

then the limit of A in (2) is different for different continuous x processes. In a univariate model with second-order Markov x (for example, one with

$\mathrm{cov}[x(t), x(t-s)] = (1+\theta|s|)\mathrm{e}^{-\theta|s|}\ \mathrm{var}[x(t)]$, the limiting discrete time model, as τ goes to zero, is

$$\begin{aligned} y(t) = a\{&-0.02X(t+4) + 0.06X(t+3) - 0.22X(t+2) \\ &+ 0.80X(t+1) - 0.80X(t-1) + 0.22X(t-2) \\ &- 0.06X(t-3) + 0.02X(t-4)\} + U(t) \end{aligned} \tag{7}$$

(see Sims, 1971).

This result is not as strange as it may look. The coefficients on X sum to zero and are anti-symmetric about zero. Nonetheless, (7) is far from the naive approximation which simply replaces the derivative operator with the first difference operator. In fact, if the estimation equation were constrained to involve only positive lags of X, the limiting form would be

$$\begin{aligned} Y(t) = \alpha\{&1.27X(t) - 1.161X(t-1) + 0.43X(t-2)\} \\ &- 0.12X(t-3) + 0.03X(t-4) - 0.01X(t-5) + U(t). \end{aligned} \tag{8}$$

The naive approximation of (3) by $Y(t) = \alpha[X(t) - X(t) - 1] + U(t)$ is valid only in the sense that, if this form is imposed on the discrete model a priori, the least squares estimate of α will converge to its true value. If the resulting estimated model is tested for fit against (8) or (7), it will be rejected.

Although the underlying model involves only the contemporaneous derivative of x, (8) and (7) both involve fairly long lags in X. If x paths have higher than first-order derivatives (for example, if they are generated by a third-order stochastic differential equation) the lag distributions in (8) and (7) are replaced by still higher-order limiting forms. Thus, different continuous time processes for x which all imply differentiable time paths produce different limiting discrete A. Here the fact that the time unit becomes small relative to the rate of variation in x does not justify the assumption that approximation of continuous by discrete models is innocuous. In particular, the notion that discrete differencing can approximate derivatives is potentially misleading.

It should not be surprising that the discrete time models may not do well in approximating a continuous time model in which derivatives appear. Nonetheless, empirical and theoretical work which ignores this point is surprisingly common.

If a is an ordinary function, there is still chance for error despite Geweke's result. His result implies only that the mean square deviation of a from A is small. This does not require that individual $A(t/\tau)$'s converge to the corresponding $a(t)$ values. For example, in a model where x is univariate and $a(t) = 0,\ t<0,\ a(0) = 1,\ a(s)$ continuous on $[0, \infty]$, the limiting value for $A(0)$ is 0.5, not 1.0. Thus, if $a(t) = e^{-\theta t}$ on $[0, \infty)$, making a monotone decreasing over that range, $A(t)$ will not be monotone decreasing. It will instead rise between $t = 0$ and $t = 1$. This is not unreasonable on reflection: the discrete lag distribution gives a value at $t = 0$ which averages the continuous time distribution's behaviour on either side of $t = 0$. It should therefore not be surprising that monotonicity of a does not necessarily imply monotonicity of A, but the point is ignored in some economic research.

Another example of possible confusion arises from the fact that, if the x process has differentiable paths, $a(t) = 0$ for $t<0$ does not imply $A(t) = 0$ for $t<0$. The mean-square approximation result implies that when the time unit is small the sum of squares of coefficients on $X(t-s)$ for negative s must be small relative to the sum of squares on $X(t-s)$ for positive s, but the first few lead coefficients will generally be non-zero and will not go to zero as the time interval goes to zero. This would lead to mistaken conclusions about Granger causal priority in large samples, if significance tests were applied naively.

Geweke's exploration of multivariate models shows that the possibilities for confusing results are more numerous and subtle in that case. In particular, there are ways by which poor approximation of $\alpha_j(s)$ by $A_j(s/\tau)$ in some s interval (for example, around $s = 0$) can lead to contamination of the estimates of other elements of the A matrix, even though they correspond to x_j's and a_j's that in a univariate model would not raise difficulties.

In estimation of a dynamic prediction model for a single vector y, such as a vector autoregression (VAR) or dynamic stochastic general equilibrium model (DSGE), the question for approximation theory becomes whether the continuous time dynamics for y, summarized in a Wold moving average representation

$$y(t) = a^{*}u(t) \tag{9}$$

has an intuitively transparented connection to the corresponding discrete time Wold representation

$$Y(t) = A^{*}U(t). \tag{10}$$

In discrete time the $U(t)$ of the Wold representation is the one-step-ahead prediction error, and in continuous time $u(t)$ also represents new information about y arriving at t. There are two related sub-questions. Is the A function the same shape as the a function; and is the U vector related in a natural way to the u vector? The u vector is a continuous time white noise, so that U cannot possibly be a simple discrete sampling of u.

If y is stationary and has an autoregressive representation, then $U(t) = A^{-1*}a^{*}u_t$, with the expression interpreted as convolution in continuous time, but with A^{-1} putting discrete weight on integers. The operator connecting U and u is then $A^{-1*}a$. There are cases where the connection between continuous and discrete time representations is intuitive. For example, if $a(s) = \exp(-Bs)$ (with the exponentiation interpreted as a matrix exponential in a multivariate case), then

$$U(t) = \int_0^1 e^{-Bs}u(t-s)ds \tag{11}$$

and $A(s) = a(s)$ at integers. This is a more intuitive and precise matching than in any case we examined above for projection of one variable on another. If $a(0)$ is full rank and right-continuous at zero and if $a(s)$ is differentiable at all $s>0$, then a similar intuitively simple matching of A to a arises when the time unit is small enough.

However, non-singularity of $a(0)$ rules out differentiability of time paths for y. When time paths for y, or some elements of it, are differentiable, no simple intuitive matching between A and a arises as the time unit shrinks.

There is one clear pattern in the difference in shape between A and a that stands in contrast to the case of distributed lag projection considered above. If both the continuous time and the discrete time moving average representations are fundamental, then by definition the one-step-ahead prediction error in $y(t)$ based on $y(t-s)$, $s \geq 1$ is

$$\int_0^1 a(s)u(t-s)ds, \tag{12}$$

while the one-step-ahead prediction error in $Y(t)$ based on $Y(t-s)$, $s = 1, 2, \ldots$ is $A_0 Y(t)$. Now the information set we use in forecasting based on the past of Y at integer values alone is smaller than the information set based on all past values of y, so the one-step-ahead error based on the discrete data alone must be larger. If we normalize in the natural way to give U an identity covariance matrix and to make $\text{var}(g^* u(t)) = \int g(s)g'(s)$ (so u is a unit white noise vector), then it must emerge that

$$A_0 A_0 \geq \int_0^1 a(s)a(s)'ds, \tag{13}$$

where the inequality is interpreted as meaning that the left-hand-side matrix minus the right-hand-side matrix is positive semi-definite. In other words, the initial coefficient in the discrete MAR will always be as big or bigger than the average over (0, 1) of the coefficients in the continuous MAR. This tendency of the discrete MAR to seem to have a bigger instant response to innovations is proportionately larger the smoother a is near zero.

More detailed discussion of these points, together with numerous examples, appears in Marcet (1991).

Estimation and continuous time modelling

How can one proceed if one has a model like, say, (6), to which a discrete time model is clearly not a good approximation? The only possibility is to introduce explicitly a model for how x behaves between discrete time intervals, estimating this jointly with (6) from the available data. Doing so converts (6) from a single-equation to a multiple-equation model. That is, the device of treating x as 'given' and non-stochastic cannot work because an important part of the error term in the discrete model arises from the error in approximating a^*x by A^*X. Furthermore, because separating the approximation error component of U from the component due to u is essential, one would have to model serial correlation in u explicitly. The model could take the form

$$\begin{bmatrix} y(t) \\ x(t) \end{bmatrix} = \begin{bmatrix} c(s) & a^*b(s) \\ 0 & b(s) \end{bmatrix} * \begin{bmatrix} w(t) \\ v(t) \end{bmatrix}, \tag{14}$$

where w and v are white noise processes fundamental (in the terminology of Rosanov, 1967), for y and x. To give b and c a convenient parametric form, one might suppose them rational, so that (14) can be written as a differential equation, that is,

$$P(D)\,y(t) = P(D)a^*x(t) + w(t) \tag{15}$$

$$Q(D)x(t) = v(t), \tag{16}$$

where P and Q are finite-order polynomials in the derivative operator, $Q^{-1}(D)v = b^*v$, and $P^{-1}(D)w = c^*w$.

A discrete time model derived explicitly from a continuous time model is likely to be nonlinear at least in parameters and therefore to be more difficult to handle than a more naive discrete model. However with modern computing power, such models are usable. Bergstrom (1983) provides a discussion of estimating continuous time constant coefficient linear stochastic differential equation systems from discrete data, the papers in the book (1976) he edited provide related discussions, and Hansen and Sargent (1991), in some of their own chapters of that book, discuss estimation of continuous time rational expectations models from discrete data.

Estimating stochastic differential equation models from discrete data has recently become easier with the development of Bayesian Markov chain Monte Carlo (MCMC) methods. Though implementation details vary across models, the basic idea is to approximate the diffusion equation

$$dy_t = a(y_t)dt + b(y_t)dW_t\,, \tag{17}$$

where W_t is a Wiener process, by

$$y_t = e^{-a(y_{t-\delta})}y_{t-\delta} + b(y_{t-\delta})\varepsilon_t\,. \tag{18}$$

Such an approximation can be quite inaccurate unless δ is very small. But one can in fact choose δ very small, much smaller than the time interval at which data are observed. The values of y_t at times between observations are of course unknown, but if they are simply treated as unknown 'parameters it may be straightforward to sample from the joint posterior distribution of the y's at non-observation times and the unknown parameters of the model. The Gibbs sampling version of MCMC samples alternately from conditional posterior distributions of blocks of parameters. Here, sampling from the distribution of y at non-observation dates conditioning on the values of model parameters is likely to be easy. If the model has a tractable form, it will also be easy to sample from the posterior distribution of the parameters conditional on all the y values, both observed and unobserved. Application of these general ideas to a variety of financial models is discussed in Johannes and Polson (2006).

Another approach that has become feasible with increased computing power is to develop numerical approximations to the distribution of $y_{t+\delta}$ conditional on data through time t. Aït-Sahalia (2007) surveys methods based on this approach.

Modelling in continuous time does not avoid the complexities of connecting discrete time data to continuous time reality – it only allows us to confront them directly. One reason this is so seldom done despite its technical feasibility is that it forces us to confront the weakness of economic theory in continuous time. A model like (15)–(16) makes an assertion about how many times y and x are differentiable, and a mistake in that assertion can result in error as bad as the mistake of ignoring the time aggregation problem. Economic theory does not have much to say about the degree of differentiability of most aggregate macroeconomic time series. When the theory underlying the model has no believable restrictions to place on fine-grained dynamics, it may be better to begin the modelling effort in discrete time. As is often true when models are in some respect under-identified, it is likely to be easier to begin from a normalized reduced form (in the case the discrete time model) in exploring the range of possible interpretations generated by different potential identifying assumptions.

Recent developments in financial economics have produced one area where there are continuous time economic theories with a solid foundation. Stochastic differential equations (SDEs) provide a convenient and practically useful framework for modelling asset prices. These SDE models imply non-differentiable time paths for prices, and it is known (Harrison, Pitbladdo and Schaefer, 1984) that differentiable time paths for asset prices would imply arbitrage opportunities, if there were no transactions costs or bounds on the frequency of transactions.

However, there are in fact transactions costs and bounds on transactions frequencies, and no-arbitrage models for asset prices break down at very fine, minute-by-minute, time scales. Successful behavioural modelling of these fine time scales requires a good theory of micro-market structure, which is still work in progress.

It is worthwile noting that a process can have non-differentiable paths without producing white noise residuals at any integer order of differentiation: for example, a model satisfying (3) with $a(s) = s^{0.5}e^{-s}$. Such a process has continuous paths with unbounded variation and is not a semimartingale. That is, it is not the sum of a martingale and a process with bounded variation, and therefore cannot be generated from an integer-order SDE. Similarly, if $a(s) = s^{-0.5}e^{-s}$, the process has non-differentiable paths but is nonetheless not a semimartingale. The existence of such non-semimartingale processes and their possible applications to financial modelling is discussed in Sims and Maheswaran (1993).

CHRISTOPHER A. SIMS

See also **time series analysis.**

Bibliography

Aït-Sahalia, Y. 2007. Estimating continuous-time models using discretely sampled data. In *Advances in Economics and Econometrics, Theory and Applications. Ninth World Congress*, vol. 3, ed. R. Blundell, T. Persson and W.K. Newey. Cambridge: Cambridge University Press.

Bergstrom, A.R., ed. 1976. *Statistical Inference in Continuous Time Economic Models*. Amsterdam: North-Holland.

Bergstrom, A.R. 1983. Gaussian estimation of structural parameters in higher order continuous time dynamic models. *Econometrica* 51, 117–52.

Geweke, J. 1978. Temporal aggregation in the multiple regression model. *Econometrica* 46, 643–62.

Hansen, L.P. and Sargent, T.J., eds. 1991. *Rational Expectations Econometrics*. Boulder and Oxford: Westview Press.

Harrison, J.M., Pitbladdo, R. and Schaefer, S.M. 1984. Continuous price processes in frictionless markets have infinite variation. *Journal of Business* 57, 353–65.

Johannes, M. and Polson, N. 2006. MCMC methods for continuous-time financial econometrics. In *Handbook of Financial Econometrics*, ed. Y. Aït-Sahalia and L.P. Hansen. Amsterdam: North-Holland.

Marcet, A. 1991. Temporal aggregation of economic time series. In *Rational Expectations Econometrics*, ed. L.P. Hansen and T.J. Sargent. Boulder and Oxford: Westview Press.

Rozanov, Yu.A. 1967. *Stationary Random Processes*, trans A. Feinstein. San Francisco, Cambridge, London, Amsterdam: Holden-Day.

Sims, C.A. 1971. Approximate specifications in distributed lag models. In Proceedings of the 38th Session, *Bulletin of the International Statistical Institute* 44, Book 1.

Sims, C.A. and Maheswaran, S. 1993. Empirical implications of arbitrage-free asset markets. In *Models, Methods and Applications of Econometrics*, ed. P.C.B. Phillips. Oxford: Blackwell.

data filters

Economic models are by definition incomplete representations of reality. Modellers typically abstract from many features of the data in order to focus on one or more components of interest. Similarly, when confronting data, empirical economists must somehow isolate features of interest and eliminate elements that are a nuisance from the point of view of the theoretical models they are studying. Data filters are sometimes used to do that.

For example, Figure 1 portrays the natural logarithm of US GDP. Its dominant feature is sustained growth, but business cycle modellers often abstract from this feature in order to concentrate on the transient ups and downs. To relate business cycle models to data, empirical macroeconomists frequently filter the data prior to analysis to remove the growth component. Until the 1980s, the most common way to do that was to estimate and subtract a deterministic linear trend. Linear de-trending is conceptually unattractive, however, because it presupposes that all shocks are neutral in the long run. While some disturbances – such as those to monetary policy – probably are neutral in the long run, others probably are not. For instance, a technical innovation is likely to remain relevant for production until it is superseded by another, later technical innovation.

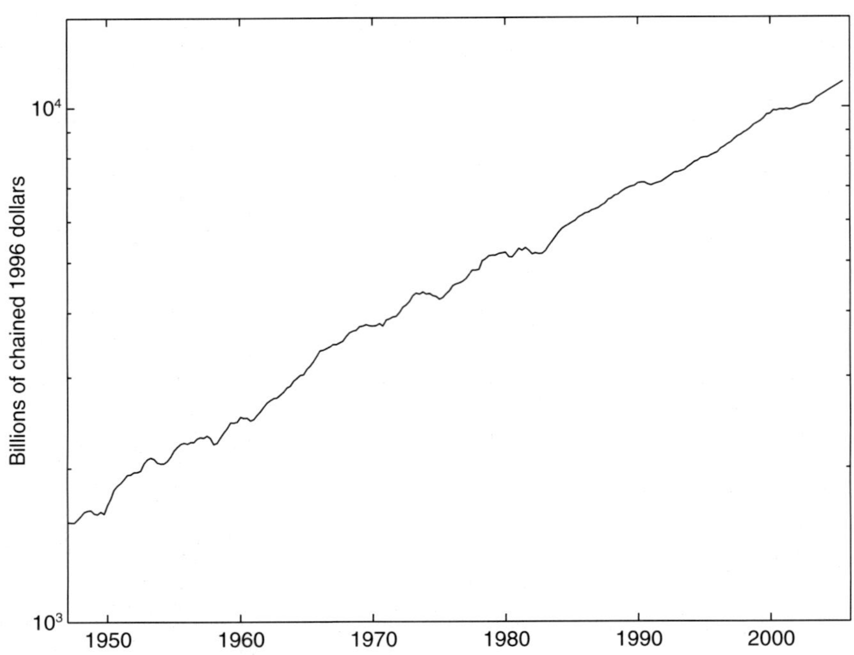

Figure 1 Real US GDP, 1947–2006. *Source*: Federal Reserve Economic Database.

The desire to model permanent shocks in macroeconomic time series led to the development of a variety of stochastic de-trending methods. For example, Beveridge and Nelson (1981) define a stochastic trend in terms of the level to which a time series is expected to converge in the long run. Blanchard and Quah (1989) adopt a more structural approach, enforcing identifying restrictions in a vector autoregression that separate permanent shocks that drive long-run movements from the transitory disturbances which account for cyclical fluctuations.

Another popular way to measure business cycles involves application of band-pass and high-pass filters. Engle (1974) was one of the first to introduce band-pass filters to economics. In the business cycle literature, the work of Hodrick and Prescott (1997) and Baxter and King (1999) has been especially influential. Figure 2 illustrates measures of the business cycle that emerge from the Baxter–King and Hodrick–Prescott filters.

In this article, I describe how data filters work and explain the theoretical rationale that lies behind them. I focus on the problem of measuring business cycles because that is one of the principal areas of application. Many of the issues that arise in this context are also relevant for discussions of seasonal adjustment. For a review of that literature, see Fok, Franses and Paap (2006).

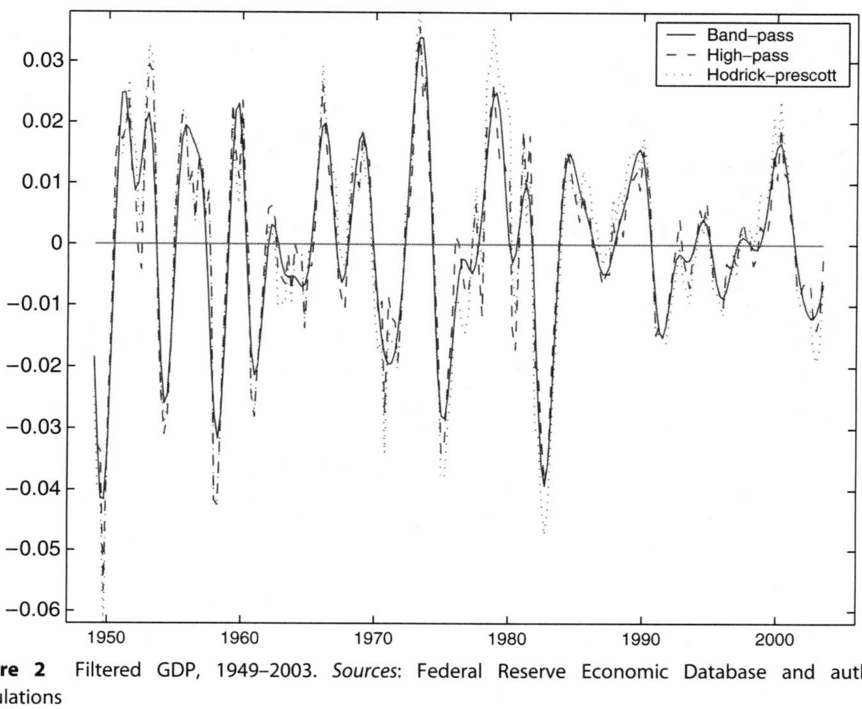

Figure 2 Filtered GDP, 1949–2003. *Sources*: Federal Reserve Economic Database and author's calculations

How data filters work

The starting point is the Cramer representation theorem. Cramer's theorem states that a covariance stationary random variable x_t can be expressed as

$$x_t - \mu_x = \int_{-\pi}^{\pi} \exp(i\omega t) dZ_x(\omega), \tag{1}$$

where μ_x is the mean, t indexes time, $i = \sqrt{-1}$, ω represents frequency, and $dZ_x(\omega)$ is a mean zero, complex-valued random variable that is continuous in ω. The complex variate $dZ_x(\omega)$ is uncorrelated across frequencies, and at a given frequency its variance is proportional to the spectral density $f_{xx}(\omega)$. If we integrate the spectrum across frequencies, we get the variance of x_t,

$$\sigma_x^2 = \int_{-\pi}^{\pi} f_{xx}(\omega) d\omega. \tag{2}$$

This theorem provides a basis for decomposing x_t and its variance by frequency. It is perfectly sensible to speak of long- and short-run variation by identifying the long run with low-frequency components and the short run with high-frequency oscillations. High frequency means that many complete cycles occur within a given time span, while low frequency means the opposite.

Baxter and King (1999) define a business cycle in terms of the periodic components $dZ_x(\omega)$. They partition x_t into three pieces: a trend, a cycle, and irregular fluctuations. Inspired by the NBER business cycle chronology, they say the business cycle consists of periodic components whose frequencies lie between 1.5 and 8 years per cycle. Those whose cycle length is longer than 8 years are identified with the trend, and the remainder are consigned to the irregular component.

The units for ω are radians per unit time. A more intuitive measure of frequency is units of time per cycle, which is given by the transformation $\lambda = 2\pi/\omega$. Often we work with quarterly data. To find the ω corresponding to a cycle length of 1.5 years, just set $\lambda_h = 6$ quarters per cycle and solve for $\omega_h = 2\pi/6 = \pi/3$. Similarly, the frequency corresponding to a cycle length of 8 years is $\omega_l = 2\pi/32 = \pi/16$. Baxter and King define the interval $[\pi/16, \pi/3]$ as 'business cycle frequencies'. The interval $[0, \pi/16)$ corresponds to the trend, and $(\pi/3, \pi]$ defines irregular fluctuations. One nice feature of the Baxter–King filter is that it can be easily adjusted to accommodate data sampled monthly or annually, just be resetting ω_l and ω_h.

To extract the business cycle component, we need to weigh the components $dZ_x(\omega)$ in accordance with Baxter and King's definition and integrate across frequencies,

$$x_t^B = \int_{-\pi}^{\pi} B(\omega) \exp(i\omega t) dZ_x(\omega), \tag{3}$$

where

$$\begin{aligned} B(\omega) &= 1 \quad \text{for } \omega \in [\pi/16, \pi/3] \text{ or } [-\pi/3, -\pi/16], \\ &= 0 \quad \text{otherwise} \end{aligned} \tag{4}$$

In technical jargon, $B(\omega)$ is an example of a 'band-pass' filter: the filter passes periodic components that lie within a pre-specified frequency band and eliminates everything else. The Baxter–King filter suppresses all fluctuations that are too long or short to be classified as part of the business cycle and allows the remaining elements to pass through without alteration.

Many economists are more comfortable working in time domain, and for that purpose it is helpful to express the cyclical component as a two-sided moving average,

$$x_t^B = \sum_{j=-\infty}^{\infty} \beta_j (x_{t+j} - \mu_x). \tag{5}$$

The lag coefficients can be found by solving

$$\beta_j = \frac{1}{2\pi} \int_{-\pi}^{\pi} B(\omega) \exp(i\omega j) d\omega. \tag{6}$$

The solution is

$$\begin{aligned} \beta_0 &= \frac{\omega_h - \omega_l}{\pi}, \\ \beta_j &= \frac{\sin(\omega_h j) - \sin(\omega_l j)}{\pi j} \quad \text{for } j \neq 0. \end{aligned} \tag{7}$$

Notice that an ideal band-pass filter cannot be implemented in actual data samples because it involves infinitely many leads and lags. In practice, economists approximate x_t^B with finite-order moving averages,

$$x_t^B \doteq \sum_{j=-n}^{n} \tilde{\beta}_j (x_{t+j} - \mu_x). \tag{8}$$

Baxter and King (1999) and Christiano and Fitzgerald (2003) analyse how to choose the lag weights $\tilde{\beta}_j$ in order to best approximate the ideal measure for a given n.

For real-time applications, the two-sided nature of the filter is a drawback because the current output of the filter depends on future values of x_{t+j}, which are not yet available. Kaiser and Maravall (2001) address this problem by supplementing the filter with an auxiliary forecasting model such as a vector autoregression or univariate ARIMA model, replacing future x_{t+j} with forecasted values. This substantially reduces the approximation error near the end of samples.

That the filter is two-sided is also relevant for models that require careful attention to the timing of information. Economic hypotheses can often be formulated as a statement that some variable z_t should be uncorrelated with any variable known in period $t-1$ or earlier. These hypotheses can be examined by testing for absence of Granger causation from a collection of potential predictors to z_t. The output of a two-sided filter should never be included among those predictors, however, for that would put information about present and future conditions on the right-hand side of the regression and bias the test towards a false finding of Granger causation. Similar comments apply to the choice of instruments in generalized-method-of-moments problems. For applications like these, one-sided filters are needed in order to respect the integrity of the information flow.

While Baxter and King favour a three-part decomposition, other economists prefer a two-part classification in which the highest frequencies also count as part of the business cycle. The trend component is still defined in terms of fluctuations lasting more than eight years, but the cyclical component now consists of all oscillations lasting eight years or less. To construct this measure, we define a new filter $H(\omega)$ such that

$$\begin{aligned} H(\omega) &= 1 \quad \text{for } \omega \in [\pi/16, \pi] \text{ or } [-\pi, -\pi/16], \\ &= 0 \quad \text{otherwise.} \end{aligned} \tag{9}$$

This is known as a 'high-pass' filter because it passes all components at frequencies higher than some pre-specified value and eliminates everything else. If we use this filter in the Cramer representation, we can extract a new measure of the business cycle by computing

$$x_t^H = \int_{-\pi}^{\pi} H(\omega) \exp(i\omega t) dZ_x(\omega). \tag{10}$$

Once again, this corresponds to a two-sided, infinite-order moving average of the original series x_t,

$$x_t^H = \int_{j=-\infty}^{\infty} \gamma_j (x_{t+j} - \mu_x), \tag{11}$$

with lag coefficients $\gamma_0 = 1 - \omega_l/\pi$ and $\gamma_j = -\sin(\omega_l j)/\pi j$. As before, this involves infinitely many leads and lags, so an approximation is needed to make it work. The approximation results of Baxter and King (1999) and Christiano and Fitzgerald (2003) apply here as well.

Hodrick and Prescott (1997) also seek a two-part decomposition of x_t. They proceed heuristically, identifying the trend τ_t and the cycle c_t by minimizing the variance of the cycle subject to a penalty for variation in the second difference of the trend,

$$\min_{\{\tau_t\}} \left\{ \sum_{t=-\infty}^{\infty} [(x_t - \tau_t)^2 + \varphi(\tau_{t+1} - 2\tau_t + \tau_{t-1})^2] \right\}. \tag{12}$$

The Lagrange multiplier φ controls the smoothness of the trend component. After experimenting with US data, Hodrick and Prescott set $\varphi = 1600$, a choice still used in most macroeconomic applications involving quarterly data. After differentiating (12) with respect to τ_t and rearranging the first-order condition, one finds that c_t can be expressed as an infinite-order, two-sided moving average of x_t,

$$c_t = HP(L)x_t = \frac{\varphi(1-L)^2(1-L^{-1})^2}{1+\varphi(1-L)^2(1-L^{-1})^2} x_t, \tag{13}$$

where L is the lag operator. Although Hodrick and Prescott's derivation is heuristic, King and Rebelo (1993) demonstrate that $HP(L)$ can be interpreted rigorously as an approximation to a high-pass filter with a cut-off frequency of eight years per cycle. The close connection between the two filters is also apparent in Figure 2, which shows that high-pass and Hodrick–Prescott filtered GDP are highly correlated.

Data filters for measuring of business cycles?

While data filters are very popular, there is some controversy about whether they represent appealing definitions of the business cycle. For one, there is a disconnect between the theory and macroeconomic applications, for the theory applies to stationary random processes and applications involve non-stationary variables. This is not critical, however, because the time-domain filters $\beta(L)$, $\gamma(L)$, and $HP(L)$ all embed difference operators, so business cycle components are stationary even if x_t has a unit root.

A more fundamental criticism concerns the fact that the Baxter–King definition represents a deterministic vision of the business cycle. According to a theorem of Szego, Kolmogorov, and Krein, the prediction error variance can be expressed as

$$\sigma_\varepsilon^2 = 2\pi \exp\left[\frac{1}{2\pi}\int_{-\pi}^{\pi} \log f_{BC}(\omega) d\omega\right], \tag{14}$$

where $f_{BC}(\omega)$ is the spectrum for the business-cycle component (see Granger and Newbold 1986, pp. 135–6). For an ideal band-pass filter, the spectrum of x_t^B is

$$f_{BC}(\omega) = |B(\omega)|^2 f_{xx}(\omega). \tag{15}$$

Since $B(\omega) = 0$ outside of business cycle frequencies, it follows that $f_{BC}(\omega) = 0$ on a measurable interval of frequencies. But then eq. (14) implies $\sigma_\varepsilon^2 = 0$, which means that x_t^B is perfectly predictable from its own past. The same is true of measures based on ideal high-pass filters. A variable that is perfectly predictable based on its own history is said to be 'linearly deterministic'. Thus, according to the Baxter–King definition, the business cycle is linearly deterministic.

In practice, of course, measured cycles are not perfectly predictable because actual filters only approximate the ideal. But this means that innovations in measured cycles are due solely to approximation errors in the filter, not to something intrinsic in the concept. The better the approximation, the closer the measures are to determinism.

How to square this deterministic vision with stochastic general equilibrium models is not obvious. Engle (1974), Sims (1993) and Hansen and Sargent (1993) suggest one rationale. They were interested in estimating models that are well specified at some frequencies but mis-specified at others. Engle studied linear regressions and showed how to estimate parameters by band-spectrum regression. This essentially amounts to running regressions involving band-pass filtered data, but band-pass filtering induces serial correlation in the residuals, and Engle showed how to adjust for this when calculating standard errors and other test statistics. He also developed methods for diagnosing mis-specification on particular frequency bands.

Sims (1993) and Hansen and Sargent (1993) are interested in fitting a rational-expectations model of the business cycle to data that contain seasonal fluctuations. They imagine that the model abstracted from seasonal features, as is common in practice, and they wonder whether estimates could be improved by filtering the data with a narrow band-pass filter centred on seasonal frequencies. They find that seasonal filtering does help, because otherwise parameters governing business cycle features would be distorted to fit unmodelled seasonal fluctuations. Filtering out the seasonals lets the business cycle parameters fit business cycle features.

Business cycle modellers also frequently abstract from trends, and Hansen and Sargent conjectured that the same rationale would apply to trend filtering. Cogley (2001) studies this conjecture but finds disappointing results. The double-filtering strategy common in business cycle research (which applies the filter to both the data and the model) has no effect on periodic terms in a Gaussian log likelihood, so it is irrelevant for estimation. The seasonal analogy (which filters the data but not the model) also fails, but for a different reason. The key assumption underlying the work of Engle, Sims, and Hansen and Sargent is that specification errors are confined to a narrow frequency band whose location is known a priori. That is true of the seasonal problem but not of the trend problem. Contrary to intuition, trend-specification errors spread throughout the frequency domain and are not quarantined to low frequencies. That difference explains why the promising results on seasonality do not carry over to trend filtering.

Finally, some economists question whether filter-based measures capture an important feature of business cycles. Beveridge and Nelson (1981) believe that trend reversion is a defining characteristic of the business cycle. They say that expected growth should be higher than average at the trough of a recession because agents can look forward to a period of catching up to compensate for past output losses. By the same token, expected growth should be lower than average at the peak of an expansion. Cochrane (1994) confirms that this is a feature of US business cycles by studying a vector autoregression for consumption and GDP.

Cogley and Nason (1995) consider what would happen if x_t were a random walk with drift. For a random walk, expected growth is constant regardless of whether the level is a local maximum or minimum. Because it lacks the catching-up feature, many economists would say that a random walk is acyclical. Nevertheless, when the Hodrick–Prescott filter is applied to a random walk, a large and persistent cycle emerges. Thus the Hodrick–Prescott filter can create a business cycle even if no trend reversion is present in the original data. Cogley and Nason call this a spurious cycle. Furthermore, the problem is not unique to the Hodrick–Prescott filter; Benati (2001), Murray (2003) and Osborn (1995) document similar results for band-pass filters and for other approximations to high-pass filters.

Conclusion

Christiano and Fitzgerald remark that data filters are not for everyone. They are certainly convenient for constructing rough and ready measures of the business cycle, and they produce nice pictures when applied to US data. But some economists worry about the spurious cycle problem, especially in applications to business cycle models where the existence and properties of business cycles are points to be established. In much of that literature, attention has shifted away from replicating properties of filtered data to matching the shape of impulse response functions.

TIMOTHY COGLEY

See also **seasonal adjustment; spectral analysis; structural vector autoregressions; trend/cycle decomposition.**

Bibliography

Baxter, M. and King, R. 1999. Measuring business cycles: approximate band-pass filters for economic time series. *Review of Economics and Statistics* 81, 575–93.

Benati, L. 2001. Band-pass filtering, cointegration, and business cycle analysis. Working Paper No. 142, Bank of England.

Beveridge, S. and Nelson, C. 1981. A new approach to decomposition of economic time series into permanent and transitory components with particular attention to measurement of the business cycle. *Journal of Monetary Economics* 7, 151–74.

Blanchard, O. and Quah, D. 1989. The dynamic effects of aggregate demand and supply disturbances. *American Economic Review* 79, 655–73.

Christiano, L. and Fitzgerald, T. 2003. The band pass filter. *International Economic Review* 44, 435–65.

Cochrane, J. 1994. Permanent and transitory components of GNP and stock prices. *Quarterly Journal of Economics* 109, 241–65.

Cogley, T. 2001. Estimating and testing rational expectations models when the trend specification is uncertain. *Journal of Economic Dynamics and Control* 25, 1485–525.

Cogley, T. and Nason, J. 1995. Effects of the Hodrick–Prescott filter on trend and difference stationary time series: implications for business cycle research. *Journal of Economic Dynamics and Control* 19, 253–78.

Engle, R. 1974. Band-spectrum regression. *International Economic Review* 15, 1–11.

Fok, D., Franses, P. and Paap, R. 2006. Comparing seasonal adjustment methods. In *Palgrave Handbook of Econometrics: Volumel; Econometric Theory*, ed. T. Mills and K. Patterson. Basingstoke: Palgrave Macmillan.

Granger, C. and Newbold, P. 1986. *Forecasting Economic Time Series*. New York: Academic Press.

Hansen, L. and Sargent, T. 1993. Seasonality and approximation errors in rational expectations models. *Journal of Econometrics* 55, 21–55.

Hodrick, R. and Prescott, E. 1997. Postwar U.S. business cycles: an empirical investigation. *Journal of Money Credit and Banking* 29, 1–16.

Kaiser, R. and Maravall, A. 2001. *Measuring Business Cycles in Economic Time Series*. New York: Springer-Verlag.

King, R. and Rebelo, S. 1993. Low-frequency filtering and real business cycles. *Journal of Economic Dynamics and Control* 17, 207–31.

Murray, C. 2003. Cyclical properties of Baxter–King filtered time series. *Review of Economics and Statistics* 85, 472–76.

Osborn, D. 1995. Moving average detrending and the analysis of business cycles. *Oxford Bulletin of Economics and Statistics* 57, 547–58.

Sims, C. 1993. Rational expectations modeling with seasonally adjusted data. *Journal of Econometrics* 55, 9–19.

equilibrium-correction models

1. Introduction

An equilibrium is a state from which there is no inherent tendency to change. Since we deal with stochastic processes, the equilibrium is the expected value of the variable in an appropriate representation, since that is the state to which the process would revert in the absence of further shocks. Then, we define an equilibrium-correction model (EqCM) as one (*a*) which has a well-defined equilibrium, and (*b*) in which adjustment takes place towards that equilibrium. A key aspect of an EqCM is that deviations from its expected value are attenuated, and eventually eliminated if no additional outside influences impinge. As such, equilibrium-correction models are a very broad class, comprising all regressions, autoregressions, autoregressive-distributed lag (ADL) models, linear simultaneous equations, vector autoregressions (VARs), vector equilibrium-correction systems based on cointegration (VEqCMs), dynamic stochastic general equilibrium systems (DSGEs), autoregressive conditional heteroscedastic processes as in Engle (1982) (ARCH), and generalized ARCH (GARCH, see Bollerslev, 1986) processes among others. Their formulation (in levels or differences) determines the equilibrium to which they converge (level or steady state). For example, a random walk without drift is a non-stationary process in levels, but is stationary in differences (its non-integrated representation), and has an expectation of zero, so the differences equilibrium corrects to zero.

We first address the broad nature of the equilibrium-correction class in Section 2, then review the history of equilibrium-correction model formulation in Section 3, and consider its links to cointegration in Section 4. The roles of cointegration and equilibrium correction in economic forecasting are examined in Section 5, in particular the non-robustness of EqCMs to location shifts in the underlying equilibria, and consequently their proneness to forecast failure. Section 6 concludes.

2. The equilibrium-correction class

Often it is not realized that the model being used is a member of the equilibrium-correction class, so this section establishes that the models listed above are indeed in the EqCM class. The properties of the class are partly specific to the precise model, but primarily generic, as Section 5 emphasizes. We consider six cases.

2.1 Regression as an equilibrium-correction model

Consider a conditional linear equation of the form in (1) for $t = 1, \ldots, T$:

$$y_t = \beta_0 + \sum_{i=1}^{k} \beta_i z_{i,t} + \varepsilon_t = \beta_0 + \beta' \mathbf{z}_t + \varepsilon_t \qquad (1)$$

with $\varepsilon_t \sim \mathsf{IN}[0, \sigma_\varepsilon^2]$ (normally and independently distributed, mean zero, variance σ_ε^2) independently of the past and present of the k regressors $\{\mathbf{z}_t\}$. Then:

$$\mathsf{E}[(y_t - \beta_0 - \beta'\mathbf{z}_t)|\mathbf{z}_t] = 0 \tag{2}$$

defines the conditional equilibrium, where adjustment to that equilibrium is instantaneous as entailed by (1). Re-expressing (1) in differences ($\Delta x_t = x_t - x_{t-1}$ for any x) and lagged deviations from (2) delivers the (isomorphic) EqCM formulation:

$$\Delta y_t = \beta'\Delta\mathbf{z}_t - (y_{t-1} - \beta_0 - \beta'\mathbf{z}_{t-1}) + \varepsilon_t \tag{3}$$

where the feedback coefficient is -1. Then (3) is an EqCM where the equilibrium-correction term is $(y_{t-1} - \beta_0 - \beta'\mathbf{z}_{t-1})$. Notice that differencing is a linear transformation, not an operator, in any setting beyond a scalar time series.

The existence of (2) does not require that y_t and $\mathbf{z}_t$ are stationary, provided the linear combination is; and could hold, for example, for growth rates rather than the original levels if y_t and $\mathbf{z}_t$ were differences of those original variables.

2.2 Autoregressive-error models as equilibrium-corrections

Even extending a static regression like (1) by (say) a first-order autoregressive error as in:

$$y_t = \beta_0 + \beta'\mathbf{z}_t + u_t \quad \text{where} \quad u_t = \rho u_{t-1} + \varepsilon_t \tag{4}$$

leads to:

$$y_t = \beta_0 + \beta'\mathbf{z}_t + \rho(y_{t-1} - \beta_0 - \beta'\mathbf{z}_{t-1}) + \varepsilon_t$$

or:

$$\Delta y_t = \beta'\Delta\mathbf{z}_t + (\rho - 1)(y_{t-1} - \beta_0 - \beta'\mathbf{z}_{t-1}) + \varepsilon_t \tag{5}$$

showing that the common-factor model class (see Sargan, 1980; Hendry and Mizon, 1978) is also a restricted equilibrium-correction mechanism, constrained by the impact effects (from $\Delta\mathbf{z}_t$) being the same as the long-run effects (from $\mathbf{z}_{t-1}$).

2.3 ADLs as equilibrium-correction models

A first-order autoregressive distributed-lag (ADL) model is:

$$y_t = \beta_0 + \beta_1'\mathbf{z}_t + \beta_2 y_{t-1} + \beta_3'\mathbf{z}_{t-1} + \varepsilon_t \quad \text{where} \quad \varepsilon_t \sim \mathsf{IN}[0, \sigma_\varepsilon^2]. \tag{6}$$

The error $\{\varepsilon_t\}$ on (6) is an innovation against the available information, and its serial independence is part of the definition of the model, whereas normality and homoscedasticity are just for convenience. The condition $|\beta_2| < 1$ is needed to ensure a levels' equilibrium solution: Ericsson (2007) provides an extensive discussion. We consider (6) for both stationary and integrated $\{\mathbf{z}_t\}$, the latter denoting that some of the $\mathbf{z}_t$ have unit roots in their levels representations, but are stationary in differences.

First, under stationarity, taking expectations in (6) where $\mathsf{E}[y_t] = y^*$ and $\mathsf{E}[\mathbf{z}_t] = \mathbf{z}^*$ $\forall t$:

$$\mathsf{E}[(1 - \beta_2)y_t - \beta_0 - (\beta_1 + \beta_3)'\mathbf{z}_t] = 0 \tag{7}$$

so:

$$y^* = \frac{\beta_0}{1-\beta_2} + \frac{1}{1-\beta_2}(\beta_1 + \beta_3)'\mathbf{z}^* = \kappa_0 + \kappa_1'\mathbf{z}^*. \tag{8}$$

Since many economic theories have long-run partial equilibria like (8), they could be modelled by this class. Transforming (6) to differences and the equilibrium-correction term $(y - \kappa_0 - \kappa_1'\mathbf{z})_{t-1}$ delivers:

$$\Delta y_t = \beta_1'\Delta\mathbf{z}_t + (\beta_2 - 1)(y_{t-1} - \kappa_0 - \kappa_1'\mathbf{z}_{t-1}) + \varepsilon_t. \tag{9}$$

The immediate impact of a change in $\mathbf{z}_t$ on y_t is β_1, and the rapidity with which Δy_t converges to zero, which is its equilibrium outcome under stationarity, depends on the magnitude of $(\beta_2 - 1) < 0$; when both changes and ε_t are zero (their expectations), (7) results.

When y_t and $\mathbf{z}_t$ are integrated of order 1 (denoted I(1)), so are stationary in differences, the reformulation in (9) remains valid provided $|\beta_2| < 1$ in which case $(y_t - \kappa_0 - \kappa_1'\mathbf{z}_t)$ is a cointegration relation, as discussed in Section 4. Let $\mathrm{E}[\Delta\mathbf{z}_t] = \delta$ (say) so $\mathrm{E}[\Delta y_t] = \kappa_1'\delta = g_y$ where $\mathrm{E}[y_t - \kappa_1'\mathbf{z}_t] = \mu$, then taking expectations in (9) using (7):

$$g_y = \beta_1'\delta + (\beta_2 - 1)(\mu - \kappa_0) \tag{10}$$

and subtracting (10) from (9) delivers:

$$\Delta y_t = g_y + \beta_1'(\Delta\mathbf{z}_t - \delta) + (\beta_2 - 1)(y_{t-1} - \kappa_1'\mathbf{z}_{t-1} - \mu) + \varepsilon_t. \tag{11}$$

Re-specifying deterministic terms as in (11) plays an important role in EqCMs, both by helping to orthogonalize the regressors, and because of the pernicious effects of shifts in μ, a topic addressed in Section 5. It is so well known that the standard error of the mean of an IID random variable is the standard deviation of the data divided by the square root of the sample size that it hardly bears reiterating: except that it is somehow almost always ignored in this context. The standard error of the intercept in an EqCM equation like (11) should, therefore, be $\hat{\sigma}_\varepsilon/\sqrt{T}$ but is often a hundred times larger in reported empirical models, revealing a highly collinear specification (a similar comment applies to VARs). Moreover, a check on the model formulation follows from using sample means to estimate δ and μ, then checking that g_y has a sensible value, which may be given by theory (for example, no autonomous inflation, so $g_y = 0$).

Finally, if $\beta_2 = 1$, (9) equilibrium corrects in differences. An autoregression is the special case where $\beta_1 = \beta_3 = 0$, so is also an EqCM; and partial adjustment is another special case where now $\beta_3 = 0$.

2.4 GARCH as an equilibrium-correction model

As a fourth example, consider a non-integrated GARCH(1, 1) process for ε_t, where $\mathrm{E}[\varepsilon_t^2|I_{t-1}] = \sigma_t^2$ when I_{t-1} denotes past information, and:

$$\sigma_t^2 = \omega + \alpha\varepsilon_{t-1}^2 + \theta\sigma_{t-1}^2 \tag{12}$$

with $0<\alpha<1$, $0<\theta<1$ and $0<\alpha+\theta<1$. Let $\sigma_t^2 = \varepsilon_t^2 - \nu_t$, where $\mathsf{E}[\nu_t] = 0$, then:

$$\varepsilon_t^2 = \omega + (\alpha+\theta)\varepsilon_{t-1}^2 + \nu_t - \theta\nu_{t-1} \tag{13}$$

where the equilibrium is:

$$\sigma_\varepsilon^2 \equiv \mathsf{E}[\varepsilon_t^2] = \frac{\omega}{1-(\alpha+\theta)}. \tag{14}$$

Substituting $\omega = (1-(\alpha+\theta))\sigma_\varepsilon^2$ from (14) into the equation for σ_t^2:

$$\Delta\sigma_t^2 = (\theta-1)(\sigma_{t-1}^2 - \sigma_\varepsilon^2) + \alpha(\varepsilon_{t-1}^2 - \sigma_\varepsilon^2). \tag{15}$$

Thus, the change in the conditional variance σ_t^2 responds less than proportionally ($\theta<1$) to the previous disequilibrium between the conditional variance and the long-run variance, perturbed by the zero-mean discrepancy between the previous squared disturbance ε_{t-1}^2 and the long-run variance σ_ε^2, so the model equilibrium corrects to σ_ε^2, consistent with (14). ARCH is simply a special case.

2.5 VARs as equilibrium-correction models

The fifth example is an n-dimensional VAR with m lags and an innovation error $\varepsilon_t \sim \mathsf{IN}_n[0, \mathbf{\Omega}_\varepsilon]$:

$$\mathbf{x}_t = \pi + \sum_{i=1}^{m} \mathbf{\Pi}_i x_{t-i} + \varepsilon_t \tag{16}$$

where the nm eigenvalues of the polynomial $|\mathbf{I}_n - \sum_{i=1}^{m}\mathbf{\Pi}_i L^i|$ in L determine the characteristics of the time series. If all the eigenvalues are inside the unit circle, (16) is stationary (when all the parameters are constant and the initial conditions also satisfy the process). In that case, $\mathbf{\Gamma} = (I_n - \sum_{i=1}^{m}\mathbf{\Pi}_i)$ is invertible and has all its eigenvalues inside the unit circle, so the process equilibrium corrects to $\psi = \mathbf{\Gamma}^{-1}\pi$. To illustrate for $m=2$, (16) can be expressed as:

$$\Delta\mathbf{x}_t = (\mathbf{\Pi}_1 - \mathbf{I}_n)\Delta\mathbf{x}_{t-1} - \mathbf{\Gamma}(\mathbf{x}_{t-2} - \psi) + \varepsilon_t \tag{17}$$

where $\mathsf{E}[\Delta x_t] = 0$ by stationarity, so $\mathsf{E}[x_t - \psi] = 0$ is indeed the equilibrium to which x_t converges in the absence of further shocks. Conversely, if all the eigenvalues are unity, x_t is I(1) with $\mathbf{\Gamma} = 0$ in (17), so does not equilibrium correct in levels, but does so in the differences (unless their polynomial has further unit roots, making the process doubly integrated, I(2)). Finally, for a combination of eigenvalues inside and on the unit circle, $\boldsymbol{\Gamma}$ has reduced rank $0<r<n$ equal to the number of non-unit eigenvalues, so can be expressed as $\mathbf{\Gamma} = \alpha\beta'$ where α and β also have rank r. Then π in (16) can be decomposed into the unconditional growth rate of x_t, denoted γ, and $\alpha\mu$ such that in place of (17), we have:

$$\Delta\mathbf{x}_t = \gamma + (\mathbf{\Pi}_1 - \mathbf{I}_n)(\Delta\mathbf{x}_{t-1} - \gamma) - \alpha(\beta'\mathbf{x}_{t-2} - \mu) + \varepsilon_t \tag{18}$$

so that $\mathsf{E}[\beta' x_t - \mu] = 0$ and the system converges to that equilibrium when the original variables are I(1), hence $\beta' x_{t-2}$ is an I(0) process which equilibrium corrects to μ. At the same time, Δx_t is an I(0) process which equilibrium corrects to γ, noting that $\beta'\gamma = 0$, whereas x_t drifts.

Linear simultaneous equations systems of time series are a restriction on a VAR, so are also EqCMs.

2.6 DSGEs as equilibrium-correction models

As a final brief example, well-defined general equilibrium systems have equilibria. Using Taylor-series expansions around the steady-state values of the discretized representation of a system of differential equations, Bårdsen, Hurn and Lindsay (2004) show that any dynamic system with a steady-state solution has a linear EqCM representation. Thus, they argue that linearizations of DSGEs imply linear EqCM representations. In principle, these could be in terms of changes only, corresponding to a steady-state path. More usually, level solutions result.

3. Historical overview

Equilibrium-correction models are a special case of the general class of proportional, derivative and integral control mechanisms, so have a long pedigree in that arena: for economics examples, see Phillips (1954); Phillips and Quenouille (1960); and Whittle (1963), with the links summarized in Salmon (1988). Explicit examples of EqCMs are presented in Sargan (1964) and were popularized by Davidson et al. (1978), although they were called 'error-correction mechanisms' (ECMs) by those authors. The major developments underlying cointegration in Engle and Granger (1987) established its isomorphism with equilibrium correction for integrated processes, leading to an explosion in the application of EqCMs and the development of a formal analysis of vector EqCM systems in Johansen (1988; 1995). We now review the two stages linking control mechanisms with error correction, then that with equilibrium correction.

3.1 Error correction and control mechanisms

Phillips (1954; 1957), in particular, pioneered the application of control methods for macroeconomic stabilization, specifically techniques for derivative, proportional and integral control servomechanisms. In this form of control, a target (say an unemployment rate of five per cent) is to be achieved by adjusting an instrument (say government expenditure), and changes to the instrument, its level, and cumulative past errors may need to be included in the rule to stabilize the target.

That approach is a precursor to the well-known linear-quadratic model in which one optimizes a quadratic function of departures from target trajectories for a linear dynamic system over a finite future horizon (see, for example, Holt et al., 1960; Preston and Pagan, 1982). For example, consider the quadratic cost function C_H which penalizes the deviations of a variable x_{t+j} from a pre-specified target trajectory $\{x^*_{t+j}\}$ subject to costs of adjustment from changes $\Delta x_t = x_t - x_{t-1}$ over an H-period horizon commencing at time t:

$$C_H = \sum_{j=0}^{H} c_{t+j} = \sum_{j=0}^{H} \frac{1}{2}\left[(x_{t+j} - x^*_{t+j})^2 + \alpha(\Delta x_{t+j})^2\right]. \tag{19}$$

To minimize c_{t+j} at time $t+j$, differentiate with respect to x_{t+j}, noting the intertemporal link that $\Delta x_{t+j+1} = x_{t+j+1} - x_{t+j}$ also depends on x_{t+j}, which yields (ignoring the end point for simplicity):

$$\frac{\partial C_H}{\partial x_{t+j}} = \frac{\partial c_{t+j}}{\partial x_{t+j}} + \frac{\partial c_{t+j+1}}{\partial x_{t+j}} = x_{t+j} - x^*_{t+j} + \alpha(\Delta x_{t+j}) - \alpha(\Delta x_{t+j+1}), \tag{20}$$

so equating to zero for a minimum for any j, and hence for $j = 0$:

$$x_t - x^*_t + \alpha\Delta x_t - \alpha\Delta x_{t+1} = 0.$$

Expressed as a polynomial in leads and lags in the operator L (for $\alpha \neq 0$):

$$(L^{-1} - (2 + \alpha^{-1}) + L)x_t = (L^{-1} - \lambda_2)(1 - \lambda_1 L)x_t = -\frac{x^*_t}{\alpha}. \tag{21}$$

The polynomial in (21) has roots λ_1 and λ_2 with a product of unity (so they are inverses, with λ_1 inside and λ_2 outside the unit circle) and a sum of $(2 + \alpha^{-1})$. Inverting the first factor $(L^{-1} - \lambda_2)$, using $(1/\lambda_2) = \lambda_1 < 1$ and expanding the last term as a power series in L^{-1}expresses x_t as a function of lagged x_s and current and future values of x^*_{t+k}:

$$(1 - \lambda_1 L)x_t = \frac{\lambda_1}{\alpha}(1 + \lambda_1 L^{-1} + \lambda_1^2 L^{-2} + \cdots)x^*_t = \frac{\lambda_1}{\alpha}\sum_{k=0}^{\infty} \lambda_1^k x^*_{t+k}. \tag{22}$$

Since $(1 - \lambda_1) = \lambda_1/\alpha(1 - \lambda_1)$, let:

$$x^{**}_t = (1 - \lambda_1)\sum_{k=0}^{\infty} \lambda_1^k x^*_{t+k} \tag{23}$$

denote the 'ultimate' target (scaled so that the weights sum to unity as in, for example, Nickell, 1985) then from (22) using (23), for $t < H$:

$$\begin{aligned}\Delta x_t &= -(1 - \lambda_1)(x_{t-1} - x^{**}_t) \\ &= (1 - \lambda_1)\Delta x^{**}_t - (1 - \lambda_1)(x_{t-1} - x^{**}_{t-1}).\end{aligned} \tag{24}$$

Thus, x_t adjusts to changes in the ultimate target, and to the previous error from that target, and is an EqCM when $-1 < \lambda_1 < 1$. Mistakes in plans, errors in expectations, and relations between the ultimate target and its determinants all need to be modelled for an operational rule. To hit a moving target requires a feedforward rule, and the role of $\alpha(\Delta x_{t+j})^2$ in (19) is to penalize the controller from making huge changes to x_t when doing so. However, it is difficult to imagine real world adjustment costs being proportional to changes, which in any case then depend on the specification of x_t as logs, levels, proportions or even changes (see, for example, Nickell, 1985). Moreover, the entire class is partial adjustment, as (24) shows.

For 1-period optimization (so $H = 0$: see, for example, Hendry and Anderson, 1977), only the end point is relevant, so (20) delivers the planned value x_t^p as a function of $x_t^{**} = x_t^*$:

$$x_t^p - x_{t-1} = \frac{1}{1-\alpha}(x_t^* - x_{t-1}) = \rho(x_t^* - x_{t-1}). \tag{25}$$

When the error on the plan is $\varepsilon_t = x_t - x_t^p$, where $\mathsf{E}[x_t^p \varepsilon_t] = 0$ under rationality, and $x_t^* = \beta' \mathbf{z}_t$ (say), (25) becomes:

$$\begin{aligned} \Delta x_t &= \rho(\beta'\mathbf{z}_t - x_{t-1}) + \varepsilon_t \\ &= \rho\beta'\Delta\mathbf{z}_t - \rho(x_{t-1} - \beta'\mathbf{z}_{t-1}) + \varepsilon_t. \end{aligned}$$

This is a partial adjustment again. The static regression in Section 2.1 has a more restrictive dynamic structure, but otherwise the properties of the ADL in Section 2.3 can vary over a wide range (see Hendry, 1995, Ch. 6).

3.2 From error correction to equilibrium correction

The model in Sargan (1964) was explicitly an ECM for wages and prices (w_t and p_t denote their respective logs), building on previous models of wage and price inflation written as:

$$\Delta w_t = \beta_0 + \beta_1 \Delta p_t + \beta_2 \Delta w_{t-1} + \varepsilon_t. \tag{26}$$

When $\mathsf{E}[\varepsilon_t] = 0$ and the differenced variables are stationary with means $\mathsf{E}[\Delta w_t] = \dot{w}$ and $\mathsf{E}[\Delta p_t] = \dot{p}$, then the long-run steady-state solution to (26) is:

$$\dot{w} = \frac{\beta_0 + \beta_1 \dot{p}}{1 - \beta_2}.$$

As formulated, (26) does not establish any relationship between the levels w_t and p_t, hence these could drift apart. Since economic agents are concerned about the level of real wages, $w_t - p_t$, Sargan postulated the equilibrium:

$$(w - p)_{e,t} = \delta_0 + \delta_1 \Delta p_t + \delta_2' \mathbf{z}_t, \tag{27}$$

where $\mathbf{z}_t$ denotes a vector of additional variables, such as unemployment (u), productivity (q) and political factors. The disequilibrium is:

$$v_t = w_t - p_t - \delta_0 - \delta_1 \Delta p_t - \delta_2' \mathbf{z}_t \tag{28}$$

and, to re-establish equilibrium whenever the levels drift apart, he used the explicit adjustment equation:

$$\Delta w_t = \alpha(w_{t-1} - p_{t-1} - (w - p)_{e,t-1}) = \alpha v_{t-1}. \tag{29}$$

If a relation like (28) is well defined with v_t being I(0) when the levels are I(1), so the differences are I(0), then w_t forms a non-integrated combination with p_t and $\mathbf{z}_t$ so these variables are cointegrated (see, among many others, Engle and Granger, 1987; Phillips and Loretan, 1991; Banerjee, et al., 1993).

A less restricted specification than (26) entails including the levels terms $(w - p)_{t-1}$ and $\mathbf{z}_{t-1}$ (and their differences), so if contemporaneous variables are excluded:

$$\begin{aligned}\Delta w_t = {} & \pi_0 + \pi_1 \Delta p_{t-1} + \pi_2 \Delta w_{t-1} - \pi_3 (w - p)_{t-1} + \pi_4' \mathbf{z}_{t-1} \\ & + \pi_5' \Delta \mathbf{z}_{t-1} + u_t.\end{aligned} \tag{30}$$

When $\pi_3 \neq 0$, the long-run levels equilibrium solution to (30) matching (27) is $(\phi_4 = \pi_4/\pi_3)$:

$$\mathrm{E}[w - p - \phi_4' \mathbf{z}] = f(\dot{w}, \dot{p}, \dot{\mathbf{z}}).$$

The model in (30) has both derivative and proportional control (e.g., Δp_{t-1} and $(w - p)_{t-1}$) following up Phillips (1954; 1957) (see Salmon, 1982). The proportional mechanism ensures the disequilibrium adjustment, based on the (possibly detrended) log-ratio of two nominal levels (see, for example, Bergstrom, 1962). The equivalent of g_y in Section 2.3 should be $\dot{p} + \dot{q}$ in (30) to avoid having ‘autonomous wage inflation’ independent of all economic forces.

The long-run stability of the ‘great ratios’ in Klein (1953) was often implicitly assumed to justify such transformations, but had come under question (see, for example, Granger and Newbold, 1977, and the discussion in Hendry, 1977), although Hendry and Mizon (1978) had argued that what mattered was that the errors in (30) were stationary, not that all the variables were stationary. Granger (1981) related the type of model in (30) to cointegration, and Granger (1986) showed the important new result that one of Δw_t or Δp_t must depend on the equilibrium correction if w_t and p_t were cointegrated: the assumption in (29) is that w_t adjusts to the disequilibrium. If both variables w_t and p_t adjust to the disequilibrium, then p_t is not weakly exogenous for the $\{\pi_i\}$ (see Phillips and Loretan, 1991; Hendry, 1995). It is primarily because of cointegration that equilibrium-correction models like (30) have proved a popular specification. Engle and Granger (1987) showed that cointegration and proportional EqCM were equivalent, linking time-series approaches more closely with econometric modelling. Davidson and Hall (1991) also linked VARs as in Section 2.5 to target relations as discussed in Section 3.1 using cointegration analysis, so we now turn to the topic of cointegration in more detail.

4. Equilibrium-correction and cointegration

4.1 From the ADL to a VAR

To complete (6), a process is needed for $\{\mathbf{z}_t\}$. Let:

$$\mathbf{z}_t | y_{t-1}, \mathbf{z}_{t-1} \sim \mathsf{N}_k[\pi_{20} + \pi_{21} y_{t-1} + \pi_{22}\mathbf{z}_{t-1}, \boldsymbol{\Omega}_{zz}]. \tag{31}$$

Given (31), the joint distribution is the first-order VAR:

$$\begin{pmatrix} y_t \\ \mathbf{z}_t \end{pmatrix} | y_{t-1}, \mathbf{z}_{t-1} \sim \mathsf{N}_{k+1}\left[\begin{pmatrix} \pi_{10} \\ \pi_{20} \end{pmatrix} + \begin{pmatrix} \pi_{11} & \pi_{12}' \\ \pi_{21} & \pi_{22} \end{pmatrix}\begin{pmatrix} y_{t-1} \\ \mathbf{z}_{t-1} \end{pmatrix}, \begin{pmatrix} \sigma_{11} & \sigma_{12}' \\ \sigma_{12} & \boldsymbol{\Omega}_{zz} \end{pmatrix}\right]. \tag{32}$$

Consequently, to match (6):

$$\begin{aligned} \mathsf{E}[y_t|\mathbf{z}_t, y_{t-1}, \mathbf{z}_{t-1}] &= \pi_{10} + \pi_{11} y_{t-1} + \pi_{12}\mathbf{z}_{t-1} \\ &\quad + \sigma_{12}'\boldsymbol{\Omega}_{zz}^{-1}(\mathbf{z}_t - \pi_{20} - \pi_{21} y_{t-1} - \pi_{22}\mathbf{z}_{t-1}), \end{aligned} \tag{33}$$

so $\beta_0 = (\pi_{10} - \phi'\pi_{20})$, $\beta_1 = \phi$, $\beta_2 = \pi_{11} - \phi'\pi_{21}$ and $\beta_3 = (\pi_{12} - \phi'\pi_{22})$ when $\phi = \boldsymbol{\Omega}_{zz}^{-1}\sigma_{12}$, and $\sigma_\varepsilon^2 = \sigma_{11} - \sigma_{12}'\boldsymbol{\Omega}_{zz}^{-1}\sigma_{12}$. When $\mathbf{z}_t$ is weakly exogenous for $(\beta_0, \ldots, \beta_3')$, the model in (31) can be ignored when analysing (6); also $\pi_{21} = 0$ then ensures the strong exogeneity of $\mathbf{z}_t$ for $(\beta_0, \ldots, \beta_3')$.

Sufficient conditions for stationarity of (32) are that all the eigenvalues λ_i of the matrix of the $\{\pi_{ij}\}$ are inside the unit circle, but a more realistic setting allows for unit roots in π_{22}. On that basis, we now investigate the properties of the VAR in (32) letting $\mathbf{x}_t' = (y_t : \mathbf{z}_t')$ as in (16).

4.2 Cointegration

Linear combinations of I(1) processes are usually I(1) as well: differencing is still needed to remove the unit root. Sometimes integration cancels between series to yield an I(0) outcome and thereby deliver cointegration. Cointegrated processes in turn define a 'long-run equilibrium trajectory' for the economy, departures from which induce 'equilibrium correction' to move the economy back towards its path. A rationale for integrated–cointegrated data is that economic agents use fewer equilibrium corrections than there are variables they need to control. We can see that effect as follows.

Consider the bivariate VAR:

$$\begin{aligned} x_{1,t} &= \pi_{10} + \pi_{11} x_{1,t-1} + \pi_{12} x_{2,t-1} + \varepsilon_{1,t} \\ x_{2,t} &= \pi_{20} + \pi_{21} x_{1,t-1} + \pi_{22} x_{2,t-1} + \varepsilon_{2,t}, \end{aligned} \tag{34}$$

where $(\varepsilon_{1,t}, \varepsilon_{2,t})$ are bivariate independent normal. To determine when the system is I(1) and if so, whether or not some linear combinations of variables are cointegrated, rewrite (34) as:

$$\begin{pmatrix} \Delta x_{1,t} \\ \Delta x_{2,t} \end{pmatrix} = \begin{pmatrix} \pi_{10} \\ \pi_{20} \end{pmatrix} + \begin{pmatrix} (\pi_{11} - 1) & \pi_{12} \\ \pi_{21} & (\pi_{22} - 1) \end{pmatrix}\begin{pmatrix} x_{1,t-1} \\ x_{2,t-1} \end{pmatrix} + \begin{pmatrix} \varepsilon_{1,t} \\ \varepsilon_{2,t} \end{pmatrix} \tag{35}$$

or as (a special case of (18)):

$$\Delta \mathbf{x}_t = \pi + \mathbf{\Pi} \mathbf{x}_{t-1} + \varepsilon_t. \tag{36}$$

Three cases are of interest. First $\mathbf{\Pi} = 0$, so (36) is a vector random walk without any levels relationships, and so $\mathbf{x}_t$ is I(1) with $\Delta \mathbf{x}_t$ being I(0) and equilibrium correcting to π. Secondly, if Π has full rank, then $\mathbf{x}_t$ is I(0) and equilibrium corrects to $\mathbf{\Pi}^{-1}\pi$. The most interesting case is when Π is reduced rank so can be expressed as:

$$\mathbf{\Pi} = \alpha\beta' = \begin{pmatrix} \alpha_{11} \\ \alpha_{12} \end{pmatrix} \begin{pmatrix} \beta_{11} & \beta_{12} \end{pmatrix},$$

where we will normalize $\beta_{11} = 1$. Then in (35):

$$\begin{aligned} \begin{pmatrix} \Delta x_{1,t} \\ \Delta x_{2,t} \end{pmatrix} &= \begin{pmatrix} \pi_{10} \\ \pi_{20} \end{pmatrix} + \begin{pmatrix} \alpha_{11} \\ \alpha_{12} \end{pmatrix} \begin{pmatrix} 1 & \beta_{12} \end{pmatrix} \begin{pmatrix} x_{1,t-1} \\ x_{2,t-1} \end{pmatrix} + \begin{pmatrix} \varepsilon_{1,t} \\ \varepsilon_{2,t} \end{pmatrix} \\ &= \begin{pmatrix} \pi_{10} \\ \pi_{20} \end{pmatrix} + \begin{pmatrix} \alpha_{11} \\ \alpha_{12} \end{pmatrix} \left(x_{1,t-1} + \beta_{12} x_{2,t-1} \right) + \begin{pmatrix} \varepsilon_{1,t} \\ \varepsilon_{2,t} \end{pmatrix} \end{aligned} \tag{37}$$

which is an EqCM with $(x_{1,t-1} + \beta_{12}x_{2,t-1})$ stationary. Thus, cointegration entails EqCM and vice versa when the feedback relation is I(0). However, prior to Granger (1981) the EqCM literature did not visualize a single cointegration relation affecting several variables, and thereby making them integrated, but instead just took the non-stationarity of the observed data as due to the behaviour of the non-modelled variables. Consequently, system cointegration 'endogenizes' data integrability in a consistent way, and so represents a significant step forward. The extensive literature on cointegration analysis also addresses most of the estimation and formulation issues that arise when seeking to conduct inference in integrated-cointegrated processes: much of this is summarized in Hendry and Juselius (2001), to which the interested reader is referred for bibliographic perspective.

5. Equilibrium correction and forecast failure

Recent research on the impact of structural breaks, particularly location shifts, on cointegrated processes has emphasized the need to distinguish equilibrium correction, which operates successfully only within regimes, from error correction, which stabilizes in the face of other non-stationarities (see, for example, Clements and Hendry, 1995). The assumptions concerning the stationarity, or otherwise, of the entity to be controlled in Section 3.1 were rarely explicitly stated, but suggest an implicitly stationary system (or perhaps steady-state growth). In such a setting, equilibrium-correction or cointegration relationships prevent the levels of the variables from 'drifting apart', and so improve the properties of forecasts.

Practical work, however, must allow the data generation process to be non-stationary both from unit roots (that is, I(1) or possibly I(2)) and from a lack of time invariance. When data processes are non-stationary even after differencing and cointegration, equilibrium-correction mechanisms tend to suffer from forecast failure, defined as a significant deterioration in forecast performance relative to in-sample behaviour. Since most empirical model forms are members of the EqCM class, this is a serious practical problem.

To illustrate, reconsider the special case of (18) with just one lag, written as:

$$\Delta \mathbf{x}_t = \gamma + \alpha(\beta' \mathbf{x}_{t-1} - \mu) + \varepsilon_t. \tag{38}$$

The shift of interest here is $\nabla\mu^* = \mu^* - \mu$, where μ^* denotes the post-break equilibrium mean (reasonable magnitude shifts in γ, α and Ω_ε rarely entail forecast failure). Denote the forecast origin as time T, then following a change to μ^* immediately after forecasting, the next outcome is:

$$\Delta \mathbf{x}_{T+1} = \gamma + \alpha(\beta' \mathbf{x}_T - \mu^*) + \varepsilon_{T+1} = \gamma + \alpha(\beta' \mathbf{x}_T - \mu) + \varepsilon_{T+1} - \alpha\nabla\mu^* \tag{39}$$

where $-\alpha\Delta\mu^*$ is the unanticipated break, and becomes the mean forecast error for known parameters. Importantly, the 1-step ahead forecast at $T+1$ using an unchanged model suffers the same mistake:

$$\mathrm{E}[\Delta \mathbf{x}_{T+2} - (\gamma + \alpha(\beta' \mathbf{x}_{T+1} - \mu))] = -\alpha\nabla\mu^* \tag{40}$$

so the shift in the equilibrium mean induces systematic mis-forecasting. The impact on multi-step forecasts of the levels is even more dramatic, as the mean forecast error increases at every horizon, eventually converging to $\alpha(\beta'\alpha)^{-1}\nabla\mu^*$, which can be very large (see Clements and Hendry, 1999). Thus, EqCMs are a non-robust forecasting device in the face of equilibrium-mean shifts, a comment which therefore applies to all members of this huge class of model, including GARCH (as noted earlier), where the pernicious shift is in the unconditional variance σ_ε^2 in (15).

To avoid forecast failure, more adaptive methods merit consideration. One generic approach to improving robustness to location shifts is to difference the forecasting device (although that may well worsen the impact of large measurement errors at the forecast origin). Differencing can be before estimation, as in a double-differenced VAR, or after, as in differencing the estimated EqCM to eliminate the equilibrium mean and growth intercept. Such devices perform as badly as the EqCM in terms of forecast biases when a break occurs after forecasts are announced (see Clements and Hendry, 1999), and have a larger error variance. The key difference is their performance when forecasting after a break has already occurred, in which case the EqCM continues to perform badly (as shown above in (48)), but a DEqCM becomes relatively immune to the earlier break. Taking (47) as an example, an additional difference yields:

$$\Delta^2 \mathbf{x}_{T+1} = \Delta(\gamma - \alpha\mu^*) + \alpha\beta'\Delta \mathbf{x}_T + \Delta\varepsilon_{T+1} = -\alpha\Delta\mu^* + \alpha\beta'\Delta \mathbf{x}_T + \Delta\varepsilon_{T+1}$$

so there is no benefit when forecasting immediately after the break (as $\Delta\mu^* = \nabla\mu^*$), whereas (48) becomes:

$$\Delta^2 \mathbf{x}_{T+2} = \alpha\beta'\Delta \mathbf{x}_{T+1} + \Delta\varepsilon_{T+2}$$

since $\Delta\mu^* = 0$. Thus, there is no longer any systematic failure. The same comment applies to double-differenced devices, although Hendry (2006) shows how to improve these while retaining robustness.

A further consequence is that, when a location shift is not modelled, since most econometric estimators minimize mis-fitting, the coefficients of dynamic models will be driven towards unity, which induces differencing to convert a location shift into a 'blip'. Thus, estimates that apparently manifest 'slow adjustment' may just reflect unmodelled breaks.

An alternative approach to avoiding forecast failure would be to construct a genuine error-correction model, adjusting more or less rapidly to wherever the target variable moves: for example, exponentially weighted moving averages do so for some processes. In essence, either the dynamics must ensure correction or the target implicit in the econometric model must move when the regime alters. This last result also explains why models in differences are not as susceptible to certain forms of structural break as equilibrium-correction systems (again see Clements and Hendry, 1999), and in turn helps to account for many of the findings reported in the forecasting competitions literature. When the shift in question is a change in a policy regime, Hendry and Mizon (2005) suggest approaches to merging robust forecasts with policy models.

6. Conclusion

Equilibrium-correction models have a long pedigree as an 'independent' class, related to optimal control theory. However, their isomorphism with cointegrated relationships has really been the feature that has ensured their considerable popularity in empirical applications. In both cases, part of the benefit from the EqCM specification came from expressing variables in the more orthogonalized forms of differences and equilibrium-correction terms, partly from the resulting insights into both short-run and long-run adjustments, partly from discriminating between the different components of the deterministic terms, and partly from 'balancing' regressors of the same order of integration, namely I(0).

Unfortunately, science is often two steps forward followed by one back, and that backwards step came from an analysis of EqCMs when forecasting in the face of structural breaks. Unmodelled shifts in the equilibrium mean (and less so in the growth rate) induce forecast failure, making EqCMs a non-robust device with which to forecast when data processes are prone to breaks, as many empirical studies suggest they are (see, for example, Stock and Watson, 1996). Since cointegration hopefully captures long-run causal relations, and ties together the levels of I(1) variables, eliminating its contribution should not be undertaken lightly, hence the suggestion in Section 5 of using the differenced version of the estimated EqCM for forecasting.

DAVID F. HENDRY

See also **cointegration.**

Financial support from the ESRC under Professorial Research Fellowship RES051270035 is gratefully acknowledged, as are helpful comments from Gunnar Bårdsen, Julia Campos, Jennifer Castle, Mike Clements, Søren Johansen and Graham Mizon.

Bibliography

Bårdsen, G., Hurn, S. and Lindsay, K.A. 2004. Linearizations and equilibrium correction models. *Studies in Nonlinear Dynamics and Econometrics* 8(4), Article 5.

Banerjee, A., Dolado, J.J., Galbraith, J.W. and Hendry, D.F. 1993. *Co-integration, Error Correction and the Econometric Analysis of Non-Stationary Data.* Oxford: Oxford University Press.

Bergstrom, A.R. 1962. A model of technical progress, the production function and cyclical growth. *Economica* 29, 357–70.

Bollerslev, T. 1986. Generalised autoregressive conditional heteroskedasticity. *Journal of Econometrics* 51, 307–27.

Clements, M.P. and Hendry, D.F. 1995. Macro-economic forecasting and modeling. *Economic Journal* 105, 1001–13.

Clements, M.P. and Hendry, D.F. 1999. *Forecasting Non-stationary Economic Time Series.* Cambridge, MA: MIT Press.

Davidson, J.E.H. and Hall, S. 1991. Cointegration in recursive systems. *Economic Journal* 101, 239–51.

Davidson, J.E.H., Hendry, D.F., Srba, F. and Yeo, J.S. 1978. Econometric modelling of the aggregate time-series relationship between consumers' expenditure and income in the United Kingdom. *Economic Journal* 88, 661–92.

Engle, R.F. 1982. Autoregressive conditional heteroscedasticity, with estimates of the variance of United Kingdom inflation. *Econometrica* 50, 987–1007.

Engle, R.F. and Granger, C.W.J. 1987. Cointegration and error correction: representation, estimation and testing. *Econometrica* 55, 251–76.

Ericsson, N.R. 2007. *Econometric Modeling.* Oxford: Oxford University Press.

Granger, C.W.J. 1981. Some properties of time series data and their use in econometric model specification. *Journal of Econometrics* 16, 121–30.

Granger, C.W.J. 1986. Developments in the study of cointegrated economic variables. *Oxford Bulletin of Economics and Statistics* 48, 213–28.

Granger, C.W.J. and Newbold, P. 1977. The time series approach to econometric model building. In *New Methods in Business Cycle Research*, ed. C.A. Sims. Minneapolis: Federal Reserve Bank of Minneapolis.

Hendry, D.F. 1977. On the time series approach to econometric model building. In *New Methods in Business Cycle Research*, ed. C.A. Sims. Minneapolis: Federal Reserve Bank of Minneapolis.

Hendry, D.F. 1995. *Dynamic Econometrics.* Oxford: Oxford University Press.

Hendry, D.F. 2006. Robustifying forecasts from equilibrium-correction models. *Journal of Econometrics* 135, 399–426. Special issue in honor of Clive Granger.

Hendry, D.F. and Anderson, G.J. 1977. Testing dynamic specification in small simultaneous systems: an application to a model of building society behaviour in the United Kingdom. In *Frontiers in Quantitative Economics*, vol. 3, ed. M.D. Intriligator. Amsterdam: North-Holland.

Hendry, D.F. and Juselius, K. 2001. Explaining cointegration analysis: Parts I and II. *Energy Journal* 21, 1–42; 22, 75–120.

Hendry, D.F. and Mizon, G.E. 1978. Serial correlation as a convenient simplification, not a nuisance: a comment on a study of the demand for money by the Bank of England. *Economic Journal* 88, 549–63.

Hendry, D.F. and Mizon, G.E. 2005. Forecasting in the presence of structural breaks and policy regime shifts. In *Identification and Inference for Econometric Models: Essays in Honor of Thomas Rothenberg*, ed. D.W.K. Andrews and J.H. Stock. Cambridge: Cambridge University Press.

Holt, C., Modigliani, F., Muth, J.F. and Simon, H. 1960. *Planning Production, Inventories and Work Force.* Englewood Cliffs, NJ: Prentice-Hall.
Johansen, S. 1988. Statistical analysis of cointegration vectors. *Journal of Economic Dynamics and Control* 12, 231–54.
Johansen, S. 1995. *Likelihood-based Inference in Cointegrated Vector Autoregressive Models.* Oxford: Oxford University Press.
Klein, L.R. 1953. *A Textbook of Econometrics.* Evanston, IL: Row, Peterson and Company.
Nickell, S.J. 1985. Error correction, partial adjustment and all that: an expository note. *Oxford Bulletin of Economics and Statistics* 47, 119–30.
Phillips, A.W.H. 1954. Stabilization policy in a closed economy. *Economic Journal* 64, 290–333.
Phillips, A.W.H. 1957. Stabilization policy and the time form of lagged response. *Economic Journal* 67, 265–77.
Phillips, P.C.B. and Loretan, M. 1991. Estimating long-run economic equilibria. *Review of Economic Studies* 58, 407–36.
Phillips, A.W.H. and Quenouille, M.H. 1960. Estimation, regulation and prediction in interdependent dynamic systems. *Bulletin de l'Institut de Statistique* 37, 335–43.
Preston, A.J. and Pagan, A.R. 1982. *The Theory of Economic Policy.* Cambridge: Cambridge University Press.
Salmon, M. 1982. Error correction mechanisms. *Economic Journal* 92, 615–29.
Salmon, M. 1988. Error correction models, cointegration and the internal model principle. *Journal of Economic Dynamics and Control* 12, 523–49.
Sargan, J.D. 1964. Wages and prices in the United Kingdom: a study in econometric methodology (with discussion). In *Econometric Analysis for National Economic Planning*, vol. 16 of *Colston Papers*, ed. P.E. Hart, G. Mills and J.K. Whitaker. London: Butterworth.
Sargan, J.D. 1980. Some tests of dynamic specification for a single equation. *Econometrica* 48, 879–97.
Sims, C.A., ed. 1977. *New Methods in Business Cycle Research.* Minneapolis: Federal Reserve Bank of Minneapolis.
Stock, J.H. and Watson, M.W. 1996. Evidence on structural instability in macroeconomic time series relations. *Journal of Business and Economic Statistics* 14, 11–30.
Whittle, P. 1963. *Prediction and Regulation by Linear Least-Square Methods.* Princeton: D. Van Nostrand.

forecasting

Decisions in the fields of economics and management have to be made in the context of forecasts about the future state of the economy or market. As decisions are so important as a basis for these fields, a great deal of attention has been paid to the question of how best to forecast variables and occurrences of interest. There are several distinct types of forecasting situations, including event timing, event outcome, and time-series forecasts. Event timing is concerned with the question of when, if ever, some specific event will occur, such as the introduction of a new tax law, or of a new product by a competitor, or of a turning point in the business cycle. Forecasting of such events is usually attempted by the use of leading indicators, that is, other events that generally precede the one of interest. Event outcome forecasts try to forecast the outcome of some uncertain event that is fairly sure to occur, such as finding the winner of an election or the level of success of a planned marketing campaign. Forecasts are usually based on data specifically gathered for this purpose, such as a poll of likely voters or of potential consumers. There clearly should be a positive relationship between the amount spent on gathering the extra data and the quality of the forecast achieved.

A time series x_t is a sequence of values gathered at regular intervals of time, such as daily stock market closing prices, interest rates observed weekly, or monthly unemployment levels. Irregularly recorded data, or continuous time sequences may also be considered but are of less practical importance. When at time n (now), a future value of the series, x_{n+h}, is a random variable where h is the forecast horizon. It is usual to ask questions about the conditional distribution of x_{n+h} given some information set I_n, available now from which forecasts will be constructed. Of particular importance are the conditional mean

$$f_{n,h} = E[x_{n+h}|I_n]$$

and variance, $V_{n,h}$. The value of $f_{n,h}$ is a point forecast and represents essentially the best forecast of the most likely value to be taken by the variable x at time $n+h$. With a normality assumption, the conditional mean and variance can be used together to determine an interval forecast, such as an interval within which $x_{n,h}$ is expected to fall with 95 per cent confidence. An important decision in any forecasting exercise is the choice of the information set I_n. It is generally recommended that I_n include at least the past and present of the individual series being forecast, $x_{n-j,j\geq 0}$. Such information sets are called *proper*, and any forecasting models based upon them can be evaluated over the past. An I_n that consists just of x_{n-j}, provides a univariate set so that future x_i are forecast just from its own past. Many simple time-series forecasting methods are based on this information set and have proved to be successful. If I_n includes several explanatory variables, one has a multivariate set. The choice of how much past data to use and which explanatory variables to include is partially a personal one, depending on one's knowledge of the series being forecast, one's levels of belief about the correctness of any economic theory that is available, and on data availability. In general terms, the more useful are the explanatory variables that are included in I_n, the better the forecast that will result. However, having many series allows for a confusing number of alternative model specifications that are possible so that using too much data could quickly lead to diminishing marginal returns in terms of forecast quality. In practice, the data to be used in I_n will often be partly determined by the length of the forecast horizon. If h is small, a short-run forecast is being made and this may concentrate on frequently varying explanatory variables. Short-term forecasts of savings may be based on interest rates, for example. If h is large so that long-run forecasts are required, then slowly changing, trending explanatory variables may be of particular relevance. A long-run forecast of electricity demand might be largely based on population trends, for example. What is considered short run or long run will usually depend on the properties of the series being forecast. For very long forecasts, allowances would have to be made for technological change as well as changes in demographics and the economy. A survey of the special and separate field of technological forecasting can be found in Martino (1993) with further discussion in Martino (2003).

If decisions are based on forecasts, it follows that an imperfect forecast will result in a cost to the decision-maker. For example, if $f_{n,h}$ is a point forecast made at time n, of x_{n+h}, the eventual forecast error will be

$$e_{n.h} = x_{n,h} - f_{n,h},$$

which is observed at time $n + h$. The cost of making an error e might be denoted as $C(e)$, where $C(e)$ is positive with $C(0) = 0$. As there appears to be little prospect of making error-free forecasts in economics, positive costs must be expected, and the quality of a forecast procedure can be measured as the expected or average cost resulting from its use. Several alternative forecasting procedures can be compared by their expected costs and the best one chosen. It is also possible to compare classes of forecasting models, such as all linear models based on a specific, finite information set, and to select the optimum model by minimizing the expected cost. In practice the true form of the cost function is not known for decision sequences, and in the univariate forecasting case a pragmatically useful substitute to the real $C(e)$ is to assume that it is well approximated by ae^2 for some positive a. This enables least-squares statistical techniques to be used when a model is estimated and is the basis of a number of theoretical results including that the optimal forecast of x_{n+h} based on I_n is just the conditional mean of $x_{n,h}$. Machina and Granger (2006) have considered cost functions generated by decision makers and then find implications for their utility functions. This is just one component of considerable developments in the area of evaluation of forecasts; see West (2006) and Timmermann (2006), for example.

When using linear models and a least-square criterion, it is easy to form forecasts under an assumption that the model being used is a plausible generating mechanism for the series of interest. Suppose that a simple model of the form

$$x_t = \alpha x_{t-1} + \beta y_{t-2} + \varepsilon_t$$

is believed to be adequate where ε_t is a zero-mean, white noise (unforecastable) series. When at time n, according to this model, the next value of x will be generated by

$$x_{n+1} = \alpha x_n + \beta y_{n-1} + \varepsilon_{n+1}.$$

The first two terms are known at time n, and the last term is unforecastable. Thus

$$f_{n,1} = \alpha x_n + \beta y_{n-1}$$

and

$$e_{n,1} = \varepsilon_{n+1}.$$

x_{n+2}, the following x, will be generated by

$$x_{n+2} = \alpha x_{n+1} + \beta y_n + \varepsilon_{n+2}.$$

The first of these terms is not known at time n, but a forecast is available for it, αf_n; the second term is known at time n, and the third term is not forecastable, so that

$$f_{n,2} = \alpha f_{n,1} + \beta y_n$$

and

$$e_{n,2} = \varepsilon_{n+2} + \alpha(x_{n+1} - f_{n,1}) = \varepsilon_{n+2} + \alpha\varepsilon_{n+1}.$$

To continue this process for longer forecast horizons, it is clear that forecasts will be required for y_{n+h-2}. The forecast formation rule is that one uses the model available as though it is true, asks how a future x_{n+h} will be generated, uses all known terms as they occur, and replaces all other terms by optimal forecasts. For non-linear models this rule can still be used, but with the additional complication that the optimum forecast of a function of x is not the same function of the optimum forecast of x.

The steps involved in forming a forecast include deciding exactly what is to be forecast, the forecast horizon, the data that is available for use, the model forms or techniques to be considered, the cost function to be used in the evaluation procedure, and whether just one single forecast would be produced or several alternatives. It is good practice to decide on the evaluation to be used before starting a sequence of forecasts. If there are several alternative forecasting methods involved, a weighted combination of the available forecasts is both helpful for evaluation and can often provide a superior forecast.

The central problem in practical forecasting is choosing the model from which the forecasts will be derived. If a univariate information set is used, it is natural to consider the model developed in the field of time-series analysis. A class of models that has proved to be successful in short-term forecasting is the autoregressive (AR) model class. If a series is regressed on itself up to p lags, the result is an AR(p) model. These models were originally influenced by Box and Jenkins (1970) as a particularly relevant subclass of their ARMA (p, q) models, which involve moving average components. The number of lags in an AR(p) can be chosen using a selection criterion; the most used are the Bayes information criterion (BIC) and the less conservative Akaike information criterion (AIC).

The natural extension was to vector autoregressive models. Later, when it was realized that many series in macroeconomics and finance had the property of being integrated, and so contained stochastic trends, the natural multivariate form was the error-correction model. It is quite often found that error-correction models improve forecasts, but not inevitably. There are a variety of ways of building models with many predictive variables, including those with unobserved components and using special data, such as survey expectations, real-time macro data, and seasonal components.

In recent years the linear models have been joined by a variety of nonlinear forms (see Terasvita, 2006), including switching models and neural networks as well as linear models with time varying coefficients estimated using Kalman filters.

Traditionally, forecasters concentrated on the mean of the predictive distribution. Towards the end of the 20th century considerable attention was given to forecasting the variance of the distribution, particularly in the financial area, often using Engle's (1995) ARCH model or one of its many generalizations (see the survey by Andersen et al., 2006). Recently forecasts of the whole distribution have become more common in practice, both in finance and in macroeconomics: see Corradi and Swanson (2006) for a recent discussion. These forecasts will include discussions of quantiles, and the use of copulas gives a way into multivariate distribution forecasts. The topics mentioned in this paragraph are covered by chapters in Elliot, Granger and Timmermann (2006).

CLIVE W.J. GRANGER

Bibliography

Andersen, T., Bollerslev, T., Christoffersen, P. and Diebold, F. 2006. Volatility and correlation forecasting. In Elliot, Granger and Timmermann (2006).

Box, G. and Jenkins, G. 1970. *Time Series Analysis, Forecasting and Control.* San Francisco: Holden Day.

Clements, M. and Hendry, D. 2003. *Forecasting Economic Time Series.* Cambridge: Cambridge University Press.

Corradi, V. and Swanson, N.R. 2006. Predictive density evaluation. In Elliot, Granger and Timmermann 2006).

Elliott, G., Granger, C.W.J. and Timmermann, A., eds. 2006. *Handbook of Economic Forecasting.* Amsterdam: North-Holland.

Engle, R.F. 1995. *ARCH: Selected Readings.* Oxford: Oxford University Press.

Granger, C.W.J. and Newbold, P. 1987. *Forecasting Economic Time Series.* 2nd edn. New York: Academic Press.

Machina, M. and Granger, C.W.J. 2006. Forecasting and decision theory. In Elliot, Granger and Timmermann (2006).

Martino, J.P. 1993. *Technological Forecasting for Decision Making.* 2nd edition. New York: McGraw Hill.

Martino, J.P. 2003. A review of selected recent advances in technological forecasting. *Technological Forecasting and Social Change* 70, 719–33.

Terasvita, T. 2006. Forecasting economic variables with nonlinear models. In Elliott, Granger and Timmermann (2006).

Timmermann, A. 2006. Forecast combinations. In Elliott, Granger and Timmermann (2006).

West, K. 2006. Forecast evaluation. In Elliott, Granger and Timmermann (2006).

fractals

The word 'fractal' was coined by the French mathematician Benoît Mandelbrot (1982) to characterize a wide class of highly irregular scale-invariant objects. It originates from the Latin adjective *fractus*, meaning 'broken' or 'fragmented'. The defining characteristic of fractals is that their degree of irregularity remains the same at all scales. This invariance permits parsimonious modelling of complex objects, and has been useful for analysing a wide variety of natural phenomena. The entry reviews the use of fractals in economics and finance, and more specifically their application in the statistical modelling of asset returns, which has been a remarkably active field since the early 1960s.

Consider the price $P(t)$ of a financial asset, such as a stock or a currency, and let $p(t)$ denote its logarithm. The process $p(t)$ is said to be *self-similar* if there exists a constant $H>0$ such that for every set of instants $t_1 \leq \ldots \leq t_k$ and for every $\lambda > 0$, the vector $\{p(\lambda t_1), \ldots, p(\lambda t_k)\}$ has the same distribution as $\lambda^H\{p(t_1), \ldots, p(t_k)\}$, that is,

$$\{p(\lambda t_1), \ldots, p(\lambda t_k)\} \overset{d}{=} \{\lambda^H p(t_1), \ldots, \lambda^H p(t_k)\}. \quad (1)$$

The constant H is called the *self-similarity* index.

Three classes of self-similar processes have been widely used in finance: the Brownian motion, Lévy-stable processes and the fractional Brownian motion, which are successively discussed. The Brownian motion (Bachelier, 1900), with self-similarity index $H = 1/2$, pervades modern financial theory and notably the Black–Merton–Scholes approach to continuous time valuation. Its lasting success arises from several appealing properties, including tractability and consistency with the financial concepts of no-arbitrage and market efficiency.

The stable processes of Paul Lévy (1924) are characterized by thicker tails than the Brownian motion. They are thus more likely to accommodate the outliers exhibited by financial series, as was pointed out by Mandelbrot in a series of seminal papers (for example, 1963). The increments of Lévy-stable processes are stationary and have stable distributions, where stability refers to invariance under linear combinations (see Samorodnitsky and Taqqu, 1994). Tails are Paretian:

$$\mathbb{P}\{p(\Delta t) > x\} \sim c\,x^{-\alpha} \quad \text{as } x \to +\infty,$$

with index $\alpha = 1/H \in (0; 2)$. The variance of a Lévy-stable process is infinite, which is at odds with both empirical evidence and mean-variance asset pricing. Furthermore, stable processes have independent increments and thus cannot account for volatility clustering.

The fractional Brownian motion (Kolmogorov, 1940; Mandelbrot, 1965; Mandelbrot and Van Ness, 1968) with $H > 1/2$ is a self-similar process with strongly dependent returns. Increments are stationary, correlated, and normally distributed. Their autocorrelation declines at the hyperbolic rate

$$Cov[r(t); r(t+n)] \sim c\ (2H-1)n^{2H-2} \quad \text{as } n \to \infty,$$

where $r(t) = p(t) - p(t - \Delta t)$ denotes the return on a time interval of fixed length Δt. Hyperbolic autocorrelation is the defining property of long-memory processes, whose use in economics was advanced by the discrete-time fractional integration approach of Granger and Joyeux (1980). While research on long memory has generally been very fruitful in economics (see Baillie, 1996, for a review), the fractional Brownian motion rarely represents a practical model of asset prices. Specifically, long memory in returns is both empirically inaccurate in most markets (Lo, 1991) and inconsistent with arbitrage-pricing in continuous time (Maheswaran and Sims, 1993). There is, however, abundant evidence of long memory in the *volatility* of returns (for example, Dacorogna et al., 1993; Ding, Granger and Engle, 1993).

In all the above self-similar processes, returns observed at various frequencies have identical distributions up to a scalar renormalization:

$$p(t + \lambda\Delta t) - p(t) \ \underline{\underline{d}} \ \lambda^H p(\Delta t).$$

Most financial series, however, are not exactly self-similar, but have thicker tails and are more peaked in the bell at shorter horizons. This observation is consistent with the economic intuition that high-frequency returns are either large if new information has arrived, or close to zero otherwise. Thus, self-similar processes do not capture in a single model the most salient features of asset returns.

A partial solution to these difficulties is provided by the multifractal model of asset returns (MMAR; Calvet, Fisher and Mandelbrot, 1997; Calvet and Fisher, 2002a). This approach builds on multifractal measures (Mandelbrot, 1974), which are constructed by the iterative random reallocation of mass within a time interval. The MMAR extends multifractals from measures to diffusions. The asset price is specified by compounding a Brownian motion with an independent random time-deformation:

$$p(t) = B[\theta(t)],$$

where θ is the cumulative distribution of a multifractal measure $\theta(t) = \mu[0, t]$. Returns are uncorrelated and the price p is a martingale in MMAR, which precludes arbitrage. The time deformation induces sharp outliers in returns and long memory in volatility. The MMAR also captures nonlinear changes in the return density with the time horizon (Lux, 2001).

The price p inherits highly heterogeneous time-variations from the multifractal measure. Its sample paths are continuous but can be more irregular than a Brownian motion at some instants. Specifically, the local variability of a sample path at a given date t is characterized by the local Hölder exponent $\alpha(t)$, which heuristically satisfies

$$|p(t + dt) - p(t)| \sim c_t(dt)^{\alpha(t)} \quad \text{as } dt \to 0.$$

Traditional jump diffusions impose that $\alpha(t)$ be equal to 0 at points of discontinuity, and to 1/2 otherwise. In a multifractal process, however, the exponent $\alpha(t)$ takes a *continuum* of values in any time interval.

Asset returns at different frequencies satisfy the moment-scaling rule:

$$\mathbb{E}[|p(\Delta t)|^q] = c_q(\Delta t)^{\tau(q)+1},$$

which holds for every (finite) moment q and time interval Δt. These moment restrictions represent the basis of estimation and testing (Calvet, Fisher and Mandelbrot, 1997; Calvet and Fisher, 2002a; 2002b; Lux, 2004). The MMAR provides a well-defined stochastic framework for the analysis of moment-scaling, which has generated extensive interest in econophysics (for example, LeBaron, 2001). The multifractal model is also related to recent econometric research on power variation, which interprets return moments at various frequencies in the context of traditional jump-diffusions (for examples, Andersen et al., 2001; Barndorff-Nielsen and Shephard, 2004).

Despite its appealing properties, the MMAR is unwieldy for econometric applications because of two features of the underlying measure: (*a*) the recursive reallocation of mass on an entire time-interval does not fit well with standard time series tools; and (*b*) the limiting measure contains a residual grid of instants that makes it non-stationary.

The Markov-switching multifractal (MSM) resolves these difficulties by constructing a fully stationary volatility process that evolves stochastically through time (Calvet and Fisher, 2001; 2004). MSM builds a bridge between multifractality and regime-switching, which permits the application of Bayesian filtering and maximum likelihood estimation to a multifractal process. Volatility is driven by the first-order Markov state vector $M_t = (M_{1,t}; M_{2,t}; \ldots; M_{k,t}) \in \mathbb{R}^k_+$, whose components have unit mean and heterogeneous persistence levels. In discrete time, returns are specified as

$$r_t = \sigma(M_{1,t}M_{2,t}\ldots M_{k,t})^{1/2}\varepsilon_t, \qquad (2)$$

where σ is a positive constant and $\{\varepsilon_t\}$ are independent standard Gaussians.

Volatility components follow independent Markov processes that are identical except for time scale. Given the volatility state M_t, the next-period multiplier $M_{k,t+1}$ is drawn from a fixed distribution M with probability γ_k, and is otherwise left unchanged.

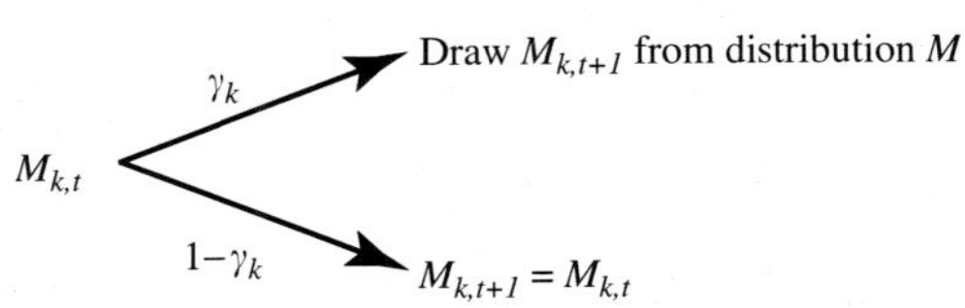

Components differ in their transition probabilities γ_k but not in their marginal distribution M. The transition probabilities are tightly specified by $\gamma_k = 1-(1-\gamma_1)^{(b^{k-1})}$, which is approximately geometric at low frequency: $\gamma_k \sim \gamma_1 b^{k-1}$. In empirical applications, a unique scalar m_0 typically determines the distribution M. The return process (2) is then specified by the four parameters $(m_0, \sigma, b, \gamma_1)$. Since the number of frequencies $\bar{k}$ can be arbitrarily large, MSM provides a tight specification of a high-dimensional state space. The approach conveniently extends to continuous time (Calvet and Fisher, 2001) or a multivariate setting (Calvet, Fisher and Thompson, 2006).

When M has a discrete distribution, the state space is finite and MSM defines a stochastic volatility model with a closed-form likelihood. It then bypasses the estimation problems of traditional stochastic volatility settings based on smooth autoregressive transitions. On the other hand when M has a continuous (for example, lognormal) distribution, estimation can proceed by simulated method of moments (Calvet and Fisher, 2002b), generalized method of moments (Lux, 2004), or simulated likelihood via a particle filter (Calvet, Fisher and Thompson, 2006).

MSM tends to substantially outperform traditional models both in and out of sample. Calvet and Fisher (2004) thus report considerable gains in exchange rate volatility forecasts at horizons of 10 to 50 days as compared with GARCH-type processes. Lux (2004) obtains similar results with lognormal MSM using linear predictions. Furthermore, bivariate MSM compares favourably with multivariate GARCH under criteria such as the likelihood function, integral transforms and value-at-risk (Calvet, Fisher and Thompson, 2006).

The integration of multifrequency models into asset pricing is now at the forefront of current research. Calvet and Fisher (2005a) thus introduce a parsimonious equilibrium set-up in which regime shifts of heterogeneous durations affect the volatility of dividend news. The resulting return process is endogenously skewed and has significantly higher likelihood than the classic Campbell and Hentschel (1992) specification. Calvet and Fisher (2005b) similarly illustrate the potential of MSM for building parsimonious multifrequency jump-diffusions.

LAURENT E. CALVET

See also **regime switching models.**

Bibliography

Andersen, T., Bollerslev, T., Diebold, F. and Labys, P. 2001. The distribution of realized exchange rate volatility. *Journal of the American Statistical Association* 96, 42–55.

Bachelier, L. 1900. Théorie de la spéculation. *Annales Scientifiques de l'Ecole Normale Supérieure* 17, 21–86.

Baillie, R. 1996. Long memory processes and fractional integration in econometrics. *Journal of Econometrics* 73, 5–59.

Barndorff-Nielsen, O. and Shephard, N. 2004. Power and bipower variation with stochastic volatility and jumps. *Journal of Financial Econometrics* 2, 1–37.

Calvet, L. and Fisher, A. 2001. Forecasting multifractal volatility. *Journal of Econometrics* 105, 27–58.

Calvet, L. and Fisher, A. 2002a. Multifractality in asset returns: theory and evidence. *Review of Economics and Statistics* 84, 381–406.

Calvet, L. and Fisher, A. 2002b. Regime-switching and the estimation of multifractal processes. Working Paper. Harvard University and University of British Columbia.

Calvet, L. and Fisher, A. 2004. How to forecast long-run volatility: regime-switching and the estimation of multifractal processes. *Journal of Financial Econometrics* 2, 49–83.

Calvet, L. and Fisher, A. 2005a. Multifrequency news and stock returns. Working Paper No. 11441. Cambridge, MA: NBER.

Calvet, L. and Fisher, A. 2005b. Multifrequency jump diffusions: an equilibrium approach. Working Paper. HEC School of Management and University of British Columbia.

Calvet, L., Fisher, A. and Mandelbrot, B. 1997. A multifractal model of asset returns. Discussion Papers 1164–1166. Cowles Foundation, Yale University.

Calvet, L., Fisher, A. and Thompson, S. 2006. Volatility comovement: a multifrequency approach. *Journal of Econometrics* 131, 179–215.

Campbell, J. and Hentschel, L. 1992. No news is good news: an asymmetric model of changing volatility in stock returns. *Journal of Financial Economics* 31, 281–318.

Dacorogna, M., Müller, U., Nagler, R., Olsen, R. and Pictet, O. 1993. A geographical model for the daily and weekly seasonal volatility in the foreign exchange market. *Journal of International Money and Finance* 12, 413–38.

Ding, Z., Granger, C. and Engle, R. 1993. A long memory property of stock returns and a new model. *Journal of Empirical Finance* 1, 83–106.

Granger, C. and Joyeux, R. 1980. An introduction to long memory time series models and fractional differencing. *Journal of Time Series Analysis* 1, 15–29.

Kolmogorov, A. 1940. Wienersche Spiralen und einige andere interessante Kurven im Hilbertschen raum. *Comptes Rendus de l'Académie des Sciences de l'URSS* 26, 115–18.

LeBaron, B. 2001. Stochastic volatility as a simple generator of apparent financial power laws and long memory. *Quantitative Finance* 1, 621–31.

Lévy, P. 1924. Théorie des erreurs: la loi de Gauss et les lois exceptionnelles. *Bulletin de la Société Mathématique de France* 52, 49–85.

Lo, A. 1991. Long memory in stock market prices. *Econometrica* 59, 1279–313.

Lux, T. 2001. Turbulence in financial markets: the surprising explanatory power of simple cascade models. *Quantitative Finance* 1, 632–40.

Lux, T. 2004. The Markov-switching multifractal model of asset returns: GMM estimation and linear forecasting of volatility. Working paper. Kiel University.

Maheswaran, S. and Sims, C. 1993. Empirical implications of arbitrage-free asset markets. In *Models, Methods and Applications of Econometrics*, ed. P. Phillips. Oxford: Blackwell.

Mandelbrot, B. 1963. The variation of certain speculative prices. *Journal of Business* 36, 394–419.

Mandelbrot, B. 1965. Une classe de processus stochastiques homothétiques à soi. *Comptes Rendus de l'Académie des Sciences de Paris* 260, 3274–77.

Mandelbrot, B. 1974. Intermittent turbulence in self-similar cascades: divergence of high moments and dimension of the carrier. *Journal of Fluid Mechanics* 62, 31–58.

Mandelbrot, B. 1982. *The Fractal Geometry of Nature*. New York: Freeman.

Mandelbrot, B. and Van Ness, J. 1968. Fractional Brownian motion, fractional noises and applications. *SIAM Review* 10, 422–37.

Samorodnitsky, G. and Taqqu, M. 1994. *Stable Non-Gaussian Random Processes*. New York: Chapman and Hall.

functional central limit theorems

Central limit theorems guarantee that the distributions of properly normalized sums of certain random variables are approximately normal. In many cases, however, a more detailed analysis is necessary. When testing for structural constancy in models, we might be interested in the temporal evolution of our sums. So for random variables X_i we are interested in analysing the behaviour of

$$\frac{1}{\sqrt{N}}\sum_{i=1}^{t} X_i \tag{1}$$

as a function of t for $t \leq N$. It is convenient to normalize the time, too, and consider for $0 \leq z \leq 1$

$$\frac{1}{\sqrt{N}}\sum_{i=1}^{Tz} X_i. \tag{2}$$

Another popular application is the asymptotic behaviour of the empirical distribution function or its multivariate generalizations, though we will only briefly discuss it.

'Functional limit theorems' are generalizations of the classical central limit theorem (CLT). Instead of analysing random variables with values in **R**, I deal with random variables in more general spaces. Here I discuss only one specific example, namely the analysis of the properly normalized partial sums of random variables. In order to do so, I will first sketch the necessary concepts concerning the topology of the spaces involved. In particular, I want to demonstrate the necessity of using spaces and metrics which, at the first glance, may not look that plausible. The results are well known and can be found in many textbooks. A classical reference is Billingsley (1999). Another introduction in this field, more geared towards econometricians, is Davidson (1994).

Foundations: metric spaces and convergence in distribution

A common framework, allowing us to formulate more general limit theorems, assumes that our 'random variables' take values in so-called 'Polish spaces', which are just metric spaces which are separable and complete. So let us assume that we have given such a space E, with a metric $d(.,.)$ on it, so that there exists a countable dense subset and that the space is complete (that is, every Cauchy sequence converges). Examples are the finite-dimensional spaces with the usual distance. The space $C[0, 1]$ of all continuous functions from $[0, 1]$ to **R** (the set of real numbers), endowed with the metric

$$d_M(x, y) = \max_{0 \leq t \leq 1} |x(t) - y(t)|. \tag{3}$$

Let us assume that we have random variables X_n, X with values in E: Then we define convergence in distribution of X_n to X

$$X_n \to^D X \tag{4}$$

if and only if for all bounded, continuous functions ϕ from E to $\mathbf{R}$,

$$E\phi(X_n) \to E\phi(X). \tag{5}$$

We can easily see that, in the special case of the space E being the set $\mathbf{R}$, our definition here is a generalization of the familiar concept of convergence in distribution. An 'invariance principle' is simply a statement convergence in distribution of random variables in a complex space.

If we have given a statement like (4), then for continuous ϕ and large n we can approximate the distribution of $\phi(X_n)$ by the distribution of $\phi(X)$. As an example, assume our underlying space is $C[0, 1]$ (defined above), and our distance is given by (3). Suppose we have $X_n \to^D X$. We can easily see that the functions attaching each $z \in C[0, 1]$ $\max_{0 \leq t \leq 1} z(t)$ or $\int_0^1 z(t)^2 dt$ are continuous with respect to our metric. Hence we can immediately conclude that

$$\max_{0 \leq t \leq 1} X_n(t) \to \max_{0 \leq t \leq 1} X(t) \tag{6}$$

or

$$\int_0^1 X_n(t)^2 dt \to \int_0^1 X(t)^2 dt, \tag{7}$$

where '$\to$' stands for the usual convergence in distribution of real-valued random variables. Sometimes it is, however, burdensome to establish continuity for some functionals, or we might even be forced to consider discontinuous functionals. In this kind of situation the following theorem is helpful. Since we only work in separable, metric spaces a function ϕ defined on a general metric space E is continuous at a point $x \in E$ if for all $x_n \to x \phi(x_n) \to \phi(x)$. Otherwise the ϕ is called discontinuous in x, and let D_ϕ be the set of all points where ϕ is discontinuous. Now assume we have some random elements X_n, X and

$$X_n \to^D X.$$

Then we have the following theorem.

Theorem 1 *Suppose that*

$$P([X \in D_\phi]) = 0.$$

Then

$$\phi(X_n) \to^D \phi(X).$$

If the discontinuities of ϕ are a null set with respect to the limiting distribution, the distributions of $\phi(X_n)$ can be approximated better and better by the distribution of $\phi(X)$.

In any case, the usefulness of functional limit theorems depends on the set of continuous functions associated with our space. On the one hand, a metric with 'many' continuous functions will allow us to establish many limiting relationships like (6) or (7). On the other hand, it will be harder to establish convergence, since we have to show the relation (5) for *more* functions ϕ. Hence we have to compromise.

The space *D*[0, 1]

The first and most important application of functional limit theorems is the analysis of partial sums. When dealing with normalized sums like (1), (2) we encounter the first problem: we can easily let the time t or z be a continuous variable. but then the sum (1), (2) is a discontinuous function. Hence we have to look at spaces more general than $C[0, 1]$. One such space is the space $D[0, 1]$, defined to be the space of all bounded functions f which have only 'jumps' as discontinuities: at every time z the limits to the right and left of f ($f(z+0)$ and $f(z-0)$ exist).

Next we have to define a distance between the functions f, g from $D[0, 1]$. The first candidate, namely the supremum-norm (3), has the disadvantage that the corresponding space is not separable: consider for each $a \in (0, 1)$ the functions f_a defined as

$$f_a(z) = \begin{cases} 0 & \text{if } z < a \\ 1 & \text{if } z \geq a. \end{cases} \tag{8}$$

Then we can easily see that in the supremum norm (3), the distance between f_a and f_b is equal to 1. Since we have more than countable real numbers in (0, 1), we cannot have a countable dense subset.

A distance better suited to this space is the so-called Skorohod metric. Let us first define the set Λ to be the set of all functions from [0, 1] to [0, 1] which are monotonically increasing, continuous, and map 0 and 1 into 0 and 1, respectively. Then define

$$d_S(f, g) = \inf_{\lambda \in \Lambda} \sup_z (|f(z) - g(\lambda(z))| + |z - \lambda(z)|).$$

The Skorohod distance is related to the maximal distance. The main difference, however, is that we do not compare the functions f and g for the same values. The Skorohod metric allows us to 'bend' the argument a little. This rather small modification has enormous consequences. The corresponding space is separable: that is, there exists a countable dense subset. The metric itself is not complete (that is, there exist Cauchy sequences which do not converge). There exists, however, an equivalent metric (that is, a metric which determines the same open sets, neighbourhoods, convergent subsequences, continuous functions,...) which is complete. Moreover, we can easily see that

$$d_S(f, g) \leq d_M(x, y),$$

so convergence in the maximum distance implies convergence in the Skorohod metric.

The next question is the set of continuous functions. We can easily see that some of the usual candidates, like for example the functional mapping each f to $\sup_{0 \leq z \leq 1} f(z)$, are continuous. The functional mapping f to $f(z)$, however, is for $0<z<1$ *not* continuous. Hence theorem 1 will come in handy.

The most important types of limiting processes will all have continuous trajectories. Hence, the class of functionals covered by theorem 1 contains all functionals which are continuous in $C[0, 1]$. For establishing this continuity, we have an interesting criterion.

Theorem 2 *Suppose we have $f \in C[0, 1]$, and $f_n \in D[0, 1]$ so that $f_n \to f$ in the Skorohod metric. Then we have convergence in the supremum metric* (3), *too.*

This result may explain the usefulness of $D[0, 1]$. On the one hand, the metric on $D[0, 1]$ is weak enough to allow for separability. This has, however, the drawback that it is hard to establish continuity of a function in the general case.

If the limiting random element lies with probability 1 in $C[0, 1]$, however, it is easy to check the requirements of theorem 1 for a function $\phi : D[0, 1] \to R$. One only has to show that $\phi(f_n) \to \phi(f)$ if $f_n \to f$ *uniformly*, which is much easier to handle.

Examples for limit theorems

In this section, I want to bring some examples of functional limit theorems. Together with the discussion above, they can be used as 'building blocks' for the derivation of general limit theorems.

The first functional limit theorem is one of the most important, namely, the functional limit theorem for martingale differences. This theorem is of utmost importance in many statistical applications: the scores of the conditional likelihood functions are martingale differences. Furthermore, the theorem is quite general. It only assumes a Lindeberg condition (which is quite similar to the case of the classical central limit theorem) and some kind of normalization condition. The role of the standard normal distribution is played by the 'standard random walk' W. W is a random element with values in $C[0, 1]$ (that is, a random function) with the following properties:

- $W(0) = 0$.
- W is 'Gaussian'. All finite-dimensional marginal distributions $(W(z_0), \ldots, W(z_k))$ are Gaussian with expectation 0.
- The covariance of $W(z_1)$ and $W(z_2)$ is $\min(z_1, z_2)$.

A quite tedious but well known proof shows that there exists such a random element, and that its distribution (the induced probability measure on $C[0, 1]$) is unique. Moreover, it is easy to show that W has all the properties associated with a random walk: its increments are independent from past values.

Theorem 3 (McLeish, 1974): *Suppose we have given a triangular array of random variables $X_{i,n}$. $1 \leq i \leq n$, together with some adapted σ-algebras $\mathfrak{F}_{i,n}$ so that*

$$E(X_{i,n}/\mathfrak{F}_{i-1,n}) = 0.$$

Furthermore assume that the following two conditions are satisfied:

1. *The 'norming condition' is satisfied:*

$$\sum_{i \leq nz} E(X_{i,n}^2/\mathfrak{F}_{i-1,n}) \to z$$

uniformly in probability as $n \to \infty$.

2. *The 'conditional Lindeberg condition' is fulfilled: for all* $\varepsilon > 0$

$$\sum_{i \leq n} E(X_{i,n}^2 I_{[|X_{i,n}| > \varepsilon]}/\mathfrak{F}_{i-1,n}) \to 0$$

in probability as $n \to \infty$.

Then let us introduce the random elements S_n, *defined by*

$$S_n(z) = \sum_{i \leq nz} X_{i,n}$$

for $0 \leq z \leq 1$. *Then the* S_n *converge in distribution to a standard random walk W.*

Another important class of processes are stationary processes X_n, $n \in \boldsymbol{Z}$. In general, we will not even have a CLT. If, however, the conditional expectation of X_n given the X_0, $X_{-1}, \ldots$ decreases sufficiently fast, we will have a functional limit theorem, analogous to Gordin's theorem. Let us define the normalized, partial sums S_n by

$$S_n(z) = \frac{1}{\sqrt{n}} \sum_{i \leq nz} X_i.$$

Furthermore we will use the $\boldsymbol{L}_2$-norm of random variables: Define $\|X\| = \sqrt{EX^2}$.

Theorem 4 (Peligrad and Utev, 2005): *Assume that we have given a stationary process* X_i *so that*

$$\sum \frac{1}{n} \|E(S_n(1)/X_0, X_{-1}, \ldots)\| < \infty,$$

for all i

$$\frac{1}{n} \sum X_t X_{t-1} \to E(X_t X_{t-i})$$

and

$$\sigma^2 = \sum_{i \in \mathbf{Z}} E(X_t X_{t-i}) < \infty.$$

Then

$$\frac{1}{\sigma}S_n$$

converges in distribution to a standard random walk W.

These two theorems should only act as illustrations for functional limit theorems. Especially for stationary processes, more general theorems are available. A good survey about recent results can be found in Merlevede, Peligrad and Utev (2006).

Conclusion

This short introduction article should serve only as an introduction to functional limit theorems. Over the years, a rich theory has developed unifying many aspects of the limiting behaviour of functions of random variables. In particular, I would like to mention the limiting theorems for empirical distribution functions and their generalizations (see for example van der Vaart and Wellner, 1996, for a survey, and Andrews and Pollard, 1994, for dependent random variables). These results can be used to derive 'uniform' central limit theorems.

WERNER PLOBERGER

See also **central limit theorems.**

Bibliography

Andrews, D.W.K. and Pollard, D. 1994. An introduction to functional central limit theorems for dependent stochastic processes. *Revue internationale de statistique* 62, 119–32.

Billingsley, P. 1999. *Convergence of Probability Measures*, 2nd edn. New York: Wiley-Interscience.

Davidson, J. 1994. *Stochastic Limit Theory: An Introduction for Econometricians.* Oxford: Oxford University Press.

McLeish, D.L. 1974. Dependent central limit theorems and invariance principles. *Annals of Probability* 2, 620–8.

Merlevede, F., Peligrad, M. and Utev, S. 2006. Recent advances in invariance principles for stationary sequences. *Probability Surveys* 3, 1–36.

Peligrad, M. and Utev, S. 2005. A new maximal inequality and invariance principle for stationary sequences. *Annals of Probability* 33, 789–815.

van der Vaart, A. and Wellner, J.A. 1996. *Weak Convergence and Empirical Processes.* Berlin: Springer.

generalized method of moments estimation

1. Introduction

Generalized method of moments (GMM) refers to a class of estimators constructed from the sample moment counterparts of population moment conditions (sometimes known as orthogonality conditions) of the data generating model. GMM estimators have become widely used, for the following reasons:

1. GMM estimators have large sample properties that are easy to characterize. A family of such estimators can be studied simultaneously in ways that make asymptotic efficiency comparisons easy. The method also provides a natural way to construct tests which take account of both sampling and estimation error.
2. In practice, researchers find it useful that GMM estimators may be constructed without specifying the full data generating process (which would be required to write down the maximum likelihood estimator). This characteristic has been exploited in analysing partially specified economic models, studying potentially misspecified dynamic models designed to match target moments, and constructing stochastic discount factor models that link asset pricing to sources of macroeconomic risk.

Books with good discussions of GMM estimation with a wide array of applications include: Cochrane (2001), Arellano (2003), Hall (2005), and Singleton (2006). For a theoretical treatment of this method see Hansen (1982) along with the self-contained discussions in the books. See also Ogaki (1993) for a general discussion of GMM estimation and applications, and see Hansen (2001) for a complementary article that, among other things, links GMM estimation to related literatures in statistics. For a collection of recent methodological advances related to GMM estimation, see the journal issue edited by Ghysels and Hall (2002). While some of these other references explore the range of substantive applications, in what follows we focus more on the methodology.

2. Set-up

As we will see, formally there are two alternative ways to specify GMM estimators, but they have a common starting point. Data are a finite number of realizations of the process $\{x_t : t = 1, 2, \ldots\}$. The model is specified as a vector of moment conditions:

$$Ef(x_t, \beta_0) = 0$$

where f has r coordinates and β_0 is an unknown vector in a parameter space $\mathbb{P} \subset \mathbb{R}^k$. To achieve identification we assume that on the parameter space $\mathbb{P}$

$$Ef(x_t, \beta) = 0 \text{ if, and only if } \beta = \beta_0. \qquad (1)$$

The parameter β_0 is typically not sufficient to write down a likelihood function. Other parameters are needed to specify fully the probability model that underlies the data generation. In other words, the model is only partially specified.

Examples include:

(a) linear and nonlinear versions of instrumental variables estimators as in Sargan (1958; 1959), and Amemiya (1974);
(b) rational expectations models as in Hansen and Singleton (1982), Cumby, Huizinga and Obstfeld (1983), and Hayashi and Sims (1983);
(c) security market pricing of aggregate risks as described, for example, by Cochrane (2001), Singleton (2006) and Hansen et al. (2007);
(d) matching and testing target moments of possibly misspecified models as described by, for example, Christiano and Eichenbaum (1992) and Hansen and Heckman (1996).

Regarding example (a), many related methods have been developed for estimating correctly specified models, dating back to some of the original applications in statistics of method-of-moments-type estimators. The motivation for such methods was computational. See Hansen (2001) for a discussion of this literature and how it relates to GMM estimation. With advances in numerical methods, the fully efficient maximum likelihood method and Bayesian counterparts have become much more tractable. On the other hand, there continues to be an interest in the study of dynamic stochastic economic models that are misspecified because of their purposeful simplicity. Thus moment matching remains an interesting application for the methods described here. Testing target moments remains valuable even when maximum likelihood estimation is possible (for example, see Bontemps and Meddahi, 2005).

2.1 Central limit theory and martingale approximation

The parameter dependent average

$$g_N(\beta) = \frac{1}{N}\sum_{t=1}^{N} f(x_t, \beta)$$

is featured in the construction of estimators and tests. When the law of large numbers is applicable, this average converges to the $Ef(x_t, \beta)$. As a refinement of the identification condition:

$$\sqrt{N} g_N(\beta_0) \Rightarrow \text{Normal}(0, V) \tag{2}$$

where $\Rightarrow$ denotes convergence in distribution and V is a covariance matrix assumed to be nonsingular. In an iid data setting, V is the covariance matrix of the random vector $f(x_t, \beta_0)$. In a time series setting:

$$V = \lim_{N\to\infty} NE[g_N(\beta_0) g_N(\beta_0)'], \tag{3}$$

which is the long-run counterpart to a covariance matrix.

Central limit theory for time series is typically built on martingale approximation (see Gordin, 1969; Hall and Heyde, 1980). For many time series models, the martingale approximators can be constructed directly and there is specific structure to the V matrix. A leading example is when $f(x_t, \beta_0)$ defines a conditional moment restriction. Suppose that x_t, $t=0, 1, \ldots$ generates a sigma algebra $\mathscr{F}_t$, $E[|f(x_t, \beta_0)|^2] < \infty$ and

$$E[f(x_{t+\ell}, \beta_0)|\mathscr{F}_t] = 0$$

for some $\ell \geq 1$. This restriction is satisfied in models of multi-period security market pricing and in models that restrict multi-period forecasting. If $\ell = 1$, then g_N is itself a martingale; but when $\ell > 1$ it is straightforward to find a martingale m_N with stationary increments and finite second moments such that

$$\lim_{N\to\infty} E[|g_N(\beta_0) - m_N(\beta_0)|^2] = 0,$$

where $|\cdot|$ is the standard Euclidean norm. Moreover, the lag structure may be exploited to show that the limit in (3) is

$$V = \sum_{j=-\ell+1}^{\ell-1} E[f(x_t, \beta_0)f(x_{t+j}, \beta_0)']. \tag{4}$$

(The sample counterpart to this formula is not guaranteed to be positive semidefinite. There are a variety of ways to exploit this dependence structure in estimation in constructing a positive semidefinite estimate. See Eichenbaum, Hansen and Singleton, 1988, for an example.) When there is no exploitable structure to the martingale approximator, the matrix V is the spectral density at frequency zero.

$$V = \sum_{j=-\infty}^{\infty} E[f(x_t, \beta_0)f(x_{t+j}, \beta_0)'].$$

2.2 Minimizing a quadratic form

One approach for constructing a GMM estimator is to minimize the quadratic form:

$$b_N = \arg\min_{\beta\in\mathbb{P}} g_N(\beta)' W g_N(\beta)$$

for some positive definite *weighting matrix* W. Alternative weighting matrices W are associated with alternative estimators. Part of the justification for this approach is that

$$\beta_0 = \arg\min_{\beta\in\mathbb{P}} Ef(x_t, \beta)' W Ef(x_t, \beta).$$

The GMM estimator mimics this identification scheme by using a sample counterpart.

There are a variety of ways to prove consistency of GMM estimators. Hansen (1982) established a uniform law of large numbers for random functions when the data generation is stationary and ergodic. This uniformity is applied to show that

$$\sup_{\beta \in \mathbb{P}} |g_N(\beta) - E[f(x_t, \beta)]| = 0$$

and presumes a compact parameter space. The uniformity in the approximation carries over directly the GMM criterion function $g_N(\beta)'Wg_N(\beta)$. See Newey and McFadden (1994) for a more complete catalogue of approaches of this type.

The compactness of the parameter space is often not ignored in applications, and this commonly invoked result is therefore less useful than it might seem. Instead, the compactness restriction is a substitute for checking behaviour of the approximating function far away from β_o to make sure that spurious optimizers are not induced by approximation error. This tail behaviour can be important in practice, so a direct investigation of it can be fruitful. For models with parameter separation:

$$f(x, \beta) = Xh(\beta)$$

where X is an $r \times m$ matrix constructed from x and h is a one-to-one function mapping $\mathbb{P}$ into subset of $\mathbb{R}^m$, there is an alternative way to establish consistency (see Hansen, 1982 for details). Models that are either linear in the variables or models based on matching moments that are nonlinear functions of the underlying parameters can be written in this separable form.

The choice of $W = V^{-1}$ receives special attention, in part because

$$Ng_N(\beta)'V^{-1}g_N(\beta) \Rightarrow \chi^2(r).$$

While the matrix V is typically not known, it can be replaced by a consistent estimator without altering the large sample properties of b_N. When using martingale approximation, the implied structure of V can often be exploited as in formula (4). When there is no such exploitable structure, the method of Newey and West (1987b) and others can be employed that are based on frequency-domain methods for time series data.

For asset pricing models there are other choices of a weighting matrix motivated by considerations of misspecification. In these models with parameterized stochastic discount factors, the sample moment conditions $g_N(\beta)$ can be interpreted as a vector of pricing errors associated with the parameter vector β. A feature of $W = V^{-1}$ is that, if the sample moment conditions (the sample counterpart to a vector pricing errors) happened to be the same for two models (two choices of β), the one for which the implied asymptotic covariance matrix is larger will have a smaller objective. Thus there is a *reward* for parameter choices that imply variability in the underlying central limit approximation. To avoid such a reward, it is also useful to compare models or parameter values in other ways. An alternative weighting matrix is constructed by minimizing the least squares distance between the parameterized stochastic discount factor and one among the family of discount factors that correctly price the assets. Equivalently, parameters or models are selected on the basis of the maximum pricing error among constant weighted portfolios with payoffs that have common magnitude (a unit second moment). See Hansen and Jagannathan (1997) and Hansen, Heaton and Luttmer (1995) for this and related approaches.

2.3 Selection matrices

An alternative depiction is to introduce a *selection matrix* A that has dimension $k \times r$ and to solve the equation system:

$$Ag_N(\beta) = 0$$

for some choice of β, which we denote b_N. The selection matrix A reduces the number of equations to be solved from r to k. Alternative selection matrices are associated with alternative GMM estimators. By relating estimators to their corresponding selection matrices, we have a convenient device for studying simultaneously an entire family of GMM estimators. Specifically, we explore the consequence of using alternative subsets of moment equations or more generally alternative linear combinations of the moment equation system. This approach builds on an approach of Sargan (1958; 1959) and is most useful for characterizing limiting distributions. The aim is to study simultaneously the behaviour of a family of estimators. When the matrix A is replaced by a consistent estimator, the asymptotic properties of the estimator are preserved. This option expands considerably the range of applicability, and, as we will see, is important for implementation.

Since alternative choices of A may give rise to alternative GMM estimators, index alternative estimators by the choice of A. In what follows, replacing A by a consistent estimator does not alter the limiting distribution. For instance, the first-order conditions from minimizing a quadratic form can be represented using a selection matrix that converges to a limiting matrix A. Let

$$D = E\left[\frac{\partial f(x_t, \beta_0)}{\partial \beta}\right].$$

Two results are central to the study of GMM estimators:

$$\sqrt{N}(b_N - \beta_0) \approx -(AD)^{-1}A\sqrt{N}g_N(\beta_0) \tag{5}$$

and

$$\frac{1}{\sqrt{N}}g_N(b_N) \approx [I - D(AD)^{-1}D]\sqrt{N}g_N(\beta_0). \tag{6}$$

Both approximation results are expressed in terms of $\sqrt{N}g_N(\beta_0)$, which obeys a central limit theorem, see (2). These approximation results are obtained by standard local methods. They require the square matrix AD to be nonsingular. Thus, for there to exist a valid selection matrix, D must have full column rank k. Notice from (6) that the sample moment conditions evaluated at b_N have a degenerate distribution.

Pre-multiplying by A makes the right-hand side zero. This is to be expected because linear combinations of the sample moment conditions are set to zero in estimation.

In addition to assess the accuracy of the estimator (approximation (5)) and to validate the moment conditions (approximation (6)), Newey and West (1987a) and Eichenbaum, Hansen and Singleton (1988) show how to use these and related approximations to devise tests of parameter restrictions. (Their tests imitate the construction of the likelihood ratio, Lagrange multiplier and the Wald tests familiar from likelihood inference methods.)

Next we derive a sharp lower bound on the asymptotic distribution of a family of GMM estimators indexed by the selection matrix A. For a given A, the asymptotic covariance matrix for a GMM estimator constructed using this selection is:

$$cov(A) = (AD)^{-1}AVA'(D'A')^{-1}.$$

A selection matrix in effect over-parameterizes a GMM estimator, as can be seen from this formula. Two such estimators with selection matrices of the form A and BA for a nonsingular matrix B imply

$$cov(BA) = cov(A)$$

because the same linear combinations of moment conditions are being used in estimation. Thus without loss of generality we may assume that $AD = I$. With this restriction we may imitate the proof of the famed Gauss–Markov Theorem to show that

$$D'V^{-1}D \leq cov(A) \tag{7}$$

and that the lower bound on left is attained by any $\tilde{A}$ such that $\tilde{A} = BD'V^{-1}$ for some nonsingular B. The quadratic form version of a GMM estimator typically satisfies this restriction when W_N is a consistent estimator of V^{-1}. This follows from the first-order conditions of the minimization problem.

To explore further the implications of this choice, factor the inverse covariance matrix V^{-1} as $V^{-1} = \Lambda'\Lambda$ and form $\Delta = \Lambda D$. Then

$$V^{-1}D(D'V^{-1}D)^{-1}D'V^{-1} = \Lambda'[\Delta(\Delta'\Delta)^{-1}\Delta']\Lambda.$$

The matrices $\Delta(\Delta'\Delta)^{-1}\Delta'$ and $I–\Delta(\Delta'\Delta)^{-1}\Delta'$ are each idempotent and

$$\begin{bmatrix} [I - \Delta(\Delta'\Delta)^{-1}\Delta'] \\ \Delta(\Delta'\Delta)^{-1}\Delta' \end{bmatrix} \sqrt{N}\Lambda g_N(\beta_0)$$
$$\rightarrow \text{Normal}\left(\begin{bmatrix} 0 \\ 0 \end{bmatrix}, \begin{bmatrix} I - \Delta(\Delta'\Delta)^{-1}\Delta' & 0 \\ 0 & \Delta(\Delta'\Delta)^{-1}\Delta' \end{bmatrix} \right).$$

The first coordinate block is an approximation for $\sqrt{N}\Lambda g_N(b_N)$ and the sum of the two coordinate blocks is $\sqrt{N}\Lambda g_N(\beta_o)$. Thus we may decompose the quadratic form

$$\begin{aligned} N[g_N(\beta_0)]'V^{-1}g_N(\beta_0) \approx\, & N[g_N(b_N)]'V^{-1}g_N(b_N) \\ & + N[g_N(\beta_0)]'V^{-1}D(D'V^{-1}D)^{-1}D'V^{-1}g_N(\beta_0). \end{aligned} \tag{8}$$

where the two terms on the right-hand side are distributed as independent chi-square. The first has r degrees of freedom and the second one has $r-k$ degrees of freedom.

3. Implementation using the objective function curvature

While the formulas just produced can be used directly using consistent estimators of V and D in conjunction with the relevant normal distributions, looking directly at the curvature of the GMM objective function based on a quadratic form is also revealing. Approximations (5) and (6) give guidance on how to do this.

For a parameter vector β let $V_N(\beta)$ denote an estimator of the long-run covariance matrix. Given an initial consistent estimator b_N, suppose that $V_N(b_N)$ is a consistent estimator of V and

$$D_N = \frac{1}{N}\sum_{t=1}^{N}\frac{\partial f(x_t, b_N)}{\partial\beta}.$$

Then use of the selection $A_N = D_N'[V_N(b_N)]^{-1}$ attains the efficiency bound for GMM estimators. This is the so-called two-step approach to GMM estimation. Repeating this procedure, we obtain the so-called iterative estimator. (There is no general argument that repeated iteration will converge.) In the remainder of this section we focus on a third approach, resulting in what we call the continuous-updating estimator. This is obtained by solving:

$$\min_{\beta\in\mathbb{P}} L_N(\beta)$$

where

$$L_N(\beta) = N[g_N(\beta)]'[V_N(\beta)]^{-1}g_N(\beta).$$

Let b_N denote the minimized value. Here the weighting matrix varies with β.

Consider three alternative methods of inference that look at the global properties of the GMM objective $L_N(\beta)$:

(a) $\{\beta\in\mathbb{P} : L_N(\beta)\le C\}$ where C is a critical value from a $\chi^2(r)$ distribution.
(b) $\{\beta\in\mathbb{P} : L_N(\beta) - L_N(b_N)\le C\}$ where C is a critical value from a $\chi^2(k)$ distribution.
(c) Choose a *prior* π. Mechanically, treat $-\frac{1}{2}L_N(\beta)$ as a log-likelihood and compute

$$\frac{\exp[-\frac{1}{2}L_N(\beta)]\pi(\beta)}{\int\exp[-\frac{1}{2}L_N(\beta)]\pi(\tilde{\beta})d\tilde{\beta}}.$$

Method (a) is based on the left-hand side of (8). It was suggested and studied in Hansen, Heaton and Luttmer (1995) and Stock and Wright (2000). As emphasized by Stock and Wright, it avoids using a local identification condition (a condition that the matrix D have full column rank). On the other hand, it combines evidence about the parameter as reflected by the curvature of the objective with overall evidence about the model. A misspecified model will be reflected as an empty confidence interval.

Method (b) is based on the second term on right-hand side of (8). By translating the objective function, evidence against the model is netted out. Of course it remains important to consider such evidence because parameter inference may be hard to interpret for a misspecified model. The advantage of (b) is that the degrees of freedom of the chi-square distribution are reduced from r to k. Extensions of this approach to accommodate nuisance parameters were used by Hansen and Singleton (1996) and Hansen, Heaton and Luttmer (1995). The decomposition on the right-hand side of (8) presumes that the parameter is identified locally in the sense that D has full column rank, guaranteeing that the $D'V^{-1}D$ is nonsingular. Kleibergen (2005) constructs an alternative decomposition based on a weaker notion of identification that can be used in making statistical inferences.

Method (c) was suggested by Chernozhukov and Hong (2003). It requires an integrability condition which will be satisfied by specifying a uniform distribution π over a compact parameter space. The resulting histograms can be sensitive to the choice of this set or more generally to the choice of π. All three methods explore the global shape of the objective function when making inferences. (The large sample justification remains local, however.)

4. Backing off from efficiency

In what follows we give two types of applications that are not based on efficient GMM estimation.

4.1 Calibration-verification

An efficient GMM estimator selects the best linear combination among a set of moment restrictions. Implicitly a test of the over-identifying moment conditions examines whatever moment conditions are not used in estimation. This complicates the interpretation of the resulting outcome. Suppose instead there is one set of moment conditions for which we have more confidence and are willing to impose for the purposes and calibration or estimation. The remaining set of moment conditions are used for the purposes of verification or testing. The decision to use only a subset of the available moment conditions for purposes of estimation implies a corresponding loss in efficiency. See Christiano and Eichenbaum (1992) and Hansen and Heckman (1996) for a discussion of such methods for testing macroeconomic models.

To consider this estimation problem formally, partition the function f as:

$$f(x,\beta) = \begin{bmatrix} f^{[1]}(x,\beta) \\ f^{[2]}(x,\beta) \end{bmatrix}$$

where $f^{[1]}$ has r_1 coordinates and $f^{[2]}$ has $r - r_1$ coordinates. Suppose that $r_1 \geq k$ and that β is estimated using an A matrix of the form:

$$A = [A_1 \quad 0],$$

and hence identification is based only on

$$A_1 E f^{[1]}(x_t, \beta) = 0.$$

This is the so-called calibration step. Let b_N be the resulting estimator.

To verify or test the model we check whether $g_N^{[2]}(b_N)$ is close to zero as predicted by the moment implication:

$$E f^{[2]}(x_t, \beta_0) = 0.$$

Partition the matrix D of expected partial derivatives as:

$$D = \begin{bmatrix} D_1 \\ D_2 \end{bmatrix}$$

where D_1 is r_1 by k and D_2 is $r - r_1$ by k. Here we use limit approximation (6) to conclude that

$$\sqrt{N} g_N^{[2]}(b_N) \approx [-D_2(A_1 D_1)^{-1} A_1 \quad I] \sqrt{N} g_N(\beta_0),$$

which has a limiting normal distribution. A chi-square test can be constructed by building a corresponding quadratic form of $r - r_1$ asymptotically independent standard normally distributed random variables. (When r_1 exceeds k it is possible to improve the asymptotic power by exploiting the long-run covariation between $f^{[2]}(x_t, \beta_o)$ and linear combination of $f^{[1]}(x_t, \beta_o)$ not used in estimation. This can be seen formally by introducing a new parameter $\gamma_o = E[f^{[2]}(x_t, \beta)]$ and using the GMM formulas for efficient estimation of β_o and γ_o.)

4.2 Sequential estimation

Sequential estimation methods have a variety of econometric applications. For models of sample selection see Heckman (1976), and for related methods with generated regressors see Pagan (1984). For testing asset pricing models, see Cochrane (2001, chs 12 and 13).

To formulate this problem in a GMM setting, partition the parameter vector as

$$\beta = \begin{bmatrix} \beta^{[1]} \\ \beta^{[2]} \end{bmatrix}$$

where $\beta^{[1]}$ has k_1 coordinates. Partition the function f as:

$$f(x,\beta) = \begin{bmatrix} f^{[1]}(x,\beta^{[1]}) \\ f^{[2]}(x,\beta) \end{bmatrix}$$

where $f^{[1]}$ has r_1 coordinates and $f^{[2]}$ has $r-r_1$ coordinates. Notice that the first coordinate block only depends on the first component of the parameter vector. Thus the matrix d is block lower triangular:

$$D = \begin{bmatrix} D_{11} & 0 \\ D_{21} & D_{22} \end{bmatrix}$$

where

$$D_{ij} = E\left[\frac{\partial f^{[i]}(x_t,\beta_0)}{\partial \beta^{[j]}}\right].$$

A sequential estimation approach exploits the triangular structure of the moment conditions as we now describe. The parameter $\beta_0^{[1]}$ is estimable from the first partition of moment conditions. Given such an estimator, $b_N^{[1]}$, $\beta_0^{[2]}$ is estimable from the second partition of moment conditions. Estimation error in the first stage alters the accuracy of the second stage estimation, as I now illustrate.

Assume now that $r_1 \geq k_1$. Consider a selection matrix that is block diagonal:

$$A = \begin{bmatrix} A_{11} & 0 \\ 0 & A_{22} \end{bmatrix}$$

where A_{11} has dimension k_1 by r_1 and A_{22} has dimension $k-k_1$ by $r-r_1$. It is now possible to estimate $\beta_o^{[1]}$ using the equation system:

$$A_{11} g_N^{[1]}(\beta^{[1]}) = 0$$

or a method that is asymptotically equivalent to this. Let $b_N^{[1]}$ be the solution. This initial estimation may be done for simplicity or because these moment conditions are embraced with more confidence. Given this estimation of $\beta_o^{[1]}$, we seek an estimator $b_N^{[2]}$ of $\beta_0^{[2]}$ by solving:

$$A_{22} g_N^{[2]}(b_N^{[1]}, \beta^{[2]}) = 0.$$

To proceed, we use this partitioning and apply (5) to obtain the limiting distribution for the estimator $b_N^{[2]}$. Straightforward matrix calculations yield,

$$\sqrt{N}\left(b_N^{[2]} - \beta_0^{[2]}\right) \approx -(A_{22}D_{22})^{-1} A_{22}\left[-D_{21}(A_{11}D_{11})^{-1}A_{11} \quad I\right]\sqrt{N} g_N(\beta_0). \qquad (9)$$

This formula captures explicitly the impact of the initial estimation of $\beta_0^{[1]}$ on the subsequent estimation of $\beta_0^{[2]}$. When D_{21} is zero an adjustment is unnecessary.

Consider next a (second-best) efficient choice of selection matrix A_{22}. Formula (9) looks just like formula (5) with A_{22} replacing A, D_{22} replacing D and a particular linear combination of $g_N(\beta_0)$. The matrix used in this linear combination 'corrects' for the estimation error associated with the use of an estimator $b_N^{[1]}$ instead of the unknown true value $\beta_0^{[1]}$. By imitating our previous construction of an asymptotically efficient estimator, we construct the (constrained) efficient choice of A_{22} given A_{11}:

$$A_{22} = B_{22}(D_{22})'\left(\begin{bmatrix}-D_{21}(A_{11}D_{11})^{-1}A_{11} & I\end{bmatrix} V \begin{bmatrix}-[D_{21}(A_{11}D_{11})^{-1}A_{11}]' \\ I\end{bmatrix}\right)^{-1}$$

for some nonsingular matrix B_{22}. An efficient estimator can be implemented in the second stage by solving:

$$\min_{\beta^{[2]}} g_N^{[2]}(b_N^{[1]}, \beta^{[2]})' W_N g_N^{[2]}(b_N^{[1]}, \beta^{[2]})$$

for $V_N^{[2]}$ given by a consistent estimator of

$$V^{[2]} = \left(\begin{bmatrix}-D_{21}(A_{11}D_{11})^{-1}A_{11} & I\end{bmatrix} V \begin{bmatrix}-[D_{21}(A_{11}D_{11})^{-1}A_{11}]' \\ I\end{bmatrix}\right)^{-1}$$

or by some other method that selects (at least asymptotically) the same set of moment conditions to use in estimation. Thus we have a method that adjusts for the initial estimation of $\beta^{[1]}$ while making efficient use of the moment conditions $Ef^{[2]}(x_t, \beta) = 0$.

As an aside, notice the following. Given an estimate $b_N^{[1]}$, the criterion-based methods of statistical inference described in Section 3 can be adapted to making inferences in this second stage in a straightforward manner.

5. Conditional moment restrictions

The bound (7) presumes a finite number of moment conditions and characterizes how to use these conditions efficiently. If we start from the conditional moment restriction:

$$E[f(x_{t+\ell}, \beta_0)|\mathscr{F}_t] = 0,$$

then in fact there are many moment conditions at our disposal. Functions of variables in the conditioning information set can be used to extend the number of moment conditions. By allowing for these conditions, we can improve upon the asymptotic efficiency bound for GMM estimation. Analogous conditional moment restrictions arise in cross-sectional settings.

For a characterizations and implementations appropriate for cross-sectional data, see Chamberlain (1986) and Newey (1993), and for characterizations and implementations in a time series settings see Hansen (1985; 1993), and West (2001). The characterizations are conceptually interesting but reliable implementation is more challenging. A related GMM estimation problem is posed and studied by Carrasco and Florens (2000) in which there is a pre-specified continuum of moment conditions that are available for estimation.

6. Conclusion

GMM methods of estimation and inference are adaptable to a wide array of problems in economics. They are complementary to maximum likelihood methods and their Bayesian counterparts. Their large sample properties are easy to characterize. While their computational simplicity is sometimes a virtue, perhaps their most compelling use is in the estimation of partially specified models or of misspecified dynamic models designed to match a limited array of empirical targets.

LARS PETER HANSEN

See also **Bayesian methods in macroeconometrics.**

I greatly appreciate comments from Lionel Melin, Monika Piazzesi, Grace Tsiang and Francisco Vazquez-Grande. This material is based upon work supported by the National Science Foundation under Award Number SES0519372.

Bibliography

Amemiya, T. 1974. The nonlinear two-stage least-squares estimator. *Journal of Econometrics* 2, 105–10.

Arellano, M. 2003. *Panel Data Econometrics*. New York: Oxford University Press.

Bontemps, C. and Meddahi, N. 2005. Testing normality: a GMM approach. *Journal of Econometrics* 124, 149–86.

Carrasco, M. and Florens, J.P. 2000. Generalization of GMM to a continuum of moment conditions. *Econometric Theory* 20, 797–834.

Chamberlain, G. 1986. Asymptotic efficiency in estimation with conditional moment restrictions. *Journal of Econometrics* 34, 305–34.

Chernozhukov, V. and Hong, H. 2003. An MCMC approach to classical estimation. *Journal of Econometrics* 115, 293–346.

Christiano, L.J. and Eichenbaum, M. 1992. Current real business cycle theories and aggregate labor market fluctuations. *American Economic Review* 82, 430–50.

Cochrane, J. 2001. *Asset Pricing*. Princeton: Princeton University Press.

Cumby, R.E., Huizinga, J. and Obstfeld, M. 1983. Two-step two-stage least squares estimation in models with rational expectations. *Journal of Econometrics* 21, 333–5.

Eichenbaum, M.S., Hansen, L.P. and Singleton, K.J. 1988. A time series analysis of representation agent models of consumption and leisure choice under uncertainty. *Quarterly Journal of Economics* 103, 51–78.

Ghysels, E. and Hall, A. 2002. Editors' Introduction to *JBES* twentieth anniversary issue on generalized method of moments estimation. *Journal of Business and Economic Statistics* 20, 441.

Gordin, M.I. 1969. The central limit theorem for stationary processes. *Soviet Mathematics Doklady* 10, 1174–6.

Hall, A.R. 2005. *Generalized Method of Moments*. New York: Oxford University Press.
Hall, P. and Heyde, C.C. 1980. *Martingale Limit Theory and Its Application*. Boston: Academic Press.
Hansen, L.P. 1982. Large sample properties of generalized method of moments estimators. *Econometrica* 50, 1029–54.
Hansen, L.P. 1985. A method for calculating bound on asymptotic covariance matrices of generalized method of moments estimators. *Journal of Econometrics* 30, 203–38.
Hansen, L.P. 1993. Semiparametric efficiency bounds for linear time-series models. In *Models, Methods and Applications of Econometrics: Essays in Honor of A.R. Bergstrom*, ed. P.C.B. Phillips. Cambridge, MA: Blackwell.
Hansen, L.P. 2001. Method of moments. In *International Encyclopedia of the Social and Behavior Sciences*. New York: Elsevier.
Hansen, L.P., Heaton, J. and Luttmer, E. 1995. Econometric evaluation of asset pricing models. *Review of Financial Studies* 8, 237–74.
Hansen, L.P. and Heckman, J.J. 1996. The empirical foundations of calibration. *Journal of Economic Perspectives* 10(1), 87–104.
Hansen, L.P. and Jagannathan, R. 1997. Assessing specification errors in stochastic discount factor models. *Journal of Finance* 52, 557–90.
Hansen, L.P. and Singleton, K.J. 1982. Generalized instrumental variables of nonlinear rational expectations models. *Econometrica* 50, 1269–86.
Hansen, L.P. and Singleton, K.J. 1996. Efficient estimation of linear asset pricing models with moving average errors. *Journal of Business and Economic Statistics* 14, 53–68.
Hansen, L.P., Heaton, J.C., Lee, J. and Roussanov, N. 2007. Intertemporal substitution and risk aversion. In *Handbook of Econometrics*, vol. 6A, ed. J. Heckman and E. Leamer. Amsterdam: North-Holland.
Hayashi, F. and Sims, C. 1983. Nearly efficient estimation of time-series models with predetermined, but not exogenous, instruments. *Econometrica* 51, 783–98.
Heckman, J.J. 1976. The common structure of statistical methods of truncation, sample selection, and limited dependent variables and a simple estimator of such models. *Annals of Economic and Social Measurement* 5, 475–92.
Kleibergen, F. 2005. Testing parameters in GMM without assuming that they are identified. *Econometrica* 73, 1103–23.
Newey, W. 1993. Efficient estimation of models with conditional moment restrictions. In *Handbook of Statistics*, vol. 11, ed. G.S. Maddala, C.R. Rao, and H.D. Vinod. Amsterdam: North-Holland.
Newey, W. and McFadden, D. 1994. Large sample estimation and hypothesis testing. In *Handbook of Econometrics*, vol. 4, ed. R. Engle and D. McFadden. Amsterdam: North-Holland.
Newey, W.K. and West, K.D. 1987a. Hypothesis testing with efficient method of moments estimation. *International Economic Review* 28, 777–87.
Newey, W.K. and West, K.D. 1987b. A simple, positive semi-definite, heteroskedasticity and autocorrelation consistent covariance matrix. *Econometrica* 55, 703–8.
Ogaki, M. 1993. Generalized method of moments: econometric applications. In *Handbook of Statistics*, vol. 11, ed. G.S. Maddala, C.R. Rao and H.D. Vinod. Amsterdam: North-Holland.
Pagan, A.R. 1984. Econometric issues in the analysis of models with generated regressors. *International Economic Review* 25, 221–47.
Sargan, J.D. 1958. The estimation of economic relationships using instrumental variables. *Econometrica* 26, 393–415.
Sargan, J.D. 1959. The estimation of relationships with autocorrelated residuals by the use of instrumental variables. *Journal of the Royal Statistical Society: Series B* 21, 91–105.

Singleton, K.J. 2006. *Empirical Dynamic Asset Pricing: Model Specification and Econometric Assessment*. Princeton: Princeton University Press.

Stock, J.H. and Wright, J.H. 2000. GMM with weak identification. *Econometrica* 68, 1055–96.

West, K.D. 2001. On optimal instrumental variables estimation of stationary time series models. *International Economic Review* 42, 1043–50.

Granger–Sims causality

Granger–Sims causality is based on the fundamental axiom that 'the past and present may cause the future, but the future cannot cause the past' (Granger, 1980, p. 330). A variable x then is said to cause a variable y if at time t the variable x_t helps to predict the variable y_{t+1}. While predictability in itself is merely a statement about stochastic dependence, it is precisely the axiomatic imposition of a temporal ordering that allows us to interpret such dependence as a causal connection. The reason is that correlation is a symmetric concept with no indication of a direction of influence, while 'the arrow of time imposes the structure necessary' (Granger, 1980, p. 349) to interpret correlations in a causal way.

1. Definitions

A more precise definition of Granger causality can be given as follows. Assume that all relevant current information is measured in a vector $\chi_t = (y_t, x_t, z_t)$ which is observed at equally spaced discrete points in time t. (The assumption of discrete time can be relaxed but is maintained here for expositional convenience. Almost all empirical work is in the context of discrete time models.) Denote by Y_s^t all information contained in $y_j, j = \{s, \ldots, t\}$ with equivalent definitions for X_s^t and Z_s^t. (In more formal terms, the expression 'information contained in' can be replaced by 'sigma field generated by'; see Florens and Mouchart, 1982, for precise definitions.) Then x_t does not Granger cause y_{t+1} if

$$P(y_{t+1} \in A | Y_{-\infty}^t, X_{-\infty}^t, Z_{-\infty}^t) = P(y_{t+1} \in A | Y_{-\infty}^t, Z_{-\infty}^t) \qquad \text{(CI)}$$

for all t and for any set A for which the conditional probabilities are well defined. It is worth noting that no assumptions of stationarity are needed for this definition. It is common to use the shorthand notation $y_{t+1} \perp X_{-\infty}^t | Y_{-\infty}^t, Z_{-\infty}^t, \forall t$, which states that y_{t+1} and $X_{-\infty}^t$ are independent conditional on $Y_{-\infty}^t, Z_{-\infty}^t$. This form of the definition, which imposes a conditional independence restriction on the joint distribution of the process χ_t (see Dawid, 1979, for a formal definition and alternative equivalent representations), is similar to the definition used by Granger (1980). It is more general than the original formulations of Granger (1963; 1969), which are based on prediction error variances: Let $E(y_{t+1}|X_{-\infty}^t)$ denote the optimal predictor of y_{t+1} based on $X_{-\infty}^t$, $\varepsilon_{t+1} = y_{t+1} - E(y_{t+1}|X_{-\infty}^t)$ is the prediction error and $\sigma^2(y_{t+1}|X_{-\infty}^t)$ the variance of ε_{t+1}. Then, according to Granger (1969), x_t does not cause y_{t+1} if

$$\sigma^2(y_{t+1}|Y_{-\infty}^t, X_{-\infty}^t, Z_{-\infty}^t) = \sigma^2(y_{t+1}|Y_{-\infty}^t, Z_{-\infty}^t). \qquad \text{(PV)}$$

The conditional independence definition (CI) has the advantage that it does not depend on a particular risk function and is easier to relate to other definitions of causality in stochastic environments such as Suppes (1970) and Rubin (1974). The formulation based on conditional independence was later used and refined in theoretical work by Chamberlain (1982), Florens and Mouchart (1982; 1985), Bouissou, Laffont and Vuong (1986) and Holland (1986). The advantage of the prediction error variance definition, (PV) which goes back to Wiener (1956), on the other hand, is that it is easier to implement in statistical tests and has consequently received considerable attention in applied work.

In an influential paper, Sims (1972) showed in the context of covariance stationary processes and restricted to linear predictors that in the case of a bivariate system $\chi_t' = (x_t, y_t)$ the definitions of Granger (1963; 1969) are equivalent to parameter restrictions of the moving average or distributed lag representations of χ_t. When χ_t is covariance stationary it can be represented as

$$x_t = \sum_{j=0}^{\infty} a_j u_{t-j} + \sum_{j=0}^{\infty} b_j v_{t-j}$$

$$y_t = \sum_{j=0}^{\infty} c_j u_{t-j} + \sum_{j=0}^{\infty} d_j v_{t-j}$$

where a_j, b_j, c_j and d_j are constants and u_t and v_t are mutually uncorrelated white noise processes. Sims (1972) shows that the condition 'x_t does not Granger cause $y^t + 1$' is equivalent to c_j or d_j being chosen identically zero for all j. Furthermore, if χ_t has an autoregressive representation, then $x_t = \sum_{j=0}^{\infty} \pi_j y_{t-j} + \varepsilon_t$ where the π_j are parameters and ε_t is an unobservable innovation. Strict exogeneity of y_t in the context of this model is defined as the condition that y_t and ε_s are independent for all values of t and s. Sims then shows that the residuals from a projection of x_t onto y_{t-j}, $j \geq 0$ are uncorrelated with all past and future y_t if an only if x_t does not Granger cause y_{t+1}. In other words, it follows that the condition 'x_t does not Granger cause y_{t+1}' is equivalent to the condition that y_t is strictly exogenous. The relationship between Granger non-causality and strict exogeneity, first discovered by Sims (1972), is further discussed by Engle, Hendry and Richard (1983). Hosoya (1977) shows that this relationship continues to hold in a bivariate setting when processes are not necessarily stationary and have deterministic components, a situation not originally consider by Granger (1969). The strict exogeneity restrictions discussed in Sims (1972) have become known in the literature as 'Sims non-causality'. In more general terms, and in the context of the process χ_t, the absence of Sims causality of x_t for y_{t+1} is defined as a conditional independence restriction where

$$P(x_t \in A | Y_{-\infty}^{\infty}, X_{-\infty}^{t-1}, Z_{-\infty}^{t}) = P(x_t \in A | Y_{-\infty}^{t}, X_{-\infty}^{t-1}, Z_{-\infty}^{t}) \qquad \text{(S)}$$

for all t or equivalently as $Y_{t+1}^{\infty} \perp x_t | Y_{-\infty}^{t}, X_{\infty}^{t-1}, Z_{-\infty}^{t}$, $\forall t$. This definition appears in Florens (2003) and Angrist and Kuersteiner (2004) and is closely related to Chamberlain (1982) except that here additional information contained in $Z_{-\infty}^{t}$ is allowed for.

The equivalence between the Granger and Sims notions of causality extends beyond the prediction criterion for covariance stationary processes to definitions of non-causality based on conditional independence restrictions, but, as discussed below, in models with time-varying covariates z_t Sims causality is more appealing and easier to link to the potential-outcomes causality concepts widely used to analyse randomized trials and quasi-experimental studies. Sims (1977, p. 30) emphasizes the use of strict exogeneity restrictions for model identification in structural models and notes that these restrictions can often be related to decision rules of economic agents. Granger non-causality, on the other hand, is a restriction of the reduced form. Structural vector autoregressions (VARs) provide an example of this difference where Sims non-causality imposes restrictions on the impulse response function of structural innovations (Sims, 1986, p. 9, discusses the nature of structural innovations and mentions explicitly that they may include random fluctuations of policies) while Granger non-causality imposes restrictions on the reduced form VAR representation of the model. That the two are in general not equivalent will be discussed in Section 3.

2. Motivation and interpretation of Granger–Sims causality

In order to understand the significance of the original definition of Granger–Sims causality in Granger (1963; 1969) and Sims (1972), it is useful to briefly review the preceding debate in the literature. Simon (1953) defines causality as 'properties of a scientist's model'. Asymmetric functional relationships between variables are given causal interpretations but Simon emphasizes that no notion of time is needed for this definition. In the context of linear systems of equations Simon's definition of causal relationships is equivalent to a block recursive structure of the equations. Simon (1953, p. 65) emphasizes the need for a priori knowledge about the system to identify the block recursive structure. Wold (1954) and Strotz and Wold (1960) strengthen this definition of causality to require a model to be fully recursive. If the model has three variables *a*, *b* and *c* then recursiveness means that *a* can be solved without knowing *b* and *c*, the solution for *b* generally depends on *a*, and the solution for *c* depends on both *a* and *b*. Such a system then is interpreted to have a causal relationship where *a* causes *b* and *a* and *b* both cause *c*. Wold (1954, p. 166) writes: 'The relationship is then defined as causal if it is theoretically permissible to regard the variables as involved in a fictive controlled experiment.'

The idea is that, because *a* does not depend on *b* an *c*, it can in principle be controlled or changed by an experimenter. In the terminology of Wold (1954, p. 166), *a* is the cause and *b* and *c* are the 'effect variables'. Wold (1954) discusses the distinction between randomized experiments in the sense of R. A. Fisher and non-experimental observations. In the latter case causal interpretations depend on 'subject-matter theory' (Wold, 1954, p. 170) to identify recursive structures. In other words, a priori assumptions about the structure of the model replace randomized experiments as the source of identification. This approach remains popular to this day for the identification of structural VARs where recursive relationships based on a casual appeal to economic theory are imposed. Orcutt (1952) discusses causal chains or triangular systems but also mentions the possibility of using temporal structures to identify causal links. Basmann (1963) criticizes the recursive identification schemes of Wold and argues in favour of more general structural economic models to identify causal relationships.

Both Granger and Sims voice scepticism about identifying restrictions which can in general not be tested but have to be accepted a priori. Granger (1980, p. 335) writes,

> [i]f these assumptions are correct or can be accepted as being correct, these definitions may have some value. However, if the assumptions are somewhat doubtful, these definitions do not prove to be useful.

And Sims (1972, p. 544; see also 1977, p. 33) notes,

> [i]f one is willing to identify causal ordering with Wold's causal chain form for a multivariate model, and if enough identifying restrictions are available in addition to those specifying the causal chain form, one can test a particular causal ordering as a set of overidentifying restrictions. The conditions allowing such a test are seldom met in practice, however.

The great advantage of Granger's definition of causality is that it is directly testable from observed data. Granger (1969) gives operational definitions and discusses testable parameter restrictions in linear time series specifications. The strict exogeneity restrictions derived by Sims (1972) are particularly revealing of the power of the flow of time as the identifying force behind uncovering causal links between time series. On the assumption that x_t is strictly exogenous, then, if x_t does not cause y_t it must be conditionally independent of future outcomes y_{t+j}. On the other hand, if conditional correlations between future outcomes y_t and current values of x_t are detected, they must be due to a causal influence of x_t on y_t, because, by assumption, the possibility of a causal link between events determining y_{t+j} that lie in the future and current observations of x_t have been excluded and thus cannot be the source of the observed correlation.

As Granger (1963; 1980) notes, the notion of causality has a long and controversial tradition in philosophy. Some treatments discussing relationships to econometric and statistical practice include Pearl (2000) and Hoover (2001). Holland (1986) discusses the causality definitions of David Hume, John Stuart Mill and Suppes (1970) in the context of Rubin's (1974) causal model. Holland points out that Hume's criteria for causality include the axiom of temporal precedence as well as the requirement of a 'constant conjunction' between cause and effect. (See Hoover, 2001, p. 8, for more discussion of Hume's concept of 'constant conjunction'.) Suppes (1970) proposes a probabilistic theory of causality where he replaces constant conjunction with the requirement that

> one event is the cause of another if the appearance of the first event is followed with high probability by the appearance of the second, and there is no third event that we can use to factor out the probability relationship between the first and second events. (Suppes, 1970, p. 10)

The definition of Suppes has some parallels to Granger's definition, notably the requirement of temporal succession, the fact that there are no restrictions on what can be a cause and the fact that causes are defined through their effect on the conditional distribution of the effect variable. Finally, Holland attributes the idea of identifying causal effects through experimentation to Mill. Experimentation has since played a central role in the statistical analysis of causality, although Granger (1980, p. 329) mentions it only in passing and does not rely on it in his definition of causality. An important consequence of an experimental concept of causality is that, as Holland (1986, p. 954) writes, 'causes are only those things that could, in principle, be treatments in experiments'. As discussed in Section 4, this is a critical difference to the concept of Granger causality which does not restrict possible causes.

Feigl (1953) discusses various aspects of the definition of causality that have appeared historically in philosophy and attempts to extract what he calls a 'purified' definition of causality. Zellner's (1979) critique of the concept of Granger causality is centered on Feigl's definition according to which '[t]he clarified (purified) concept of causation is defined in terms of predictability according to a law' (Feigl, 1953, p. 408). Feigl (1953, p. 417) continues to note that '[p]rediction may be analyzed as a form of deductive inference from inductive premises (laws, hypotheses, theories) with the help of descriptions or existential hypotheses'. Zellner (1979, p. 12) writes: 'predictability without a law or set of laws, or as econometricians might put it, without theory, is not causation.' In other words, the causality concept put forward by Feigl is based on a priori theoretical assumptions used to generate predictions, while the Granger–Sims notion of causality replaces these a priori restrictions with the axiom of temporal priority. Feigl (1953, p. 417) notes that causal relationships can be defined even when cause and effect are contemporaneous. According to Feigl, a more important distinction between cause and effect lies in the controllability of the cause as opposed to the effect, which leads Feigl to recommend experimental methods as the best way to identify causal factors. In light of Feigl's work Zellner's main critique of Granger–Sims causality is that it is not based on economic theory to identify causes. (Leamer, 1985, also strongly rejects the idea of conducting causal inference without relying on a priori theory.) In a reply to Zellner, Sims (1979, p. 105) notes that Feigl's definition is at least as ambiguous as the term 'causality' itself and that it is so general that it encompasses many other definitions of causality.

Zellner also criticizes three more specific features of Granger causality. First, the requirement that the information set needed to define relevant conditional distributions contain all available information makes the definition non-operational. Zellner (1979, p. 33) writes, 'Granger does not explicitly mention the important role of economic laws in defining the set of "all relevant information"' and emphasizes that additional assumptions beyond statistical criteria are necessary to implement tests for Granger non-causality. Second, the limitation to stochastic phenomena and the assumption or axiom of temporal priority of a cause is unnecessarily restrictive compared with other definitions of causality, such as the one of Feigl (1953), which does not rely on these restrictions. And finally, the use of the prediction error variance as a criterion for predictability and the reliance on an optimal predictor, which according to Zellner may both not be well defined, is too restrictive. As far as this last point is concerned it should be noted that more general definitions of Granger–Sims causality proposed by Granger (1980), Chamberlain (1982) and Florens and Mouchart (1982), which are based on conditional independence restrictions of the joint distribution, do not have the problems that Zellner mentions because these distributional restrictions can be formulated for any process with well-defined conditional distributions.

Zellner further points out that economic theory can play a role in providing overidentifying restrictions that allow directions of causality to be imposed. Sims (1979) objects to this last suggestion on the grounds that a test for causality based on overidentifying restrictions is always a joint test of the correctness of such restrictions and the hypothesis of interest and is thus never conclusive. On the other hand, strict exogeneity and thus Granger non-causality, as pointed out by Sims (1977, p. 30, 33) provides overidentifying restrictions that can be tested for. The scepticism about untestable identifying restrictions is forcefully expressed in Sims (1980).

The role of economic theory in identifying parameters of interest in empirical studies remains one of the most controversial issues in econometrics and empirical economics to date. The debate over the correct definition of causality hinges on what individual researchers are prepared to assume a priori, be it restrictions on the temporal direction of cause and effect or fundamental structures that govern the interaction between economic variables. Granger (1980) does not dispute the potential usefulness of a priori theoretical restrictions in identifying causal relationships but emphasizes the potential for misleading inference should these restrictions turn out to be incorrect.

The problem of specifying the correct information set is recognized by Granger (1969), where it is suggested to restrict the set of all available information to the set of relevant information. Granger (1980) discusses a number of examples that illustrate the sensitivity of a causal relationship between two variables to additional information in the conditioning set. This problem is also mentioned by Holland (1986). It seems, however, that, even though specification issues are of great importance in applied work, this is not a fundamental limitation of the causality concept put forward by Granger. Moreover, correct specification of relevant conditioning variables is a common problem in most statistical procedures applied to economic data and thus not specific to procedures testing for causality. At the same time, the argument in favour of guidance from economic theory when designing such procedures is probably strongest when it comes to selecting the relevant variables that need to be included in the analysis, a point elaborated in more detail below.

Further problems of interpretation are discussed in Granger (1980). Simultaneity occurs if x_t causes y_{t+1} and y_t causes x_{t+1}. In a bivariate system of equations this form of feedback, as Granger (1969) defines it, typically leads to other inferential problems as discussed in Sims (1972). In particular, the lack of exogeneity in this case invalidates conventional regression methods and complicates the interpretation of reduced form parameters such as in VARs. Furthermore, Granger causality is not a transitive relationship: if x_t causes y_{t+1} and y_t causes z_{t+1} then x_t does not necessarily cause z_{t+1}. Granger (1980, p. 339) gives the following example. Assume that ε_t and η_t are mutually independent i.i.d. sequences and that $x_t = \varepsilon_t$, $y_t = \varepsilon_{t-1} + \eta_t$ and $z_t = \eta_{t-1}$. Then, because $y_t = x_{t-1} + \eta_t$ it is clear that x_t causes y_{t+1}. In the same way, y_t causes z_{t+1} but x_t does not cause z_{t+1} if the conditioning set contains only $X^t_{-\infty}$ and $Z^t_{-\infty}$, which is the typical assumption in bivariate statements of causality. At the same time, x_t does cause z_{t+1} in this example when the information set is enlarged to$(X^t_{-\infty}, Y^t_{-\infty}, Z^t_{-\infty})$ because now the innovation η_{t-1}can be recovered from $\eta_{t-1} = y_{t-1} - x_{t-2}$. This example shows that the concept of Granger–Sims causality can be sensitive to the specification of conditioning information.

3. Equivalence and non-equivalence of Granger and Sims causality

Since the original contributions of Granger (1963; 1969) and Sims (1972), which were mostly cast in terms of forecast error criteria, there has been a sizeable literature concerned with extensions of the basic definition and establishing a number of equivalence relationships. While the conditional independence formulation of Granger causality goes back at least to Granger (1980, p. 330), a formal analysis of the equivalence with a corresponding definition of Sims causality was first obtained by Chamberlain (1982) and Florens and Mouchart (1982). It turns out that the condition for Granger non-causality, which in its more general form can be stated as $y_{t+1} \perp X^t_{-\infty} | Y^t_{-\infty}, \forall t$, does imply the generalized form of Sims non-causality formulated as $Y^{\infty}_{t+1} \perp x_t | Y^t_{-\infty}, \forall t$, but the reverse implication does not hold generally. Florens and Mouchart (1982) give a counter-example of a nonlinear process where Sims non-causality holds but Granger non-causality does not hold. As Florens and Mouchart (1982) point out, the two conditional independence relationships are equivalent for Gaussian processes where lack of covariance is equivalent to independence. Chamberlain (1982) shows, on the other hand, that a generalized form of Sims non-causality, stated as $Y^{\infty}_{t+1} \perp x_t | Y^t_{-\infty}, X^{t-1}_{-\infty}$, is equivalent to $y_{t+1} \perp X^t_{-\infty} | Y^t_{-\infty}$ under a mild regularity condition limiting temporal dependence. Florens and Mouchart (1982) obtain a very similar result for slightly different definitions of the conditioning sets. General statements of this result can also be found in Bouissou, Laffont and Voung (1986). These authors define additional causality relationships: global non-causality (C) is defined as $Y^{\infty}_{t+1} \perp X^t_{-\infty} | Y^t_{-\infty}, \forall t$, Granger non-causality of order k (G_k) is defined as $Y^{t+k}_{t+1} \perp X^t_{-\infty} | Y^t_{-\infty}, \forall t$ and Sims non-causality of order k (S_k) is defined as $Y^{\infty}_{t+1} \perp X^t_{t-k+1} | Y^t_{-\infty}, X_{t-k}, \forall t$ where X_{t-k} is any subset of $X^{t-k}_{-\infty}$. It is then shown that (G_k), (S_k) and (C) are all equivalent for all k, a result that is also stated in Florens and Mouchart (1982, p. 580). Pierce and Haugh (1977) propose an alternative definition of Granger causality in the context of linear processes. If $H_{\{Y^t_0, U\}}(y_{t+1})$ is the linear projection of y_{t+1} on $\{Y^t_0, U\}$ where $\{Y^t_0, U\}$ is the closed linear span of all the variables generating Y^t_0 and the initial conditions U, then x_t does not Granger cause y_{t+1} if the innovations $y_{t+1} - H_{\{Y^t_0, U\}}(y_{t+1})$ and $x_{p+1} - H_{\{X^p_0, U\}}(x_{p+1})$ are uncorrelated for all $p \leq t$. Florens and Mouchart (1985) show that this definition is equivalent to covariance-based definitions of Granger and Sims causality under some additional regularity conditions. Generally speaking, the results of this early literature show that Granger causality between two processes x and y is equivalent to appropriate definitions of Sims causality not only in a mean squared prediction error sense but more generally in terms of restrictions on appropriate conditional distributions of the joint process. It should also be noted that these equivalence results continue to hold when both x and y are vector valued processes.

The situation changes, however, quite markedly when an additional set of covariates z is added to the analysis. In this situation Granger non-causality defined as $y_{t+1} \perp X^t_{-\infty} | Y^t_{-\infty}, Z^t_{-\infty}, \forall t$ in general is not equivalent to Sims non-causality defined as $Y^{\infty}_{t+1} \perp x_t | Y^t_{-\infty}, X^{t-1}_{\infty}, Z^t_{-\infty}, \forall t$. This result seems to have been first obtained by Dufour and Tessier (1993) for the linear case and also appears in Florens (2003) and Angrist and Kuersteiner (2004) and in the biostatistics literature in Robins, Greenland and Hu (1999). Simple examples can be constructed where x Granger-causes y but does not Sims cause y as well as cases where x Sims causes y but does not Granger cause y. In related work Lütkepohl (1993) and Dufour and Renault (1998) show that, in general, (G_1) does not imply (C) if the information set contains z.

To illustrate the result of Dufour and Tessier (1993), assume that $\chi_t = (y_t, x_t, z_t)$ and that χ_t can be represented as linear functions of present and past structural innovations e_t. To simplify the exposition assume that for $C(L) = A(L)^{-1}$ where $A(L)$ is a matrix of lag polynomials of finite order and L is the lag operator, it holds that $\chi_t = C(L)e_t$. Also assume that the diagonal blocks of $C(0)$, partitioned according to (y_t, x_t, z_t), are full rank. The reduced form VAR representation of χ_t then is $\pi(L)\chi_t = u_t$ with $u_t = C(0)e_t$ and $\pi(L) = C(0)A(L)$. As was discussed before, Sims non-causality imposes zero restrictions on off-diagonal blocks of $C(L)$ while Granger non-causality imposes zero restrictions on corresponding off-diagonal blocks of $\pi(L)$. Now note that when χ_t contains only y_t and x_t, the partitioned inverse formula implies that off-diagonal blocks of $A(L)$ are zero if and only if corresponding blocks of $C(L)$ are zero. Because the latter can hold only if corresponding blocks of $C(0)$ are zero, it follows that $\pi(L)$ has zero off-diagonal blocks if and only if corresponding blocks of $C(L)$ are zero. This is the result of Sims (1972). On the other hand, when $\chi_t = (y_t, x_t, z_t)$ the partitioned inverse formula for matrices partitioned into three blocks shows that $C(L)$ having off-diagonal blocks no longer implies that corresponding blocks of $A(L)$ are necessarily zero as well. Thus the equivalence between Sims and Granger causality no longer holds when additional time varying covariates are included in the analysis. In Section 4, applications in monetary economics are discussed where this situation arises naturally.

4. The connection between Granger–Sims causality and potential outcomes

The notion of causality that has become standard in micro-econometrics is based on Rubin's (1974) concept of potential outcomes, which at its core uses experimental variation to identify causal relationships. The potential outcomes model has been extended to and applied in observational studies. Observational studies are situations where no experimental assignments of actions were used. Examples are medical trials where experiments might be unethical or many economic policy questions where experiments may be unethical or too expensive to carry out. The importance of experimental evidence was recognized in econometrics dating back to Haavelmo (1944). Wold (1954) similarly discusses controlled experiments as a way to uncover causal relationships. Orcutt's (1952, p. 305) notion of causality is closely related to the idea of potential outcomes and is defined in terms of consequences of actions:

> Thus when we say that *A* is the cause of *B*, we often mean that if *A* varies, then *B* will be different in a specified way from what it *would have been* if *A* had not varied.

Orcutt (1952, p. 309) goes on to discuss policy actions as a substitute for unavailable experimental evidence to identify causal relationships in observational data, an idea explored in more detail below.

For expositional purposes, assume that a certain action or treatment D can be either given or not given to individual i. The causal question in this context is whether the treatment has an effect on an outcome variable of interest measured by y. It is convenient to define $D_i = 1$ if the treatment is given and $D_i = 0$ if the treatment is not given. The potential outcome $y_i(0)$ is defined as the outcome for individual i that would have occurred if the treatment had not been given and $y_i(1)$ as the outcome that would have occurred in the case the treatment had been given. The absence of causality of D for y then is defined as the situation where $y_i(0) = y_i(1)$. This condition is referred to as the 'strong null hypothesis of no causal effect'. Usually this condition cannot be directly tested because $y_i(0)$ and $y_i(1)$ are not both observed for the same individual. Instead, the observed measurement takes the form

$$y_i = D_i y_i(1) + (1 - D_i) y_i(0) \qquad \text{(PO)}$$

Potential outcomes may depend on a list of covariates z_i. Covariates capture characteristics of the outcome variable that are not directly related to the experiment but that need to be taken into account when assessing the outcome. An identification condition is needed to proceed to testable restrictions. Formally one imposes the condition $y_i(0), y_i(1) \perp D_i | z_i$, which is sometimes referred to as selection on observables. This condition is automatically satisfied if D_i is randomly assigned in an experiment. In observational studies the condition essentially states that actions by individuals or policymakers cannot be based on unobservable information. The 'selection on observables' condition implies that

$$P(y_i(j) \in A | z_i) = P(y_i(j) \in A | D_i, z_i)$$

for all j and the null hypothesis implies that

$$P(y_i \in A | D_i, z_i) = P(y_i \in A | z_i)$$

which is identical to Granger's condition of no causal effect, a result that is discussed in Holland (1986). The power of the identifying restriction lies in the fact that it is formulated independently of the null hypothesis of no causal relationship. To be more precise, the identifying restriction imposes conditional independence of D from $y_i(0)$ and $y_i(1)$ but not from y_i. The latter holds only when the null is true. This is an important difference from Sims (1977, p. 30), who writes that 'Causality is an important identifying restriction on dynamic behavioral relations'. In other words, Sims imposes the null to identify certain structural models. Another important difference between this model of causality and the less specific definition of Granger–Sims causality is that the form of the causal link between D and y is of a simple functional form specified in (PO). This particular structure is what allows the interpretation of measured correlations as causal links.

Identification conditions thus lead to testable implications of Rubin's potential outcome framework that are identical to the Granger–Sims definition of non-causality. Nevertheless, causality in Rubin's context is closely related to experiments and counterfactuals: causal effects of a treatment are measured by comparing unobservable counterfactuals under treatment and non-treatment. On the other hand, Granger–Sims causality does not rely on the notion of treatment. It has been applied to studying such phenomena as the temporal link between interest rates and inflation, variables that are endogenously determined and where it is hard to imagine that an experiment or even a policy intervention is available for causal inference. Orcutt's idea of using policy variation can, however, still be implemented if instead of market interest rates one focuses (in the United States) on the federal funds target rate, a variable that is directly set by the policymaker. Under the additional assumption that all systematic aspects of the policy depend on observable information, it is possible to generate pseudo-experimental variation even in the interest rate example. These ideas are now explored in more detail.

The potential outcomes framework in its original form is in many ways too limited to be directly applicable to macroeconomic questions of causality where the Granger–Sims concept of causality has been mostly applied. The two main limitations of the potential outcomes approach are that it does not allow for dynamic treatments or policies and that usually the stable unit treatment assumption of Rubin (1980) is imposed. The latter rules out general equilibrium effects of treatments and is not satisfied in a macroeconomic context. Angrist and Kuersteiner (2004) propose an extension of the potential outcomes framework that overcomes these limitations (also related is the work of White, 2006). Consider an economy that is described by $\chi_t = (y_t, D_t, z_t)$ where y is a vector of outcome variables, D is a vector of policy variables and z is a vector of relevant covariates not already included in y. Potential outcomes $y_{t,j}(d)$ are defined as values the outcome variable $y_{+,j}$ would have taken if at time t the policy had been set to $D_t = d$. It is probably useful to discuss the nature of the potential outcome $y_{t,j}(d)$ in a context where one has a dynamic general equilibrium model describing the evolution of the process χ_t as a system of stochastic difference equations. Then $y_{t,j}(d)$ has to be thought of as a specific solution of that model indexed against a specific decision rule d of some policymaker. It is helpful, but not necessary, to assume that the model has a strong solution, in the sense of stochastic process theory, such that the strong null of no causal effect can be represented as the restriction $y_{t,j}(d) = y_{t,j}(d')$ for all possible values of d. It should be clear from this description that $y_{t,j}(d)$ is a possibly highly complex function of all the inputs that go into the model, including policy decisions taken at times different from t. Solving for $y_{t,j}(d)$ explicitly is not necessary for the definition of a causal link between D and y, a feature that is very much in the spirit of Granger and Sims. All that is needed is an identifying restriction that allows us to interpret observed correlations as causal links. A sufficient condition is

$$y_{t,j}(d) \perp D_t | Y^t_{-\infty}, D^{t-1}_{-\infty}, Z^t_{-\infty}, \forall t, \ \forall j > 0. \tag{ID}$$

Under the sharp null of no causal effect where $y_{t,j}(d) = y_{t+j}$ it then follows immediately that $y_{t+j} \perp D_t | Y^t_{-\infty}, D^{t-1}_{-\infty}, Z^t_{-\infty}$. This is the same as the condition for Sims non-causality. As discussed earlier, it is generally not equivalent to Granger non-causality. The form of the testable restriction depends critically on the form of (ID). At least in cases where D is a decision variable of a policymaker, this restriction leading to Sims causality seems plausible. This can be seen easily in the context of linear models where the identification assumption leading to Sims non-causality is identical to the restriction that policy innovations are independent of all future innovations affecting the outcome variables.

In order to better understand the nature of the identifying assumption (ID) it is useful to consider a specific example. The notion of Granger causality was applied to the question of a causal link between monetary policy and real economic activity, starting with Sims (1972), and thus has a long tradition in the empirical macro literature. Most of the early empirical literature has investigated this question using linear regressions of some measure of monetary aggregates or interest rates and various measures of real economic activity. In an important methodological contribution, Romer and Romer (1989) use information from the minutes of the Federal Open Markets Committee to classify US Federal Reserve policy into times of purely anti-inflationary monetary tightening and other periods. Times of tight monetary policy are called Romer dates. The idea then is to measure average economic activity following Romer dates and to compare these measurements to average economic activity at other times. While the argument that Romer dates are exogenous has been criticized (see for example Hoover and Perez, 1994; Shapiro, 1994; Leeper, 1997), the basic premise of the approach of Romer and Romer (1989) remains valid. It is to use a behavioural theory or policy rule for a policymaker to construct policy innovations which serve as exogenous variation that can be used to evaluate the effectiveness of the policy in question (Jorda, 2005, emphasizes the importance of exogenous variation to identify causal relationships). Angrist and Kuersteiner (2004) analyse the consequences of allowing for additional covariates z in the policy model to capture information about nominal macroeconomic variables such as the inflation rate. These variables are clearly relevant for policy decisions of the Federal Reserve and thus constitute relevant conditioning information in the sense of Granger (1969). At the same time these nominal variables are not part of the null hypothesis of no causal effects on real economic activity and thus cannot be subsumed into the y-process. As discussed earlier, under these circumstances reduced form regressions based on Granger's notion of non-causality cannot be used to test for Sims non-causality.

Generally speaking, a model of the policymaker, in this case the Federal Reserve, is a conditional probability distribution $P(D_t \in A|Y^t_{-\infty}, D^{t-1}_{-\infty}, Z^t_{-\infty})$. An example of a policy model for the Federal Reserve in the context of the Romer and Romer data was developed by Shapiro (1994). The fundamental identifying assumption then is that this model is correct, especially in the sense that the conditional probability of D_t does not depend on $y_{t,j}(d)$. This condition will be satisfied when two criteria are met. All relevant information that the policymaker used to decide on the policy D_t is included in the model and the problem at hand is of a nature where the policymaker does not foresee the future. This is the way Granger's fundamental axiom of 'the arrow of time' plays a role to provide identifying assumptions in this setting. If these conditions are met then all random deviations of the policy D_t from what is predicted by the model are conditionally independent of $y_{t,j}(d)$. Random deviations could be due to the variation over time in policymakers' beliefs about the workings of the economy, decision-makers' tastes and goals, political factors, and the temporary pursuit of objectives other than changes in the outcomes of interest (for example, monetary policy that targets exchange rates instead of inflation or unemployment), and finally harder-to-quantify factors such as the mood and character of decision-makers. It is then precisely these random deviations from prescribed policies that help to identify causal links. In other words, it is not the systematic or predictable policy changes that are helpful to answer causal questions but the deviations from prescribed rules. The reason for this is that the causal model used here does not provide enough structure to disentangle causal links from endogenously varying policy variables. The situation is quite similar to the analysis of impulse response functions in structural VAR models where identification is driven by the independence of structural innovations. Impulse response analysis can thus be viewed as a special case of the potential outcomes model when χ_t is a linear process.

The potential outcome framework has the advantage that it focuses on exogenous variation and puts the identification discussion at the centre of causal inference. It helps to clarify the source of identifying variation in an analysis of Granger–Sims causality. A priori arguments for the identifying exogeneity restrictions can be based on institutional settings such as the introduction of new legislation, on procedural details as in the Romer and Romer (1989) example, or on behavioural models derived from economic theory. At the same time the potential outcome approach to identification is limited in the sense that its most natural areas of application lie in the analysis of policy effectiveness. It is less suited to analyse causality between variables that are jointly determined in equilibrium.

A point made by Granger (1980) is also relevant here. The analysis of causality is not necessarily relevant for the analysis of controlled processes. To illustrate the issue, consider the linearized Lucas model of McCallum (1984), where in an overlapping generations framework price setting happens in isolated markets based on local information. McCallum shows that random innovations to the money stock affect unemployment because agents cannot completely distinguish between real price changes in their markets and price changes due to variation in the supply of money. A test of conditional independence between money and employment for data generated by this model would find evidence of a causal relationship in the sense of Sims. At the same time, any attempt by the monetary authority to systematically exploit this relationship through a systematic policy rule would fail in this model because agents fully incorporate predictable actions of the policymaker and do not respond to nominal changes in prices. This example shows that a statistical definition of causality may indicate the existence of a causal relationship that does not lend itself to policy intervention and control. Whether individual researchers are willing to call such a finding a causal relationship hinges upon their notion of causality and is likely to be controversial.

The situation is reversed in some models where the monetary authority can fully control output through appropriate policy rules. For the purpose of illustration consider the model by Rudebusch and Svensson, which 'consists of an aggregate supply equation that relates inflation to an output gap and an aggregate demand equation that relates output to a short term interest rate' (1999, p. 205). Monetary policy is conducted by setting the nominal interest rate, and affects output and inflation with a one period lag. In this model it is possible for the monetary authority to fully stabilize output such that deviations from a fixed steady state level are serially uncorrelated. On the assumption that the policy rule is augmented with an independently distributed policy innovation (this assumption is necessary for statistical identification of test procedures, see Sims, 1977, p. 39), it follows that a test for Granger causality will not be able to reject the null of no causal relationship running from interest rates to output. At the same time, a test for Granger non-causality of output for interest rates will be rejected because the interest rate setting rule depends on past output. In this example, the direction of causality in the statistical sense of Granger goes in the opposite direction of what the model indicates.

On the other hand, a test of the conditional independence restriction (S) for Sims non-causality of interest rates for output will be rejected, thus revealing the influence of the policymaker on output. Sims (1977, p. 36) considers a similar reversal of the direction of Granger causality in models where a policy variable is controlled. While Sims considers a bivariate model where both Granger and Sims causality are equivalent, the model of Rudebusch and Svensson considered here has three equations, which explains why the concepts of Granger and Sims causality do not lead to the same conclusions.

A test of (S) in this model is thus able to identify the direction of causality even when variables are controlled, at least when the test is based on the assumption that the policy model is correctly specified and the policy innovation is thus identified. However, even in this case, the measured causal effects are those of random deviations from the policy rule. As discussed earlier, attempts to exploit these effects with systematic policy actions may not be feasible due to the reactions of rationally forward-looking agents.

A related issue is the problem of analysing causal effects of systematic changes in the policy rule, a problem discussed in Sims (1977, p. 30). Without additional structure such questions seem to be hard to address, and it remains an open question to what extent evidence gained from causal inference based on notions of Granger–Sims causality can be used to investigate them.

5. Summary

This article explores the notion of Granger–Sims causality as a concept of statistical predictability. The definition is appealing because it does not require a priori theoretical restrictions but rather is formulated in terms of a directly testable implication on the distribution of observed data. The simplicity of this approach to causality has led to extensive applications in areas such as macro-econometrics, where notions of causality that rely on the possibility of carrying out experiments are difficult to apply. Difficulties with a purely statistical concept of causality, however, arise when it comes to interpreting the nature of detected causal relationships. Without additional assumptions regarding the exogeneity of one or more input variables, it seems difficult to link the statistical causality concept with the more fundamental distinction between cause and effect. The latter distinction is fundamental to the analysis of controllability of outcome variables and thus central to many questions in the social sciences. As discussed above, there is clearly a distinction between a causal link between two variables and the possibility of controlling an output by manipulating certain inputs. Equilibrium effects which are at the core of economic analysis may, for example, pre-empt policy changes through the agent's rational anticipation of just these policy changes. Perceived causal relationships thus may not be exploitable for policy purposes even if they can be reliably identified in the history of an economy. The analysis of causality and controllability in dynamic equilibrium models thus remains a central topic of research.

G.M. KUERSTEINER

See also **continuous and discrete time models; structural vector autoregressions.**

Bibliography

Angrist, J. and Kuersteiner, G. 2004. Semiparametric causality tests using the policy propensity score. Working Paper No. 10975. Cambridge, MA: NBER.

Basmann, R. 1963. The causal interpretation of non-triangular systems of economic relations. *Econometrica* 31, 439–48.

Bouissou, M., Laffont, J.-J. and Voung, Q. 1986. Tests of noncausality under Markov assumptions for qualitative panel data. *Econometrica* 54, 395–414.

Chamberlain, G. 1982. The general equivalence of Granger and Sims causality. *Econometrica* 50, 569–81.

Dawid, A. 1979. Conditional independence in statistical theory. *Journal of the Royal Statistical Society*, Series B 41, 1–31.

Dufour, J.-M. and Renault, E. 1998. Short run and long run causality in time series: theory. *Econometrica* 66, 1099–125.

Dufour, J.-M. and Tessier, D. 1993. On the relationship between impulse response analysis, innovation accounting and Granger causality. *Economics Letters* 42, 327–33.

Engle, R., Hendry, D. and Richard, J. 1983. Exogeneity. *Econometrica* 51, 277–304.

Feigl, H. 1953. Notes on causality. In *Readings in the Philosophy of Science*, ed. H. Feigl and M. Brodbeck. New York: Appleton-Century-Crofts, Inc.

Florens, J.-P. 2003. Some technical issues in defining causality. *Journal of Econometrics* 112, 127–8.

Florens, J.-P. and Mouchart, M. 1982. A note on noncausality. *Econometrica* 50, 583–91.

Florens, J.-P. and Mouchart, M. 1985. A linear theory for noncausality. *Econometrica* 53, 157–76.

Granger, C. 1963. Economic processes involving feedback. *Information and Control* 6, 28–48.

Granger, C. 1969. Investigating causal relations by econometric models and cross-spectral methods. *Econometrica* 37, 424–38.

Granger, C. 1980. Tests for causation – a personal viewpoint. *Journal of Economic Dynamics and Control* 2, 329–52.

Haavelmo, T. 1944. The probability approach in econometrics. *Econometrica* 12 (supplement), iii–vi, 1–115.

Holland, P. 1986. Statistics and causal inference. *Journal of the American Statistical Association* 81, 945–60.

Hoover, K. 2001. *Causality in Macroeconomics*. Cambridge: Cambridge University Press.

Hoover, K. and Perez, S. 1994. Post hoc ergo propter once more: an evaluation of 'does monetary policy matter?' in the spirit of James Tobin. *Journal of Monetary Economics* 34, 47–73.

Hosoya, Y. 1977. On the Granger condition for non-causality. *Econometrica* 45, 1735–6.

Jorda, O. 2005. Estimation and inference of impulse responses by local projections. *American Economic Review* 95, 162–82.

Leamer, E. 1985. Vector autoregressions for causal inference? *Carnegie-Rochester Conference Series on Public Policy* 22, 255–303.

Leeper, E. 1997. Narrative and VAR approaches to monetary policy: common identification problems. *Journal of Monetary Economics* 40, 641–57.

Lütkepohl, H. 1993. Testing for causation between two variables in higher dimensional VAR models. In *Studies in Applied Econometrics*, ed. H. Schneeweiss and K. Zimmerman. Heidelberg: Springer-Verlag.

McCallum, B. 1984. A linearized version of Lucas's neutrality model. *Canadian Journal of Economics* 17, 138–45.

Orcutt, G. 1952. Actions, consequences, and causal relations. *Review of Economics and Statistics* 34, 305–13.

Pearl, J. 2000. *Causality*. Cambridge: Cambridge University Press.

Pierce, D. and Haugh, L. 1977. Causality in temporal systems. *Journal of Econometrics* 5, 265–93.

Robins, J., Greenland, S. and Hu, F. 1999. Estimation of the causal effect of a time-varying exposure on the marginal mean of a repeated binary outcome. *Journal of the American Statistical Association* 94, 687–700.

Romer, C. and Romer, D. 1989. Does monetary policy matter? A new Test in the spirit of Friedman and Schwartz. In *NBER Macroeconomics Annual 1989*, ed. O. Blanchard and S. Fischer. Cambridge, MA: MIT Press.

Rubin, D. 1974. Estimating causal effects of treatments in randomized and nonrandomized studies, *Journal of Educational Psychology* 66, 688–701.

Rubin, D. 1980. Randomization analysis of experimental data: the Fisher randomization test. Comment. *Journal of the American Statistical Association* 75, 591–93.

Rudebusch, G. and Svensson, L. 1999. Policy rules for inflation targeting. In *Monetary Policy Rules*, ed. J. Taylor. Chicago: University of Chicago Press.

Shapiro, M. 1994. Federal Reserve policy: cause and effect. In *Monetary Policy*, ed. G. Mankiew. Chicago: University of Chicago Press.

Simon, H. 1953. Causal ordering and identifiability. In *Studies in Econometric Method*, ed. W. Hood and T. Koopmans. Cowles Commission Monograph No. 14. New York: John Wiley.

Sims, C. 1972. Money, income and causality. *American Economic Review* 62, 540–52.

Sims, C. 1977. Exogeneity and causal ordering in macroeconomic models. In *New Methods in Business Cycle Research: Proceedings of a Conference*, ed. C. Sims. Minneapolis: Federal Reserve Bank of Minneapolis.

Sims, C. 1979. A comment on the papers by Zellner and Schwert. *Carnegie-Rochester Conference Series on Public Policy* 10, 103–8.

Sims, C. 1980. Macroeconomics and reality. *Econometrica* 48, 1–48.

Sims, C. 1986. Are forecasting models usable for policy analysis? *Federal Reserve Bank of Minneapolis Quarterly Review* 10(1), 2–16.

Strotz, R. and Wold, H. 1960. Recursive versus nonrecursive systems: an attempt at synthesis. *Econometrica* 28, 417–27.

Suppes, P. 1970. *A Probabilistic Theory of Causality.* Amsterdam: North-Holland.

White, H. 2006. Time-series estimation of the effects of natural experiments. *Journal of Econometrics* 135, 527–66.

Wiener, N. 1956. The theory of prediction. In *Modern Mathematics for the Engineer, Series 1*, ed. E. Beckenback. New York: McGraw-Hill.

Wold, H. 1954. Causality and econometrics. *Econometrica* 22, 162–77.

Zellner, A. 1979. Causality and econometrics, policy and policy making. *Carnegie-Rochester Conference Series on Public Policy* 10, 9–54.

heteroskedasticity and autocorrelation corrections

Heteroskedasticity and autocorrelation consistent (HAC) covariance matrix estimation refers to calculation of covariance matrices that account for conditional heteroskedasticity of regression disturbances and serial correlation of cross products of instruments and regression disturbances. The heteroskedasticity and serial correlation may be of unknown form. HAC estimation is integral to empirical research using generalized method of moments (GMM) estimation (Hansen, 1982). In this article I summarize results relating to HAC estimation, with emphasis on practical rather than theoretical aspects.

The central issue is consistent and efficient estimation of what is called a 'long-run variance', subject to the constraint that the estimator is positive semidefinite in finite samples. Positive semidefiniteness is desirable since the estimator will be used to compute standard errors and test statistics. To fix notation, let h_t be a $q \times 1$ stationary mean zero random vector. Let Γ_j denote the $q \times q$ autocovariance of h_t at lag q, $\Gamma_j \equiv Eh_t h_{t-j}'$; of course, $\Gamma_j = \Gamma_{-j}'$. The long run variance of h_t is the $q \times q$ matrix

$$S = \sum\nolimits_{j=-\infty}^{\infty} \Gamma_j = \Gamma_0 + \sum\nolimits_{j=1}^{\infty} (\Gamma_j + \Gamma_j{}'). \tag{1}$$

Apart from a factor of 2π, the symmetric matrix S, which I assume to be positive definite, is the spectral density of h_t at frequency zero. As discussed below, techniques for spectral density estimation are central to HAC estimation. (For an arbitrary stationary process, the sum in the right-hand side of (1) may not converge, and may not be positive definite even if it does converge. But here and throughout I assume unstated regularity conditions. As well, I use formulas that allow for relatively simple notation, for example assuming covariance stationarity even when that assumption can be relaxed. The cited papers may be referenced for generalizations and for technical conditions.)

To illustrate how estimation of S figures into covariance matrix estimation, consider the following simple example. As in Hansen and Hodrick (1980), let us suppose that we wish to test the 'rationality' of a scalar variable x_t as an n period ahead predictor of a variable y_{t+n+1}, for $n \geq 0$: the null hypothesis is $E_t y_{t+n+1} = x_t$, where E_t denotes expectations conditional on the information set used by market participants. The variable x_t might be the expectation of y_{t+n+1} reported by a survey, or it might be a market determined forward rate. Let u_t denote the expectational error: $u_t = y_{t+n+1} - E_t y_{t+n+1} = y_{t+n+1} - x_t$. (The expectational error u_t, which is not realized until period $t + n + 1$, is dated t to simplify notation.)

One can test one implication of the hypothesis that x_t is the expectation of y_{t+n+1} by regressing y_{t+n+1} on a constant and x_t, and checking whether the coefficient on the constant term is zero and that on x_t is 1:

$$y_{t+n+1} = \beta_0 + \beta_1 x_t + u_t \equiv X_t{}'\beta + u_t; \quad H_0 : \beta = (0, 1)'. \tag{2}$$

Under the null, $EX_tu_t = 0$, so least squares is a consistent estimator. As well, X_tu_t follows a moving average process of order n. Thus the asymptotic variance of the least squares estimator of β is $(EX_tX_t')^{-1}S(EX_tX_t')^{-1}$, where $S = \Gamma_0 + \sum_{j=1}^{n}(\Gamma_j + \Gamma_j')$, $\Gamma_j \equiv EX_tu_t(X_{t-j}u_{t-j})'$. This example maps into the notation used in (1) with $h_t = X_tu_t$, $q = 2$ and a known upper bound to the number of non-zero autocovariances of h_t. Clearly one needs to estimate EX_tX_t' and S to conduct inference. A sample average of X_tX_t' can be used to estimate EX_tX_t'. If $n = 0$, so that h_t is serially uncorrelated, $S = EX_tu_t(X_tu_t)'$ and estimation of S is equally straightforward; White's (1980) heteroskedasticity consistent estimator can be used. The subject at hand considers ways to estimate S when h_t is serially correlated. I note in passing that one cannot sidestep estimation of $S$by applying generalized least squares. In this example and more generally, generalized least squares is inconsistent. See Hansen and West (2002).

To discuss estimation of S, let us describe a more general set-up. In GMM estimation, h_t is a $q \times 1$ orthogonality condition used to identify a k-dimensional parameter vector β. The orthogonality condition takes the form

$$h_t = Z_tu_t \tag{3}$$

for a $q \times \ell$ matrix of instruments Z_t and an $\ell \times 1$ vector of unobservable regression disturbances u_t. The vector of regression disturbances depends on observable data through β, $u_t = u_t(\beta)$. In the example just given, $q = 2$, $\ell = 1$, $Z_t = X_t$, $u_t(\beta) = y_{t+n+1} - X_t'\beta$. The example just given is overly simple in that the list of instruments typically will not be identical to right-hand side variables, and the model may be nonlinear. For a suitable $k \times q$ matrix D, the asymptotic variance of the GMM estimator of β takes the form DSD' (for example, $D = (EX_tX_t')^{-1}$ in the example just given). In an overidentified model (that is, in models in which the dimension of the orthogonality condition q is greater than the number of parameters k) the form D takes depends on a certain weighting matrix. Let $h_{t\beta}$ be the $q \times k$ matrix $\partial h_t/\partial\beta$. When the weighting matrix is chosen optimally, $D = (Eh_{t\beta}'S^{-1}Eh_{t\beta})^{-1}Eh_{t\beta}'S^{-1}$ and the asymptotic variance DSD' simplifies to $(Eh_{t\beta}'S^{-1}Eh_{t\beta})^{-1}$. The optimal weighting matrix is one that converges in probability to S, and thus the results about to be presented are relevant to efficient estimation as well as to hypothesis testing. In any event, the matrix $Eh_{t\beta}$ typically is straightforward to estimate; the question is how to estimate S. This will be the focus of the remainder of the discussion.

We have sample of size $T$and sample counterparts to u_t and h_t, call them $\hat{u}_t = u_t(\hat{\beta})$ and $\hat{h}_t = h_t(\hat{\beta})$. Here, $\hat{\beta}$ is a consistent estimate of β. In the least squares example given above, $\hat{u}_t$ is the least squares residual, $\hat{u}_t = y_{t+n+1} - X_t'\hat{\beta}$, and $\hat{h}_t = X_t\hat{u}_t = X_t(y_{t+n-1} - X_t'\hat{\beta})$. One path to consistent estimation of S involves consistent estimation of the autocovariances of h_t. The natural estimator is a sample average,

$$\hat{\Gamma}_j = T^{-1}\sum\nolimits_{t=j+1}^{T} \hat{h}_t \hat{h}_{t-j}' \quad \text{for} \quad j \geq 0. \tag{4}$$

For given j, (4) is a consistent $(T \rightarrow \infty)$ estimator of Γ_j.

I now discuss in turn several possible estimators, or classes of estimators, of S: (1) the truncated estimator; (2) estimators applicable only when h_t follows a moving average (MA) process of known order; (3) an autoregressive spectral estimator; (4) estimators that smooth autocovariances; (5) some recent work, on estimators that might be described as extensions or modifications of ones the estimators described in (4).

1. The truncated estimator

Suppose first that it is known a priori that the autocovariances of h_t are zero after lag n, as is the case in the empirical example above. A natural estimator of S is one that replaces population objects in (1) with sample analogues. This is the *truncated* estimator:

$$\hat{S}_{TR} = \hat{\Gamma}_0 + \sum_{j=1}^{n}(\hat{\Gamma}_j + \hat{\Gamma}_j'). \tag{5}$$

In the more general case in which $\Gamma_j \neq 0$ for all j, the truncated estimator is consistent if the truncation point $n \rightarrow \infty$ at a suitable rate. Depending on exact technical conditions, the rate may be $n/T^{1/2} \rightarrow 0$ or $n/T^{1/4} \rightarrow 0$ (Newey and West, 1987). The truncated estimator need not, however, yield a positive semidefinite estimate. With certain plausible data generating processes, simulations indicate that it will not be p.s.d. in a large fraction of samples (West, 1997). Hence this estimator is not used much in practice.

2. Estimators applicable only when h_t follows an MA process of known order n

Such a process for h_t holds in studies of rationality (as illustrated above) and in the first order conditions from many rational expectations models (for example, Hansen and Singleton, 1982).

Write the Wold representation of h_t as $h_t = e_t + \Theta_1 e_{t-1} + \cdots + \Theta_n e_{t-n}$. Here, e_t is the $q \times 1$ innovation in h_t. Let Ω denote the $q \times q$ variance covariance matrix of e_t. Then it is well known (for example, Hamilton, 1994, p. 276) that

$$S = (I + \Theta_1 + \cdots + \Theta_n)\Omega(I + \Theta_1 + \cdots + \Theta_n)' \tag{6}$$

Suppose that one fits an MA(n) process to $\hat{h}_t$, and plugs the resulting estimates of the Θ_i and Ω into the formula for S. Clearly the resulting estimator is $T^{1/2}$ consistent and positive semidefinite. Nevertheless, to my knowledge this estimator has not been used, presumably because of numerical difficulties in estimating multivariate moving average processes.

Two related estimators have been proposed that impose a smaller computational burden. Hodrick (1992) and West (1997) suggest an estimator that requires fitting an MA(n) to the vector of regression residuals $\hat{u}_t$ (or, in Hodrick's, 1992, application, using MA coefficients that are known a priori). The computational burden of such MA estimation will typically be considerably less than that of MA estimation of the h_t process, because the dimension of u_t is usually much smaller than that of h_t. For example, $\hat{u}_t$ will be a scalar in a single equation application, regardless of the number of orthogonality conditions captured in h_t. Write the estimated MA process for $\hat{u}_t$ as $\hat{u}_t = \hat{\in}_t + \hat{\psi}_1\hat{\in}_{t-1} + \cdots + \hat{\psi}_n\hat{\in}_{t-n}$, where the $\hat{\psi}_j$ are $\ell \times \ell$. (Note that $\in_t$, the $\ell \times 1$ innovation in u_t, is not the same as e_t, the $q \times 1$ innovation in h_t.) Then a $T^{1/2}$ consistent and positive semidefinite estimator of S is

$$\hat{S}_{MA-\ell} = T^{-1}\sum_{t=1}^{T-n} \hat{d}_{t+n}\hat{d}_{t+n}', \; \hat{d}_{t+n} = (Z_t + Z_{t+1}\hat{\psi}_1 + \cdots + Z_{t+n}\hat{\psi}_n)\hat{\in}_t, \tag{7}$$

where, again, Z_t is the $q \times \ell$ matrix of instruments (see eq. (3)).

Eichenbaum, Hansen and Singleton (1988) and Cumby, Huizanga and Obstfeld (1983) propose a different strategy that avoids the need to estimate a moving average process for either u_t or h_t. They suggest estimating the parameters of $\hat{h}_t$'s autoregressive representation, and inverting the autoregressive weights to obtain moving average weights. Call the results $\hat{\Theta}_1, \ldots, \hat{\Theta}_n$, with $\hat{\Omega}$ the estimate of the innovation variance–covariance matrix. The resulting estimator $\hat{S} = (I + \hat{\Theta}_1 + \cdots + \hat{\Theta}_n)\hat{\Omega}(I + \hat{\Theta}_1 + \cdots + \hat{\Theta}_n)'$ is positive semidefinite by construction. The rate at which it converges to S depends on the rate at which the order of the autoregression is increased.

3. Autoregressive estimators

Den Haan and Levin (1997) propose and evaluate an autoregressive spectral estimator. Suppose that h_t follows a (possibly) infinite-order vector autoregression (VAR)

$$h_t = \sum_{j=1}^{\infty} \Phi_j h_{t-j} + e_t, \quad Ee_te_t' = \Omega. \tag{8}$$

Then (Hamilton, 1994, p. 237)

$$S = \left(I - \sum_{j=1}^{\infty}\Phi_j\right)^{-1} \Omega \left(I - \sum_{j=1}^{\infty}\Phi_j\right)^{-1'}. \tag{9}$$

The idea is to approximate this quantity via estimates from a finite-order VAR in $\hat{h}_t$. Write the estimate of a VAR in $\hat{h}_t$ of order p as

$$\hat{h}_t = \hat{\Phi}_1\hat{h}_{t-1} + \cdots + \hat{\Phi}_p\hat{h}_{t-p} + \hat{e}_t, \; \hat{\Omega} = T^{-1}\sum_{t=p+1}^{T} \hat{e}_t\hat{e}_t'. \tag{10}$$

Then the estimator of $\hat{S}$ is

$$\hat{S}_{AR} = \left(I - \sum_{j=1}^{p} \hat{\Phi}_j\right)^{-1} \hat{\Omega} \left(I - \sum_{j=1}^{p} \hat{\Phi}_j\right)^{-1'}. \tag{11}$$

Den Haan and Levin (1997, Section 3.5) conclude that if p is chosen by BIC, and some other technical conditions hold, then this estimator converges at a rate very near $T^{1/2}$ (the exact rate depends on certain characteristics of the data). A possible problem in practice with this estimator (as well as with the estimator described in the final paragraph of Section 2, which also requires estimates of a VAR in $\hat{h}_t$) is that it may require estimation of many parameters and inversion of a large matrix. Den Haan and Levin therefore suggest judiciously parametrizing the autoregressive process, for example by using the BIC criterion equation-by-equation for each of the q elements of $\hat{h}_t$.

4. Estimators that smooth autocovariances

In practice, the most widely used class of estimators is one that relies on smoothing of autocovariances. Andrews (1991), building on the literature on estimation of spectral densities, established a general framework for analysis. Andrews considers estimators that can be written

$$\hat{S} = \hat{\Gamma}_0 + \sum_{j=1}^{T-1} k_j(\hat{\Gamma}_j + \hat{\Gamma}_j') \tag{12}$$

for a series of kernel weights $\{k_j\}$ that obey certain properties. For example, to obtain a consistent estimator, we need k_j near zero (or perhaps identically zero) for values of j near $T-1$, since autocovariances at large lags are estimated imprecisely, while $k_j \rightarrow 1$ for each j is desirable for consistency. We would also like the choice of k_j to ensure positive definiteness.

The two most commonly used formulas for the kernel weights are:

$$\text{Bartlett : for some } m \geq 0 : \quad k_j = 1 - [j/(m+1)] \text{ for } j \leq m, k_j = 0 \quad \text{for } j > m. \tag{13a}$$

$$\text{Quadratic spectral (QS) : for some } m > 0, \quad \text{and with } x_j = j/m : \quad k_j = [25/12\pi^2 x_j^2] \times \{[\sin(6\pi x_j/5)/(6\pi x_j/5)] - \cos(6\pi x_j/5)\}. \tag{13b}$$

If we let $z_j = 6\pi x_j/5$, the QS formula for k_j can be written in more compact form as $(3/z_j^2)\{[\sin(z_j)/z_j] - \cos(z_j)\}$. Call the resulting estimators $\hat{S}_{BT}$ and $\hat{S}_{QS}$. For example,

$$\hat{S}_{BT} = \hat{\Gamma}_0 + \sum_{j=1}^{m} [1 - j/(m+1)](\hat{\Gamma}_j + \hat{\Gamma}_j'). \tag{14}$$

The vast literature on spectral density estimation suggests many other possible kernel weights. For conciseness, I consider only the Bartlett and QS kernels.

To operationalize these estimators, one needs to choose the lag truncation parameter or bandwidth m. I note that for both kernels, consistency requires $m \to \infty$ as $T \to \infty$, even if h_t follows an MA process of known finite order, as in the example given above. Thus one should not set m to be the number of non-zero autocovariances. Subject to possible problems with positive definiteness, setting $m = n$ is fine for the truncated estimator (5) but not for estimators that use nontrivial weights $\{k_j\}$.

Andrews shows that maximizing the rate at which $\hat{S}$ converges to S requires that m increase as a suitable function of sample size, with the 'suitable function' varying with kernel. For the Bartlett and QS, the maximal rates of convergence are realized when

$$\begin{aligned} &\text{Bartlett} : m = \gamma T^{1/3} (\text{or } m = (\text{integer part of } \gamma T^{1/3})) \\ &\text{for some } \gamma \neq 0, \quad \text{QS} : m = \gamma T^{1/5} \text{ for some } \gamma \neq 0, \end{aligned} \tag{15}$$

in which case $\hat{S}_{BT}$ converges to S at rate $T^{1/3}$ and the mean squared error in estimation of S goes to zero at rate $T^{2/3}$; the comparable figures for QS are $T^{2/5}$ and $T^{4/5}$. Since both estimators are nonparametric, they converge at rates slower than $T^{1/2}$; since faster convergence is better, the QS rate is preferable to that of the Bartlett. Indeed, Andrews (1991), drawing on Priestley (1981), shows that for a certain class of kernel weights $\{k_j\}$, the mean squared error of QS rate is optimal in the following sense: a $T^{4/5}$ rate on the asymptotic mean squared error is the fastest that can be achieved if one wants to ensure a positive definite $\hat{S}$, and within the class of kernels that achieve the $T^{4/5}$ rate, the QS has the smallest possible asymptotic mean squared error.

As a practical matter, the formulas in (15) have merely pushed the question of choice of m to one of choice of γ; putting arbitrary γ in (15) yields convergence that is as fast as possible, but different choices of γ lead to different asymptotic mean squared errors. The choice of γ that is optimal from the point of view of asymptotic mean squared error is a function of the data (Hannan, 1970, p. 286). Let $S^{(0)} = \sum_{j=-\infty}^{\infty} \Omega_j (= S)$; $S^{(1)} = \sum_{j=-\infty}^{\infty} |j| \Omega_j$; $S^{(2)} = \sum_{j=-\infty}^{\infty} j^2 \Omega_j$. For scalar $(q = 1)$ S optimal choices are:

$$\text{Bartlett} : \gamma = 1.1447[S^{(1)}/S^{(0)}]^{2/3}; \quad \text{QS} : \gamma = 1.3221[S^{(2)}/S^{(0)}]^{2/5}. \tag{16}$$

(See Andrews, 1991, for the derivation of these formulas.)

Andrews (1991), Andrews and Monahan (1992) and Newey and West (1994) proposed feasible data dependent to procedures to estimate γ, for vector as well as scalar h_t. Rather than exposit the general case, I will describe two 'cookbook' procedures that have been offered as reasonable starting points in empirical work. One procedure relies on Andrews (1991) and Andrews and Monahan (1992), and assumes the QS kernel and estimation of γ via parametric models. The second relies on Newey and West (1994), and assumes a Bartlett kernel and nonparametric estimation of γ.

I emphasize that both papers present more general results than are presented here; both allow the researcher to (for example) use any one of a wide range of kernels.

Let there be a $q \times 1$ vector of weights $w=(w_1, w_2, \ldots, w_q)'$ whose elements tells us how to weight the various elements of S with respect to mean squared error. The weights might be sample dependent, and den Haan and Levin (1997) argue that there are benefits to certain sample-dependent weights, but a simple choice proposed by both papers is: w_i=0 if the corresponding element of h_t is a cross product of a constant term and a regression disturbance, otherwise w_i=1. Andrews's loss function is the normalized expectation of $\sum_{i=1}^{q} w_i(S_{ii} - \hat{S}_{ii})^2$, while Newey and West's loss function is the normalized expectation of $[w'(\hat{S} - S)w]^2$; the normalization is $T^{4/5}$ for QS and $T^{2/3}$ for Bartlett.

Both procedures begin with using a vector autoregression to prewhiten, and end with re-colouring. The basic justification for prewhitening and re-colouring is that simulation evidence indicates that this improves finite sample performance.

1. Prewhitening: Estimate a vector autoregression in $\hat{h}_t$, most likely of order 1. Call the residuals $\hat{h}_t^\dagger$

$$\hat{h}_t = \hat{A}\hat{h}_{t-1} + \hat{h}_t^\dagger, \quad \hat{A} = \sum_{t=2}^{T} \hat{h}_t \hat{h}_{t-1}' \left(\sum_{t=2}^{T} \hat{h}_{t-1} \hat{h}_{t-1}' \right)^{-1}. \tag{17}$$

2. Let $\hat{\Gamma}_j^\dagger$ denote the jth autocovariance of the VAR residual $\hat{h}_t^\dagger$, $\hat{\Gamma}_j^\dagger = (T-1)^{-1}\sum_{t=2+j}^{T} \hat{h}_t^\dagger \hat{h}_{t-j}^{\dagger\prime}$. Using $\{\hat{\Gamma}_j^\dagger\}$ (rather than $\{\hat{\Gamma}_j\}$ [the autocovariances of $\hat{h}_t$]), and choosing m optimally as described in steps 2a or 2b below, construct an estimate of the long run variance of the residual of the VAR just estimated. Call the result $\hat{S}^\dagger$.

2a. Andrews and Monahan (1992): Fit a univariate AR(1) to each of the q elements of $\hat{h}_t^\dagger$. Call the resulting estimate of the AR coefficient and variance of the residual $\hat{\rho}_i$ and $\hat{\sigma}_i^2$. Compute

$$\begin{aligned} \hat{s}_2 &= \sum_{i=1}^{q} w_i(4\hat{\rho}_i^2\hat{\sigma}_i^4)/(1-\hat{\rho}_i)^8, \quad \hat{s}_0 = \sum_{i=1}^{q} w_i\hat{\sigma}_i^4/(1-\hat{\rho}_i)^4, \hat{\gamma}_{QS} \\ &= 1.3221[\hat{s}_2/\hat{s}_0]^{1/5}, \quad \hat{m}_{QS} = \hat{\gamma}_{QS}T^{1/5}. \end{aligned} \tag{18}$$

Then plug $\hat{m}_{QS}$ into formula (13b). Call the result $\hat{k}_j$. Compute $\hat{S}^\dagger = \hat{\Gamma}_0^\dagger + \sum_{j=1}^{T-1}\hat{k}_j(\hat{\Gamma}_j^\dagger + \hat{\Gamma}_j^{\dagger\prime}$.

2b. Newey and West (1994): Set n = integer part of $12(T/100)^{2/9}$. Compute

$$\begin{aligned} \hat{s}^{(1)} &= w'\,\hat{\Gamma}_0 w + 2\sum_{i=1}^{n} iw'\,\Gamma_i^\dagger w, \hat{s}^{(0)} = 2\sum_{i=1}^{n} w'\,\hat{\Gamma}_i^\dagger w, \hat{\gamma}_{BT} \\ &= 1.1447[\hat{s}^{(1)}/\hat{s}^{(0)}]^{2/3}, \hat{m}_{BT} = \text{integer part of } \hat{\gamma}_{BT}T^{1/3}. \end{aligned} \tag{19}$$

Then compute $\hat{S}^\dagger$ according to (14), using $\hat{m}_{BT}$.

3. Re-colouring: compute $\hat{S} = (I - \hat{A})^{-1}\hat{S}^{\dagger}(I - \hat{A})^{-1\,\prime}$.

These two recipes for estimates of S can serve as a starting point for experimentation for alternative choices of m and alternative kernels.

What is the simulation evidence on behaviour of these and other proposed estimators? In answering this question, I focus on sizing of test statistics and accuracy of confidence interval coverage: accuracy in estimation of S is desirable mainly insofar as it leads to accuracy of inference using the relevant variance–covariance matrix. The simulations in papers cited in this article suggest the following. First, no one estimator dominates others. This means in particular that the rate of convergence is not a sufficient statistic for performance in finite samples. The truncated estimator often and the autoregressive estimator sometimes perform more poorly than the slower converging QS estimator, which in turn sometimes performs more poorly than the still slower converging Bartlett estimator. Second, given that one decides to use QS or Bartlett, performance generally though not always is improved if one prewhitens and uses a data-dependent bandwidth as described in the recipes above. Third, the QS and Bartlett estimators tend to reject too much in the presence of positive serial correlation in h_t, and have what I read as a DGP dependent rejection rate (sometimes over-reject, sometimes under-reject) in the presence of negative serial correlation in h_t. The truncated estimator is much likelier to fail to be positive semidefinite in the presence of negative than positive serial correlation. Finally, the performance of all estimators leaves much to be desired. Plausible data-generating processes and sample sizes can lead to serious mis-sizing of any given estimator. Nominal 0.05 tests can have empirical size as low as 0.01 and higher than 0.25.

5. Some recent work

Because simulation studies have yielded disappointing performance, ongoing research aims to develop better estimators. I close by summarizing a few of many recently published papers.

1. I motivated my topic by observing that consistent estimation of S is a natural element of consistent estimation of the variance–covariance matrix of a GMM estimator. Typically we estimate the variance–covariance matrix because we wish to construct confidence intervals or conduct hypothesis tests. A recent literature has evaluated inconsistent estimators that lead to well-defined test statistics, albeit statistics with non-standard critical values. These estimators set lag truncation (or bandwidth) equal to sample size. For example, for the Bartlett estimator, these estimators set $m = T-1$ (see Kiefer, Vogelsang and Bunzel, 2000; Kiefer and Vogelsang, 2002). Simulation evidence indicates that the non-standard statistics may be better behaved than standard statistics. Jansson (2004) provides a theoretical rationale for improved performance in a special case, with more general results in Kiefer and Vogelsang (2005). Phillips, Sun and Jin (2006; 2007) propose a related approach, which under some assumptions will yield statistics with standard critical values.

2. Politis and Romano (1995) propose what they call a 'trapezoidal' kernel. A trapezoidal kernel is a combination of the truncated and Bartlett kernels. For given truncation lag m, let $x_j = j/(m+1)$. Then for some c, $0 < c < 1$, the trapezoidal weights satisfy: $k_j = 1$ if $0 \leq x_j \leq c$, $k_j = (x_j - 1)/(c - 1)$ for $c < x_j \leq 1$. Thus for $0 \leq j \leq c(m+1)$, the autocovariances receive equal weight, as in the truncated kernel; for $c(m+1) < j \leq m+1$, the weights on the autocovariances decline linearly to zero, as in the Bartlett kernel. Such kernels have the advantage that, like the truncated kernel, their convergence is rapid (near $T^{1/2}$). They share with the truncated kernel the possibility of not being positive semidefinite. The authors argue, however, that these kernels are better behaved in finite samples than is the truncated kernel.

3. Xiao and Linton (2002) propose 'twicing' kernels. Operationally, one first computes an estimate such as one of those described in Section 4. One also constructs a multiplicative bias correction by smoothing periodogram ordinates via a 'twiced' kernel. For a properly chosen bandwidth and kernel, the mean squared error of the estimator is of order $T^{8/9}$ (versus $T^{4/5}$ for the QS and $T^{2/3}$ for the Bartlett, absent any corrections). As well, Hirukawa's (2006) version of the Xiao and Linton estimator is positive semidefinite by construction. (The rate results for this estimator and that described in the previous paragraph do not contradict Andrews's, 1991, optimality result for the QS kernel, because these procedures fall outside the class considered by Andrews.)

KENNETH D. WEST

See also **generalized method of moments estimation; spectral analysis; time series analysis.**

I thank Steven Durlauf, Masayuki Hirukawa and Tim Vogelsgang for helpful comments, and the National Science Foundation for financial support.

Bibliography

Andrews, D.W.K. 1991. Heteroskedasticity and autocorrelation consistent covariance matrix estimation. *Econometrica* 59, 817–58.

Andrews, D.W.K. and Monahan, J.C. 1992. An improved heteroskedasticity and autocorrelation consistent covariance matrix. *Econometrica* 60, 953–66.

Cumby, R.E., Huizanga, J. and Obstfeld, M. 1983. Two step, two-stage least squares estimation in models with rational expectations. *Journal of Econometrics* 21, 333–55.

den Haan, W.J. and Levin, A.T. 1997. A practitioner's guide to robust covariance matrix estimation. In *Handbook of Statistics: Robust Inference*, vol. 15, ed. G. Maddala and C. Rao. New York: Elsevier.

Eichenbaum, M.S., Hansen, L.P. and Singleton, K.J. 1988. A time series analysis of representative agent models of consumption and leisure choice under uncertainty. *Quarterly Journal of Economics* 103, 51–78.

Hamilton, J. 1994. *Time Series Analysis*. Princeton: Princeton University Press.

Hannan, E.J. 1970. *Multiple Time Series*. New York: Wiley.

Hansen, B.E. and West, K.D. 2002. Generalized method of moments and macroeconomics. *Journal of Business and Economic Statistics* 20, 460–9.

Hansen, L.P. 1982. Large sample properties of generalized method of moments estimators. *Econometrica* 50, 1029–54.

Hansen, L.P. and Hodrick, R.J. 1980. Forward exchange rates as optimal predictors of future spot rates: an econometric analysis. *Journal of Political Economy* 96, 829–53.

Hansen, L.P. and Singleton, K.J. 1982. Generalized instrumental variables estimation of nonlinear rational expectations models. *Econometrica* 50, 1269–86.

Hirukawa, M. 2006. A modified nonparametric prewhitened covariance estimator. *Journal of Time Series Analysis* 27, 441–76.

Hodrick, R.J. 1992. Dividend yields and expected stock returns: alternative procedures for inference and measurement. *Review of Financial Studies* 5, 357–86.

Jansson, M. 2004. The error in rejection probability of simple autocorrelation robust tests. *Econometrica* 72, 937–46.

Kiefer, N.M. and Vogelsang, T.J. 2002. Heteroskedasticity-autocorrelation robust standard errors using the Bartlett kernel without truncation. *Econometrica* 70, 2093–5.

Kiefer, N.M. and Vogelsang, T.J. 2005. A new asymptotic theory for heteroskedasticity-autocorrelation robust tests. *Econometric Theory* 21, 1130–64.

Kiefer, N.M., Vogelsang, T.J. and Bunzel, H. 2000. Simple robust testing of regression hypotheses. *Econometrica* 68, 695–714.

Newey, W.K. and West, K.D. 1987. A simple, positive semidefinite, heteroskedasticity and autocorrelation consistent covariance matrix. *Econometrica* 55, 703–08.

Newey, W.K. and West, K.D. 1994. Automatic lag selection in covariance matrix estimation. *Review of Economic Studies* 61, 631–54.

Phillips, P.C.B., Sun, Y. and Jin, S. 2006. Spectral density estimation and robust hypothesis testing using steep origin kernels without truncation. *International Economic Review* 47, 837–94.

Phillips, P.C.B., Sun, Y. and Jin, S. 2007. Long run variance estimation and robust regression testing using sharp origin kernels with no truncation. *Journal of Statistical Planning and Inference* 137, 985–1023.

Politis, D.N. and Romano, J.P. 1995. Bias-corrected nonparametric spectral estimation. *Journal of Time Series Analysis* 16, 67–103.

Priestley, M.B. 1981. *Spectral Analysis and Time Series.* New York: Academic Press.

West, K.D. 1997. Another heteroskedasticity and autocorrelation consistent covariance matrix estimator. *Journal of Econometrics* 76, 171–91.

White, H. 1980. A heteroskedasticity consistent covariance matrix estimator and a direct test for heteroskedasticity. *Econometrica* 48, 817–38.

Xiao, Z. and Linton, O. 2002. A nonparametric prewhitened covariance estimator. *Journal of Time Series Analysis* 23, 215–50.

impulse response function

Sims (1980) questioned the way classical simultaneous equations models were specified and identified. He argued in particular that the exogeneity assumptions for some of the variables are often problematic. As an alternative he advocated the use of vector autoregressive (VAR) models for macroeconometric analysis. These models have the form

$$y_t = A_1 y_{t-1} + \cdots + A_p y_{t-p} + u_t,$$

where $y_t = (y_{1t}, \ldots, y_{Kt})'$ (the prime denotes the transpose) is a vector of K observed variables of interest, the A_i's are $(K \times K)$ parameter matrices, p is the lag order and u_t is an error process which is assumed to be white noise with zero mean, that is, $E(u_t) = 0$, the covariance matrix, $E(u_t u_t') = \Sigma_u$, is time invariant and the u_t's are serially uncorrelated or independent. There are usually also deterministic terms such as constants, seasonal dummies or polynomial trends. These terms are neglected here because they are not of interest in what follows. The relations between the variables in a VAR model are difficult to see directly from the parameter matrices. Therefore, *impulse response functions* have been proposed as tools for interpreting VAR models.

A VAR model can be written more compactly as $A(L)y_t = u_t$, where the lag or back-shift operator L is defined such that $Ly_t = y_{t-1}$ and $A(L) = I_K - A_1 L - \cdots - A_p L^p$ is a matrix polynomial in the lag operator. If the polynomial in z defined by $\det A(z)$ has all its roots outside the complex unit circle, the process is stationary and has a Wold moving average (MA) representation

$$y_t = A(L)^{-1} u_t = u_t + \sum_{i=1}^{\infty} \Phi_i u_{t-i}. \tag{1}$$

In this framework impulse response analysis may be based on the counterfactual experiment of tracing the marginal effect of a shock to one variable through the system by setting one component of u_t to one and all other components to zero and evaluating the responses of the y_t's to such an impulse as time goes by. These impulse responses are just the elements of the Φ_i matrices. Because the u_t's are the one-step ahead forecast errors of the system, the resulting functions are sometimes referred to as *forecast error impulse responses* (for example, Lütkepohl, 2005, Section 2.3.2).

Such a counterfactual experiment may not properly reflect the actual responses of an economic system of interest because the components of u_t are instantaneously correlated, that is, Σ_u may not be a diagonal matrix. In that case, forecast error impulses are just not the kinds of impulses that occur in practice, because an impulse in one variable is likely to be accompanied by an impulse in another variable and should not be considered in isolation. Therefore, *orthogonalized impulse responses* are often considered in this context. They are obtained from (1) by choosing some matrix

B such that $BB' = \Sigma_u$ or such that $B^{-1}\Sigma_u B'^{-1}$ is a diagonal matrix and defining $\varepsilon_t = B^{-1}u_t$. Substituting in (1) gives

$$y_t = B\varepsilon_t + \sum_{i=1}^{\infty} \Theta_i \varepsilon_{t-i}, \tag{2}$$

where $\Theta_i = \Phi_i B$, $i = 1, 2, \ldots$. The ε_t's have a diagonal or even a unit covariance matrix and are hence contemporaneously uncorrelated (orthogonal). Thus, ε_t shocks may give a more realistic picture of the reactions of the system. The problem is, however, that the matrix B is not unique and many different orthogonal shocks exist. Thus, identifying restrictions based on non-sample information are necessary to find the unique impulses of interest which represent the actual responses of the system to shocks that occur in practice. These considerations have led to what is known as *structural VAR (SVAR) models* and *structural impulse responses.*

SVAR models

Various types of restrictions have been considered for identifying the structural innovations or, equivalently, for finding a unique or at least locally unique B matrix. For example, using a triangular B matrix obtained from a Choleski decomposition of Σ_u is quite popular (for example, Sims, 1980; Christiano, Eichenbaum and Evans, 1996). Choosing a lower-triangular matrix amounts to setting up a recursive system with a so-called *Wold causal ordering* of the variables. One possible interpretation is that an impulse in the first variable can have an instantaneous impact on all other variables as well, whereas an impulse in the second variable can also have an instantaneous effect on the third to last variables but not on the first one, and so on. Because such a causal ordering is sometimes difficult to defend, other types of restrictions have also been proposed. Examples are:

1. Instantaneous effects of some shocks on certain variables may be ruled out. In other words, zero restrictions are placed on B just as in the Choleski decomposition approach. The zero restrictions do not have to result in a triangular B matrix, however.
2. Identification is achieved by imposing restrictions on the instantaneous relations of the variables. In this case a structural form model of the type $A_0 y_t = A_1 y_{t-1} + \cdots + A_p y_{t-p} + \varepsilon_t$ may be considered and typically linear restrictions are imposed on A_0. Usually the elements on the main diagonal of A_0 will be normalized to unity. The restrictions on A_0 imply restrictions for $B = A_0^{-1}$. For example, if A_0 is triangular, then so is B.
3. It is also possible to set up a model in the form $A_0 y_t = A_1 y_{t-1} + \cdots + A_p y_{t-p} + B\varepsilon_t$ and impose restrictions on both A_0 and B to identify structural shocks. Combining restrictions on B with those on the instantaneous effects on the observed variables results in the so-called *AB*-model of Amisano and Giannini (1997).
4. There may be prior information on the long-run effects of some shocks. In this case restrictions may be placed on $B + \sum_{i=1}^{\infty} \Theta_i = A(1)^{-1}B$ (for example, Blanchard

and Quah, 1989). For instance, demand shocks may be assumed to have no accumulated long-run effects on some variable (in their case output). In fact, distinguishing between shocks with permanent and transitory effects is perhaps done more naturally in models which allow for integrated variables. They will be discussed later.

5. Sign restrictions may be imposed on the impulse responses (for example, Canova and De Nicoló, 2003; Uhlig, 2005), that is, one may want to require that certain shocks have positive or negative effects on certain variables. For example, a restrictive monetary shock should reduce the inflation rate.

Integrated and cointegrated variables

If the VAR operator has unit roots, that is, $\det A(z) = 0$ for $z = 1$, then the variables have stochastic trends. Variables with such trends are called integrated. They can be made stationary by differencing. Moreover, they are called cointegrated if stationary linear combinations exist. If the VAR model contains integrated and cointegrated variables, impulse response analysis can still be performed as for stationary processes. For the latter processes the Φ_i's go to zero for $i \to \infty$ and, hence, the marginal response to an impulse to a stationary process is transitory, that is, the effect goes to zero as time goes by. In contrast, some impulses have permanent effects in cointegrated systems. In fact, in a K-dimensional system with $r<K$ cointegration relations, at least $K - r$ of the K shocks have permanent effects and at most r shocks have transitory effects (King et al, 1991; Lütkepohl, 2005, Ch. 9). These facts open up the possibility to find identifying restrictions for the structural innovations by taking into account the cointegration properties of the system.

Estimation of impulse responses

Estimation of reduced form and structural form parameters of VAR processes is usually done by least squares, maximum likelihood or Bayesian methods. Estimates of the impulse responses are then obtained from the VAR parameter estimates. Suppose the VAR coefficients are contained in a vector $\boldsymbol{\alpha}$ and denote its estimator by $\hat{\boldsymbol{\alpha}}$. Any specific impulse response coefficient θ is a (nonlinear) function of $\boldsymbol{\alpha}$ and may be estimated as $\hat{\theta} = \theta(\hat{\boldsymbol{\alpha}})$. If $\hat{\boldsymbol{\alpha}}$ is asymptotically normal, that is, $\sqrt{T}(\hat{\boldsymbol{\alpha}} - \boldsymbol{\alpha}) \xrightarrow{d} \mathcal{N}(0, \Sigma_{\hat{\boldsymbol{\alpha}}})$, then, under general conditions, $\hat{\theta}$ is also asymptotically normally distributed, $\sqrt{T}(\hat{\theta} - \theta) \xrightarrow{d} \mathcal{N}(0, \sigma_\theta^2)$. The variance of the asymptotic distribution is $\sigma_\theta^2 = \frac{\partial\theta}{\partial\boldsymbol{\alpha}'}\Sigma_{\hat{\boldsymbol{\alpha}}}\frac{\partial\theta}{\partial\boldsymbol{\alpha}}$. Here $\partial\theta/\partial\alpha$ denotes the vector of first order partial derivatives of θ with respect to the elements of $\boldsymbol{\alpha}$ (see Lütkepohl, 1990, for the precise expressions). This result can be used for setting up asymptotic confidence intervals for impulse responses in the usual way.

Asymptotic normality of $\hat{\theta}$ requires that σ_θ^2 is non-zero, which follows if $\Sigma_{\hat{\boldsymbol{\alpha}}}$ is non-singular and $\partial\theta/\partial\boldsymbol{\alpha} \neq 0$. In general the covariance matrix $\Sigma_{\hat{\boldsymbol{\alpha}}}$ will not be non-singular for cointegrated systems, for example. Moreover, the impulse responses generally consist of sums of products of the VAR coefficients and, therefore, the partial derivatives will also be sums of products of such coefficients. Consequently, the partial derivatives will also usually be zero in parts of the parameter space. Thus, $\sigma_\theta^2 = 0$ may hold and, hence, $\hat{\theta}$ may actually converge at a faster rate than $\sqrt{T}$ in parts of the parameter space (cf. Benkwitz, Lütkepohl and Neumann, 2000).

Even under ideal conditions where the asymptotic theory holds, it may not provide a good guide for small sample inference. Therefore, bootstrap methods are often used to construct confidence intervals for impulse responses (for example, Kilian, 1998; Benkwitz, Lütkepohl and Wolters, 2001). If one uses these methods, deriving explicit forms of the analytical expressions for the asymptotic variances of the impulse response coefficients can be avoided. Unfortunately, bootstrap methods generally do not overcome the problems due to zero variances in the asymptotic distributions of the impulse responses. In fact, they may provide confidence intervals which do not have the desired coverage level even asymptotically (Benkwitz, Lütkepohl and Neumann, 2000).

Confidence bands for impulse response functions can also be constructed with Bayesian methods (for example, Koop, 1992). Prior information on the VAR parameters or the impulse responses can in that case be considered. It is not uncommon to report confidence intervals for individual impulse response coefficients and connecting them to get a confidence band around an impulse response function. This approach has been criticized by Sims and Zha (1999), who propose likelihood-characterizing error bands instead.

Extensions

There are a number of extensions to the models and impulse response functions considered so far. For example, all observed variables are treated as endogenous. A main criticism regarding problematic exogeneity assumptions in classical simultaneous equations models is thereby accounted for. On the other hand, this approach often results in heavily parameterized models and imprecise estimates. Therefore, it is occasionally desirable to classify some of the variables as exogenous or consider partial models where we condition on some of the variables which remain unmodelled. In this case one may be interested in tracing the effects of changes in the exogenous or unmodelled variables on the endogenous variables. The resulting impulse response functions are often referred to as dynamic multipliers in the literature on simultaneous equations (see Lütkepohl, 2005, for an introductory treatment). The inference problems related to these quantities are similar to those discussed earlier for VAR impulse responses.

It was also acknowledged in the related literature that finite order VAR models are at best good approximations to the actual data generation processes of multiple time series. Therefore, inference for impulse responses was also considered under the assumption that finite order VAR processes are fitted to data generated by infinite order processes (for example, Lütkepohl, 1988; Lütkepohl and Saikkonen, 1997).

Impulse responses associated with linear VAR models have the property of being time invariant and their shape is invariant to the size and direction of the impulses. These features make it easy to represent the reactions of the variables to impulses hitting the system in a small set of graphs. Such responses are often regarded as unrealistic in practice, where, for instance, a positive shock may have a different effect from a negative shock or the effect of a shock may depend on the state of the system at the time when it is hit. Hence, the linear VAR models are too restrictive for some analyses. These problems can be resolved by considering nonlinear models. Although nonlinear models have their attractive features for describing economic systems or phenomena, their greater flexibility makes them more difficult to interpret properly. In fact, it is not obvious how to define impulse responses of nonlinear models in a meaningful manner. Gallant, Rossi and Tauchen (1993) proposed so-called *conditional moment profiles* which may give useful information on important features of nonlinear multiple time series models. For example, one may consider quantities of the general form $E[g(y_{t+h})|y_t + \xi, \Omega_{t-1}] - E[g(y_{t+h})|y_t, \Omega_{t-1}]$, $h = 1, 2, \ldots$, where $g(\cdot)$ denotes some function of interest, ξ represents the impulses hitting the system at time t, and $\Omega_{t-1} = (y_{t-1}, y_{t-2}, \ldots)$ denotes the history of the variables at time t. In other words, the conditional expectation of some quantity of interest, given the history of y_t in period t, is compared to the conditional expectation that is obtained if a shock ξ occurs at time t. For example, defining $g(y_{t+h}) = [y_{t+h} - E(y_{t+h}|\Omega_{t+h-1})][y_{t+h} - E(y_{t+h}|\Omega_{t+h-1})]'$ results in conditional volatility profiles, which may be compared to a baseline profile obtained for a specific history of the process and a zero impulse. Clearly, in general the conditional moment profiles depend on the history Ω_{t-1} as well as the impulse ξ. Similar quantities were also considered by Koop, Pesaran and Potter (1996), who called them *generalized impulse responses* (see also Pesaran and Shin, 1998).

Although these quantities may be interesting to look at, they depend on t, h, and ξ. Hence, there is a separate impulse response function for each given t and ξ. In empirical work it will therefore be necessary to summarize the wealth of information in the conditional moment profiles in a meaningful way – for instance, by considering summary statistics. In practice, an additional obstacle is that the actual data generation process is unknown and estimated models are available at best. In that case, the conditional moment profiles or generalized impulse responses will be estimates, and it would be useful to have measures for their sampling variability. It is not clear how this additional information may be computed and presented in the best way in practice.

HELMUT LÜTKEPOHL

See also **cointegration; structural vector autoregressions; vector autoregressions.**

Bibliography

Amisano, G. and Giannini, C. 1997. *Topics in Structural VAR Econometrics*, 2nd edn. Berlin: Springer.

Benkwitz, A., Lütkepohl, H. and Neumann, M. 2000. Problems related to bootstrapping impulse responses of autoregressive processes, *Econometric Reviews* 19, 69–103.

Benkwitz, A., Lütkepohl, H. and Wolters, J. 2001. Comparison of bootstrap confidence intervals for impulse responses of German monetary systems. *Macroeconomic Dynamics* 5, 81–100.

Blanchard, O. and Quah, D. 1989. The dynamic effects of aggregate demand and supply disturbances. *American Economic Review* 79, 655–73.

Canova, F. and De Nicoló, G. 2003. On the sources of business cycles in the G-7. *Journal of International Economics* 59, 77–100.

Christiano, L., Eichenbaum, M. and Evans, C. 1996. The effects of monetary policy shocks: evidence from the flow of funds. *Review of Economics and Statistics* 78, 16–34.

Gallant, A., Rossi, P. and Tauchen, G. 1993. Nonlinear dynamic structures. *Econometrica* 61, 871–907.

Kilian, L. 1998. Small-sample confidence intervals for impulse response functions. *Review of Economics and Statistics* 80, 218–30.

King, R., Plosser, C., Stock, J. and Watson, M. 1991. Stochastic trends and economic fluctuations. *American Economic Review* 81, 819–40.

Koop, G. 1992. Aggregate shocks and macroeconomic fluctuations: a Bayesian approach. *Journal of Applied Econometrics* 7, 395–411.

Koop, G., Pesaran, M. and Potter, S. 1996. Impulse response analysis in nonlinear multivariate models. *Journal of Econometrics* 74, 119–47.

Lütkepohl, H. 1988. Asymptotic distribution of the moving average coefficients of an estimated vector autoregressive process. *Econometric Theory* 4, 77–85.

Lütkepohl, H. 1990. Asymptotic distributions of impulse response functions and forecast error variance decompositions of vector autoregressive models. *Review of Economics and Statistics* 72, 116–25.

Lütkepohl, H. 2005. *New Introduction to Multiple Time Series Analysis.* Berlin: Springer-Verlag.

Lütkepohl, H. and Saikkonen, P. 1997. Impulse response analysis in infinite order cointegrated vector autoregressive processes. *Journal of Econometrics* 81, 127–57.

Pesaran, M. and Shin, Y. 1998. Generalized impulse response analysis in linear multivariate models. *Economics Letters* 58, 17–29.

Sims, C. 1980. Macroeconomics and reality. *Econometrica* 48, 1–48.

Sims, C. and Zha, T. 1999. Error bands for impulse responses. *Econometrica* 67, 1113–55.

Uhlig, H. 2005. What are the effects of monetary policy on output? Results from an agnostic identification procedure. *Journal of Monetary Economics* 52, 381–419.

Kalman and particle filtering

The Kalman and particle filters are algorithms that recursively update an estimate of the state and find the innovations driving a stochastic process given a sequence of observations. The Kalman filter accomplishes this goal by linear projections, while the particle filter does so by a sequential Monte Carlo method.

Since both filters start with a state-space representation of the stochastic processes of interest, Section 1 presents the state-space form of a dynamic model. Section 2 introduces the Kalman filter and Section 3 develops the particle filter. For extended expositions of this material, see Doucet, de Freitas, and Gordon (2001), Durbin and Koopman (2001), and Ljungqvist and Sargent (2004).

1. The state-space representation of a dynamic model

A large class of dynamic models can be represented by a state-space form:

$$X_{t+1} = \varphi(X_t, W_{t+1}; \gamma) \tag{1}$$

$$Y_t = g(X_t, V_t; \gamma). \tag{2}$$

This representation handles a stochastic process by finding three objects: a vector that describes the position of the system (a *state*, $X_t \in X \subset R^l$) and two functions, one mapping the state today into the state tomorrow (*the transition equation*, (1)) and one mapping the state into observables, Y_t (*the measurement equation*, (2)). An iterative application of the two functions on an initial state X_0 generates a fully specified stochastic process. The variables W_{t+1} and V_t are independent *i.i.d.* shocks. A realization of T periods of observables is denoted by $y^T \equiv \{y_t\}_{t=1}^T$ with $y^0 = \{\phi\}$. Finally, γ, which belongs to the set $\Upsilon \subset R^n$, is a vector of parameters. To avoid stochastic singularity, we assume that $\dim(W_t) + \dim(V_t) \geq \dim(Y_t)$ for all t.

This framework can accommodate cases in which the dimensionality of the shocks is zero, where the shocks have involved structures, or where some or all of the states are observed. Also, at the cost of heavier notation, we could deal with more general problems. For example, the state could be a function or a correspondence, and the transition equation a functional operator. The basic ideas are, however, identical.

The transition and measurement equations may come from a statistical description of the process or from the equilibrium dynamics of an economic model. For example, dynamic stochastic general equilibrium models can be easily written in state-space form with the transition and measurement equations formed by the policy functions that characterize the optimal behaviour of the agents of the model. This observation tightly links modern dynamic macroeconomics with the filtering tools presented in this article.

It is important to note that there are alternative timing conventions for the state-space representation of a dynamic model and that, even while the timing convention is kept constant, the same model can be written in different state-space forms. All of those representations are equivalent, and the researcher should select the form that best fits her needs.

2. The Kalman filter

The Kalman filter deals with state-space representations where the transition and measurement equations are linear and where the shocks to the system are Gaussian. The procedure was developed by Kalman (1960) to transform ('filter') some original observables y_t into Wold innovations a_t and estimates of the state x_t. With the innovations, we can build the likelihood function of the dynamic model. With the estimates of the states, we can forecast and smooth the stochastic process.

We begin with the state-space system defined by the transition equation:

$$x_{t+1} = Ax_t + G\omega_{t+1}, \omega_{t+1} \sim \mathcal{N}(0, Q)$$

and the measurement equation:

$$y_t = Cx_t + \upsilon_t, \upsilon_t \sim \mathcal{N}(0, R)$$

where A, G, C, Q, and R are known matrices.

There are different ways to derive and interpret the Kalman filter, including an explicitly Bayesian one. We follow a simple approach based on linear least-square projections. The reader will enhance her understanding with the more general expositions in Durbin and Koopman (2001) and Ljungqvist and Sargent (2004).

Let $x_{t|t-1} = E(x_t|y^{t-1})$ be the best linear predictor of x_t given the history of observables until $t-1$, i.e., y^{t-1}. Let $y_{t|t-1} = E(y_t|y^{t-1}) = Cx_{t|t-1}$ be the best linear predictor of y_t given y^{t-1}. Let $x_{t|t} = E(x_t|y^t)$ be the best linear predictor of x_t given the history of observables until t, i.e., y^t. Let $\Sigma_{t|t-1} \equiv E((x_t - x_{t|t-1})(x_t - x_{t|t-1})'|y^{t-1})$ be the predicting error variance-covariance matrix of x_t given y^{t-1}. Finally, let $\Sigma_{t|t} \equiv E((x_t - x_{t|t})(x_t - x_{t|t})'|y^t)$ be the predicting error variance-covariance matrix of x_t given y^t.

How does the Kalman filter work? Let's assume we have $x_{t|t-1}$ and $y_{t|t-1}$, that is, an estimate of the state and a forecast of the observable given y^{t-1}. Then, we observe y_t. Thus, we want to revise our linear predictor of the state and obtain an estimate, $x_{t|t}$, that incorporates the new information. Note that $x_{t+1|t} = Ax_{t|t}$ and $y_{t+1|t} = Cx_{t+1|t}$, so we can go back to the first step and wait for the y_{t+1} next period. Therefore, the key of the Kalman filter is to obtain $x_{t|t}$ from $x_{t|t-1}$ and y_t.

We do so with the formula:

$$x_{t|t} = x_{t|t-1} + K_t(y_t - y_{t|t-1}) = x_{t|t-1} + K_t(y_t - Cx_{t|t-1}),$$

that is, our new value $x_{t|t}$ is equal to $x_{t|t-1}$ plus the difference between the actual y_t and the forecasted $y_{t|t-1}$, times a matrix K_t, called the Kalman gain. Durbin and Koopman (2001) derive this formula from probabilistic foundations. Ljungqvist and Sargent (2004) find it through an application of a Gram–Schmidt orthogonalization procedure.

Then, if we choose K_t to minimize $\Sigma_{t|t}$, we get $K_t = \Sigma_{t|t-1}C'(C\Sigma_{t|t-1}C' + R)^{-1}$. This expression shows the determinants of K_t. If we made a big mistake forecasting $x_{t|t-1}$ using past information ($\Sigma_{t|t-1}$ large), we give a lot of weight to the new information (K_t large). Also, if the new information is noisy (R large), we give a lot of weight to the old prediction (K_t small).

Now, note that $\Sigma_{t|t} \equiv E((x_t - x_{t|t})(x_t - x_{t|t})'|y^t) = \Sigma_{t|t-1} - K_tC\Sigma_{t|t-1}$. Therefore, from $x_{t|t-1}$, $\Sigma_{t|t-1}$, and y_t, we compute $x_{t|t}$ and $\Sigma_{t|t}$ using K_t. Also, we derive $\Sigma_{t+1|t} = A\Sigma_{t|t}A' + GQG'$, $x_{t+1|t} = Ax_{t|t}$, and $y_{t+1|t} = Cx_{t+1|t}$.

We collect all the previous steps. We start with some estimates of the state $x_{t|t-1}$, the observables $y_{t|t-1}$, and the variance-covariance matrix $\Sigma_{t|t-1}$. Then we observe y_t and compute $x_{t+1|t}$, $y_{t+1|t}$, and $\Sigma_{t+1|t}$.

Thus, the Kalman filter can be recursively written as follows:

- $y_{t|t-1} = Cx_{t|t-1}$
- $K_t = \Sigma_{t|t-1}C'(C\Sigma_{t|t-1}C' + R)^{-1}$
- $\Sigma_{t|t} = \Sigma_{t|t-1} - K_tC\Sigma_{t|t-1}$
- $x_{t|t} = x_{t|t-1} + K_t(y_t - Cx_{t|t-1})$
- $\Sigma_{t+1|t} = A\Sigma_{t|t}A' + GQG'$
- $x_{t+1|t} = Ax_{t|t}$.

The differences between the observable and its forecast, $a_t = y_t - y_{t|t-1} = y_t - Cx_{t|t-1}$ are, by construction, Wold innovations. Moreover, since the system is linear and Gaussian, a_t is normally distributed with zero mean and variance $C\Sigma_{t|t-1}C' + R$. That is why the Kalman filter is a whitening filter: it takes as an input a correlated sequence y^T and it produces a sequence of white noise innovations a_t.

With this last result, we write the likelihood function of $y^T = \{y_t\}_{t=1}^T$ as:

$$\log L(y^T|A,G,C,Q,R) = \sum_{t=1}^{T} \log L(y_t|y^{t-1}A,G,C,Q,R)$$
$$= -\sum_{t=1}^{T}\left[\frac{N}{2}\log 2\pi + \frac{1}{2}\log|C\Sigma_{t|t-1}C' + R| + \frac{1}{2}\sum_{t=1}^{T} a_t'(C\Sigma_{t|t-1}C' + R)^{-1}a_t\right].$$

This likelihood is one of the most important results of the Kalman filter. With it, we can undertake statistical inference in the dynamic model, both with maximum likelihood and with Bayesian approaches.

An important step in the Kalman filter is to set the initial conditions $x_{1|0}$ and $\Sigma_{1|0}$. If we consider stationary stochastic processes, the standard approach is to set $x_{1|0} = x^*$ and $\Sigma_{1|0} = \Sigma^*$ such that $x^* = Ax^*$ and

$$\Sigma^* = A\Sigma^*A' + GQG' = [I - A \otimes A]^{-1} vec(GQG').$$

Non-stationary time series require non-informative prior conditions for $x_{1|0}$. This approach, called the diffuse initialization of the filter, begins by postulating that $x_{1|0}$ is equal to:

$$x_{1|0} = \tau + \Phi\delta + G\omega_0, \omega_0 \sim \mathcal{N}(0, Q) \text{ and } \delta \sim \mathcal{N}(0, \kappa I_q)$$

where τ is given and Φ and G are formed by columns of the identity matrix such that $\Phi G' = 0$. This structure allows for some elements of $x_{1|0}$ to have a known joint distribution, while, by letting $\kappa \to \infty$, to formalize ignorance with respect to other elements. Clearly, $x_{1|0} = E(x_{1|0}) = \tau$. To determine the initial variance, we expand $\Sigma_{1|0} = \kappa\Phi\Phi' + GQG'$ as a power series of κ^{-1} and take $\kappa \to \infty$ to find the dominant term of the expansion. Durbin and Koopman (2001) provide details.

The Kalman filter can also be applied for smoothing, that is, to obtain $x_{t|T}$, an estimate of x_t given the whole history of observables, that is, y^T. Smoothing is of interest when the state x_t has a structural interpretation of its own. Since smoothing uses more information than filtering, the predicting error variance covariance matrix of x_t given y^T will be smaller than $\Sigma_{t|t-1}$. Finally, we note that the Kalman filtering problem is the dual of the optimal linear regulator problem.

3. The particle filter

The Kalman filter relies on the linearity and normality assumptions. However, many models in which economists are interested are nonlinear and/or non-Gaussian. How can we undertake the forecast, smoothing, and estimation of dynamic models when any of those two assumptions are relaxed?

Sequential Monte Carlo methods, in particular the particle filter, reproduce the work of the Kalman filter in those nonlinear and/or non-Gaussian environments. The key difference is that, instead of deriving analytic equations as the Kalman filter does, the particle filter uses simulation methods to generate estimates of the state and the innovations. If we apply the particle filter to a linear and Gaussian model, we will obtain the same likelihood (as the number of simulations grows) that we would if we used the Kalman filter. Since it avoids simulations, the Kalman filter is more efficient in this linear and Gaussian case.

We present here only the basic particle filter. Doucet, de Freitas and Gordon (2001) discuss improvements upon the basic filter. Fernández-Villaverde and Rubio-Ramírez (2007) show how this particle filter can be implemented to estimate dynamic stochastic general equilibrium models. Our goal is to evaluate the likelihood function of a sequence of realizations of the observable y^T implied by a stochastic process at a parameter value γ:

$$L(y^T; \gamma) = p(y^T; \gamma). \tag{3}$$

Our first step is to factor the likelihood function as:

$$\begin{aligned} p(y^T; \gamma) &= \prod_{t=1}^{T} p(y_t | y^{t-1}; \gamma) \\ &= \prod_{t=1}^{T} \iint p(y_t | W^t, X_0, y^{t-1}; \gamma) \times p(W^t, X_0 | y^{t-1}; \gamma) dW^t dX_0, \end{aligned} \tag{4}$$

where X_0 is the initial state of the model and the p's represent the relevant densities. In general, the likelihood function (4) cannot be computed analytically. The particle filter uses simulation methods to estimate it.

Before introducing the filter, we assume that, for all γ, x_0, w^t, and t, the following system of equations:

$$X_1 = \varphi(x_0, w_1; \gamma)$$
$$y_m = g(X_m, V_m; \gamma) \qquad \text{for } m = 1, 2, \ldots t$$
$$X_m = \varphi(X_{m-1}, w_m; \gamma) \quad \text{for } m = 2, 3, \ldots t$$

has a unique solution, (v^t, x^t), and we can evaluate $p(v^t; \gamma)$. This assumption implies that we can evaluate the conditional densities $p(y_t|w^t, x_0, y^{t-1}; \gamma)$ for all γ, x_0, w^t, and t. Then, we have:

$$p(y_t|w^t, x_0, y^{t-1}; \gamma) = |dy(v_t; \gamma)| p(v_t; \gamma)$$

for all γ, x_0, w^t, and t, where $|dy(v_t; \gamma)|$ stands for the determinant of the Jacobian of y_t with respect to V_t evaluated at v_t.

Conditional on having N draws of $\{\{x_0^{t|t-1,i}, w^{t|t-1,i}\}_{i=1}^N\}_{t=1}^T$ from the sequence of densities $\{p(W^t, X_0|y^{t-1}; \gamma)\}_{t=1}^T$, the law of large numbers implies that the likelihood function ((4)) can be approximated by:

$$p(y^T; \gamma) \simeq \prod_{t=1}^{T} \frac{1}{N} \sum_{i=1}^{N} p(y_t|w^{t|t-1,i}, x_0^{t|t-1,i}, y^{t-1}; \gamma).$$

This observation shows that the problem of evaluating the likelihood ((4)) is equivalent to the problem of drawing from $\{p(W^t, X_0|y^{t-1}; \gamma)\}_{t=1}^T$. Since the algorithm does not require any assumption about the distribution of the shocks except the ability to evaluate $p(y_t|w^t, x_0, y^{t-1}; \gamma)$, either analytically or by simulation, we can deal with models with a rich specification of non-Gaussian innovations. But, how do we sample from $\{p(W^t, X_0|y^{t-1}; \gamma)\}_{t=1}^T$?

Let $\{x_0^{t-1,i}, w^{t-1,i}\}_{i=1}^N$ be a sequence of N *i.i.d.* draws from $p(W^{t-1}, X_0|y^{t-1}; \gamma)$. Let $\{x_0^{t|t-1,i}, w^{t|t-1,i}\}_{i=1}^N$ be a sequence of N *i.i.d.* draws from $p(W^t, X_0|y^{t-1}; \gamma)$. We call each draw $(x_0^{t,i}, w^{t,i})$ a *particle* and the sequence $\{x_0^{t,i}, w^{t,i}\}_{i=1}^N$ a *swarm of particles*. Also, define the weights:

$$q_t^i = \frac{p(y_t|w^{t|t-1,i}, x_0^{t|t-1,i}, y^{t-1}; \gamma)}{\sum_{i=1}^N p(y_t|w^{t|t-1,i}, x_0^{t|t-1,i}, y^{t-1}; \gamma)}. \tag{5}$$

The next proposition shows how to use $p(W^t, X_0|y^{t-1}; \gamma)$, the weights $\{q_t^i\}_{i=1}^N$, and importance sampling to draw from $p(W^t, X_0|y^t; \gamma)$:

Proposition 1 *Let* $\{x_0^{t|t-1,i}, w^{t|t-1,i}\}_{i=1}^N$ *be a draw from* $p(W^t, X_0|y^{t-1}; \gamma)$. *Let the sequence* $\{\tilde{x}_0^i, \tilde{w}^i\}_{i=1}^N$ *be a draw with replacement from* $\{x_0^{t|t-1,i}, w^{t|t-1,i}\}_{i=1}^N$ *where* q_t^i *is the* *probability of* $(x_0^{t|t-1,i}, w^{t|t-1,i})$ *being drawn* $\forall i$. *Then* $\{\tilde{x}_0^i, \tilde{w}^i\}_{i=1}^N$ *is a draw from* $p(W^t, X_0|y^t;\gamma)$.

Then, with a draw $\{x_0^{t|t-1,i}, w^{t|t-1,i}\}_{i=1}^N$ from $p(W^t, X_0|y^{t-1};\gamma)$, we get a draw $\{x_0^{t,i}, w^{t,i}\}_{i=1}^N$ from $p(W^t, X_0|y^t;\gamma)$ and we generate a sequence of particles $\{\{x_0^{t|t-1,i}, w^{t|t-1,i}\}_{i=1}^N\}_{t=1}^T$ from the sequence $\{p(W^t, X_0|y^{t-1};\gamma)\}_{t=1}^T$. Given some initial conditions, we can recursively apply the idea of the previous proposition as summarized by the algorithm:

Step 0, Initialization: Set $t \rightsquigarrow 1$. Initialize $p(W^{t-1}, X_0|y^{t-1};\gamma) = p(X_0;\gamma)$.
Step 1, Prediction: Sample N values $\{x_0^{t|t-1,i}, w^{t|t-1,i}\}_{i=1}^N$ from the conditional density $p(W_1^t, X_0|y^{t-1};\gamma) = p(W_t;\gamma)p(W^{t-1}, X_0|y^{t-1};\gamma)$.
Step 2, Filtering: Assign to each draw $(x_0^{t|t-1,i}, w^{t|t-1,i})$ the weight q_t^i as defined in (5).
Step 3, Sampling: Sample N times with replacement from $\{x_0^{t|t-1,i}, w^{t|t-1,i}\}_{i=1}^N$ with probabilities $\{q_t^i\}_{i=1}^N$. Call each draw $(x_0^{t,i}, w^{t,i})$. If $t < T$ set $t \rightsquigarrow t+1$ and go to Step 1. Otherwise stop.

With the algorithm's output $\{\{x_0^{t|t-1,i}, w_1^{t|t-1,i}\}_{i=1}^N\}_{t=1}^T$, we obtain the estimate of the states in each period and compute the likelihood:

$$p(y^T;\gamma) \simeq \frac{1}{N}\left(\prod_{t=1}^{T}\frac{1}{N}\sum_{i=1}^{N} p(y_t|w^{t|t-1,i}, x_0^{t|t-1,i}, y^{t-1};\gamma)\right).$$

The **sampling step** is the heart of the algorithm. If we skip it and weight each draw in $\{x_0^{t|t-1,i}, w^{t|t-1,i}\}_{i=1}^N$ by $\{Nq_t^i\}_{i=1}^N$, we have a sequential importance sampling. The problem with this approach is that it diverges as t grows. The reason is that, as $t \to \infty$, all the sequences become arbitrarily far away from the true sequence of states (the true sequence being a zero measure set), and the sequence that happens to be closer dominates all the remaining sequences in weight. In practice, after a few steps only one sequence has a non-zero weight. Through resampling, we eliminate this problem as we keep (and multiply) those sequences that do not diverge from the true one.

The algorithm outlined above is not the only procedure to evaluate the likelihood of nonlinear and/or non-Gaussian dynamic models. However, the alternatives, such as the extended Kalman filter, the Gaussian sum approximations, or grid-based filters, are of limited use, and many, such as the extended Kalman filter, fail asymptotically. Consequently, the particle filter is the most efficient and robust procedure to undertake inference for nonlinear and/or non-Gaussian models, and we will witness many applications of this filter in economics in future years.

JESÚS FERNÁNDEZ-VILLAVERDE

See also **Bayesian methods in macroeconometrics; state space models.**

Bibliography

Doucet, A., de Freitas, N. and Gordon, N. 2001. *Sequential Monte Carlo Methods in Practice.* New York: Springer Verlag.

Durbin, J. and Koopman, S.J. 2001. *Time Series Analysis by State Space Methods*. Oxford: Oxford University Press.
Fernández-Villaverde, J. and Rubio-Ramírez, J.F. 2007. Estimating macroeconomic models: a likelihood approach. *Review of Economic Studies* 74, 1059–87.
Kalman, R.E. 1960. A new approach to linear filtering and prediction problems. *Transactions of the ASME – Journal of Basic Engineering* D82, 35–45.
Ljungqvist, L. and Sargent, T.J. 2004. *Recursive Macroeconomic Theory*, 2nd edn. Cambridge, MA: MIT Press.

law(s) of large numbers

When we have a large number of independent replications of a random experiment, we observe that the frequency of the outcomes can be very well approximated by the probabilities of the corresponding events. The profits of many commercially successful enterprises – like casinos or insurance companies – are based on random events obeying some laws.

Mathematically, this idea was first formulated by Jacob Bernoulli, for experiments with only two outcomes ('Bernoulli experiments'). The terminology 'law of large numbers' was introduced by S.D. Poisson in 1835.

In the most basic version, LLN (the standard abbreviation for 'law(s) of large numbers') describes results of the following type. We assume that we have given a sequence of random variables $X_1, X_2, \ldots$ We say we have a LLN if

$$\frac{1}{N}(X_1 + \cdots + X_N) \tag{1}$$

converges for $N \to \infty$, preferably to a constant.

For stating our results, we have to state the nature of the convergence in our LLN and impose some restrictions on the X_i. The more we restrict our X_i, the stronger our convergence results will be.

The weak law of large numbers

The 'weak law of large numbers' states that averages like (1) converge in a 'weak' sense (like for example convergence in probability) to a limit. In most cases, the requirements for the random variables involved are not very restrictive. A typical weak LLN is the following theorem.

Theorem 1 *Assume that the random variables X_i satisfy*

$$EX_i = 0, \tag{2}$$

$$\sup EX_i^2 < \infty \tag{3}$$

and

$$\lim_{M\to\infty} \sup_{|i-j|>M} |EX_iX_j| < \infty. \tag{4}$$

Then for $N \to \infty$

$$\frac{1}{N}(X_1 + \cdots + X_N) \to^P 0,$$

where $\to^P$ denotes convergence in probability.

Our random variables have to be centred, of bounded variance, and condition (4) requires that the correlation of random variables 'far apart' converges to zero uniformly. This is a very general and important result. Another advantage is the simplicity of its proof: it is an elementary task to show that the variance of the average converges to zero. Then the theorem is an immediate consequence of Chebyshev's inequality. Moreover, the assumptions of the theorem can easily be checked, and only depend on the second moments of the X_i.

The strong law of large numbers

In some cases, we want to have more than convergence in probability of the averages. For this purpose, we have strong laws of large numbers. We do need, however, stricter requirements. The following theorem is a typical strong LLN. A more stringent discussion of this type of theorems

Theorem 2 *Assume that the random variables X_i satisfy* (2), (3). *Let $\mathfrak{F}_i$ be an increasing sequence of σ-algebras (for example $\mathfrak{F}_{i-1} \subset \mathfrak{F}_i$) so that X_i is $\mathfrak{F}_i$ –measurable. Then let us assume that*

$$E(X_i/\mathfrak{F}_{i-1}) = 0. \tag{5}$$

Then

$$\frac{1}{N}(X_1 + \cdots + X_N) \to 0 \ P - \textit{almost surely}.$$

Heuristically, we can interpret $\mathfrak{F}_i$ as information available at time i. Then (5) postulates that we cannot predict X_i given the information at time $i - 1$. One important special case where (5) is fulfilled is the case of independent. In this case, we can choose $\mathfrak{F}_i$ to be the σ-algebra generated by $X_1, \ldots, X_i$. Then, assuming the X_i to be independent, we have $E(X_i/\mathfrak{F}_{i-1}) = E(X_i)$.

Hence (5) is more general than the requirement of independence, but still far more restrictive than (4).

Ergodic theorems

We can easily see that (5) implies that our X_i are uncorrelated. In many applications, this requirement is unrealistic. Fortunately, there is a theory guaranteeing convergence of sums like (1) at least for stationary processes X_i. A process X_i, $i \in \mathbf{Z}$ is called (strictly) stationary if for all $n \in \mathbf{Z}$ the distributions of $(X_1, X_2, \ldots, X_m)$ and $(X_{n+1}, X_2, \ldots, X_{n+m})$ are the same. To describe the limits of our process, we need to introduce the transition operator T: This operator is a mapping defined on the space of random variables measurable with respect to the σ-algebra generated by the X_i, $i \in \mathbf{Z}$. For random variables

$$Y = f(X_{t_1}, X_{t_2}, \ldots, +X_{t_n}) \tag{6}$$

we define the random variable TY by

$$TY = f(X_{t_1+1}, X_{t_2+1}, \dots, X_{t_n+1}). \tag{7}$$

So the transition operator T shifts every random variable 'one step in the future'. (T can be considered as the inverse of the usual lag operator). One can show that the definition based on (6), (7) can be uniquely extended to the space of all X_i, $i \in Z$ measurable random variables. Then an event A is called *invariant* if

$$TI_A = I_A \text{ almost surely,}$$

where I_A is the indicator of the event A. It can be easily seen that the invariant events form a σ-algebra, which we denote by $\mathfrak{F}$. Then the ergodic theorem states that

$$\lim_{n\to\infty} \frac{1}{n} \sum_{i=1}^{n} X_i = E(X_i/\mathfrak{F}). \tag{8}$$

(Since we are taking the conditional expectation with respect to $\mathfrak{F}$, it can easily be seen that $E(X_1/\mathfrak{F}) = E(X_2/\mathfrak{F}) = \dots$).

The ergodic theorem is included in most of advanced textbooks on probability theory (see, for example, Billingsley, 1995). A more detailed exposition can be found in Gray (2007).

We now can take various conclusions from our theorem. First of all, we can regardless of the nature of the σ-algebra $\mathfrak{F}$ conclude that the limit of $\frac{1}{n}\sum_{i=1}^{n} X_i$ *exists.* In econometric theory, one often postulates the existence of limits of certain averages (that is, in regression theory we often assume that $\lim_{n\to\infty}\frac{1}{n}\sum_{i=1}^{n} x_i x_i'$ exists). In case of stationary processes, the theorem here makes assumptions of this type very plausible.

If the σ-algebra $\mathfrak{F}$ is trivial (that is, consists only of events of probability 0 and 1), then the right-hand side of (8) is constant. One sufficient criterion for this property is that the process is a causal function of i.i.d. random variables. So if

$$X_i = f(e_i, e_{i-1}, \dots)$$

where e_i are i.i.d., $\mathfrak{F}$ is trivial.

Applications and uniform laws of large numbers

For many statistical applications, we need stronger results. As a first example, consider the asymptotic of the maximum likelihood estimator. As a simplest case, let us discuss the case of i.i.d. random variables X_i, distributed according to densities f_θ for parameters $\theta \in \Theta$, and let θ_0 be the true parameter. Then the LLN guarantees that for every fixed θ

$$\frac{1}{n}\sum \ln(f_\theta(X_i) \to \int \ln(f_\theta) f_{\theta_0}, \tag{9}$$

and the function on the right-hand side is maximized if $\theta = \theta_0$. Since the maximum likelihood estimator maximizes the right-hand side, it seems reasonable to exploit this relation for a proof of consistency of the maximum likelihood estimator. The LLN guarantees only convergence for *fixed* θ, from our LLN we cannot say anything about the limiting behaviour of

$$\sup_{\theta \in \Theta} \ln\left(\frac{1}{n}\sum \ln(f_\theta(X_i)\right).$$

This problem would go away if one could establish that the convergence in (9) is *uniform* in θ. This strategy was first realized in a path breaking paper by A. Wald (Wald, 1949), where he first established the consistency of the maximum likelihood estimator. Today the techniques are a little more sophisticated. Nevertheless, consistency proofs for M-estimators still rely to good extend on Wald's idea.

Another application of uniform LLN is the consistency of 'plug-in' estimators. In many cases, the asymptotic variance of certain estimators can be expressed as a function of the expectations of certain random functions, possibly depending on the parameter to be estimated (for example, the well-known 'sandwich formula' derived by H. White; see for example Hayashi, 2000). A standard strategy is to estimate the parameter, then replace the expectation by an average (and hope that – due to the LLN – average and expectation are close together) and use the estimated parameter as an argument. One can easily see that only a uniform law of large numbers can justify procedures of this type.

Fortunately, there exist a lot of criteria to establish uniform laws of large numbers. For most cases of interest to econometricians, the papers by Andrews (1992) and Pőtscher and Prucha (1989) will be sufficient.

A more general and abstract theory can be found in van der Vaart and Wellner (1996). These theories allow us also to estimate the cumulative distribution function of random variables directly. Suppose we have given random variables $X_i, \ldots, X_n$. Then the empirical distribution function F_n is defined as

$$F_n(x) = \frac{1}{n}\sum_{i=1}^{n} I(X_i \leq x)$$

(that is F_n jumps $1/n$ in X_i and is constant in between the jumps). Then the theorem of Glivenko-Cantelli (see van der Vaart and Wellner, 1996) states that if the X_i are i.i.d. with cumulative distribution function F, then

$$\sup|F_n(x) - F(x)| \to 0.$$

It should be noted that there are generalizations to multivariate or even more general X_i. In these cases, however, one has to use slightly more sophisticated techniques. Instead of the 'empirical distribution function', one has to use the 'empirical measure' (a random measure, which puts mass $1/n$ in the points X_i, and instead of the maximum difference of the distribution functions one has to consider the maximal difference of the measures over certain classes ('VC-classes').

WERNER PLOBERGER

Bibliography

Andrews, D.W.K. 1992. Generic uniform convergence. *Econometric Theory* 8, 241–57.

Billingsley, P. 1995. *Probability and Measure*, 3rd edn. New York: Wiley.

Gray, R.M. 2007. *Probability, Random Processes, and Ergodic Properties*. Online. Available at http://ee.stanford.edu/~gray/arp.html, accessed 29 April 2007.

Hall, P. and Heyde, C.C. 1980. *Martingale Limit Theory and its Application*. San Diego: Academic Press.

Hayashi, F. 2000. *Econometrics*. Princeton: Princeton University Press.

Pőtscher, B.M. and Prucha, I.R. 1989. A uniform law of large numbers for dependent and heterogeneous data processes. *Econometrica* 57, 675–83.

van der Vaart, A.W. and Wellner, J.A. 1996. *Weak Convergence and Empirical Processes*. New York: Springer.

Wald, A. 1949. Note on the consistency of the maximum likelihood estimate. *Annals of Mathematical Statistics* 20, 595–601.

long memory models

Much analysis of economic and financial time series focuses on stochastic modelling. Deterministic sequences, based on polynomials and dummy variables, can explain some trending or cyclic behaviour, but residuals typically exhibit serial dependence. Stochastic components have often been modelled by stationary, weakly dependent processes: parametric models include stationary and invertible autoregressive moving average (ARMA) processes, while a non-parametric approach usually focuses on a smooth spectral density. In many cases, however, we need to allow for a greater degree of persistence or 'memory'. This is characterized by stationary time series whose autocorrelations are not summable or whose spectral densities are unbounded, or by non-stationary series evolving over time. The latter are partly covered by unit root processes, but considerably greater flexibility is possible.

Basic models

Early empirical evidence of slowly decaying autocorrelations emerged long ago, in analyses of astronomical, chemical, agricultural and hydrological data, and then in economics and finance. A stationary parametric model which attracted early interest is 'fractional noise'. Let x_t, $t = 0, \pm 1, \ldots$, be a covariance stationary discrete time process, so its autocovariance $cov(x_t, x_{t+u})$ depends only on u, and thus may be denoted by γ_u. Then fractional noise x_t has autocovariance

$$\gamma_u = \gamma_0\{|u+1|^{2d+1} - 2|u|^{2d+1} + |u-1|^{2d+1}\}, \quad u = 0, \pm 1, \ldots, \tag{1}$$

where the parameter d is called the 'memory parameter', and satisfies $-\frac{1}{2} < d < \frac{1}{2}$. When $d = 0$ (1) implies that $\gamma_u = 0$ for $u \neq 0$, so x_t is white noise. But if $0 < d < \frac{1}{2}$, we have

$$\gamma_u \sim 2d(d+\frac{1}{2})\gamma_0|u|^{2d-1}, \quad \text{as } |u| \to \infty, \tag{2}$$

where '$\sim$' means that the ratio of left- and right-hand sides tends to one. It follows from (2) that γ_u does decrease with lag u, but so slowly that

$$\sum_{u=-\infty}^{\infty} \gamma_u = \infty. \tag{3}$$

In the frequency domain, when x_t has a spectral density $f(\lambda)$, $\lambda \in (-\pi, \pi)$ given by

$$f(\lambda) = (2\pi)^{-1} \sum_{u=-\infty}^{\infty} \gamma_u \cos(u\lambda), \lambda \in (-\pi, \pi),$$

the property (3) is equivalent to

$$f(0) = \infty, \tag{4}$$

and more precisely a fractional noise process x_t has spectral density satisfying

$$f(\lambda) \sim C\lambda^{-2d}, \quad \text{as } \lambda = 0+. \tag{5}$$

In general we can regard (3) and (4) as basic indicators of a 'long memory' process x_t, and (2) and (5) as providing more detailed description of autocorrelation structure at long lags, or spectral behaviour at low frequencies. By contrast, if x_t were a stationary ARMA, γ_u would decay exponentially and $f(\lambda)$ would be analytic at all frequencies. The structure (5) is similar to Granger's (1966) 'typical spectral shape of an economic variable'.

The model (1) is connected with the physical property of 'self-similarity', and, so far as economic and financial data are concerned, found early application in work of Mandelbrot (1972) and others. However, (1) imposes a very rigid structure, with autocorrelations decaying monotonically and depending on a single parameter. In addition, though a formula for $f(\lambda)$ corresponding to (1) can be written down, it is complicated, and (1) does not connect well mathematically with other important time series models, and does not lend itself readily to forecasting.

An alternative class of 'fractionally integrated' processes leads to a satisfactory resolution of these concerns. This is conveniently expressed in terms of the lag operator L, where $Lx_t = x_{t-1}$. Given the formal expansion

$$(1-s)^d = \sum_{j=0}^{\infty} \frac{\Gamma(j+d)}{\Gamma(d)\Gamma(j+1)} s^j,$$

we consider generating x_t from a zero-mean stationary sequence u_t, $t = 0, \pm 1, \ldots$, by

$$(1-L)^d (x_t - \mu) = v_t, \tag{6}$$

where $\mu = Ex_t$ and

$$|d| < \frac{1}{2}.$$

If v_t has absolutely summable autocorrelations, that satisfy some mild additional conditions, both the properties (2) and (5) hold. In the simplest case of (6), v_t is a white noise sequence. Then γ_u decays monotonically when $d \in (0, \frac{1}{2})$ and indeed behaves very much like (1). This model may have originated in Adenstedt (1974), though he stressed the case $d \in (-\frac{1}{2}, 0)$, where x_t is said to have 'negative dependence' or 'antipersistence'. Taking v_t to be a stationary and invertible ARMA process, with autoregressive order p and moving average order q, gives the FARIMA (p, d, q) process of Granger and Joyeux (1980). In principle, the short memory process v_t in (6) can be specified in any number of ways so as to yield (2) and/or (5); a process satisfying this condition is sometimes called $I(d)$.

Statistical inference

Given observations x_t, $t = 1, \ldots, n$ there is interest in estimating d. If v_t has parametric autocorrelation, as when x_t is a FARIMA (p, d, q), one can form a Gaussian maximum likelihood estimate of d and any other parameters. This estimate has the classical properties of being $n^{1/2}$ – consistent and asymptotically normal and efficient. Computationally somewhat more convenient estimates, called Whittle estimates, have the same asymptotic properties. Indeed, for standard FARIMA (p, d, q) parameterizations, say, the estimates of d and of ARMA coefficients have asymptotic variance matrix that is unaffected by many departures from Gaussianity. Though these asymptotic properties are of the same type as one obtains for estimates of short memory processes, such as ARMAs, their proof is considerably more difficult (see Fox and Taqqu, 1986), due to the spectral singularity (4). In econometrics, generalized method of moments (GMM) estimation has become very popular, and GMM estimates have been proposed for long memory models. However, unless a suitable weighting is used, they are not efficient under Gaussianity, are not more robust asymptotically to non-Gaussianity, and are not even asymptotically normal when $d > \frac{1}{4}$.

If the parametric autocorrelation is mis-specified, for example if in the FARIMA (p, d, q) p or q are chosen too small or both are chosen too large, then the procedures described in the previous paragraph will generally produce inconsistent estimates of d, as well as of other parameters. Essentially, the attempt to model the short memory component of x_t damages estimation of the long memory component. This difficulty can be tackled by a 'semiparametric' approach, if one regards the local or asymptotic specifications (2) or (5) as the model, and estimates d using only information in low frequencies or in long lags. Frequency domain versions are by far the more popular here, having the nicest asymptotic statistical properties. In the log periodogram estimate of d, logged periodograms are regressed on a logged local approximation to $f(\lambda)$, over the m Fourier frequencies closest to the origin (Geweke and Porter-Hudak, 1983), m having the character of a bandwidth number similar to those used in smoothed nonparametric functional estimation. An alternative approach optimizes a local Whittle function, again based on the lowest m Fourier frequencies (Künsch, 1987). In the asymptotics for both types of estimate (see Robinson, 1995a; 1995b) m must increase with n, but more slowly (to avoid bias); both the log periodogram and local Whittle estimates are $m^{1/2}$ – consistent and asymptotically normal, with the latter the more efficient (though it is computationally more onerous, being only implicitly defined). Because both converge more slowly than estimates of correctly specified parametric models, a larger amount of data may be necessary for estimates to be reasonably precise. Moreover, estimates are sensitive to the choice of m. However, automatic and other rules are available for determining m; and semiparametric methods of estimating memory parameters have become very popular not only because of the robust character of the asymptotic results, but because of their relative simplicity.

The long memory processes we have been discussing exhibit an excess of low frequency power (5). But one can also consider parametric or semiparametric models for a spectral density with one or more poles at non-zero frequencies. These models can be used to describe seasonal or cyclic behaviour (see Arteche and Robinson, 2000).

It is also possible to estimate the unknown location of a pole, that is, cycle (see Giraitis, Hidalgo and Robinson, 2001).

Nonlinear models

In non-Gaussian series, not all information is contained in first and second moments. In particular, in many financial series observations x_t may appear to have little or no autocorrelation, but instantaneous nonlinear functions, such as squares x_t^2, exhibit long memory behaviour. We can develop models to describe such phenomena. For example, let

$$x_t = \varepsilon_t h_t, \tag{7}$$

where x_t is a sequence of independent and identically distributed random variables with unit variance, whereas h_t is a stationary autocorrelated sequence, such that ε_s and h_t are independent for all s, t. Then for all $u \neq 0$, $cov = (x, x_{t+u}) = 0$ but $cov(x_t^2, x_{t+u}^2) = cov(h_t^2, h_{t+u}^2)$, which in general can be non-zero. In particular, if h_t^2 has long memory, so has x_t^2. In a more fundamental modelling we can take h_t to be a nonlinear function of an underlying long memory Gaussian processes, with the functional form of h determining the extent of any long memory in h_t^2; these issues were discussed in some generality by Robinson (2001). The models form a class of long memory stochastic volatility models, whose estimation has been discussed by Hurvich, Moulines and Soulier (2005), for example.

The fractional class (6) can be modified or extended to describe a wide class of nonstationary behaviour. For $d \geq \frac{1}{2}$ the variance of x_t (6) explodes, but we can consider truncated versions such as

$$x_t = (1 - L)^{-d}\{v_t 1(t \geq 1)\}$$

where $1(\cdot)$ is the indicator function, or

$$x_t = (1 - L)^{-k}\{w_t 1(t \geq 1)\}$$

for integer $k \geq 1$, where w_t is a stationary $I(c)$ process, $|c| < \frac{1}{2}$, and $d = k + c$. In either case we might call x_t a (nonstationary) $I(d)$ process, for $d \geq \frac{1}{2}$. Both models include the unit root case $d = 1$ that has proved so popular in econometrics. However, the fractional class $I(d)$, for real-valued d, bridges the gap between short memory and unit root processes, allowing also for the possibility of arbitrarily long memory d. The 'smoothness' of the $I(d)$ family is associated with classical asymptotic theory, which is not found in autoregressive based models around a unit root. Robinson (1994) showed that Lagrange multiplier tests for the value of d, and any other parameters, have asymptotic null χ^2 distributions for all real d. Also, under nonstationary suitably modified parametric and semiparametric methods of estimating d, extending those for the stationary case, tend still to be respectively $n^{1/2}-$ and $m^{1/2}$ – consistent, and asymptotically normal, unlike, say, the lag-one sample autocorrelation of a unit root series.

Multivariate models

Often in economics and finance we are concerned with a vector of jointly dependent series, so x_t is vector-valued. Such series can be modelled, either parametrically or semiparametrically, to have long memory, with different elements of x_t possibly having different memory parameters, and being stationary or nonstationary. Methods of statistical inference developed for the univariate case can be extended to such settings. However, multivariate data introduces the possibility of (fractional) cointegration, where a linear combination of x_t (the cointegrating error) can have smaller memory parameter than the elements of x_t. Cointegration has been extensively developed for the case x_t is $I(1)$ and cointegrating errors are $I(0)$, and methods developed for this case can fail to detect fractional cointegration. Moreover, it is possible for stationary series, not only nonstationary ones, to be fractionally cointegrated, as seems relevant in financial series. In either case, methods of analysing cointegration that allow memory parameters of observables and cointegrating errors to be unknown (see, for example, Hualde and Robinson, 2004) afford considerable flexibility.

P.M. ROBINSON

See also **central limit theorems; time series analysis.**

Research supported by ESRC Grant R000239936.

Bibliography

Adenstedt, R. 1974. On large-sample estimation for the mean of a stationary random sequence. *Annals of Statistics* 2, 1095–107.

Arteche, J. and Robinson, P. 2000. Semiparametric inference in seasonal and cyclic long memory processes. *Journal of Time Series Analysis* 21, 1–25.

Fox, R. and Taqqu, M.S. 1986. Large sample properties of parameter estimates of strongly dependent stationary Gaussian time series. *Annals of Statistics* 14, 517–32.

Geweke, J. and Porter-Hudak, S. 1983. The estimation and application of long memory time series models. *Journal of Time Series Analysis* 4, 221–38.

Giraitis, L., Hidalgo, J. and Robinson, P. 2001. Gaussian estimation of parametric spectral density with unknown pole. *Annals of Statistics* 29, 987–1023.

Granger, C. 1966. The typical spectral shape of an economic variable. *Econometrica* 34, 150–67.

Granger, C. and Joyeux, R. 1980. An introduction to long memory time series models and fractional differencing. *Journal of Time Series Analysis* 1, 15–39.

Hualde, J. and Robinson, P. 2004. Semiparametric estimation of fractional cointegration. Mimeo, London School of Economics.

Hurvich, C., Moulines, E. and Soulier, P. 2005. Estimating long memory in volatility. *Econometrica* 73, 1283–328.

Künsch, H. 1987. Statistical aspects of self-similar processes. In *Proceedings of the First World Congress of the Bernoulli Society*, vol. 1, ed. Y. Prohorov and V. Sazonov. Utrecht: VNU Science Press.

Mandelbrot, D. 1972. Statistical methodology for non-periodic cycles: from the covariance to R/S analysis. *Annals of Economic and Social Measurement* 1, 259–90.

Robinson, P. 1994. Efficient tests of nonstationary hypotheses. *Journal of the American Statistical Association* 89, 1420–37.

Robinson, P. 1995a. Log-periodogram regression of time series with long range dependence. *Annals of Statistics* 23, 1048–72.
Robinson, P. 1995b. Gaussian semiparametric estimation of long range dependence. *Annals of Statistics* 5, 1630–61.
Robinson, P. 2001. The memory of stochastic volatility models. *Journal of Econometrics* 101, 192–218.

nonlinear time series analysis

Since the early 1980s, there has been a growing interest in stochastic nonlinear dynamical systems of the form

$$x_{t+1} = f(x_t, x_{t-1}, \ldots, x_{t-p}) + \sigma(x_t)\varepsilon_t, \tag{1}$$

where $\{x_t\}_{t=0}^{\infty}$ is a zero mean, covariance stationary process, $f : R^{p+1} \rightarrow R$, σ is the conditional volatility, and $\{\varepsilon_t\}_{t=0}^{\infty}$ is an independent and identically distributed noise process. The major recent developments in nonlinear time series are described here using this canonical model. The first section develops representation theory for a third order approximation. Nonparametric approaches follow; these rely on series expansions of the general model. Ergodic properties including path dependence and dimension are considered next. I then consider two widely utilized parametric models, piecewise linear models of f and autoregressive models for volatility. I conclude with a discussion of hypothesis testing and forecasting.

Volterra expansion

There is no general causal representation for nonlinear time series as in the linear case. Series approximations rely on the *Volterra expansion*,

$$x_{t+1} \simeq f(0) + \sum_{i=1}^{p} f_{i_1} x_{t-i_1} + \sum_{i_1=1}^{p} \sum_{i_2=i_1}^{p} f_{i_1 i_2} x_{t-i_1} x_{t-i_2} + \sum_{i_1=1}^{p} \sum_{i_2=i_1}^{p} \sum_{i_3=i_2}^{p} f_{i_1 i_2 i_3} x_{t-i_1} x_{t-i_2} x_{t-i_3} + \cdots \tag{2}$$

Brockett (1976) shows any continuous map over $[0, T]$ can be approximated by a finite Volterra series. Mittnik and Mizrach (1992) examine forecasts using generalized polynomial expansions like (2). Potter (2000) shows that in the cubic case, a one-sided Wold-type representation in terms of white noise v_t can be obtained,

$$x_{t+1} \simeq \sum_{i=1}^{\infty} g_{i_1} v_{t-i_1} + \sum_{i_1=1}^{\infty} \sum_{i_2=i_1}^{\infty} g_{i_1 i_2} v_{t-i_1} v_{t-i_2} + \sum_{i_1=1}^{\infty} \sum_{i_2=i_1}^{\infty} \sum_{i_3=i_2}^{\infty} g_{i_1 i_2 i_3} x_{t-i_1} x_{t-i_2} x_{t-i_3}. \tag{3}$$

Koop, Pesaran and Potter (1996) note that the impulse response functions, $E[x_{t+n}|x_t, v_t] - E[x_{t+n}|x_t]$ will depend upon the size and sign of v_t as well as the current state x_t.

I now turn to nonparametric approaches which build on approximations like 2.

Nonparametric estimation

Consider the local polynomial approximation to $f(.)$ around x_0,

$$\hat{f}(x) = \sum_{j=0}^{m} \beta_j (x - x_0)^j. \tag{4}$$

In the case $j = 0$, this corresponds to the *kernel regression* estimator of Nadaraya and Watson,

$$\hat{f}(x) = \frac{\sum_{t=1}^{T} x_{t+1} K_h(x_t - x_0)}{\sum_{t=1}^{T} K_h(x_t - x_0)}. \tag{5}$$

The K_h are *kernels*, usually functions with a support on a compact set, assigning greater weight to observations closer to x_0. h is the *bandwidth* parameter, determining the size of the histogram bin. *Nearest neighbours* estimation is the case where h is adjusted to find a fixed number of nearby observations k.

More generally, the *local linear approximation* solves,

$$\alpha_0, \beta_0 \min \left(x_{t+1} - \alpha_0 - \beta_0(x_t - x_0)\right)^2 K_h(x_t - x_0). \tag{6}$$

The estimator (5) corresponds to the case where the only regressor in (6) is the constant term.

The application of these methods in the time series case is a fairly recent development. Conditions for consistency and asymptotic normality rely on *mixing conditions* where the dependence between x_{t+j} and x_t becomes negligible as j grows large.

A closely related approach involves the use of a *recurrent neural network*,

$$\Psi_i(x_t, h_{t-1}) = \Psi(\gamma_{i0} + \gamma_{i1} x_t + \sum_{k=1}^{r} \delta_{ik} h_{k,t-k}),$$

$$x_{t+1} = \Phi(\beta_0 + \sum_{i=1}^{p} \beta_i \Psi_i(x_t, h_{t-1})). \tag{7}$$

Kuan, Hornik and White (1994) provide convergence results for bounded Ψ (most commonly the logistic) as p grows large.

A popular approach in the frequency domain is wavelets. The *discrete wavelet transform* is

$$x_{t+1} = \sum_{j=-\infty}^{\infty} \sum_{k=-\infty}^{\infty} \gamma(j,k) \Psi_{j,k}(t), \tag{8}$$

where the *mother wavelet* $\Psi(t)$,

$$\Psi_{j,k}(t) = \frac{1}{\sqrt{s_0^j}} \Psi\left(\frac{t - k\tau_0 s_0^j}{s_0^j}\right), \tag{9}$$

is parameterized by scale s_0 and translation τ, and the wavelet coefficients are given by

$$\gamma(j,k) = \langle \Psi_{j,k}(t), x(t) \rangle. \tag{10}$$

Daubechies (1992) orthonormal basis functions,

$$E[\Psi_{j,k}(t)\Psi_{m,n}(t)] = 0, \quad \forall j \neq m, k \neq n, \tag{11}$$

have received the widest application.

Even when very little is known about f or σ, nonlinear time series analysis can shed light on the long run average or *ergodic* properties of the dynamical system.

Ergodic properties

Mathematicians have known since Poincaré that even simple maps like (1) can produce very complex dynamics. The nonlinear time series literature has developed tools for estimation of ergodic properties of these systems. Denote by $Df(\bar{x})$ the Jacobian matrix of partial derivatives of (1),

$$\begin{bmatrix} \partial f_1/\partial x_1 & \cdots & \partial f_1/\partial x_p \\ \vdots & \ddots & \vdots \\ \partial f_p/\partial x_1 & \cdots & \partial f_p/\partial x_p \end{bmatrix} \tag{12}$$

evaluated at $\bar{x}$. Replacing 12 with a sample analog,

$$J_t = \begin{bmatrix} \Delta f_1/\Delta x_{1,t} & \cdots & \Delta f_1/\Delta x_{p,t} \\ \vdots & \ddots & \vdots \\ \Delta f_p/\Delta x_{1,t} & \cdots & \Delta f_p/\Delta x_{p,t} \end{bmatrix} \tag{13}$$

we compute eigenvalues V_i,

$$V_i(Q'_T Q_T) \tag{14}$$

rank ordered from $1, \ldots, p$, where

$$Q_T = J_{T-p} \cdot J_{T-p-1} \cdots J_1 \tag{15}$$

The *Lyapunov exponents* are defined for the positive eigenvalues V_i^+ as

$$T \to \infty \lim \lambda_i = \frac{1}{2(T-p)} \ln V_i^+, \tag{16}$$

and a single exponent greater than 1 characterizes a system with sensitive dependence. Popularly known as 'chaos', this property implies that dynamic trajectories become unpredictable even when the state of the system is known with certainty. Gençay and Dechert (1992) and Shintani and Linton (2004) provide methods for estimating these. Shintani and Linton (2003; 2004) reject the presence of positive Lyapunov exponents in both real output and stock returns.

The sum of the Lyapunov exponents also provides a measure of the Kolmogorov–Sinai *entropy* of the system. This tells the researcher how quickly trajectories separate. Mayfield and Mizrach (1991) estimate this time at about 15 minutes for the S&P 500 index.

A final quantity of interest is the *dimension p* of the dynamical system. Nonlinear econometricians try to estimate the dimension from a scalar m-history. A powerful result due to Takens (1981) says this can be done as long as $m \geq 2p + 1$. Diks (2004) has shown that the scaling of correlation exponents seems to be consistent with the stochastic volatility model.

A great deal of progress has been made with parametric models of (1) as well. I begin with the widely utilized piecewise linear models.

Piecewise linear models

The most widely applied parametric nonlinear time series specification has been the Markov switching model introduced by James Hamilton (1989). The function f is a piecewise linear function,

$$f(x_t) = \left\{ \begin{array}{l} \mu^{(1)} + \sum_{j=0}^{p} \phi_j^{(1)} \left(x_{t-j} - \mu^{(1)}, S_t = s_t^{(1)} \right) \\ \vdots \\ \mu^{(m)} + \sum_{j=0}^{p} \phi_j^{(m)} \left(x_t - \mu^{(m)}, S_t = s_t^{(m)} \right) \end{array} \right\}, \tag{17}$$

where the changes among states are governed by an unobservable regime switching process, $S_t = s_t^{(i)}$, $i = 1, \ldots m$, an $m \times m$ transition matrix Π, and $E[x_t|S_t = s_t^{(i)}] = \mu^{(i)}$. When S_t is unobserved, $Pr(S_t|x_{t-1})$ is nonlinear in x_{t-1}. Hamilton has shown that a two-dimensional switching model describes well the business cycle dynamics in the United States. This model has been extended to include regime dependence in volatility (Kim, 1994) and time varying transition probabilities (Filardo, 1994).

The latent state vector requires forming prior and posterior estimates of which regime you are in. The EM algorithm (Hamilton, 1990) and Bayesian Gibbs sampling methods (Albert and Chib, 1993) have proven fruitful in handling this problem. Hypothesis testing is also non-standard because under the alternative of $m - 1$ regimes, the conditional mean parameters are nuisance parameters. Hansen (1996) has explored carefully these issues.

A closely related framework is the *threshold autoregressive* (TAR) model,

$$f(x_t) = \left\{ \begin{array}{l} \left[\mu^{(1)} + \sum_{j=0}^{p} \phi_j^{(1)} \left(x_{t-j} - \mu^{(1)} \right) \right] I(q(x_{t-d,} Z_t) \leq \gamma_1) \\ \left[\mu^{(2)} + \sum_{j=0}^{p} \phi_j^{(2)} \left(x_{t-j} - \mu^{(2)} \right) \right] I(\gamma_1 < q(x_{t-d,} Z_t) \leq \gamma_2) \\ \vdots \\ \left[\mu^{(m)} + \sum_{j=0}^{p} \phi_j^{(m)} \left(x_t - \mu^{(m)} \right) \right] I(q(x_{t-d,} Z_t) > \gamma_{m-1}) \end{array} \right\}. \tag{18}$$

$I(.)$is the indicator function, and $q(x_{t-d}, Z_t)$, the regime switching variable, is assumed to be an observable function of exogenous variables Z_t and lagged x's. The integer d is known as the *delay parameter*. When q depends only upon x, the model is called *self-exciting*.

Teräsvirta (1994) has developed a two-regime version of the TAR model in which regime changes are governed by a smooth transition function $G(x_{t-d}, Z_t) : R^k \to [0, 1]$,

$$f(x_t) = G(x_{t-d}, Z_t){}_{j=0}^{p}\phi_j^{(1)}(x_{t-j} - \mu^{(1)}) + (1 - G(x_{t-d}, Z_t)){}_{j=0}^{p}\phi_j^{(2)}(x_{t-j} - \mu^{(2)}). \quad (19)$$

Luukkonen, Saikkonen and Teräsvirta (1988) have shown that inference and hypothesis testing in this model is often much simpler than in the piecewise linear models. Van Dijk and Franses (1999) have extended this model to multiple regimes. Applications of this framework have been widespread from macroeconomics (Teräsvirta and Anderson, 1992) to empirical finance (Franses and van Dijk, 2000).

Krolzig (1997) considers the multivariate case where $x_t = (x_{1,t}, x_{2,t}, \ldots, x_{k,t})'$ is $k \times 1$. Balke and Fomby (1997) introduced threshold cointegration by incorporating error correction terms into the thresholds. Koop, Pesaran and Potter (1996) develop a bivariate model of US GDP and unemployment where the threshold depends upon the depth of the recession.

I now turn to models that introduce nonlinearity through the error term.

Models of volatility

Engle and Bollerslev have introduced the generalized autoregressive conditional heteroskedasticity (GARCH) model,

$$h_t = \alpha_0 + {}_{i=1}^{q}\alpha_i\sigma^2(x_{t-i})\varepsilon_{t-i}^2 + {}_{i=1}^{p}\beta_i h_{t-i}, \quad (20)$$

where $h_t = E[(x_t - E[x_t|\Omega_{t-1}])^2|\Omega_{t-1}]$ is the *conditional variance*. This is just a Box–Jenkins model in the squared residuals of 1 of order $(max[p, q], p)$. The model is nonlinear because the disturbances are uncorrelated, but their squares are not.

The GARCH model describes the volatility clustering and heavy-tailed returns in financial market data, and has found wide application in asset pricing and risk management applications.

Volatility modelling has been motivated by the literature on options pricing. Popular alternatives to the GARCH model include the stochastic volatility (SV) model (Ghysels, Harvey and Renault, 1996), and the realized volatility approach of Andersen et al. (2003) and Barndorff-Nielsen and Shephard (2002). The discrete-time SV model takes the form,

$$x_t = \sigma_\varepsilon \exp{(h_t/2)}\varepsilon_t, \quad (21)$$

$$h_t = \beta h_{t-1} + \sigma_{ht},$$

where x_t is the demeaned log asset return, and ε_t and η_t are noise terms. Realized volatility sums high-frequency squared returns as an approximation of lower frequency volatility. Both GARCH and SV have been successful in explaining the departures from the Black–Scholes observed empirically.

The final two sections address the marginal contribution of nonlinear modelling to goodness of fit and forecasting.

Testing for linearity and Gaussianity

There is a large literature on testing the importance of the nonlinear components of a model. The most widely used test is due to Brock, Dechert, Scheinkman and LeBaron (BDSL, 1996). Their nonparametric procedure is built upon U-statistics. Serfling (1980) is a good introduction.

The first step is to form m-histories of the data,

$$x_t^m = (x_t, x_{t+1}, \dots, x_{t+m-1}), \tag{22}$$

with joint distribution $F(x_t^m)$. Introduce the kernel $h : R^m \times R^m \rightarrow R$,

$$h(x_t^m, x_s^m) = I(x_t^m, x_s^m, \varepsilon) \equiv I[\|x_t^m - x_s^m\| < \varepsilon], \tag{23}$$

where $I(.)$ is the indicator function. The *correlation integral* of Grassberger and Procaccia (1983),

$$C(m, \varepsilon) \equiv \int_X \int_X I(x_t^m, x_s^m, \varepsilon) dF(x_t^m) dF(x_s^m), \tag{24}$$

is the expected number of m-vectors in an ε neighbourhood. A U-statistic,

$$C(m, N, \varepsilon) \equiv \frac{2}{N(N-1)} \sum_{t=1}^{N-1} \sum_{s=t+1}^{N} I(X_t^m, X_s^m, \varepsilon), \tag{25}$$

is a consistent estimator of 24. BDSL demonstrate the asymptotic normality of the statistic

$$\sqrt{N} \frac{S(m, N, \varepsilon)}{\sqrt{Var[S(m, N, \varepsilon)]}} \sqrt{N} d\times \rightarrow N(0, 1), \tag{26}$$

where

$$S(m, N, \varepsilon) = C(m, N, \varepsilon) - C(1, N, \varepsilon)^m. \tag{27}$$

There is a multi-dimensional extension due to Baek and Brock (1992). De Lima (1997) explores the use of the BDSL under moment condition failure.

There is a direct relationship between nonlinear and non-Gaussian time series. In the model (1), even if the disturbance term ε_t is normal, nonlinear transformations of Gaussian noise will make x_t non-Gaussian. Testing for Gaussianity is then an instrumental part of the nonlinear time series toolkit.

Hinich (1982) has developed testing in the time domain using the *bicorrelation*,

$$\gamma(r,s)={}_{t=1}^{s}x_t x_{t+r} x_{t+s}/(N-s),\quad 0 \leq r \leq s, \tag{28}$$

and in the frequency domain using the *bispectrum*,

$$B(\omega_1,\omega_2)={}_{r=-\infty}^{\infty}{}_{s=-\infty}^{\infty}\gamma(r,s)\exp\left[-i(\omega_1 r+\omega_2 s)\right]. \tag{29}$$

For a Gaussian time series, the bicorrelation should be close to zero, and the bispectrum should be flat across all frequencies. Both tests have good power against skewed alternatives.

Ramsey and Rothman (1996) have proposed a related time domain procedure that looks for *time reversibility*,

$$F(X_t, X_{t+1}, \ldots, X_{t+r}) = F(X_{s-t}, X_{s-t-1}, \ldots, X_{s-t-r}) \tag{30}$$

for any r, s and t, where $F(.)$ is the joint distribution. This condition is stronger than stationarity because of the triple index. The authors find evidence of business cycle asymmetry using this diagnostic.

Forecasting

For many, the bottom line on nonlinear modelling is the ability to generate superior forecasts. In this respect, the results from the nonlinear literature are decidedly mixed. Harding and Pagan (2002) are prominent sceptics. Teräsvirta, van Dijk and Medeiros (2005) provide a very wide set of evidence in favour of nonlinear models.

Aside from the comparison of point forecasts from model i, $u_{i,t+1} = x_{t+1} - f_i(x_t)$, with a particular loss function $g(.)$,

$$H_0 : E[g(u_{i,t+1}) - g(u_{j,t+1})] = 0, \tag{31}$$

there has been growing interest in comparing forecast densities $p_i(x_{t+1}|f_i(x_t))$,

$$H_0 : \int [p_i(x_{t+1}|f_i(x_t)) - p_j(x_{t+1}|f_j(x_t))]dx = 0. \tag{32}$$

Corradi and Swanson (2005) provide a comprehensive overview of available tools.

BRUCE MIZRACH

See also **stochastic volatility models.**

I would like to thank Cees Diks, James Hamilton, Sebastiano Manzan, Simon Potter, Phil Rothman, Dick van Dijk and Steven Durlauf for helpful comments.

Bibliography

Albert, J. and Chib, S. 1993. Bayesian analysis via Gibbs sampling of autoregressive time series subject to Markov mean and variance shifts. *Journal of Business and Economic Statistics* 11, 1–15.

Andersen, T.G., Bollerslev, T., Diebold, F.X. and Labys, P. 2003. Modeling and forecasting realized volatility. *Econometrica* 71, 579–625.

Barndorff-Nielsen, O.E. and Shephard, N. 2002. Econometric analysis of realised volatility and its use in estimating stochastic volatility models. *Journal of the Royal Statistical Society*, Series B 64, 253–80.

Baek, B. and Brock, W.A. 1992. A nonparametric test for independence of a multivariate time series. *Statistica Sinica* 2, 137–56.

Balke, N. and Fomby, T. 1997. Threshold cointegration. *International Economic Review* 38, 627–45.

Brock, W.A., Dechert, W.D., Scheinkman, J.A. and LeBaron, B. 1996. A test for independence based on the correlation dimension. *Econometric Reviews* 15, 197–235.

Brockett, R.W. 1976. Volterra series and geometric control theory. *Automatica* 12, 167–76.

Corradi, V. and Swanson, N.R. 2005. Predictive density evaluation. In *Handbook of Economic Forecasting*, ed. C.W.J. Granger, A. Timmermann and G. Elliott. Amsterdam: North-Holland.

Daubechies, I. 1992. *Ten Lectures on Wavelets*. 2nd ed. Philadelphia: SIAM.

De Lima, P. 1997. On the robustness of nonlinearity tests due to moment condition failure. *Journal of Econometrics* 76, 251–80.

Diks, C. 2004. The correlation dimension of returns with stochastic volatility. *Quantitative Finance* 4, 45–54.

Filardo, A.J. 1994. Business cycle phases and their transitional dynamics. *Journal of Business and Economic Statistics* 12, 299–308.

Franses, P.H. and van Dijk, D. 2000. *Nonlinear Time Series Models in Empirical Finance*. New York: Cambridge University Press.

Gallant, A.R. and Tauchen, G. 1987. Seminonparametric maximum likelihood estimation. *Econometrica* 55, 363–90.

Gallant, A.R. and Tauchen, G. 1996. Which moments to match? *Econometric Theory* 12, 657–81.

Gençay, R. and Dechert, W.D. 1992. An algorithm for the n Lyapunov exponents of an n-dimensional unknown dynamical system. *Physica D* 59, 142–57.

Ghysels, E., Harvey, A. and Renault, E. 1996. Stochastic volatility. In *Handbook of Statistics* 14: *Statistical Methods in Finance*, ed. G.S. Maddala and C.R. Rao. Amsterdam: North-Holland.

Granger, C.W.J. and Hallman, J. 1991. Nonlinear transformations of integrated time series. *Journal of Time Series Analysis* 12, 207–24.

Grassberger, P. and Procaccia, J. 1983. Measuring the strangeness of strange attractors. *Physica D* 9, 189–208.

Hamilton, J.D. 1989. A new approach to the economic analysis of nonstationary time series and the business cycle. *Econometrica* 57, 357–84.

Hamilton, J.D. 1990. Analysis of time series subject to changes in regime. *Journal of Econometrics* 45, 39–70.

Hansen, B.E. 1996. Inference when a nuisance parameter is not identified under the null hypothesis. *Econometrica* 64, 413–30.

Harding, D. and Pagan, A. 2002. Dissecting the cycle: a methodological investigation. *Journal of Monetary Economics* 49, 365–81.

Hinich, M.J. 1982. Testing for Gaussianity and linearity of a stationary time series. *Journal of Time Series Analysis* 3, 169–76.

Kim, C.J. 1994. Dynamic linear models with Markov-switching. *Journal of Econometrics* 60, 1–22.

Koop, G., Pesaran, H. and Potter, S. 1996. Impulse response analysis in nonlinear multivariate models. *Journal of Econometrics* 74, 119–48.
Krolzig, H.-M. 1997. *Markov Switching Vector Autoregressions: Modelling, Statistical Inference and Application to Business Cycle Analysis*. Berlin: Springer.
Kuan, C.-M., Hornik, K. and White, H. 1994. A convergence result for learning in recurrent neural networks. *Neural Computation* 6, 620–40.
Luukkonen, R., Saikkonen, P. and Teräsvirta, T. 1988. Testing linearity against smooth transition autoregressive models. *Biometrika* 75, 491–9.
Mayfield, S. and Mizrach, B. 1991. Nonparametric estimation of the correlation exponent. *Physical Review* A 88, 5298–301.
Mittnik, S. and Mizrach, B. 1992. Parametric and seminonparametric analysis of nonlinear time series. In *Advances in GLIM and Statistical Modeling*, ed. L. Fahrmeir, B. Francis, R. Gilchrist and G. Tutz. New York: Springer.
Potter, S. 2000. Nonlinear impulse response functions. *Journal of Economic Dynamics and Control* 24, 1425–46.
Ramsey, J.B. and Rothman, P. 1996. Time irreversibility and business cycle asymmetry. *Journal of Money, Credit, and Banking* 28, 1–21.
Serfling, R.J. 1980. *Approximation Theorems of Mathematical Statistics*. New York: John Wiley.
Shintani, M. and Linton, O. 2003. Is there chaos in the world economy? A nonparametric test using consistent standard errors. *International Economic Review* 44, 331–58.
Shintani, M. and Linton, O. 2004. Nonparametric neural network estimation of Lyapunov exponents and a direct test for chaos. *Journal of Econometrics* 120, 1–33.
Takens, F. 1981. Detecting strange attractors in turbulence. In *Springer Lecture Notes in Mathematics*, vol. 898, ed. D. Rand and L.-S. Young. Berlin: Springer.
Teräsvirta, T. 1994. Specification, estimation, and evaluation of smooth transition autoregressive models. *Journal of the American Statistical Association* 89, 208–18.
Teräsvirta, T. and Anderson, H.M. 1992. Characterizing nonlinearities in business cycles using smooth transition autoregressive models. *Journal of Applied Econometrics* 7, S119–S136.
Teräsvirta, T., van Dijk, D. and Medeiros, M.C. 2005. Linear models, smooth transition autoregressions, and neural networks for forecasting macroeconomic time series: a re-examination. *International Journal of Forecasting* 21, 755–74.
Van Dijk, D. and Franses, P.H. 1999. Modeling multiple regimes in the business cycle. *Macroeconomic Dynamics* 3, 311–40.

prediction formulas

1. Introduction

This article reviews the derivation of formulas for linear least squares and robust prediction of stationary time series and geometrically discounted distributed leads of such series. The derivations employed are the classical, frequency-domain procedures employed by Whittle (1983) and Whiteman (1983), and result in nearly closed-form expressions. The formulas themselves are useful directly in forecasting, and have also found uses in economic modelling, primarily in macroeconomics. Indeed, Hansen and Sargent (1980) refer to the cross-equation restrictions connecting the time series representation of driving variables to the analogous representation for predicting the present value of such variables as the 'hallmark of rational expectations models'.

2. The Wold representation

Suppose that $\{x_t\}$ is a covariance-stationary stochastic process and assume (without loss of generality) that $Ex_t = 0$. Covariance stationarity ensures that first and second unconditional moments of the process do not vary with time. Then, by the Wold decomposition theorem (see Sargent, 1987, for an elementary exposition and proof), x_t can be represented by:

$$x_t = \sum_{j=0}^{\infty} a_j \varepsilon_{t-j} \tag{1}$$

with

$$a_0 = 1, \sum_{j=0}^{\infty} a_j^2 < \infty$$

and

$$\varepsilon_t = x_t - P(x_t | x_{t-1}, x_{t-2}, \ldots), \quad E\varepsilon_t^2 = \sigma^2$$

where $P(x_t | x_{t-1}, x_{t-2}, \ldots)$ denotes the linear least squares projection (population regression) of x_t on $x_{t-1}, x_{t-2}, \ldots$ here, 'represented by' need not mean 'generated by', but rather 'has the same variance and covariance structure as'. By construction, the 'fundamental' innovation ε_t is uncorrelated with information dated prior to t, including earlier values of the process itself: $E\varepsilon_t \varepsilon_{t-s} = 0 \ \forall s > 0$. This fact makes the Wold representation very convenient for computing predictions. The convolution in (1) is often written $x_t = A(L)\varepsilon_t$ using the polynomial $A(L) = \sum_{j=0}^{\infty} a_j L^j$ in the 'lag operator' L, where $L\varepsilon_t = \varepsilon_{t-1}$.

3. Squared-error loss optimal prediction

The optimal prediction problem under squared-error loss can be thought of as follows. Given $\{x_t\}$ with the Wold representation (1) we want to find the stochastic process y_t,

$$y_t = \sum_{j=0}^{\infty} c_j \varepsilon_{t-j} = C(L)\varepsilon_t$$

that will minimize the squared forecast error of the h-step ahead prediction

$$\min_{\{y_t\}} E(x_{t+h} - y_t)^2.$$

Equivalently, the problem can be written as

$$\min_{\{y_t\}} E(L^{-h}x_t - y_t)^2$$

or

$$\min_{\{c_j\}} E\left(L^{-h}\sum_{j=0}^{\infty} a_j\varepsilon_{t-j} - \sum_{j=0}^{\infty} c_j\varepsilon_{t-j}\right)^2. \tag{2}$$

The problem in (2) involves finding a *sequence* of coefficients in the Wold representation of the unknown prediction process y_t, and is referred to as the *time domain problem*. By virtue of the Riesz–Fisher theorem (see again Sargent, 1987, for an exposition), the time-domain problem is equivalent to a *frequency domain problem* of finding an analytic function $C(z)$ on the unit disk $|z| \leq 1$ corresponding to the 'z-transform' of the $\{c_j\}$ sequence

$$C(z) = \sum_{j=0}^{\infty} c_j z^j$$

that solves

$$\min_{C(z)\in H^2} \frac{1}{2\pi i}\oint | \; z^{-h}A(z) - C(z)|^2 \frac{dz}{z} \tag{3}$$

where H^2 denotes the Hardy space of square-integrable analytic functions on the unit disk, and $\oint$ denotes (counterclockwise) integration about the unit circle. The requirement that $C(z) \in H^2$ ensures that the forecast is causal, and contains no future values of the ε's; this is equivalent to the requirement that $C(z)$ have a well-behaved power series expansion in non-negative powers of z.

Each formulation of the problem is useful, as often one or the other will be simpler to solve. This stems from the fact that convolution in the time domain becomes multiplication in the frequency domain and vice versa. To see this, consider the two sequences $\{g_k\}_{k=-\infty}^{\infty}$ and $\{h_k\}_{k=-\infty}^{\infty}$. The convolution of $\{g_k\}$ and $\{h_k\}$ is the sequence $\{f_k\}$, in which a typical element would be:

$$f_k = \sum_{j=-\infty}^{\infty} g_j h_{k-j}.$$

The z-transform of the convolution is given by

$$\begin{aligned}
\sum_{k=-\infty}^{\infty} f_k z^k &= \sum_{k=-\infty}^{\infty} \left(\sum_{j=-\infty}^{\infty} g_j h_{k-j} \right) z^k = \sum_{k=-\infty}^{\infty} \sum_{j=-\infty}^{\infty} g_j z^j h_{k-j} z^{k-j} \\
&= \sum_{(k-j)=-\infty}^{\infty} \sum_{j=-\infty}^{\infty} g_j z^j h_{k-j} z^{k-j} = \sum_{s=-\infty}^{\infty} \sum_{j=-\infty}^{\infty} g_j z^j h_s z^s \\
&= (\text{Substituting } s = k - j) \\
&= \sum_{s=-\infty}^{\infty} h_s z^s \sum_{j=-\infty}^{\infty} g_j z^j = g(z)h(z).
\end{aligned}$$

Thus the 'z-transform' of the convolution of the sequences $\{g_k\}$ and $\{h_k\}$ is the product of the z-transforms of the two sequences.

Similarly, the z-transform of the product of two sequences is the convolution of the z-transforms:

$$\sum_{k=-\infty}^{\infty} g_k h_k z^k = \frac{1}{2\pi i} \oint g(p) h(z/p) \frac{dp}{p}.$$

To see why this is the case, note that

$$g(p)h(z/p)p^{-1} = \sum_{j=-\infty}^{\infty} g_j p^j \sum_{k=-\infty}^{\infty} h_k z^k p^{-k-1},$$

implying

$$\frac{1}{2\pi i} \oint g(p)h(z/p)p^{-1} dp = \frac{1}{2\pi i} \oint \sum_{j=-\infty}^{\infty} \sum_{k=-\infty}^{\infty} g_j h_k z^k p^{j-k-1} dp.$$

But all of the terms vanish except where $j = k$ because

$$\frac{1}{2\pi i} \oint z^k \frac{dz}{z} = 0$$

except when $k = 0$. To see why, let $z = e^{i\theta}$. As θ increases from 0 to 2π, z goes around the unit circle. So, since $dz = ie^{i\theta} d\theta$, we have that

$$\frac{1}{2\pi i} \oint z^k \frac{dz}{z} = \frac{i}{2\pi i} \oint e^{i\theta k} d\theta = \begin{cases} 1 & \text{if } k = 0 \\ \frac{1}{2\pi} \frac{1}{ik} e^{i\theta k} \Big|_0^{2\pi} = 0 & \text{otherwise.} \end{cases}$$

Thus,

$$\frac{1}{2\pi i}\oint g(p)h(z/p)p^{-1}dp = \sum_{j=-\infty}^{\infty} g_j h_j z^j \frac{1}{2\pi i}\oint \frac{dp}{p} = \sum_{j=-\infty}^{\infty} g_j h_j z^j$$

by Cauchy's Integral formula.

The frequency domain formulas can now be used to calculate moments quickly and conveniently. Consider Ex_t^2:

$$Ex_t^2 = E(A(L)\varepsilon_t)^2 = E\left(\sum_{j=0}^{\infty} A_j\varepsilon_{t-j}\right)^2 = \sigma_\varepsilon^2 \sum_{j=0}^{\infty} A_j^2. \tag{4}$$

The result in eq. (4) comes from the fact that $E\varepsilon_t\varepsilon_{t-s} = 0, \ \forall s \neq 0$. Using the product-convolution relation, we see that

$$\begin{aligned}\sum_{j=0}^{\infty} A_j^2 &= \sum_{j=0}^{\infty} A_j^2 z^j|_{z=1} \\ &= \frac{1}{2\pi i}\oint A(p)A(z/p)\frac{dp}{p}\bigg|_{z=1} \\ &= \frac{1}{2\pi i}\oint A(p)A(p^{-1})\frac{dp}{p} \\ &= \frac{1}{2\pi i}\oint |A(z)|^2\frac{dz}{z}.\end{aligned} \tag{5}$$

Returning to the prediction problem, the task is to choose $c_0, c_1, c_2, \ldots$ to

$$\min_{\{c_j\}} \frac{1}{2\pi i}\oint \left| z^{-h}A(z) - \sum_{j=0}^{\infty} c_j z^j \right|^2 \frac{dz}{z}. \tag{6}$$

The first order conditions for the optimization in expression (7) are

$$\begin{aligned} 0 &= \frac{1}{2\pi i}\oint \{z^j[z^h A(z^{-1}) - C(z^{-1})] + z^{-j}[z^{-h}A(z) - C(z)]\}\frac{dz}{z} \\ &= \frac{1}{2\pi i}\oint z^{-j}[z^{-h}A(z) - C(z)]\frac{dz}{z} - \frac{1}{2\pi i}\oint p^{-j}[p^{-h}A(p) - C(p)]\frac{dp}{p}\end{aligned} \tag{7}$$

for $j = 0, 1, 2, \ldots$, where the second integral is the result of a change of variable $p = z^{-1}$ so that $dp = -z^{-1}dz$, resulting in

$$\frac{dp}{p} = z(-z^{-2}dz) = -\frac{dz}{z}.$$

The result is that in the second integral, the direction of the contour integration is clockwise. Multiplying by -1 and integrating counterclockwise, the second integral becomes identical to the first, and we can write the set of first-order conditions as

$$0 = \frac{1}{\pi i}\oint z^{-j}[z^{-h}A(z) - C(z)]\frac{dz}{z} \quad j = 0, 1, 2, \ldots \tag{8}$$

Define $F(z)$ such that

$$F(z) = z^{-h}A(z) - C(z) = \sum_{j=-\infty}^{\infty} F_j z^j.$$

From eq. (8), it must be the case that all coefficients on non-negative powers of z equal zero:

$$F_j = 0, \quad j = 0, 1, 2, \ldots.$$

Multiplying by z^j and summing over *all* $j = 0, \pm 1, \pm 2, \ldots$, we obtain

$$F(z) = \sum_{-\infty}^{-1} \tag{9}$$

where the term on the right-hand-side of (9) represents an unknown function in negative powers of z. Thus

$$z^{-h}A(z) - C(z) = \sum_{-\infty}^{-1},$$

which is an example of a 'Wiener–Hopf' equation. Now apply the (linear) 'plussing' operator, $[\cdot]_+$, which means 'ignore negative powers of z'. The unknown function in negative powers of z is 'annihilated' by this operation, resulting in

$$\begin{aligned} C(z) &= [z^{-h}A(z)]_+ = [z^{-h}a_0 + z^{-h+1}a_1 + z^{-h+2}a_2 + \ldots]_+ \\ &= [z^0 a_h + z^1 a_{h+1} + z^2 a_{h+2} + \ldots] \\ &= \sum_{j=h}^{\infty} a_j z^{j-h} = z^{-h}A(z) - pr[z^{-h}A(z)] \end{aligned}$$

where $pr[z^{-h}A(z)]$ is the principal part of the Laurent expansion of $z^{-h}A(z)$ about $z = 0$. (The principal part of the Laurent expansion about $z = 0$ is the part involving negative powers of z.) This provides a very simple formula for computing forecasts.

3.1 AR(1) example

Suppose that $x_t = ax_{t-1} + \varepsilon_t$. This means that $A(z) = 1/(1-az)$. In this case:

$$\begin{aligned} C(z) &= [z^{-h}A(z)]_+ = [z^{-h}(1 + az + a^2z^2 + \cdots)]_+ \\ &= a^h(1 + az + a^2z^2 + \cdots) = \frac{a^h}{(1 - az)} \end{aligned}$$

and the least squares loss predictor of x_{t+h} using information dated t and earlier is

$$P_t^{LS} x_{t+h} = y_t = C(L)\varepsilon_t = C(L)A^{-1}(L)x_t = a^h x_t.$$

The forecast error is

$$x_{t+h} - a^h x_t = \varepsilon_{t+h} + a\varepsilon_{t+h-1} + \cdots + a^{h-1}\varepsilon_{t+1},$$

which is serially correlated (for $h \geq 2$), but not correlated with information dated t and earlier.

3.2 MA(1) example

Supposed that $x_t = \varepsilon_t - \alpha\varepsilon_{t-1}$, meaning $A(z)$=1$-az$. Thus,

$$C(z) = [z^{-h}A(z)] = [z^{-h}(1 - \alpha z)] = \begin{cases} \alpha \text{ if } h = 1, \\ 0 \text{ otherwise.} \end{cases}$$

So, the best one-step ahead predictor is

$$\alpha\varepsilon_t = \alpha(1 + \alpha L + \alpha^2 L^2 + \cdots)x_t$$

and the best predictor for forecasts of horizon two or more is exactly zero. For two-step-ahead (and beyond) prediction, the forecast error is x_{t+h} itself, which is serially correlated but not correlated with information dated t and earlier.

4. Least squares prediction of geometric distributed leads

A prediction problem that characterizes many models in economics involves the expectation of a discounted value. Perhaps the most common and widely studied example is the present value formula for stock prices. Abstracting from mean and trend, suppose the dividend process has a Wold representation given by

$$d_t = \sum_{j=0}^{\infty} q_j \varepsilon_{t-j} = q(L)\varepsilon_t \quad E(\varepsilon_t) = 0, \;\; E(\varepsilon_t^2) = 1. \tag{10}$$

Assuming that the constant discount factor is given by γ, we have the present value formula

$$p_t = E_t \sum_{j=0}^{\infty} \gamma^j d_{t+j} = E_t\left(\frac{q(L)}{1 - \gamma L^{-1}}\varepsilon_t\right) = E_t(p_t^*). \tag{11}$$

The least-squares minimization problem the predictor faces is to find a stochastic process p_t to minimize the expected squared prediction error $E(p_t - p_t^*)^2$. In terms of the information known at date t, the agent's task is to find a linear combination of current and past dividends, or, equivalently, of current and past dividend innovations ε_t, that is 'close' to p_t^*. Writing $p_t = f(L)\varepsilon_t$, the problem becomes one of finding the coefficients f_j in $f(L) = f_0 + f_1 L + f_2 L^2 + \cdots$ to minimize $E(f(L)\varepsilon_t - p_t^*)^2$. Using the method described in the previous section, the problem has an equivalent, frequency-domain representation

$$\min_{f(z)\in H^2} \frac{1}{2\pi i}\oint \left| \frac{q(z)}{1-\gamma z^{-1}} - f(z)\right|^2 \frac{dz}{z}. \tag{12}$$

The first-order conditions for choosing f_j are, after employing the same simplification used in (7),

$$-\frac{2}{2\pi i}\oint z^{-j}\left[\frac{q(z)}{1-\gamma z^{-1}} - f(z)\right]\frac{dz}{z} = 0, \quad j = 0, 1, 2, \ldots. \tag{13}$$

Now define

$$H(z) = \frac{q(z)}{1-\gamma z^{-1}} - f(z)$$

so that (13) becomes

$$-\frac{2}{2\pi i}\oint z^{-j}H(z)\frac{dz}{z} = 0.$$

Then multiplying by z^j and summing over all $j = 0, \pm 1, \pm 2, \ldots$ as above, we obtain

$$H(z) = \frac{q(z)}{1-\gamma z^{-1}} - f(z) = \sum_{-\infty}^{-1},$$

the Wiener–Hopf equation for this problem. Applying the plussing operator to both sides yields

$$\left[\frac{q(z)}{1-\gamma z^{-1}}\right]_+ - [f(z)]_+ = 0$$

implying

$$f(z) = \left[\frac{q(z)}{1-\gamma z^{-1}}\right]_+ = \left[\frac{zq(z)}{z-\gamma}\right]_+$$

because $f(z)$ is, by construction, one-sided in non-negative powers of z. As in the previous section,

$$[A(z)]_+ = A(z) - P(z)$$

where $P(z)$ is the principal part of the Laurent series expansion of $A(z)$. To determine the principal part of $[(z-\gamma)^{-1}zq(z)]$, note that $zq(z)$ has a well-behaved power series expansion about $z=\gamma$, where 'well-behaved' means 'involving no negative powers of $(z-\gamma)$'. Thus $[(z-\gamma)^{-1}zq(z)]$ has a power series expansion about $z=\gamma$ involving a single term in $(z-\gamma)^{-1}$:

$$\left(\frac{zq(z)}{z-\gamma}\right) = \frac{b_{-1}}{z-\gamma} + b_0 + b_1(z-\gamma)^1 + b_2(z-\gamma)^2 + \cdots.$$

The principal part here is the part involving negative powers of $(z-\gamma)$: $b_{-1}(z-\gamma)^{-1}$. To determine it, multiply both sides by $(z-\gamma)$ and evaluate what is left at $z=\gamma$ to find $b_{-1} = \gamma q(\gamma)$. Thus

$$f(z) = \left[\frac{q(z)}{1-\gamma z^{-1}}\right]_+ = \left[\frac{zq(z)}{z-\gamma}\right]_+ = \frac{zq(z)-\gamma q(\gamma)}{z-\gamma}. \tag{14}$$

The 'cross-equation restrictions' of rational expectations refer to the connection between the serial correlation structure of the driving process (here dividends) and the serial correlation structure of the expected discounted value of the driving process (here prices). That is, when dividends are characterized by $q(z)$, prices are characterized by $f(z)$, and $f(z)$ depends upon $q(z)$ as depicted in (14).

To illustrate how the formula works, suppose detrended dividends are described by a first-order autoregression; that is, that $q(L) = (1-\rho L)^{-1}$. Then

$$p_t = f(L)\varepsilon_t = \frac{Lq(L)-\gamma q(\gamma)}{L-\gamma}\varepsilon_t = \left(\frac{1}{1-\rho\gamma}\right)d_t. \tag{15}$$

It is instructive to note that, while the pricing formula (15) makes p_t the best least squares predictor of p_t^*, the prediction errors $p_t - p_t^*$ will not be serially uncorrelated. Indeed

$$\begin{aligned} p_t - p_t^* &= \gamma\left\{\frac{Lq(L)-\gamma q(\gamma)}{L-\gamma} - \frac{q(L)}{1-\gamma L^{-1}}\right\}\varepsilon_t \\ &= \frac{-\gamma^2 q(\gamma)}{L-\gamma}\varepsilon_t = -\gamma^2 q(\gamma)\frac{L^{-1}}{1-\gamma L^{-1}}\varepsilon_t \\ &= -\gamma^2 q(\gamma)\{\varepsilon_{t+1} + \gamma\varepsilon_{t+2} + \gamma^2\varepsilon_{t+3} + \ldots\}. \end{aligned}$$

Thus the prediction errors will be described by a highly persistent (γ is close to unity) first-order autoregression. But because this autoregression involves *future* ε_t's, the serial correlation structure of the errors cannot be exploited to improve the quality of the prediction of p_t^*. The reason is that the predictor 'knows' the *model* for price setting (the present value formula) and the dividend process; the best predictor $p_t = E_t p_t^*$ of p_t^* 'tolerates' the serial correlation because the (correct) model implies that it involves *future* ε_t's and therefore cannot be predicted. If one only had data on the errors (and did not know the model that generated them), they would appear (rightly) to be characterized by a first-order autoregression; fitting an AR(1) (that is, the best *linear* model) and using it to 'adjust' p_t by accounting for the serial correlation in the errors $p_t - p_t^*$ would decrease the quality of the estimate of p_t^*. The reason is the usual one that the Wold representation for $p_t - p_t^*$ is not the economic model of $p_t - p_t^*$, and (correct) models always beat Wold representations. This also serves as a reminder of circumstances under which one should be willing to tolerate serially correlated errors: when one knows the model that generated them, and the model implies that they are as small as they can be made.

5. Robust optimal prediction of time series

The squared-error loss function employed to this point is appropriate for situations in which the model (either the time series model or the economic model) is thought to be correct. But in many settings the forecaster or model builder may wish to guard against the possibility of misspecification. There are many ways to do this; an approach popular in the engineering literature and recently introduced into the economics literature by Hansen and Sargent (2007) involves behaving so as to minimize the maximum loss sustainable by using an approximating model when the truth may be something else. The 'robust' approach to this involves replacing the squared-error loss problem

$$\min_{\{C(z)\}} \frac{1}{2\pi i} \oint |z^{-h}A(z) - C(z)|^2 \frac{dz}{z}$$

with the 'min-max' problem

$$\min_{\{C(z)\}} \sup_{|z|=1} |z^{-h}A(z) - C(z)|^2,$$

so that minimizing the 'average' value on the unit circle has been replaced by minimizing the max. This problem can also be written

$$\min_{\{C(z)\}} \sup_{|z|=1} |A(z) - z^h C(z)|^2.$$

This is known as the 'minimum norm interpolation problem' and amounts to finding a function $\phi(z)$ to

$$\min ||\phi(z)||_\infty$$

subject to the restriction that the power series expansion of $\phi(z)$ matches that of $A(z)$ for the first $h-1$ powers of z. This means that the following must hold:

$$\sum_{j=0}^{h-1} \phi_j z^j = \sum_{j=0}^{h-1} a_j z^j. \tag{16}$$

Theorem 5.1 *The minimizing $\phi(z)$ function is such that $|\phi(z)|^2$ is constant on $|z| = 1$. Moreover,*

$$\phi(z) = M \prod_{j=1}^{h} \frac{z - \alpha_j}{1 - \alpha_j z}$$

where $M, \alpha_1, \alpha_2, \ldots, \alpha_n$ are chosen to ensure that (16) holds.

Proof: see Nehari (1957).

To see that $\phi(z)$ must be of the indicated form, note that the 'Blaschke factors' in the product have unit modulus:

$$\frac{z-\alpha_j}{1-\alpha_j z}\left(\frac{z^{-1}-\alpha_j}{1-\alpha_j z^{-1}}\right)=\left(\frac{z-\alpha_j}{1-\alpha_j z}\right)\left(z^{-1}z\right)\left(\frac{z^{-1}-\alpha_j}{1-\alpha_j z^{-1}}\right)$$
$$=\left(\frac{1-\alpha_j z^{-1}}{1-\alpha_j z}\right)\left(\frac{1-\alpha_j z}{1-\alpha_j z^{-1}}\right)=1,$$

so that $|\phi(z)|^2 = M^2$.

In the general h-step-ahead prediction problem, we have that

$$\phi(z)=M\prod_{j=1}^{h-1}\frac{z-\alpha_j}{1-\alpha_j z}=A(z)-z^h C(z),$$

meaning that

$$C(z)=\frac{1}{z^h}\left(A(z)-M\prod_{j=1}^{h-1}\frac{z-\alpha_j}{1-\alpha_j z}\right).$$

This is analogous to the solution in the least-squares case, but, instead of subtracting the principal part of $z^{-h}A(z)$, we subtract a different function from $z^{-h}A(z)$. Note also that because

$$M\prod_{j=1}^{h-1}\frac{z-\alpha_j}{1-\alpha_j z}$$

matches the power series expansion of $A(z)$ up to the power z^{h-1}, $C(z)$ is of the form

$$C(z)=c_0+c_1 z+c_2 z^2+\ldots$$

Finally, note that the forecast error is serially uncorrelated because $\phi(z)$ is constant on $|z|=1$.

5.1 Example. AR(1)

Let

$$A(z)=\frac{1}{1-az}.$$

For $h=1$, we see that $\phi(z)=A(z)-zC(z)$ must be constant on $|z|=1$, and that $\phi(0)=A(0)=1$. Thus, $\phi(z)=M=1$, so that

$$C(z)=\frac{A(z)-1}{z}=\frac{az}{(1-az)z}=\frac{a}{1-az},$$

which implies that the robust one-step ahead forecast is

$$y_t^R=ax_t,$$

which coincides with the best least-squares forecast. This equivalence between the robust and least-squares one-step ahead forecasts is to be expected because the best one-step-ahead least-squares forecast also has serially uncorrelated errors. For $h=2$, we have that

$$\phi(z) = \frac{M(z-\alpha)}{1-\alpha z}$$

where (again) $\phi(0)=1$, but now we also see that $\phi'(0)=a$. Thus,

$$\phi(0) = 1 = -\alpha M \;\Rightarrow\; M = -\frac{1}{\alpha},$$

and furthermore

$$\phi'(0) = a = \left.\frac{(1-\bar{\alpha}z)M - M(z-\alpha)(-\bar{\alpha})}{(1-\bar{\alpha}z)^2}\right|_{z=0} = M - M(\alpha\bar{\alpha}) = M(1-\alpha\bar{\alpha}).$$

Therefore, the solution will have the property that

$$a = -\frac{1}{\alpha}(1-\alpha\bar{\alpha}) - a\alpha = 1 - \alpha\bar{\alpha} \quad 0 = 1 + a\alpha - \alpha\bar{\alpha}.$$

That is, the roots are reciprocal pairs. Notice that the discriminant is positive ($a^2\alpha^2 + 4\alpha\bar{\alpha} > 0$), meaning that we will always have a real solution, and we choose $|\alpha| < 1$. Then, we have that

$$C(z) = \frac{1}{z^2}\left[\frac{1}{1-az} - \frac{M(z-\alpha)}{1-\alpha z}\right] = \frac{1}{z^2}\frac{1-\alpha z-(1-az)(1-\frac{1}{\alpha}z)}{(1-az)(1-\alpha z)}$$

$$= \frac{1-\alpha z - 1 + az + \frac{1}{\alpha}z - \frac{a}{\alpha}z^2}{z^2(1-az)(1-\alpha z)} = \frac{-\frac{a}{\alpha}}{(1-az)(1-\alpha z)}.$$

So, the robust prediction is given by

$$P_t^R x_{t+2} = -\frac{a}{\alpha}\sum_{j=0}^{\infty} \alpha^j x_{t-j},$$

in contrast to the least-squares prediction

$$P_t^{LS} x_{t+2} = a^2 x_t.$$

5.2 Example: MA(1)

Suppose that the process follows an MA(1), $x_t = \varepsilon_t - \beta\varepsilon_{t-1}$, and therefore $A(z)=1-\beta z$. The analysis from the previous example still holds, and all of the following are true:

$$\phi(z) = \frac{M(z-\alpha)}{1-\alpha z}$$

while

$$\phi(0) = 1 = -\alpha M \quad \Rightarrow \quad M = -\alpha^{-1}$$

and

$$\phi'(0) = -\beta = -\frac{1}{\alpha}(1 - \alpha\bar{\alpha}).$$

Therefore,

$$0 = 1 - \alpha\beta - \alpha\bar{\alpha},$$

meaning that, again, we have real roots which are reciprocal pairs and we can choose $|\alpha| < 1$. Of course, α will depend upon the value of β, and we write $\alpha(\beta)$. Thus

$$\begin{aligned} C(z) &= \frac{1}{z^2}\left[1 - \beta z - \frac{M(z - \alpha(\beta))}{(1 - \alpha(\beta)z)}\right] \\ &= \frac{1}{z^2}\left[\frac{(1 - \beta z)(1 - \alpha(\beta)z)) - M(z - \alpha(\beta))}{1 - \alpha(\beta)z}\right] \\ &= \frac{1}{z^2}\left[\frac{1 - \beta z - \alpha(\beta)z + \beta\alpha(\beta)z^2 - Mz + M\alpha(\beta)}{1 - \alpha(\beta)z}\right] \\ &= \frac{\beta\alpha(\beta)}{1 - \alpha(\beta)z}. \end{aligned}$$

Therefore, we have the robust prediction

$$P_t^R x_{t+2} = \frac{\beta\alpha(\beta)}{1 - \alpha(\beta)L}\varepsilon_t = \frac{\beta\alpha(\beta)}{1 - \alpha(\beta)L}[x_t + \beta x_{t-1} + \beta x_{t-2} + \ldots],$$

while the least-squares prediction is the standard

$$P_t^{LS} x_{t+2} = 0.$$

6. Robust prediction of geometric distributed leads

Following the excellent treatment in Kasa (2001), a robust present-value predictor fears that dividends may not be generated by the process in (10), and so, instead of choosing an $f(z)$ to minimize the average loss around the unit circle, chooses $f(z)$ to minimize the maximum loss:

$$\min_{f(z)\in H^\infty} \sup_{|z|=1} \left|\frac{q(z)}{1 - \gamma z^{-1}} - f(z)\right|^2 \Leftrightarrow \min_{f(z)\in H^\infty} \sup_{|z|=1} \left|\frac{zq(z)}{z - \gamma} - f(z)\right|^2.$$

Unlike in the least squares case (14), where $f(z)$ was restricted to the class H^2 of functions finitely square integrable on the unit circle, the restriction now is to the class of functions with finite maximum modulus on the unit circle, and the H^2 norm has been replaced by H^∞ norm.

To begin the solution process, note that there is considerable freedom in designing the minimizing function $f(z)$: it must be well-behaved (that is, must have a convergent power series in non-negative powers of z on the unit disk), but is otherwise unrestricted. Recalling the Laurent expansion

$$\frac{zq(z)}{z-\gamma} = \frac{b_{-1}}{z-\gamma} + b_0 + b_1(z-\gamma) + b_2(z-\gamma)^2 + \ldots,$$

while in the least squares case $f(z)$ was set to 'cancel' all the terms of this series except the first, here $f(z)$ will be set to do something else. Now define the Blaschke factor $B_\gamma(z) = (z-\gamma)/(1-\gamma z)$ and note that, because of the unit modulus condition, the problem can be written

$$\min_{\{f(z)\}} \sup_{|z|=1} \left| \frac{zq(z)}{1-\gamma z} - \frac{z-\gamma}{1-\gamma z} f(z) \right|^2.$$

Defining

$$T(z) = \frac{zq(z)}{1-\gamma z}$$

we have

$$\min_{f\in H^\infty} \sup_{|z|=1} |T(z) - B_\gamma(z)f(z)| \Leftrightarrow \min_{f\in H^\infty} ||T(z) - B_\gamma(z)f(z)||_\infty.$$

Define the function inside the ||'s as

$$\phi(z) = T(z) - B_\gamma(z)f(z)$$

and note that $\phi(\gamma)=T(\gamma)$. Thus the problem of finding $f(z)$ reduces to the problem of finding the smallest $\phi(z)$ satisfying $\phi(\gamma)=T(\gamma)$:

$$\min_{\phi\in H^\infty} ||\phi(z)||_\infty \text{ s.t. } \phi(\gamma) = T(\gamma)$$

Theorem 6.1 (Kasa, 2001). *The solution to (17) is the constant function $\phi(z)=T(\gamma)$.*
Proof. To see this, first note that the norm of a constant function is the modulus of the constant itself. This is written as

$$||\phi(z)||_\infty = ||T(\gamma)||_\infty = |T(\gamma)|^2. \tag{17}$$

Next, suppose that there exists another function $\Psi(z) \in H^\infty$, with $\Psi(\gamma) = T(\gamma)$ and also

$$||\Psi(z)||_\infty < ||\phi(z)||_\infty. \tag{18}$$

Recall the definition of the H^∞ norm, and using equations (17) and (18):

$$||\Psi(z)||_\infty = \sup_{|z|=1} |\Psi(z)|^2 < |T(\gamma)|^2.$$

The maximum modulus theorem states that a function f which is analytic on the disk U achieves its maximum on the boundary of the disk. That is

$$\sup_{z\in U} |f(z)|^2 \leq \sup_{z\in\partial U} |f(z)|^2.$$

Therefore, we can see that

$$\sup_{|z|<1} |\Psi(z)|^2 \leq \sup_{|z|=1} |\Psi(z)|^2 < |T(\gamma)|^2.$$

However, one of the values on the interior of the unit disk is $z = \gamma$, which can be inserted into the far left-hand-side of eq. (6) to get the result

$$|\Psi(\gamma)|^2 \leq \sup_{|z|=1} |\Psi(z)|^2 < |T(\gamma)|^2 \Rightarrow |\Psi(\gamma)|^2 < |T(\gamma)|^2.$$

This contradicts the requirement that $\Psi(\gamma) = T(\gamma)$. Therefore, we have verified that there does not exist another function $\Psi(z) \in H^\infty$ such that $\Psi(\gamma) = T(\gamma)$ and $||\Psi(z)||_\infty < ||\phi(z)||_\infty$. □

Given the form for $\phi(z)$, the form for $f(z)$ follows. After some tedious algebra, we obtain

$$f(z) = \frac{T(z) - \phi(z)}{B_\gamma(z)} = \frac{zq(z) - \gamma q(\gamma)}{z - \gamma} + \frac{\gamma^2}{1 - \gamma^2} q(\gamma)$$

which is the least squares solution plus a constant. Thus the robust cross-equation restrictions likewise differ from the least squares cross-equation restrictions. After the initial period, the impulse response function for the robust predictor is identical to that of the least squares predictor. In the initial period, the least squares impulse response is $q(\gamma)$, while the robust impulse response is larger: $q(\gamma)/(1-\gamma^2)$.

Because γ is the discount factor, and therefore close to unity, the robust impulse response can be considerably larger than that of the least squares response. Relatedly, the volatility of prices in the robust case will be larger as well. For example, in the first-order autoregressive case studied above,

$$p_t = f(L)\varepsilon_t = \frac{1}{1-\rho\gamma} d_t + \frac{\gamma^2}{(1-\gamma^2)(1-\rho\gamma)}\varepsilon_t \tag{19}$$

from which the variance can be calculated as

$$\sigma^2(p_t) = \left(\frac{1}{1-\rho\gamma}\right)^2 \sigma^2(d_t) + \frac{2\gamma^2-\gamma^4}{(1-\rho\gamma)^2(1-\gamma^2)^2}.$$

When the discount factor is large and dividends are highly persistent, the variance of the robust present value prediction can be considerably larger than that of the least squares prediction (the first term on the right alone).

Finally, recall that the least-squares present-value predictor behaved in such a way as to minimize the *variance* of the error $p_t - p_t^*$. Here, robust prediction results in an error with Wold representation

$$\begin{aligned} p_t - p_t^* &= \gamma\left\{\frac{Lq(L)-\gamma q(\gamma)}{L-\gamma} + \frac{\gamma^2}{1-\gamma^2}q(\gamma) - \frac{q(L)}{1-\gamma L^{-1}}\right\}\varepsilon_t \\ &= -\frac{\gamma q(\gamma)}{1-\gamma^2}\left\{\frac{1-\gamma L}{L-\gamma}\right\}\varepsilon_t. \end{aligned}$$

The term in braces has the form of a Blaschke factor. Applying such factors in the lag operator to a serially uncorrelated process like ε_t leaves a serially uncorrelated result; thus the robust present value predictor has behaved in such a way that the resulting errors are white noise. Of course this comes at a cost: to make the error serially uncorrelated, the robust predictor must tolerate an error variance that is larger than the least squares error variance by a factor of $a^2/(1-\gamma^2)$, which can be substantial when γ is close to unity.

CHARLES H. WHITEMAN AND KURT F. LEWIS

See also **forecasting.**

Bibliography

Hansen, L.P. and Sargent, T.J. 1980. Formulating and estimating dynamic linear rational expectations models. *Journal of Economic Dynamics and Control* 2, 7–46.

Hansen, L.P. and Sargent, T.J. 2007. *Robustness*. Princeton: Princeton University Press.

Kasa, K. 2001. A robust Hansen–Sargent prediction formula. *Economic Letters* 71, 43–8.

Nehari, Z. 1957. On bounded bilinear forms. *Annals of Mathematics* 65(1), 153–62.

Sargent, T.J. 1987. *Macroeconomic Theory.* New York: Academic Press.

Whiteman, C.H. 1983. *Linear Rational Expectations: A User's Guide.* Minneapolis: University of Minnesota Press.

Whittle, P. 1983. *Prediction and Regulation by Linear Least-square Methods*, 2nd edn. Minneapolis: University of Minnesota Press.

rational expectations

'Rational expectations' is an equilibrium concept that can be applied to dynamic economic models that have elements of 'self-reference', that is, models in which the endogenous variables are influenced by the expectations about future values of those variables held by the agents in the model. The concept was introduced and applied by John F. Muth (1960; 1961) in two articles that interpreted econometric distributed lag models. Muth used explicitly stochastic dynamic models and brought to bear his extensive knowledge of classical linear prediction theory to interpret distributed lags in terms of economic parameters. For Muth, an econometric model with rational expectations possesses the defining property that the forecasts made by agents within the model are no worse than the forecasts that can be made by the economist who has the model.

Muth's first concrete application of rational expectations was to find restrictions on a stochastic process for income that would render Milton Friedman's (1957) geometric distributed lag formula for permanent income an optimal predictor for income. Muth showed that, if the first difference of income is a first-order moving average process, then Friedman's formula is optimal for forecasting income over any horizon. The independence of this formula from the horizon makes precise the sense in which Friedman's formula extracts from past income an estimator of 'permanent' income. In working backwards from Friedman's formula to a process for income in this way, Muth touched Lucas's critique (1976). Given any distributed lag for forecasting income, one can work backwards as Muth did and discover a stochastic process for income that makes that distributed lag an optimal predictor for income over some horizon. Similarly, Sargent (1977) reverse engineered a joint inflation-money creation process that makes Cagan's (1956) adaptive expectations scheme for forecasting inflation a linear least squares forecast.

Solving a few such inverse-optimal prediction problems in the fashion of Muth and Sargent quickly reveals the dependence of a distributed lag for forecasting the future on the form of the stochastic process that is being forecast. In 1963, Peter Whittle published a book that conveniently summarized and made more accessible to economists the classical linear prediction theory that Muth had used. That book repeatedly applies the Wiener–Kolmogorov formula for the optimal j-step ahead predictor of a covariance stationary stochastic process x_t with moving average representation $x_t = c(L)\varepsilon_t$. The Wiener–Kolmogorov formula displays the dependence of the optimal distributed lag for predicting future x on the form of $c(L)$. That dependence underlies Lucas's critique of econometric policy evaluation procedures that were common when Lucas composed his critique in 1973. Those procedures had assumed that distributed lags in behavioural relations would remain invariant with respect to alterations in government policy rules, alterations that took the form of changes in $c(L)$ for government policy instruments. Although the formulas in Whittle's book were used extensively by Nerlove (1967) to work out additional examples along the lines of Muth, it was not until the writing of Lucas's critique in 1973 and its publication in 1976 that the implications for econometric practice of Muth's ideas and the prediction formulas in Whittle began to be widely appreciated.

Lucas and Prescott (1971) clarified and extended rational expectations as an equilibrium concept and also pointed the way to connecting theory with observations. They described the partial equilibrium of an industry in which there exists a fixed number of identical firms, each subject to costs of adjustment for a single factor of production, capital. The industry faces a downward sloping demand curve for its output that shifts randomly due to a demand shock that follows a Markov process. The representative firm maximizes the expected present value of its profits by choosing a contingency plan for investment. To state the firm's optimum problem, it is necessary to describe what the firm believes about the motion of variables that influence its future returns even though they are beyond the firm's control. The price of output is such an uncontrollable variable, but the demand curve for output and the hypothesis of market clearing make price a function of the capital stock in the industry as a whole. It follows that to state the firm's decision problem requires the firm's view about the law of motion of the industry-wide capital stock be stated. The representative firm's optimum problem can then be solved, yielding a law of motion for the capital stock of the representative firm in which both the individual firm's capital stock and the market-wide capital stock are both state variables. Multiplying this law of motion by the number of firms then gives the actual law of motion for capital in the industry. In this way, the firm's optimization problem and the hypothesis of market clearing induce a mapping from a perceived law of motion to an actual law of motion for the industry's capital stock. A rational expectations equilibrium is a fixed point of this mapping. By studying an artificial planning problem that maximizes consumer plus producer surplus, Lucas and Prescott pursued an indirect approach to describing conditions under which a unique fixed point exists. In this way, they formulated a recursive competitive equilibrium.

From a practical perspective, an important property of a rational expectations model is that it imposes a communism of models and expectations. If we define a model as a probability distribution over a sequence of outcomes, possibly indexed by a parameter vector, a rational expectations equilibrium asserts that the same model is shared by (1) all of the agents within the model, (2) the econometrician estimating the model, and (3) nature, also known as the data generating mechanism. Different agents might have different information, but they form forecasts by computing conditional expectations with respect to a common joint density, that is, a common model. Communism of models gives rational expectations much of its empirical power and underlies the cross-equation restrictions that are used by rational expectations econometrics to identify and estimate parameters. A related perspective is that, within models that have unique rational expectations equilibria, the hypothesis of rational expectations makes agents' expectations disappear as objects to be specified by the model-builder or to be estimated by the econometrician. Instead, they are equilibrium outcomes.

The equilibrium law of motion for capital induces a stochastic process for capital that assumes the form of a Markov process. Lucas and Prescott showed that this Markov process converges in distribution to a unique invariant distribution. That justifies an asymptotic distribution theory adequate for doing time series econometrics, in particular, a mean ergodic theorem that guarantees that sample moments converge to the corresponding population moments. Lucas and Prescott's notion of a recursive competitive equilibrium thus takes a big step towards integrating dynamic theory and econometrics because it supplies an explicit mapping from economic parameters describing preferences, technology, and information sets to the population moments of observable sequences of economic time series. The task of econometrics under rational expectations is to 'invert' this mapping by using time series data to make inferences about economic parameters.

Hansen and Sargent (1980) used linear versions of Lucas–Prescott and Brock and Mirman (1972) models as laboratories for working out econometric techniques for estimating rational expectations models. They studied both generalized method of moments (GMM) and maximum likelihood approaches. They described how desirable statistical properties including consistency and asymptotic efficiency for estimators of the model's economic parameters induce a metric for measuring distance between the sample moments and the theoretical population moments implied by the equilibrium of the model at given parameter values. Typical metrics are those associated with the generalized method of moments, a special case of which is associated with the first-order conditions for maximizing a Gaussian likelihood function. Parameter estimates are obtained by minimizing the metric with respect to the parameter values, a nonlinear minimization problem.

Econometric identification of parameters means uniqueness of the minimizer of distance between the theory and the observations. Identification is partially achieved by the rich set of cross-equation restrictions that the hypothesis of rational expectations imposes (the same parameters appear in many equations, in highly nonlinear ways). These cross-equation restrictions achieve identification in a different manner from the Cowles Commission's 'rank and order' conditions, which explicitly excluded cross-equation restrictions. Dynamic rational expectations models subvert such 'exclusion restrictions', and thereby destroy the neat division between 'supply' and 'demand' curves that underlay the 'exclusion' approach to identification.

Minimum distance estimation of a rational expectations model requires recomputing an equilibrium for each set of parameter values used during a descent with respect to the data-fitting metric. Except for linear models, Bellman's 'curse of dimensionality' makes it challenging to compute an equilibrium, so developing improved computational methods has become an important research area. Judd (1998) describes a variety of numerical approaches. Methods for computing equilibria are required not only for parameter estimation, but also for quantitatively evaluating the effects of proposed interventions, for example, new policies for setting government instruments. A new government policy implies, via the cross-equation restrictions, new laws of motion for all the endogenous variables in the models. It is no coincidence that full information estimation methods require calculations closely connected to those needed to evaluate policy.

Good computer programmes for solving and estimating complete rational expectations models have recently become available. A suite of Matlab programmes called Dynare was written by Michele Juilliard and colleagues and is available on the Internet. Dynare solves linear models as systems of expectational difference equations using methods originally described by Sargent (1979), Blanchard and Khan (1980), and Whiteman (1983). Dynare estimates models by either maximum likelihood or a Markov chain Monte Carlo procedure to construct a Bayesian posterior density over free parameters. Dynare also knows how to compute and estimate various linear and log-linear approximations to nonlinear models.

Hansen and Singleton (1982) suggested a short-cut estimation method capable of estimating the parameters of a subset of preference and technologies without computing or estimating a complete equilibrium. Their idea was to use back out parameter estimates from conditional moment restrictions implied by the first-order necessary conditions (Euler equations) for an agent's dynamic optimization problem. Hansen and Singleton pointed out that their GMM method requires special restrictions on the stochastic process of disturbances to the function being estimated, and that it typically fails to estimate enough parameters to permit evaluating many kinds of interventions. Nevertheless, its ease of use and presumed robustness to features of the environment that a researcher prefers not to specify have made it a very popular and fruitful approach.

As already mentioned, a rational expectations equilibrium is a fixed point from a perceived to an actual law of motion. It is tempting to hope that iterations on that mapping converge to a fixed point. But that is asking for too much because the mapping is not a contraction and it is easy to construct examples in which iterations diverge. Nevertheless, the mapping from a perceived to an actual law of motion plays an important role in studying how a rational expectations equilibrium can emerge as the limit point of a system of adaptive agents who use least squares on historical data to forecast the future, rather than the population moments from the equilibrium that are handed to them within a rational expectations equilibrium. By applying the theory of stochastic approximation, Marcet and Sargent (1989) and Woodford (1990) derived an ordinary differential equation (ODE) for beliefs that describe the limiting behaviour of such an adaptive system. That ODE expresses how the gap between the perceived and implied actual law of motion governs a limiting rate of change of beliefs. Necessary and sufficient conditions for convergence to a rational expectations equilibrium are stated in terms of the stability of the associated ode. These conditions have been dubbed the E-stability conditions by Evans and Honkapohja (2001) and are useful for constructing algorithms for computing rational expectations equilibria via least squares learning algorithms or direct attacks on the ordinary differential equation governing E-stability. This is in effect what Krusell and Smith (1998) do, though they do not connect their method to the learning literature.

The literature on least squares learning and adaptive learning in games (for example, Marcet and Sargent, 1989; Woodford, 1990; Fudenberg and Levine, 1998) began partly as a response to a widespread scepticism about the plausibility of the communism of expectations imposed by rational expectations. How could people possibly come to learn to share a common model with each other, the econometrician, and nature? The learning literature offers an explanation. But the learning literature falls short of implying a communism of models as extensive as the one typically imposed in macroeconomics. A meta-theorem is that, if a system of least squares agents converges, it converges to a self-confirming equilibrium (see Fudenberg and Levine, 1998; Sargent, 1999). In a self-confirming equilibrium, agents' models agree about events that occur frequently enough (infinitely often) within the equilibrium. But agents can have different subjective distributions about events that occur infrequently because they are off the equilibrium path. For those events, a law of numbers just doesn't have enough observations to work on. In a macro model, it is typically irrelevant that private agents' beliefs can be wrong off an equilibrium path because, being atomistic, all that matters for them are their conditional forecasts along an equilibrium path. But for the government, its beliefs about off-equilibrium paths events influence its choices in important ways: designing government policy is all about evaluating the effects of alternative hypothetical outcome paths, most of which will not be observed. Kreps (1998) defends the concept of self-confirming equilibrium.

Lucas and Prescott's model can be used to study aspects of the theory of policy. Their model generates a stochastic process for output, price and industry capital that exhibits recurrent but aperiodic 'cycles', as realizations of stochastic difference equations do. Thus, Lucas and Prescott's model is an alternative to the 'cobweb' mechanism for generating fluctuations in commodity markets. Two-industry versions of the model can readily be constructed to model 'corn-hog' cycles. Models along the lines of Lucas and Prescott's reveal a different perspective on these cycles than do cobweb models. Lucas and Prescott show that, despite cyclical fluctuations, the equilibrium of their model is optimal in the sense that it maximizes the expected present value of consumer surplus net of producer surplus. Therefore, unlike cobweb models, in which cycles partly reflect erroneous and readily improved upon perceptions of private agents, matters cannot be improved by government interventions designed to smooth out the cycles. Models of this kind have been calibrated to price and quantity data from markets for cattle, housing, and engineers by Rosen, Murphy and Scheinkman (1994), Topel and Rosen (1988), and Ryoo and Rosen (2004).

For studying a variety of macroeconomic questions, researchers have used what can be interpreted as a version of Lucas and Prescott's model, suitably modified and reinterpreted to apply to an aggregative economy. Brock and Mirman (1972) analysed a centralized version of such an economy that took the form of a stochastic version of a one-sector optimal growth model. The planner in their model seeks to maximize the expected discounted value of utility of consumption subject to a technology for transforming consumption over time via investment in physical capital. Brock and Mirman gave conditions under which the optimal plan for capital and consumption induces a stochastic process that converges in distribution, so that, like Lucas and Prescott's model, theirs is prepared for rigorous treatment econometrically. It is possible to decentralize Brock and Mirman's model into an equivalent economy consisting of competitive firms and households who interact in markets for labour and capital and who have rational expectations about the evolution of the wages and interest rates that they face. Decentralized versions of Brock–Mirman models have been used to construct equilibrium theories of stock prices and interest rates, typically by computing particular shadow prices associated with the planning problem (Lucas, 1978; Brock, 1982). Decentralized versions of the Brock–Mirman model form the backbone of the modern version of 'real business cycle theory' that was initiated by Kydland and Prescott (1982). Since the stochastic optimal growth model has a stochastic difference equation for capital as its equilibrium, it shares with the Lucas–Prescott model the property that it readily generates realizations for capital, output and consumption that display recurrent but aperiodic fluctuations of the kind observed in aggregate time series data. Kydland and Prescott embarked on the task of taking seriously the possibility that the preferences and technology of a small stochastic optimal growth model could be specified so that it would approximate closely the moments of a list of important aggregate economic time series for the United States. Kydland and Prescott have constructed several such models, each driven by a single unobserved shock, which they interpret as a disturbance to technology. This research strategy is charged with meaning, since it undertakes to explain aggregate time series data with a model whose equilibrium is optimal, and in which there is no government. The government is neither a contributing source to economic fluctuations nor a potential modifier of those fluctuations. Real business cycle models of this kind are capable of determining a long list of real variables, while remaining silent about all nominal variables.

But central banks are supposed to determine nominal variables, which has created an interest in adapting real business cycle models to include interactions among nominal and real variables. By directly imposing parameterized versions of wage and price inertia, Smets and Wouter (2003) and Woodford (2003) have formulated rational expectations models with enough shocks and rigidities to fit macro data well enough to be useful to research departments of leading central banks. These models can be estimated and simulated with Dynare.

The idea of rational expectations was essential for formulating the problem of time inconsistency in macroeconomics. Three ideas underlie the time consistency problem in multi-agent dynamic games and macroeconomic models: (1) the communism of models brought by rational expectations, (2) backward induction by all agents, and (3) the observation that different timing protocols generally imply different outcomes. The time inconsistency 'problem' was recognized in macroeconomics by Kydland and Prescott (1977) and Calvo (1978), who studied macro models in which a competitive economy with a representative agent confronts a benevolent government. These papers compare outcomes under two timing protocols. In one timing protocol, private agents choose sequentially but the government has a commitment technology that allows it once and for all at time zero to choose an entire history contingent sequence of actions (for example, tax rates or money supplies). In the other, the government, or a sequence of government administrations if you prefer, must choose sequentially, that is, anew each period. Outcomes under these two timing protocols typically differ, with outcomes being better under the timing protocol that allows the government to choose once and for all at time 0. The difference in outcomes shows the value of being able to commit at time 0. In the problem under commitment, among the constraints that the government faces at time 0 are a sequence of private agents' Euler equations that involve their (rational) expectations of future government actions. The equilibrium time t values of the Lagrange multipliers on these 'implementability constraints' encode the costs in terms of the government's time t continuation value of confirming the time t expectations that the government's time 0 plan had induced private agents to expect. The presence of those implementability conditions in the government's constraint set gives rise to a conflict between the preference orderings of the government and the representative agent over outcomes. That conflict is the ultimate source of the timing inconsistency problem. Recursive methods for computing the optimal plan under commitment were first suggested by Kydland and Prescott (1980) and are surveyed in Ljungqivst and Sargent (2004). These methods are used extensively in the literature on rational expectations monetary models with ad hoc inertial wages and prices that Woodford (2003) catalogues and extends.

An important literature studies whether reputation can overcome the time inconsistency problem. The finding of this literature is that, by allowing history-dependent strategies, reputation can substitute for the ability to commit if the discount factor is sufficiently close to one. This literature, which is surveyed critically in Ljungqivst and Sargent (2004), exploits the communism of expectations inherent in rational expectations.

THOMAS J. SARGENT

Bibliography

Blanchard, O.J. and Kahn, C.M. 1980. The solution of linear difference models under rational expectations. *Econometrica* 48, 1305–11.

Brock, W.A. 1982. Asset prices in a production economy. In *Economics of Information and Uncertainty*, ed. J.J. McCall. Chicago: University of Chicago Press.

Brock, W.A. and Mirman, L. 1972. Optimal economic growth under uncertainty: the discounted case. *Journal of Economic Theory* 4, 479–513.

Cagan, P. 1956. The monetary dynamics of hyperinflation. In *Studies in the Quantity Theory of Money*, ed. M. Friedman. Chicago: University of Chicago Press.

Calvo, G.A. 1978. On the time consistency of optimal policy in a monetary economy. *Econometrica* 46, 1411–28.

Evans, G.W. and Honkapohja, S. 2001. *Learning and Expectations in Macroeconomics*. Princeton: Princeton University Press.

Friedman, M. 1957. *A Theory of the Consumption Function*. Princeton: Princeton University Press.

Fudenberg, D. and Levine, D.K. 1998. *The Theory of Learning in Games*. Cambridge, MA: MIT Press.

Hansen, L.P. and Sargent, T.J. 1980. Formulating and estimating dynamic linear rational expectations models. *Journal of Economic Dynamics and Control* 2, 7–46.

Hansen, L.P. and Singleton, K.J. 1982. Generalized instrumental variables estimation of nonlinear rational expectations models. *Econometrica* 50, 1269–86.

Judd, K. 1998. *Numerical Methods in Economics*. Cambridge, MA: MIT Press.

Kreps, D.M. 1998. Anticipated utility and dynamic choice. In *Frontiers of Research in Economic Theory: The Nancy L. Schwartz Memorial Lectures, 1983–1997*, ed. D.P. Jacobs, E. Kalai and M.I. Kamien. Cambridge: Cambridge University Press.

Krusell, P. and Smith, A. 1998. Income and wealth heterogeneity in the macroeconomy. *Journal of Political Economy* 106, 867–96.

Kydland, F.E. and Prescott, E.C. 1977. Rules rather than discretion: the inconsistency of optimal plans. *Journal of Political Economy* 85, 473–91.

Kydland, F.E. and Prescott, E.C. 1980. Dynamic optimal taxation, rational expectations and optimal control. *Journal of Economic Dynamics and Control* 2, 79–91.

Kydland, F.E. and Prescott, E.C. 1982. Time to build and aggregate fluctuations. *Econometrica* 50, 1345–70.

Lucas, R.E., Jr. 1976. Econometric policy evaluation: a critique. In *The Phillips Curve and the Labor Market*, Vol. 1 of Carnegie-Rochester Conference in Public Policy. ed. K. Brunner and A. Meltzer. Amsterdam: North-Holland.

Lucas, R.E., Jr. 1978. Asset prices in an exchange economy. *Econometrica* 46, 1429–45.

Lucas, R.E., Jr. and Prescott, E.C. 1971. Investment under uncertainty. *Econometrica* 39, 659–81.

Ljungqivst, L. and Sargent, T.J. 2004. *Recursive Macroeconomic Theory*, 2nd edn. Cambridge, MA: MIT Press.

Marcet, A. and Sargent, T.J. 1989. Convergence of least squares learning mechanisms in self-referential linear stochastic models. *Journal of Economic Theory* 48, 337–68.

Muth, J.F. 1960. Optimal properties of exponentially weighted forecasts. *Journal of the American Statistical Association* 55, 299–306.

Muth, J.F. 1961. Rational expectations and the theory of the price movements. *Econometrica* 29, 315–35.

Nerlove, M. 1967. Distributed lags and unobserved components in economic time series. In *Ten Economic Studies in the Tradition of Irving Fisher*, ed. W. Fellner. New York: Wiley.

Rosen, S. and Topel, R.H. 1988. Housing investment in the United States. *Journal of Political Economy* 96, 718–40.

Rosen, S., Murphy, K.M. and Scheinkman, J.A. 1994. Cattle cycles. *Journal of Political Economy* 102, 468–92.

Ryoo, J. and Rosen, S. 2004. The engineering labor market. *Journal of Political Economy* 112, S110–S139.

Sargent, T.J. 1977. The demand for money during hyperinflations under rational expectations: I. *International Economic Review* 18, 59–82.

Sargent, T.J. 1979. *Macroeconomic Theory*. New York: Academic Press.

Sargent, T.J. 1984. Autoregressions, expectations, and advice. *American Economic Review* 74, 408–15.

Sargent, T.J. 1999. *The Conquest of American Inflation*. Princeton: Princeton University Press.

Sargent, T.J. and Wallace, N. 1976. Rational expectations and the theory of economic policy. *Journal of Monetary Economics* 2, 169–83.

Smets, F. and Wouter, R. 2003. An estimated dynamic stochastic general equilibrium model of the Euro area. *Journal of the European Economic Association* 1, 1123–75.
Whiteman, C.H. 1983. *Linear Rational Expectations Models: A Users Guide*. Minneapolis: University of Minnesota Press.
Whittle, P. 1983. *Prediction and Regulation by Linear Least-Square Methods*, 2nd edn. Minneapolis: University of Minnesota Press.
Woodford, M. 1990. Learning to believe in sunspots. *Econometrica* 58, 277–307.
Woodford, M. 2003. *Interest and Prices: Foundations of a Theory of Monetary Policy*. Princeton: Princeton University Press.

regime switching models

Many economic time series occasionally exhibit dramatic breaks in their behaviour, associated with events such as financial crises (Jeanne and Masson, 2000; Cerra and Saxena, 2005; Hamilton, 2005) or abrupt changes in government policy (Hamilton, 1988; Sims and Zha, 2006; Davig, 2004). Of particular interest to economists is the apparent tendency of many economic variables to behave quite differently during economic downturns, when underutilization of factors of production rather than their long-run tendency to grow governs economic dynamics (Hamilton, 1989; Chauvet and Hamilton, 2006). Abrupt changes are also a prevalent feature of financial data, and the approach described below is quite amenable to theoretical calculations for how such abrupt changes in fundamentals should show up in asset prices (Ang and Bekaert, 2002a; 2000b; Garcia, Luger and Renault, 2003; Dai, Singleton and Yang, 2003).

Consider how we might describe the consequences of a dramatic change in the behaviour of a single variable y_t. Suppose that the typical historical behaviour could be described with a first-order autoregression,

$$y_t = c_1 + \phi y_{t-1} + \varepsilon_t, \tag{1}$$

with $\varepsilon_t \sim N(0, \sigma^2)$, which seemed to adequately describe the observed data for $t = 1, 2, \ldots, t_0$. Suppose that at date t_0 there was a significant change in the average level of the series, so that we would instead wish to describe the data according to

$$y_t = c_2 + \phi y_{t-1} + \varepsilon_t \tag{2}$$

for $t = t_0 + 1, t_0 + 2, \ldots$ This fix of changing the value of the intercept from c_1 to c_2 might help the model to get back on track with better forecasts, but it is rather unsatisfactory as a probability law that could have generated the data. We surely would not want to maintain that the change from c_1 to c_2 at date t_0 was a deterministic event that anyone would have been able to predict with certainty looking ahead from date $t = 1$. Instead, there must have been some imperfectly predictable forces that produced the change. Hence, rather than claim that expression (1) governed the data up to date t_0 and (2) after that date, what we must have in mind is that there is some larger model encompassing them both,

$$y_t = c_{s_t} + \phi y_{t-1} + \varepsilon_t, \tag{3}$$

where s_t is a random variable that, as a result of institutional changes, happened in our sample to assume the value $s_t = 1$ for $t = 1, 2, \ldots, t_0$ and $s_t = 2$ for $t = t_0 + 1$, $t_0 + 2, \ldots$. A complete description of the probability law governing the observed data would then require a probabilistic model of what caused the change from $s_t = 1$ to $s_t = 2$. The simplest such specification is that s_t is the realization of a two-state Markov chain with

$$\Pr(s_t = j|s_{t-1} = i, s_{t-2} = k, \ldots, y_{t-1}, y_{t-2}, \ldots) = \Pr(s_t = j|s_{t-1} = i) = p_{ij}. \quad (4)$$

On the assumption that we do not observe s_t directly, but only infer its operation through the observed behavior of y_t, the parameters necessary to fully describe the probability law governing y_t are then the variance of the Gaussian innovation σ^2, the autoregressive coefficient ϕ, the two intercepts c_1 and c_2, and the two state transition probabilities, p_{11} and p_{22}.

The specification in (4) assumes that the probability of a change in regime depends on the past only through the value of the most recent regime, though, as noted below, nothing in the approach described below precludes looking at more general probabilistic specifications. But the simple time-invariant Markov chain (4) seems the natural starting point and is clearly preferable to acting as if the shift from c_1 to c_2 was a deterministic event. Permanence of the shift would be represented by $p_{22} = 1$, though the Markov formulation invites the more general possibility that $p_{22} < 1$. Certainly in the case of business cycles or financial crises, we know that the situation, though dramatic, is not permanent. Furthermore, if the regime change reflects a fundamental change in monetary or fiscal policy, the prudent assumption would seem to be to allow the possibility of it changing back again, suggesting that $p_{22} < 1$ is often a more natural formulation for thinking about changes in regime than $p_{22} = 1$.

A model of the form of (3)–(4) with no autoregressive elements ($\phi = 0$) appears to have been first analysed by Lindgren (1978) and Baum et al. (1980). Specifications that incorporate autoregressive elements date back in the speech recognition literature to Poritz (1982), Juang and Rabiner (1985), and Rabiner (1989), who described such processes as 'hidden Markov models'. Markov-switching regressions were introduced in econometrics by Goldfeld and Quandt (1973), the likelihood function for which was first correctly calculated by Cosslett and Lee (1985). The formulation of the problem described here, in which all objects of interest are calculated as a by-product of an iterative algorithm similar in spirit to a Kalman filter, is due to Hamilton (1989; 1994). General characterizations of moment and stationarity conditions for such processes can be found in Tjøstheim (1986), Yang (2000), Timmermann (2000), and Francq and Zakoïan (2001).

Econometric inference

Suppose that the econometrician observes y_t directly but can only make an inference about the value of s_t based on what we see happening with y_t. This inference will take the form of two probabilities

$$\xi_{jt} = \Pr(s_t = j|\Omega_t; \boldsymbol{\theta}) \quad (5)$$

for $j = 1, 2$, where these two probabilities sum to unity by construction. Here $\Omega_t = \{y_t, y_{t-1}, \ldots, y_1, y_0\}$ denotes the set of observations obtained as of date t, and $\boldsymbol{\theta}$ is a vector of population parameters, which for the above example would be

$\boldsymbol{\theta} = (\sigma, \phi, c_1, c_2, p_{11}, p_{22})'$, and which for now we presume to be known with certainty. The inference is performed iteratively for $t = 1, 2, \ldots, T$, with step t accepting as input the values

$$\xi_{i,t-1} = \Pr(s_{t-1} = i|\Omega_{t-1}; \boldsymbol{\theta}) \tag{6}$$

for $i = 1, 2$ and producing as output (5). The key magnitudes one needs in order to perform this iteration are the densities under the two regimes,

$$\eta_{jt} = f(y_t|s_t = j, \Omega_{t-1}; \boldsymbol{\theta}) = \frac{1}{\sqrt{2\pi}\sigma} \exp\left[-\frac{(y_t - c_j - \phi y_{t-1})^2}{2\sigma^2}\right], \tag{7}$$

for $j = 1, 2$. Specifically, given the input (6) we can calculate the conditional density of the tth observation from

$$f(y_t|\Omega_{t-1}; \boldsymbol{\theta}) = \sum_{i=1}^{2} \sum_{j=1}^{2} p_{ij}\xi_{i,t-1}\eta_{jt} \tag{8}$$

and the desired output is then

$$\xi_{jt} = \frac{\sum_{i=1}^{2} p_{ij}\xi_{i,t-1}\eta_{jt}}{f(y_t|\Omega_{t-1}; \boldsymbol{\theta})}. \tag{9}$$

As a result of executing this iteration, we will have succeeded in evaluating the sample conditional log likelihood of the observed data

$$\log f(y_1, y_2, \ldots, y_T|y_0; \boldsymbol{\theta}) = \sum_{t=1}^{T} \log f(y_t|\Omega_{t-1}; \boldsymbol{\theta}) \tag{10}$$

for the specified value of θ. An estimate of the value of θ can then be obtained by maximizing (10) by numerical optimization.

Several options are available for the value ξ_{i0} to use to start these iterations. If the Markov chain is presumed to be ergodic, one can use the unconditional probabilities

$$\xi_{i0} = \Pr(s_0 = i) = \frac{1 - p_{jj}}{2 - p_{ii} - p_{jj}}.$$

Other alternatives are simply to set $\xi_{i0} = 1/2$ or estimate ξ_{i0} itself by maximum likelihood.

The calculations do not increase in complexity if we consider an $(r \times 1)$ vector of observations $\mathbf{y}_t$ whose density depends on N separate regimes. Let $\Omega_t = \{\mathbf{y}_t, \mathbf{y}_{t-1}, \ldots, \mathbf{y}_1\}$ be the observations through date t, $\mathbf{P}$ be an $(N \times N)$ matrix whose row j, column i element is the transition probability p_{ij}, $\boldsymbol{\eta}_t$ be an $(N \times 1)$ vector whose jth element $f(\mathbf{y}_t|s_t = j, \Omega_{t-1}; \boldsymbol{\theta})$ is the density in regime j, and $\hat{\boldsymbol{\xi}}_{t|t}$ an $(N \times 1)$ vector whose jth element is $\Pr(s_t = j|\Omega_t, \boldsymbol{\theta})$ Then (8) and (9) generalize to

$$f(\mathbf{y}_t|\Omega_{t-1};\boldsymbol{\theta}) = \mathbf{1}'(\mathbf{P}\hat{\xi}_{t-t|t-1} \odot \eta_t) \tag{11}$$

$$\hat{\xi}_{t|t} = \frac{\mathbf{P}\hat{\xi}_{t-t|t-1} \odot \eta_t}{f(\mathbf{y}_t|\Omega_{t-1};\boldsymbol{\theta})} \tag{12}$$

where $\mathbf{1}$ denotes an $(N \times 1)$ vector all of whose elements are unity and $\odot$ denotes element-by-element multiplication. Markov-switching vector autoregressions are discussed in detail in Krolzig (1997). Vector applications include describing the co-movements between stock prices and economic output (Hamilton and Lin, 1996) and the tendency for some series to move into recession before others (Hamilton and Perez-Quiros, 1996). There further is no requirement that the elements of $\boldsymbol{\eta}_t$ be Gaussian densities or even from the same family of densities. For example, Dueker (1997) studied a model in which the degrees of freedom of a Student t distribution change depending on the economic regime.

One is also often interested in forming an inference about what regime the economy was in at date t based on observations obtained through a later date T, denoted $\hat{\xi}_{t|T}$. These are referred to as 'smoothed' probabilities, an efficient algorithm for whose calculation was developed by Kim (1994).

Extensions

The calculations in (11) and (12) remain valid when the probabilities in $\mathbf{P}$ depend on lagged values of $\mathbf{y}_t$ or strictly exogenous explanatory variables, as in Diebold, Lee and Weinbach (1994), Filardo (1994) and Peria (2002). However, often there are relatively few transitions among regimes, making it difficult to estimate such parameters accurately, and most applications have assumed a time-invariant Markov chain. For the same reason, most applications assume only $N = 2$ or 3 different regimes, though there is considerable promise in models with a much larger number of regimes, either by tightly parameterizing the relation between the regimes (Calvet and Fisher, 2004), or with prior Bayesian information (Sims and Zha, 2006).

In the Bayesian approach, both the parameters θ and the values of the states $\mathbf{s} = (s_1, s_2, \ldots, s_T)'$ are viewed as random variables. Bayesian inference turns out to be greatly facilitated by Monte Carlo Markov chain methods, specifically, the Gibbs sampler. This is achieved by sequentially (for $k = 1, 2, \ldots$) generating a realization $\theta^{(k)}$ from the distribution of $\boldsymbol{\theta}|\mathbf{s}^{(k-1)}, \Omega_T$ followed by a realization of $\mathbf{s}^{(k)}$ from the distribution of $\mathbf{s}|\boldsymbol{\theta}^{(k)}, \Omega_T$. The first distribution, $\boldsymbol{\theta}|\mathbf{s}^{(k-1)}, \Omega_T$, treats the historical regimes generated at the previous iteration, $s_1^{(k-1)}, s_2^{(k-1)}, \ldots, s_T^{(k-1)}$, as if fixed known numbers. Often this conditional distribution takes the form of a standard Bayesian inference problem whose solution is known analytically using natural conjugate priors. For example, the posterior distribution of ϕ given other parameters is a known function of easily calculated OLS coefficients. An algorithm for generating a draw from the second distribution, $\mathbf{s}|\boldsymbol{\theta}^{(k)}, \Omega_T$, was developed by Albert and Chib (1993).

The Gibbs sampler turns out also to be a natural device for handling transition probabilities that are functions of observable variables, as in Filardo and Gordon (1998).

It is natural to want to test the null hypothesis that there are N regimes against the alternative of $N+1$, for example when $N=1$, to test whether there are any changes in regime at all. Unfortunately, the likelihood ratio test of this hypothesis fails to satisfy the usual regularity conditions because, under the null hypothesis, some of the parameters of the model would be unidentified. For example, if there is really only one regime, the maximum likelihood estimate $\hat{p}_{11}$ does not converge to a well-defined population magnitude, meaning that the likelihood ratio test does not have the usual χ^2 limiting distribution. To interpret a likelihood ratio statistic, one instead needs to appeal to the methods of Hansen (1992) or Garcia (1998). An alternative is to rely on generic tests of the hypothesis that an N-regime model accurately describes the data (Hamilton, 1996), though these tests are not designed for optimal power against the specific alternative hypothesis of $N+1$ regimes. A test recently proposed by Carrasco, Hu and Ploberger (2004) that is easy to compute but not based on the likelihood ratio statistic seems particularly promising. Other alternatives are to use Bayesian methods to calculate the value of N implying the largest value for the marginal likelihood (Chib, 1998) or the highest Bayes factor (Koop and Potter, 1999), or to compare models on the basis of their ability to forecast (Hamilton and Susmel, 1994).

A specification where the density depends on a finite number of previous regimes, $f(\mathbf{y}_t|s_t, s_{t-1}, \ldots, s_{t-m}, \Omega_{t-1}; \boldsymbol{\theta})$, can be recast in the above form by a suitable redefinition of regime. For example, if s_t follows a 2-state Markov chain with transition probabilities $\Pr(s_t = j|s_{t-1} = i)$ and $m = 1$, one can define a new regime variable s_t^* such that $f(\mathbf{y}_t|s_t^*, \Omega_{t-1}; \boldsymbol{\theta}) = f(\mathbf{y}_t|s_t, s_{t-1}, \ldots, s_{t-m}, \Omega_{t-1}; \boldsymbol{\theta})$ as follows:

$$s_t^* = \begin{cases} 1 & \text{when } s_t = 1 \text{ and } s_{t-1} = 1 \\ 2 & \text{when } s_t = 2 \text{ and } s_{t-1} = 1 \\ 3 & \text{when } s_t = 1 \text{ and } s_{t-1} = 2 \\ 4 & \text{when } s_t = 2 \text{ and } s_{t-1} = 2 \end{cases}.$$

Then s_t^* itself follows a 4-state Markov chain with transition matrix

$$\mathbf{P}^* = \begin{bmatrix} p_{11} & 0 & p_{11} & 0 \\ p_{12} & 0 & p_{12} & 0 \\ 0 & p_{21} & 0 & p_{21} \\ 0 & p_{22} & 0 & p_{22} \end{bmatrix}.$$

More problematic are cases in which the order of dependence m grows with the date of the observation t. Such a situation often arises in models whose recursive structure causes the density of y_t given Ω_{t-1} to depend on the entire history $y_{t-1}, y_{t-2}, \ldots, y_1$ as is the case in ARMA, GARCH or state-space models. Consider for illustration a

GARCH(1, 1) specification in which the coefficients are subject to changes in regime, $y_t = h_t v_t$, where $v_t \sim N(0,1)$ and

$$h_t^2 = \gamma_{s_t} + \alpha_{s_t} y_{t-1}^2 + \beta_{s_t} h_{t-1}^2. \tag{13}$$

Solving (13) recursively reveals that the conditional standard deviation h_t depends on the full history $\{y_{t-1}, y_{t-2}, \ldots, y_0, s_t, s_{t-1}, \ldots, s_1\}$. One way to avoid this problem was proposed by Gray (1996), who postulated that, instead of being generated by (13), the conditional variance is characterized by

$$h_t^2 = \gamma_{s_t} + \alpha_{s_t} y_{t-1}^2 + \beta_{s_t} \tilde{h}_{t-1}^2 \tag{14}$$

where

$$\tilde{h}_{t-1}^2 = \sum_{i=1}^{N} \hat{\xi}_{i,t-1|t-2}(\gamma_i + \alpha_i y_{t-2}^2 + \beta_i \tilde{h}_{t-2}^2).$$

In Gray's model, h_t in (14) depends only on s_t since $\tilde{h}_{t-1}^2$ is a function of data Ω_{t-1} only. An alternative solution, due to Haas, Mittnik and Paolella (2004), is to hypothesize N separate GARCH processes whose values h_{it} all exist as latent variables at date t,

$$h_{it}^2 = \gamma_i + \alpha_i y_{t-1}^2 + \beta_i h_{i,t-1}^2 \tag{15}$$

and then simply pose the model as $y_t = h_{s_t} v_t$. Again, the feature that makes this work is the fact that h_{it} in (15) is a function solely of the data Ω_{t-1} rather than the states $\{s_{t-1}, s_{t-2}, \ldots, s_1\}$.

A related problem arises in Markov-switching state-space models, which posit an unobserved state vector $\mathbf{z}_t$ characterized by

$$\mathbf{z}_t = \mathbf{F}_{s_t}\mathbf{z}_{t-1} + \mathbf{Q}_{s_t}\mathbf{v}_t$$

with $\mathbf{v}_t \sim N(0, \mathbf{I}_n)$, with observed vectors $\mathbf{y}_t$ and $\mathbf{x}_t$ governed by

$$\mathbf{y}_t = \mathbf{H}_{s_t}'\mathbf{z}_t + \mathbf{A}_{s_t}'\mathbf{x}_t + \mathbf{R}_{s_t}\mathbf{w}_t$$

for $\mathbf{w}_t \sim N(0, \mathbf{I}_r)$. Again, the model as formulated implies that the density of $\mathbf{y}_t$ depends on the full history $\{s_t, s_{t-1}, \ldots, s_1\}$. Kim (1994) proposed a modification of the Kalman filter equations similar in spirit to the modification in (14) that can be used to approximate the log likelihood. A more common practice recently has been to estimate such models with numerical Bayesian methods, as in Kim and Nelson (1999).

JAMES D. HAMILTON

See also **nonlinear time series analysis.**

Bibliography

Albert, J. and Chib, 1993. Bayes inference via Gibbs sampling of autoregressive time series subject to Markov mean and variance shifts. *Journal of Business and Economic Statistics* 11, 1–15.

Ang, A. and Bekaert, G. 2002a. International asset allocation with regime shifts. *Review of Financial Studies* 15, 1137–87.

Ang, A. and Bekaert, G. 2002b. Regime switches in interest rates. *Journal of Business and Economic Statistics* 20, 163–82.

Baum, L., Petrie, E., Soules, G. and Weiss, N. 1980. A maximization technique occurring in the statistical analysis of probabilistic functions of Markov chains. *Annals of Mathematical Statistics* 41, 164–71.

Calvet, L. and Fisher, A. 2004. How to forecast long-run volatility: regime-switching and the estimation of multifractal processes. *Journal of Financial Econometrics* 2, 49–83.

Carrasco, M., Hu, L. and Ploberger, W. 2004. Optimal test for Markov switching. Working paper. University of Rochester.

Cerra, V. and Saxena, S. 2005. Did output recover from the Asian crisis? *IMF Staff Papers* 52, 1–23.

Chauvet, M. and Hamilton, J. 2006. Dating business cycle turning points. In *Nonlinear Time Series Analysis of Business Cycles*, ed. C. Milas, P. Rothman and D. van Dijk. Amsterdam: Elsevier.

Chib, S. 1998. Estimation and comparison of multiple change-point models. *Journal of Econometrics* 86, 221–41.

Cosslett, S. and Lee, L.-F. 1985. Serial correlation in discrete variable models. *Journal of Econometrics* 27, 79–97.

Dai, Q., Singleton, K. and Yang, W. 2003. Regime shifts in a dynamic term structure model of U.S. Treasury bonds. Working paper, Stanford University.

Davig, T. 2004. Regime-switching debt and taxation. *Journal of Monetary Economics* 51, 837–59.

Diebold, F., Lee, J.-H. and Weinbach, G. 1994. Regime switching with time-varying transition probabilities. In *Nonstationary Time Series Analysis and Cointegration*, ed. C. Hargreaves. Oxford: Oxford University Press.

Dueker, M. 1997. Markov switching in GARCH processes and mean-reverting stock-market volatility. *Journal of Business and Economic Statistics* 15, 26–34.

Filardo, A. 1994. Business cycle phases and their transitional dynamics. *Journal of Business and Economic Statistics* 12, 299–308.

Filardo, A. and Gordon, S. 1998. Business cycle durations. *Journal of Econometrics* 85, 99–123.

Francq, C. and Zakoïan, J.-M. 2001. Stationarity of multivariate Markov-switching ARMA models. *Journal of Econometrics* 102, 339–64.

Garcia, R. 1998. Asymptotic null distribution of the likelihood ratio test in Markov switching models. *International Economic Review* 39, 763–88.

Garcia, R., Luger, R. and Renault, E. 2003. Empirical assessment of an intertemporal option pricing model with latent variables. *Journal of Econometrics* 116, 49–83.

Goldfeld, S. and Quandt, R. 1973. A Markov model for switching regressions. *Journal of Econometrics* 1, 3–16.

Gray, S. 1996. Modeling the conditional distribution of interest rates as a regime-switching process. *Journal of Financial Economics* 42, 27–62.

Haas, M., Mittnik, S. and Paolella, M. 2004. A new approach to Markov-switching GARCH models. *Journal of Financial Econometrics* 2, 493–530.

Hamilton, J. 1988. Rational-expectations econometric analysis of changes in regime: an investigation of the term structure of interest rates. *Journal of Economic Dynamics and Control* 12, 385–423.

Hamilton, J. 1989. A new approach to the economic analysis of nonstationary time series and the business cycle. *Econometrica* 57, 357–84.

Hamilton, J. 1994. *Time Series Analysis*. Princeton, NJ: Princeton University Press.

Hamilton, J. 1996. Specification testing in Markov-switching time-series models. *Journal of Econometrics* 70, 127–57.

Hamilton, J. 2005. What's real about the business cycle? *Federal Reserve Bank of St. Louis Review* 87, 435–52.

Hamilton, J. and Lin, G. 1996. Stock market volatility and the business cycle. *Journal of Applied Econometrics* 11, 573–93.
Hamilton, J. and Perez-Quiros, G. 1996. What do the leading indicators lead? *Journal of Business* 69, 27–49.
Hamilton, J. and Susmel, R. 1994. Autoregressive conditional heteroskedasticity and changes in regime. *Journal of Econometrics* 64, 307–33.
Hansen, B. 1992. The likelihood ratio test under non-standard conditions. *Journal of Applied Econometrics* 7, S61–S82. Erratum, 11(1996), 195–8.
Jeanne, O. and Masson, P. 2000. Currency crises, sunspots, and Markov-switching regimes. *Journal of International Economics* 50, 327–50.
Juang, B.-H. and Rabiner, L. 1985. Mixture autoregressive hidden Markov models for speech signals. *IEEE Transactions on Acoustics, Speech, and Signal Processing* 30, 1404–13.
Kim, C. 1994. Dynamic linear models with Markov-switching. *Journal of Econometrics* 60, 1–22.
Kim, C. and Nelson, C. 1999. *State-Space Models with Regime Switching*. Cambridge, MA: MIT Press.
Koop, G. and Potter, S. 1999. Bayes factors and nonlinearity: evidence from economic time series. *Journal of Econometrics* 88, 251–81.
Krolzig, H.-M. 1997. *Markov-Switching Vector Autoregressions: Modelling, Statistical Inference, and Application to Business Cycle Analysis*. Berlin: Springer.
Lindgren, G. 1978. Markov regime models for mixed distributions and switching regressions. *Scandinavian Journal of Statistics* 5, 81–91.
Peria, M. 2002. A regime-switching approach to the study of speculative attacks: a focus on EMS crises. In *Advances in Markov-Switching Models*, ed. J. Hamilton and B. Raj. Heidelberg: Physica Verlag.
Poritz, A. 1982. Linear predictive hidden Markov models and the speech signal. *Acoustics, Speech and Signal Processing, IEEE Conference on ICASSP '82*, vol. 7, 1291–4.
Rabiner, L. 1989. A tutorial on hidden Markov models and selected applications in speech recognition. *Proceedings of the IEEE* 77, 257–86.
Sims, C. and Zha, T. 2006. Were there switches in U.S. monetary policy? *American Economic Review* 96, 54–81.
Timmermann, A. 2000. Moments of Markov switching models. *Journal of Econometrics* 96, 75–111.
Tjøstheim, D. 1986. Some doubly stochastic time series models. *Journal of Time Series Analysis* 7, 51–72.
Yang, M. 2000. Some properties of vector autoregressive processes with Markov-switching coefficients. *Econometric Theory* 16, 23–43.

seasonal adjustment

Seasonal adjustment of economic time series dates back to the 19th century and it is based on an attitude properly expressed by Jevons, who wrote:

> Every kind of periodic fluctuation, whether daily, weekly, monthly, quarterly, or yearly, must be detected and exhibited not only as a subject of a study in itself, but because we must ascertain and eliminate such periodic variations before we can correctly exhibit those which are irregular or non-periodic, and probably of more interest and importance. (1884, p. 4)

The most popular model behind seasonal adjustment in the beginning of the 20th century was either the so-called additive unobserved components (UC) model

$$X_t = T_t + C_t + S_t + I_t, \qquad t = 1, 2, \ldots, n \tag{1}$$

where the observed series X_t is divided into a trend component, T_t, a business cycle component, C_t, a seasonal component, S_t, and an irregular component, I_t, or the multiplicative UC model

$$X_t = T_t * C_t * S_t * I_t, \qquad t = 1, 2, \ldots, n \tag{2}$$

which is applied if the series is positive and the oscillations increase with the level of the series.

The definitions of the individual components could vary, but Mills (1924, p. 357) defined the trend component, T_t, as the smoothed, regular, long-term movement of the series X_t, while the seasonal component, S_t, contains fluctuations that are definitely periodic in character with a period of one year, that is, 12 months or 4 quarters. The business cycle component, C_t, is less markedly periodic, but characterized by a considerable degree of regularity with a period of more than one year, while the irregular component, I_t, has no periodicity. A detailed description of the historical development is given in Hylleberg (1986).

The rationale behind seasonal adjustments is that the unobserved components model is useful, that the components are independent, and that the components of interest are the trend and cycle components.

The assumption of independence is highly questionable, as the actual economic time series is a result of economic agents' reaction to some exogenous seasonally varying explanatory factors such as the climate, the timing of religious festivals and business practices. For typical economic agents, decisions designed to smooth seasonal fluctuations will naturally interact with non-seasonal fluctuations, since the costs of such smoothing will necessarily be interrelated through budget constraints and so forth. Therefore, not only is the independence assumption economically unreasonable, but seasonal patterns may be expected to change if economic agents change their behavioural rules.

Hylleberg (1992, p. 4) defines seasonality as

> A systematic, although not necessarily regular, intra-year movement caused by the changes of the weather, the calendar, and timing of decisions, directly or indirectly through the production and consumption decisions made by agents of the economy. These decisions are influenced by endowments, the expectations and preferences of the agents, and the production techniques available in the economy.

Such a view of seasonality is somewhat different from the views expressed by most statistical data-producing agencies. The views of the statistical offices are well represented by the arguments put forward by OECD (1999, p. vii), where the implied definition of seasonality stresses the fixed timing of certain events during the year. Likewise, they indicate that the reason for changes in the seasonal pattern is 'the trading day effect', that is, the changing number of working days in a month, the changing number of Saturdays, and movable feasts such as Easter, Pentecost, Chinese New Year and Korean Full Moon Day. Obviously, such factors do influence the seasonal pattern in economic time series, but in the longer run technical progress and economic considerations based on these will imply changes in the seasonal pattern as well. In addition, the seasonal economic time series may constitute an invaluable and plentiful source of data for testing theories about economic behaviour, as the seasonal pattern is a recurrent though changing event where the pattern, despite the changes, is somewhat easier to forecast than many other economic phenomena. (For a general discussion of seasonality and the literature, see Hylleberg, 1986. For a presentation and discussion of the results since then, see Hylleberg, 1992; Franses, 1996; Ghysels and Osborn, 2001; Brendstrup et al., 2004.)

Seasonal adjustment and treatment of the seasonal components may in practice be undertaken in two ways: simply applying the seasonally adjusted data produced by the statistical agencies, or integrating the modelling and adjustment into the econometric analysis undertaken.

Officially applied seasonal adjustment programmes

Several different methods for seasonal adjustment are in actual use, but the most popular programme is the X-12-ARIMA seasonal adjustment programme (see Findley et al., 1998) which is a further development of the popular X-11 seasonal adjustment programme (see Shiskin, Young and Musgrave, 1967; Hylleberg, 1986). Another popular seasonal adjustment programme is the TRAMO/SEATS programme developed in Gomez and Maravall (1996).

X-12-ARIMA seasonal adjustment programme

The main characteristics of the X-11 seasonal adjustment method for the monthly multiplicative model (see (2)),

$$X_t = TC_t * S_t * TD_t * H_t * I_t, \tag{3}$$

where TC_t is the combined trend-cycle component, while TD_t is the trading day component, and H_t the holiday component, is the repeated application of selected moving averages such as a 12-month centred moving average to estimate TC_t followed by an actual extraction of the estimated trend-cycle component. The extraction by the moving average filters takes place after a prior adjustment for trading days and certain holidays, and a varying seasonal pattern is taken care of by applying so-called Henderson moving averages with a 9, 13 or 23 number of terms. The Henderson trend filters are used in preference to simpler moving averages because they can reproduce polynomials of up to degree 3, thereby capturing trend-turning points.

In addition, treatment for so-called extreme observations was possible, and a refined asymmetric moving averages filter is used at the ends of the series.

In order to robustify the initial seasonally adjusted series against data revisions, the X-11 seasonal adjustment method was improved by extending the series by forecasts and backcasts from an ARIMA model before seasonally adjusting the series (see Dagum, 1980).

The X-12-ARIMA seasonal adjustment programme described in Findley et al. (1998) extends the facilities of X-11-ARIMA by adding a modelling module denoted RegARIMA, which not only facilitates modelling the processes in order to forecast and backcast the time series, but also facilitates modelling of trading day and holiday effects, detection of outlier effects, dealing with missing data, detection of sudden level changes, and detection of changes in the seasonal pattern, trading day effects and so forth. The second major improvement on the earlier programmes is the inclusion of a module for diagnostics which contains many helpful 'tests'. The third improvement is a user-friendly interface.

Although X-12 is a major improvement to X-11, it has its critics. For example, Wallis (1998) doubts that the trend estimation procedure taken over from X-11 is still the best available despite the results obtained since the mid-1970s, and he stresses the need to give the user of the adjusted numbers an indication of their susceptibility to revision.

TRAMO/SEATS seasonal adjustment programme

The main difference between the X-12 programme and the TRAMO/SEATS programme is that the former uses signal-to-noise ratios to choose between the different moving average filters available while SEATS uses signal extraction with filters derived from a time series (ARIMA) model.

The programme also contains a preadjustment programme, TRAMO, which basically performs tasks similar to RegARIMA in X-12.

The signal extraction is based on an additive model such as (1) or

$$Y_t = \mu_t + \gamma_t + \varepsilon_t, \qquad (4)$$

where μ_t is the trend-cycle component, γ_t the seasonal component, and ε_t is the irregular component. It is then assumed that the μ_t and γ_t can be modelled as two distinct ARIMA processes

$$A_C(L)\,(1-L)^d\mu_t = B_C(L)v_t \quad \text{and} \quad A_S(L)\,(1-L^s)^D\gamma_t = B_S(L)w_t \tag{5}$$

where the processes v_t, w_t and ε_t are independent, serially uncorrelated processes with zero means and variances σ_u^2, σ_w^2 and σ_ε^2, and d and D are integers, while L is the lag operator. This class of model is also called the unobserved components autoregressive integrated moving-average model (UCARIMA) by Engle (1978).

Hence, the TRAMO/SEATS programme requires the estimation of the UCARIMA parameters for each specific series, which in principle should allow computation of the correct number of degrees of freedom. This is not possible in X-12 due to the adjustments undertaken within the programme based on the characteristics of the individual series.

A discussion of the merits and drawbacks of X-12 and TRAMO/SEATS may be found in Ghysels and Osborn (2001), Hood, Ashley and Findley (2004), and several working papers from EUROSTAT (see Mazzi and Savio, 2005), which find that X-12 is slightly preferable to TRAMO/SEATS when applied to short time series – a result to be expected as the model-based approach requires more data. In fact, the main differences between the two leading competitors reflect the difference between the model-based approach of TRAMO/SEATS, which tailors a seasonal filter to each series, and the uniform filter applied by X-12 (see below). However, the model-based approach relies on a very restrictive set of models, and the uniform filter approach is not really applying the same filter, as individual characteristics like outliers, smoothness, and so on, have an influence on the filter.

Seasonal adjustment as an integrated part of the analysis

The main objective of the production of seasonally adjusted time series is to give the policy analyst or adviser easy access to a common time series data-set that has been purged of what is considered noise contaminating the series. Obviously, the application of the seasonally adjusted data may be more or less formal and meticulous, ranging from eyeball analysis to thorough econometric analysis.

However, although the application of officially seasonally adjusted data may have the advantage of saving costs, it also implies that the user runs a severe risk of not making the most effective use of the information available and – perhaps more serious – of applying a dataset distorted by the applied seasonal filter.

The possible reasons for these shortcomings are:

- the seasonal component is a noise component but
  - the wrong seasonal adjustment filter has been applied, or
  - the data have been seasonally adjusted individually without consideration being given to the fact that they are often used as input to a multivariate analysis;
- the seasonal components of different time series may be closely connected and contain valuable information across series.

Filtering the data before they are applied may of course distort the outcome of the analysis if the wrong filter is used, but even if the filter is 'correct' as seen from the individual series, the filtering may produce biased estimates of the parameters in certain cases where, for instance, a regression model is applied (see Hylleberg, 1986, p. 3). However, this result is complicated by the application of other transformations to the original series. Which filter to apply may in fact depend on the order of the applied transformations, as shown by Ghysels (1997).

Hence, in order to optimally model many economic phenomena, seasonality may need to be treated as an integrated part of an econometric analysis based on unadjusted quarterly, monthly, weekly and daily time series or panel data observations. This may be done in many different ways depending on the specific context and the set of reasonable assumptions one can make within that context.

As both X-12 and TRAMO/SEATS seasonal adjustment programmes are available to the individual researcher, they may both be applied as part of an integrated approach and their use somewhat adapted to the specific analysis. In what follows we discuss some alternative methods, of three kinds: pure noise models, time series models and economic models.

Pure noise models

The first group comprises seasonal adjustment methods which are based on the assumption that the seasonal component is noise. Thus, the group also contains the officially applied seasonal adjustment programmes presented earlier. The seasonal adjustment methods in this group are distinguished by their ability to take care of a changing seasonal component.

Seasonal dummies

The use of seasonal dummy variables to filter quarterly and monthly times series data is a very simple, straightforward and therefore popular method in econometric applications. The dummy variable method is designed to take care of a constant, stable seasonal component. The popularity of the seasonal dummy variable method is partly due to its simplicity and the flexible way it can be used either as a pre-filtering device whereby the series are regressed on a set of seasonal dummy variables and the residuals used in the final regression, or within the regression as an extension of the set of regressors with seasonal dummy variables (see Frisch and Waugh, 1933; Lovell, 1963).

Band spectrum regression and band pass filters

A natural and quite flexible way to analyse time series with a strong and somewhat varying periodic component is to perform the analysis in the frequency domain, where the time series is represented as a weighted sum of cosine and sine waves. Hence, the time series are Fourier-transformed and the seasonal filtering of the time series may take place by removing specific frequency components from the Fourier-transformed data series.

Application of such filters dates back a long time (see Hannan, 1960). Band spectrum regression is further developed and analysed by Engle (1974; 1980) and

Hylleberg (1977; 1986). The so-called real business cycle literature has since named it 'band pass filtering' (see Baxter and King, 1999).

Let us assume that we have data series with T observations in a vector y and a matrix X related by $y = X\beta + \varepsilon$, where ε is the disturbance term and β a coefficient vector. Band spectrum regression is then performed as a regression in the transformed model $A\Psi y = A\Psi X\beta + A\Psi\varepsilon$, where the transformation by the matrix Ψ is a finite Fourier transformation of the data. The transformation by the diagonal matrix A with zeros and ones on the diagonal, symmetric around the south-west north-east diagonal, is a filtering which removes the frequency components corresponding to the elements with the zeros. Hence, by an appropriate choice of zeros in the main diagonal of A the exact seasonal frequencies and possibly a band around them may be filtered from the series.

An obvious advantage of the band spectrum regression representation is that the model $A\Psi y = A\Psi X\beta + A\Psi\varepsilon$ lends itself directly to a test for the appropriate filtering, as argued in Engle (1974). In fact the test is just the well-known so-called Chow test applied to a stacked model with the null hypothesis that the parameters are the same over the different frequencies. A drawback of band spectrum regression is that the temporal relations between series may be affected in a complicated way by the two-sided filter (see Engle, 1980; Bunzel and Hylleberg, 1982).

Seasonal integration and seasonal fractional integration

A simple filter often applied in empirical econometric work is the seasonal difference filter $(1 - L^s)^d$, where s is the number of observations per year and d the number of times the filter should be applied to render the series stationary at the long run and seasonal frequencies (see Box and Jenkins, 1970).

In the unit root literature a time series is said to be integrated of order d if its d'th difference has a stationary and invertible ARMA representation. Hylleberg et al. (1990) generalize this to seasonal integration and denote, for instance, a quarterly series y_t, $t = 1, 2, \ldots, T$ represented by the model $(1 - L^4)y_t = \varepsilon_t$, $\varepsilon_t \sim$ i.i.d. $(0, \sigma^2)$ as integrated of order 1 at frequency θ since $(1 - L^4) = (1 - L)(1 + L)(1 + L^2)$ has real roots at the unit circle at the frequencies $\theta = \{0, \frac{1}{2}, [\frac{1}{4}, \frac{3}{4}]\}$, where θ is given as the share of a total circle of 2π.

Many empirical studies have applied the so-called Hylleberg, Engle, Granger, and Yoo (HEGY) test for seasonal unit roots developed by Hylleberg et al. (1990) and Engle et al. (1993) for quarterly data, extended to monthly data by Beaulieu and Miron (1993), and to daily data integrated at a period of one week by Kunst (1997). These tests are extensions of the well-known Dickey–Fuller test for a unit root at the long-run zero frequency (Dickey and Fuller, 1979) and at the seasonal frequencies (Dickey, Hasza and Fuller, 1984).

The HEGY test is the simplest and most easily applied test for seasonal unit roots. In the quarterly case the test is based on an autoregressive model $\phi(L)y_t = \varepsilon_t$, $\varepsilon_t \sim iid$ $(0, \sigma^2)$ where $\phi(L)$ is a lag polynomial with possible unit roots at frequencies $\theta = \{0, \frac{1}{2}, [\frac{1}{4}, \frac{3}{4}]\}$. A rewritten linear regression model where the possible unit roots are isolated in specific terms is

$$\begin{aligned}
\phi^*(L)y_{4t} &= \pi_1 y_{1t-1} + \pi_2 y_{2t-1} + \pi_3 y_{3t-2} + \pi_4 y_{3t-1} + \varepsilon_t \\
y_{1t} &= (1 + L + L^2 + L^3)y_t \\
y_{2t} &= -(1 - L + L^2 - L^3)y_t \\
y_{3t} &= -(1 - L^2)y_t \\
y_{4t} &= (1 - L^4)y_t.
\end{aligned} \tag{6}$$

The lag polynomial $\phi^*(L)$ is a stationary and finite polynomial by assumption. Denoting integration of order d at frequency θ by $I_\theta(d)$ we thus have $y_{1t} \sim I_0(1)$, $y_{2t} \sim I_{\frac{1}{2}}(1)$, and $y_{3t} \sim I_{[1/4,3/4]}(1)$, while $y_{1t} \sim I_{\frac{1}{2},[1/4,3/4]}(0)$, $y_{2t} \sim I_{0,[1/4,3/4]}(0)$, $y_{3t} \sim I_{0,\frac{1}{2}}(0)$, and $y_{4t} \sim I_{0,\frac{1}{2},[1/4,3/4]}(0)$, provided $y_t \sim I_{0,\frac{1}{2},[1/4,3/4]}(1)$.

The HEGY tests of the null hypothesis of a unit root are conducted by 't-value' tests on π_1 for the long-run unit root, π_2 for the semi-annual unit root, and 'F-value tests' on π_3 and π_4 for the annual unit roots. In fact, the 't-value' tests on π_1 is just the unit root test of Dickey and Fuller with a special augmentation applied. As in the Dickey–Fuller cases the statistics are not t or F distributed but have non-standard distributions, which for the 't' are tabulated in Fuller (1976) while critical values for the 'F' test are tabulated in Hylleberg et al. (1990).

As in the Dickey–Fuller case the correct lag-augmentation in the auxiliary regression (6) is crucial. The errors need to be rendered white noise in order for the size to be close to the stipulated significance level, but the use of too many lag coefficients reduces the power of the tests.

Obviously, if the data-generating process (DGP) contains a moving average component, the augmentation of the autoregressive part may require long lags (see Hylleberg, 1995) and the HEGY test may be seriously affected by autocorrelation in the errors, moving average terms with roots close to the unit circle, so-called structural breaks, and noisy data with outliers.

The existence of seasonal unit roots in the DGP implies a varying seasonal pattern where 'summer may become winter'. In most cases such an extreme situation is not logically possible, and the findings of seasonal unit roots should be taken as an indication of a varying seasonal pattern and the unit root model as a parsimonious approximation to the DGP.

Another test where the null is no unit root at the zero frequency is suggested by Kwiatkowski et al. (1992) and extended to the seasonal frequencies by Canova and Hansen (1995), and further developed by Busetti and Harvey (2003). See Hylleberg (1995) for a comparison of the Canova–Hansen test and the HEGY test. See also Taylor (2005) for a variance ratio test.

Arteche (2000) and Arteche and Robinson (2000) have extended the analysis to include non-integer values of d in the definition of a seasonally integrated process. In case d is a number between 0 and 1 the process is called fractionally seasonally integrated. The fractionally integrated seasonal process is said to have strong dependence or long memory at a frequency ω since the autocorrelations at that frequency die out at a hyperbolic rate, in contrast to the much faster exponential rate in the weak dependence case where $d = 0$. In the integrated case where $d = 1$ the autocorrelations never die out.

The difficulty with the fractional model is estimation of the parameter d. Even in the quarterly case there are three possible d parameters, and the testing procedure may become very elaborate, requiring, for instance, a sequence of clustered tests as in Gil-Alana and Robinson (1997).

Time series models

The time series models may be univariate models such as the Box–Jenkins model, unobserved components models, time varying parameter models or evolving seasonal models, or multivariate models with seasonal cointegration or periodic cointegration, or models with seasonal common features.

Univariate seasonal models

The Box–Jenkins model. In the traditional analysis of Box and Jenkins (see Box and Jenkins, 1970), the time series where s is the number of quarters, months, and so on, in the year were made stationary by application of the filters $(1 - L)$ *and/or* $(1 - L^s) = (1 - L)S(L)$, where $S(L) = (1 + L + L^2 + L^3 + \cdots + L^{s-1})$, as many times as was deemed necessary from the form of the resulting autocorrelation function. After stationarity had been obtained, the filtered series were modelled as an autoregressive moving average model (ARMA) model. Both the AR and the MA part could be modelled as consisting of a non-seasonal and seasonal lag polynomial. Hence, the so-called seasonal ARIMA model has the form

$$\phi(L)\phi_s(L^s)\,(1 - L^s)^D(1 - L)^d y_t = \theta(L)\theta_s(L^s)\varepsilon_t \tag{7}$$

where $\phi(L)$ and $\theta(L)$ are invertible lag polynomials in L, while $\phi_s(L^s)$ and $\theta_s(L^s)$ are invertible lag polynomials in L^s, and D and d integers.

In light of the results mentioned in the section on seasonal unit roots, the modelling strategy of Box and Jenkins may easily be refined to allow for situations were the non-stationarity exists only at some of the seasonal frequencies.

The 'structural' or unobserved components model. When modelling processes with seasonal characteristics, one must apply complicated and high-ordered polynomials in the ARMA representation. As an alternative to this, the unobserved components model (UC) discussed earlier was proposed. It is easily seen that the UCARIMA model is a general ARIMA model with restrictions on the parameters. Alternatively, the UC model may be specified as a so-called structural model following Harvey (1993).

The structural model is based on a very simple and quite restrictive modelling of the components of interest such as trends, seasonals and cycles. The model is often specified as (4). The trend μ_t is normally assumed to be stationary only in first or second differences, whereas the seasonal component γ_t is stationary when multiplied by the seasonal summation operator $S(L)$. In the basic structural model (BSM) the trend is specified as

$$\begin{aligned} \mu_t &= \mu_{t-1} + \beta_{t-1} + \eta_t \\ \beta_t &= \beta_{t-1} + \zeta_t \end{aligned} \tag{8}$$

where each of the error terms is independently distributed. (If $\sigma^2_\xi = 0$ this collapses to a random walk plus drift. If $\sigma^2_\eta = 0$ as well it corresponds to a model with a linear trend.) The seasonal component is specified as

$$S(L)\gamma_t = \sum_{j=0}^{n-1} \gamma_{t-j} = \omega_t \tag{9}$$

where s is the number of periods per year and where $\omega_t \sim N(0, \sigma^2_\omega)$. (This specification is known as the dummy variable form, since it reduces to a standard deterministic seasonal component if $\sigma^2_\omega = 0$. Specifying the seasonal component this way makes it slowly changing by a mechanism that ensures that the sum of the seasonal components over any s consecutive time periods has an expected value of zero and a variance that remains constant over time.) The BSM model can also be written as

$$y_t = \frac{\xi_t}{\Delta^2} + \frac{\omega_t}{S(L)} + \varepsilon_t, \tag{10}$$

where $\xi_t = \eta_t - \eta_{t-1} + \xi_{t-1}$ is equivalent to an MA(1) process. Expressing the model in the form (10) makes the connection to the UCARIMA model in (4) clear.

Estimation of the general UC model is treated in Hylleberg (1986) and estimation of the structural model is treated in Harvey, Koopman and Shephard (2004).

In the structural approach the problems of specifying the ARMA models for the components is thus avoided by a priori restrictions. Harvey and Scott (1994) argue that the type of model above, which has a seasonal component evolving relatively slowly over time, can fit most economic time series, irrespective of the apparently strong assumptions of a trend component with a unit root and a seasonal component with all possible seasonal unit roots present.

Periodic models and other time varying parameter models. The periodic model extends the non-periodic time series models by allowing the parameters to vary with the seasons. The so-called periodic autoregressive model of order h (PAR(h)) assumes that the observations in each of the seasons can be described using different autoregressive models (see Franses, 1996).

Consider a quarterly times series y_t which is observed for N years. The stationary PAR(h) quarterly model can be written as

$$y_t = \sum_{s=1}^{4} \mu_s D_{s,t} + \sum_{s=1}^{4} \phi_{1s} D_{s,t} y_{t-1} + \cdots + \sum_{s=1}^{4} \phi_{hs} D_{s,t} y_{t-h} + \varepsilon_t \tag{11}$$

with $s = 1, 2, 3, 4$, $t = 1, 2, \ldots, T = 4N$, and where $D_{s,t}$ are seasonal dummies, or as $y_t = \mu_s + \phi_{1s} y_{t-1} + \cdots + \phi_{ps} y_{t-h} + \varepsilon_t$.

It has been shown that any PAR model can be described by a non-periodic ARMA model (Osborn, 1991). In general, however, the orders will be higher than in the PAR model. For example, a PAR(1) corresponds to a non-periodic ARMA(4,3) model. Furthermore, it has been shown that estimating a non-periodic model when the true DGP is a PAR can result in a lack of ability to reject the false non-periodic model (Franses, 1996). Fitting a PAR model does not prevent the finding of a non-periodic AR process, if the latter is in fact the DGP. In practice it is thus recommended that one starts by selecting a PAR(h) model and then tests whether the autoregressive parameters are periodically varying using the method described above.

A major weakness of the periodic model is that the available sample for estimation $N = n/s$ often is too small. Furthermore, the identification of a periodic time series model is not as easy as it is for non-periodic models.

Now, let us rewrite the series y_t, $t = 1, 2, 3, \ldots T$, as $y_{s,\tau}$, where $s = 1, 2, 3, 4$ indicating the quarter, and $\tau = 1, 2, \ldots, n$ indicating the year. The PAR(1) process can then be written as

$$y_{s,\tau} = \phi_s y_{s-1,\tau} + \varepsilon_{s,\tau}, \qquad s = 1, 2, 3, 4; \ \tau = 1, 2, \ldots, n \tag{12}$$

where $y_{0,\tau} = y_{4,\tau-1}$, or in vector notation

$$\Phi(L) Y_\tau = U_\tau \tag{13}$$

where

$$\Phi(L) = \begin{bmatrix} 1 & 0 & 0 & 0 \\ -\phi_2 & 1 & 0 & 0 \\ 0 & -\phi_3 & 1 & 0 \\ 0 & 0 & -\phi_4 & 1 \end{bmatrix} + \begin{bmatrix} 0 & 0 & 0 & -\phi_1 \\ 0 & 0 & 0 & 0 \\ 0 & 0 & 0 & 0 \\ 0 & 0 & 0 & 0 \end{bmatrix} L$$

$$Y'_\tau = [y_{1,\tau}, y_{2,\tau}, y_{3,\tau}, y_{4,\tau}]$$
$$U'_\tau = [\varepsilon_{1,\tau}, \varepsilon_{2,\tau}, \varepsilon_{3,\tau}, \varepsilon_{4,\tau}]$$

with L operating on the seasons, that is, $L\ y_{s,\tau} = y_{s-1,\tau}$ and especially $L\ y_{1,\tau} = y_{0,\tau} = y_{4,\tau-1}$. The PAR(1) process in (13) is stationary provided $|\Phi(z)| = 0$ has all its roots outside the unit circle, which is the case if and only if $|\phi_1\phi_2\phi_3\phi_4| < 1$.

The model may be estimated by maximum likelihood or OLS. Testing for periodicity in (11) amounts to testing the hypothesis $H_0 : \phi_{is} = \phi_i$ for $s = 1, 2, 3, 4$ and $i = 1, 2, \ldots, p$, and this can be done with a likelihood ratio test, which is asymptotically χ^2_{3p} under the null, irrespective of any unit roots in y_t (see Boswijk and Franses, 1995).

The vector representation of the PAR model forms an effective vehicle for generating estimation and testing procedures directly from the general result for stationary vector autoregression (VAR) models, but it also creates an effective way to handle the non-stationary case and compare the periodic models with the models with seasonal integration.

In the non-stationary case, a periodically integrated process of order 1(PI(1)) is defined as a process, where there exists a quasi-difference

$$\begin{aligned} D_s y_{s,\tau} &= 1 - \alpha_s y_{s-1,\tau} \\ \alpha_1\alpha_2\alpha_3\alpha_4 &= 1 \\ \text{not all } \alpha_s &= 1, \quad s = 1,2,3,4. \end{aligned} \tag{14}$$

such that $D_s y_{s,\tau}$ has a stationary and invertible representation. Notice that the *PI(1)* process is neither an integrated $I_0(1)$ process nor a seasonally integrated $I_{0,\frac{1}{2}[1/4,\,3/4]}(1)$ process as shown in Ghysels and Osborn (2001).

The periodic models can be considered special cases of what are referred to as the time-varying parameter models (see Hylleberg, 1986). These are regression models of the form

$$\begin{aligned} Y_t &= X_t'\beta_t + u_t \\ B(L)\ (\beta_t - \beta) &= A\gamma_t + \xi_t \end{aligned} \tag{15}$$

which can be written in state-space form and estimated using the Kalman filter. However, the number of parameters is often greater than the number of observations, and in practice one may be forced to restrict the parameter space. Gersovitz and MacKinnon (1978), applying Bayesian techniques, adopt the sensible assumption that the parameters vary smoothly over the seasons.

The evolving seasonals model. The evolving seasonals model was promulgated by Hannan, Terrell and Tuckwell (1970). The model has been revitalized by Hylleberg and Pagan (1997), who show that the evolving seasonals model produces an excellent vehicle for analysing different commonly applied seasonal models as it nests many of them. The model has been used by Koop and Dijk (2000) to analyse seasonal models from a Bayesian perspective.

The evolving seasonals model for a quarterly time series is based on a representation like

$$\begin{aligned} y_t &= \alpha_{1t}\cos(\lambda_1 t) + \alpha_{2t}\cos(\lambda_2 t) + 2\alpha_{3t}\cos(\lambda_3 t) + 2\alpha_{4t}\sin(\lambda_3 t), \\ &= \alpha_{1t} + \alpha_{2t}\cos(\pi t) + 2\alpha_{3t}\cos(\pi t/2) + 2\alpha_{4t}\sin(\pi t/2), \\ &= \alpha_{1t}(1)^t + \alpha_{2t}(-1)^t + \alpha_{3t}[i^t + (-i)^t] + \alpha_{4t}[i^{t-1} + (-i)^{t-1}], \end{aligned} \tag{16}$$

where $\lambda_1 = 0$, $\lambda_2 = \pi$, $\lambda_3 = \pi/2$, $\cos(\pi t) = (-1)^t$, $2\cos(\pi t/2) = [i^t + (-i)^t]$, $2\sin(\pi t/2) = [i^{t-1} + (-i)^{t-1}]$, $i^2 = -1$, while $\alpha_{jt}, j = 1,2,3,4$; is a linear function of its own past and a stochastic term e_{jt}, $j = 1,2,3,4$: For instance,

$$\begin{aligned} \alpha_{1t} &= \rho_1\alpha_{1,t-1} + e_{1t}, \\ \alpha_{2t} &= \rho_1\alpha_{2,t-1} + e_{2t}, \\ \alpha_{3t} &= \rho_3\alpha_{3,t-2} + e_{3t}, \alpha_{4t} = \rho_4\alpha_{4,t-3} + e_{4t}. \end{aligned} \tag{17}$$

In such a model, $\alpha_{1t}(1)^t = \alpha_{1t}$ represents the trend component with the unit root at the zero frequency, $\alpha_{2t}(-1)^t$ represents the semi-annual component with the root -1, while $\alpha_{3t}[i^t + (-1)^t]\, \alpha_{4t}[i^{t-1} + (-1)^{t-1}]$ represents the annual component with the complex conjugate roots $\pm i$. In Hylleberg and Pagan (1997) it is shown that the HEGY auxiliary regression in (6) has an evolving seasonals model representation, and also that the Canova–Hansen test and the PAR(h) model may be presented in the framework of the evolving seasonals model.

Multivariate seasonal time series models

The idea that the seasonal components of a set of economic time series are driven by a smaller set of common seasonal features seems a natural extension of the idea that the trend components of a set of economic time series are driven by common trends.

If the seasonal components are seasonally integrated, the idea immediately leads to the concept of seasonal cointegration, introduced in Engle, Granger and Hallman (1989), Hylleberg et al. (1990), and Engle et al. (1993). In case the seasonal components are stationary, the idea leads to the concept of seasonal common features (see Engle and Hylleberg, 1996), while so-called periodic cointegration considers cointegration season by season (see Birchenhal et al., 1989; Ghysels and Osborn, 2001).

Seasonal cointegration. Seasonal cointegration exists at a particular seasonal frequency if at least one linear combination of series which are seasonally integrated at the particular frequency is integrated of a lower order.

Consider the quarterly case where y_t and x_t are both integrated of order 1 at the zero and at the seasonal frequencies, that is, the transformations corresponding to 6 are $\{y_{1t}, x_{1t}\} \sim I_0(1)$, $\{y_{2t}, x_{2t}\} \sim I_{\frac{1}{2}}(1)$ and $\{y_{3t}, x_{3t}\} \sim I_{[1/4,3/4]}(1)$. Cointegration at the frequency $\theta = 0$ then exists if $y_{1t} - k_1 x_{1t} \sim I_0(0)$ for some non-zero k_1, cointegration at the frequency $\theta = \frac{1}{2}$ exists if $y_{2t} - k_2 x_{2t} \sim I_{\frac{1}{2}}(0)$ for some non-zero k_2, while cointegration at the frequency $\theta = [1/4, 3/4]$ exists if $y_{2t} - k_3 x_{2t} - k_4 x_{2,t-1} \sim I_{[1/4,3/4]}(0)$ for some non-zero pair $\{k_3,\ k_4\}$. The complex unit roots at the annual frequency [1/4, 3/4] lead to the concept of polynomial cointegration, where cointegration exists if one can find at least one linear combination including a lag of the seasonally integrated series which is stationary.

In Hylleberg et al. (1990) and Engle et al. (1993), seasonal cointegration is analysed along the path set up in Engle and Granger (1987).

The well-known drawbacks of this method, especially when the number of variables included exceeds two, is partly overcome by Lee (1992), who extends the maximum likelihood (ML)-based methods of Johansen (1995) for cointegration at the long-run frequency, to cointegration at the semi-annual frequency $\theta = \frac{1}{2}$.

To adopt the ML-based cointegration analysis at the annual frequency $\theta = [1/4, 3/4]$ with the complex pair of unit roots $\pm i$ is somewhat more complicated, however. The general results may be found in Johansen and Schaumburg (1999), and Cubadda (2001) applies the results of Brillinger (1981) on the canonical correlation analysis of complex variables to obtain tests for cointegration at all the frequencies of interest, that is, at the frequencies 0 and π with the real unit roots ± 1 and at the frequency $\theta = [1/4, 3/4]$ with the complex roots $\pm i$.

Periodic cointegration. Periodic cointegration extends the notion of seasonal cointegration by allowing the coefficients in the cointegration relations to be periodic (see Ghysels and Osborn, 2001).

Consider the example given above with two quarterly time series y_t and x_t, $t = 1, 2, \ldots, T$ which are integrated of order 1 at the zero and seasonal frequencies implying that a transformation by the fourth difference $1 - L^4$ will make the two series stationary. Such series are called seasonally integrated series. Let us rewrite the series as $y_{s,\tau}$ and $x_{s,\tau}$ with $s = 1, 2, 3, 4$ indicating the quarter, and $\tau = 1, 2, \ldots, n$ indicating the year. Hence, the eight yearly series $y_{s,\tau}$, $x_{s,\tau}$ $s = 1, 2, 3, 4$ are all integrated of order 1 at the zero frequency.

Hence, full periodic cointegration exists (see Boswijk and Franses, 1995) if $y_{\tau t} - k_s x_{\tau r} \sim I_0(0)$ for some non-zero k_s, $s = 1, 2, 3, 4$, $\tau = 1, 2, 3$. In case stationarity is only obtained for some $s = 1, 2, 3, 4$, partially periodic cointegration exists.

Several interesting and useful results reviewed in Ghysels and Osborn (2001) follow:

1. Two seasonally integrated series may fully or partially periodically cointegrate.
2. Two $I_0(1)$ processes cannot be periodically cointegrated. They are either non-periodically cointegrated or not cointegrated at all.
3. If two $PI(1)$ processes cointegrate in one quarter. they cointegrate in all four quarters.

Periodic cointegration is a promising but currently not fully exploited area of research, which has the inherent problem that it requires a large sample. It is therefore not surprising that the recent advances in this area happen when data are plentiful (daily) and it is possible to restrict the model appropriately (Haldrup et al., 2007).

Common seasonal features. Although economic time series often exhibit non-stationary behaviour, stationary economic variables exist as well, especially when conditioned on some deterministic pattern such as linear trends, seasonal dummies, breaks and so on. However, a set of stationary economic times series may also exhibit common behaviour and, for instance, share a common seasonal pattern. The technique for finding such patterns, known as common seasonal features (see Engle and Hylleberg, 1996; Cubadda, 1999) is based on earlier contributions defining common features by Engle and Kozicki (1993) and Vahid and Engle (1993).

Consider a multivariate autoregression written in error correction form as

$$\Delta Y_t = \sum_{j=1}^{p} B_j \Delta Y_{t-j} + \Pi v_{t-1} + \Gamma z_t + \varepsilon_t, \quad t = 1, 2, \ldots, T, \tag{18}$$

where Y_t is $k \times 1$ vector of observations on the series of interest in period t and the error correction term is Πv_{t-1}. The vector v_t contains the cointegrating relations at the zero frequency, and the number of cointegrating relations is equal to the rank of Π. If Π has full rank equal to k the series are stationary. In the quarterly case the vector z_t is a vector of trigonometric seasonal dummies, such as $\{\cos(2\pi ht/4 + 2\pi j/T), h = 1, 2; j \in (-\delta T \leq j \leq \delta T), \sin(2\pi h4 + 2\pi j/T), h = 1, 2; j \in (-\delta T \leq j \leq \delta T), j \neq 0,$ when $h = 2\}$. The use of trigonometric dummy variables facilitates the 'modelling' of a varying seasonal pattern, since a proper choice of δ takes care of the neighboring frequencies to the exact seasonal frequencies. If δ is 0, the filter is equivalent to the usual seasonal dummy filter.

The implication of a full rank of the $k \times m$ matrix Γ, equal to min[k, m], is that different linear combinations of the seasonal dummies in z_t are needed in order to explain the seasonal behaviour of the variables in Y_t. However, if there are common seasonal features in these variables we do not need all the different linear combinations, and the rank of Π is not full. Thus, a test of the number of common seasonal features can be based on the rank of Π (see Engle and Kozicki, 1993).

The test is based on a reduced rank regression similar to the test for cointegration by Johansen (1995). Hence, the hypotheses are tested using a canonical correlation analysis between of z_t and ΔY_t, where both sets of variables are purged of the effect from the other variables in (18).

This kind of analysis has proved useful in some situations, but it is difficult to apply in cases where the number of variables is large, and the results are sensitive to the lag-augmentation as in the case of cointegration. In addition, the somewhat arbitrary nature of the choice of z_t poses difficulties.

Economic models of seasonality

Many economic time series have a strong seasonal component, and obviously economic agents must react to that. Hence, the seasonal variation in economic time series must be an integrated part of the optimizing behaviour of economic agents, and the seasonal variation in economic time series must be a result of the optimizing behaviour of economic agents, reacting to exogenous factors such as the weather, the timing of holidays, and so on.

The fact that economic agents react and adjust to seasonal movements on one hand and influence them on the other, implies that the application of seasonal data in economic analysis may widen the possibilities for testing theories about economic behaviour. The relative ease with which the agents may forecast at least some of the causes of the seasonality may be quite helpful in setting up testable models for production smoothing, for instance.

Apart from what is caused by the easiness of forecasting exogenous factors, the type of optimizing behaviour and the agents' reactions to a seasonal phenomenon may be expected not to differ fundamentally from what is happening in a non-seasonal context. However, the recurrent characteristic of seasonality may be exploited.

The economic treatment of seasonal fluctuation has been discussed in the real business cycle (RBC) literature (for example, Chatterjee and Ravikumar, 1992; Braun and Evans, 1995), working with a utility optimizing consumer faced with some feasibility constraint. However, in most of this RBC branch seasonality arises from deterministic shifts in tastes and technology. A few other papers incorporate seasonality through stochastic productivity shocks (see for example Wells, 1997; Cubadda, Savio and Zelli, 2002).

Another area is the production smoothing literature (for instance, Ghysels, 1988; Miron and Zeldes, 1988; Miron, 1996) and habit persistence (for example, Osborn, 1988), where a model for seasonality and habit persistence is presented in a life-cycle consumption model.

SVEND HYLLEBERG

See also **data filters.**

The author is grateful for helpful comments from Niels Haldrup and Steven Durlauf.

Bibliography

Arteche, J. 2000. Gaussian semiparametric estimation in seasonal/cyclical long memory time series. *Kybernetika* 36, 279–310.

Arteche, J. and Robinson, P. 2000. Semiparametric inference in seasonal and cyclical long memory processes. *Journal of Time Series Analysis* 21, 1–25.

Baxter, M. and King, R. 1999. Measuring business cycles: approximate band-pass filters for economic time series. *Review of Economics and Statistics* 81, 575–93.

Beaulieu, J. and Miron, J. 1993. Seasonal unit roots in aggregate US data. *Journal of Econometrics* 55, 305–28.

Birchenhal, C., Bladen-Howell, R., Chui, A., Osborn, D. and Smith, J. 1989. A seasonal model of consumption. *Economic Journal* 99, 837–43.

Boswijk, H. and Franses, P. 1995. Periodic cointegration: representation and inference. *Review of Economics and Statistics* 77, 436–54.

Box, G. and Jenkins, G. 1970. *Time Series Analysis, Forecasting, and Control.* San Francisco: Holden-Day.

Braun, R. and Evans, C. 1995. Seasonality and equilibrium business cycle theories. *Journal of Economic Dynamics and Control* 19, 503–31.

Brendstrup, B., Hylleberg, S., Nielsen, M., Skipper, L. and Stentoft, L. 2004. Seasonality in economic models. *Macroeconomic Dynamics* 8, 362–94.

Brillinger, D. 1981. *Time Series: Data Analysis and Theory.* San Francisco: Holden Day.

Bunzel, H. and Hylleberg, S. 1982. Seasonality in dynamic regression models: a comparative study of finite sample properties of various regression estimators including band spectrum regression. *Journal of Econometrics* 19, 345–66.

Busetti, F. and Harvey, A. 2003. Seasonality tests. *Journal of Business and Economic Statistics* 21, 421–36.

Canova, F. and Hansen, B. 1995. Are seasonal patterns constant over time? A test for seasonal stability. *Journal of Business and Economic Statistics* 13, 237–52.

Chatterjee, S. and Ravikumar, B. 1992. A neoclassical model of seasonal fluctuations. *Journal of Monetary Economics* 29, 59–86.

Cubadda, G. 1999. Common cycles in seasonal non-stationary time series. *Journal of Applied Econometrics* 14, 273–91.

Cubadda, G. 2001. Complex reduced rank models for seasonally cointegrated time series. *Oxford Bulletin of Economics and Statistics* 63, 497–511.

Cubadda, G., Savio, G. and Zelli, R. 2002. Seasonality, productivity shocks, and sectoral comovements in a real business cycle model for Italy. *Macroeconomic Dynamics* 6, 337–56.

Dagum, E. 1980. The X-11-ARIMA seasonally adjustment method. Technical Report 12-564E. Ottawa: Statistics Canada.

Dickey, D. and Fuller, W. 1979. Distribution of the estimators for autoregressive time series with a unit root. *Journal of the American Statistical Association* 74, 427–31.

Dickey, D., Hasza, D. and Fuller, W. 1984. Testing for unit roots in seasonal time series. *Journal of the American Statistical Association* 79, 355–67.

Engle, R. 1974. Band spectrum regression. *International Economic Review* 15, 1–11.

Engle, R. 1978. Estimating structural models of seasonality. In *Seasonal Analysis of Economic Time Series*, ed. A. Zellner. Washington, DC: US Census Bureau.

Engle, R. 1980. Exact maximum likelihood methods for dynamic regressions and band spectrum regressions. *International Economic Review* 21, 391–407.

Engle, R. and Granger, C. 1987. Co-integration and error correction: representation, estimation and testing. *Econometrica* 55, 251–76.

Engle, R., Granger, C. and Hallman, J. 1989. Merging short and long run forecasts: an application of seasonal cointegration to monthly electricity sales forecasting. *Journal of Econometrics* 40, 45–62.

Engle, R., Granger, C., Hylleberg, S. and Lee, H. 1993. Seasonal cointegration – the Japanese consumption function. *Journal of Econometrics* 55, 275–98.

Engle, R. and Hylleberg, S. 1996. Common seasonal features: global unemployment. *Oxford Bulletin of Economics and Statistics* 58, 615–30.

Engle, R. and Kozicki, S. 1993. Testing for common features. *Journal of Business and Economic Statistics* 11, 369–80.

Findley, D., Monsell, B., Bell, W., Otto, M. and Chen, B. 1998. New capabilities and methods of the X-12-ARIMA seasonal adjustment program. *Journal of Business and Economic Statistics* 16, 127–76.

Franses, P. 1996. *Periodicity and Stochastic Trends in Economic Time Series*. Oxford: Oxford University Press.

Frisch, R. and Waugh, F. 1933. Partial time regressions as compared with individual trends. *Econometrica* 1, 387–401.

Fuller, W. 1976. *Introduction to Statistical Time Series*. New York: John Wiley and Sons.

Gersovitz, M. and MacKinnon, J. 1978. Seasonality in regression: an application of smoothness priors. *Journal of the American Statistical Association* 73, 264–73.

Ghysels, E. 1988. A study towards a dynamic theory of seasonality for economics time series. *Journal of the American Statistical Association* 83, 68–72.

Ghysels, E. 1997. Seasonal adjustments and other data transformations. *Journal of Business and Economic Statistics* 15, 410–18.

Ghysels, E. and Osborn, D. 2001. *The Econometric Analysis of Seasonal Time Series*. Cambridge: Cambridge University Press.

Gil-Alana, L. and Robinson, P. 1997. Testing of unit root and other non-stationary hypotheses in macroeconomic time series. *Journal of Econometrics* 80, 241–68.

Gomez, V. and Maravall, A. 1996. *Programs TRAMO and SEATS*. Madrid: Banco de Espana.

Haldrup, N., Hylleberg, S., Pons, G. and Sansó, A. 2007. Common periodic correlation features and the interaction of stocks and flows in daily airport data. *Journal of Business and Economic Statistics* 25, 21–32.

Hannan, E. 1960. *Time Series Analysis*. London: Methuen.

Hannan, E., Terrell, R. and Tuckwell, N. 1970. The seasonal adjustment of economic time series. *International Economic Review* 11, 24–52.

Harvey, A. 1993. *Time Series Models*. London: Prentice Hall/Harvester Wheatsheaf.

Harvey, A. and Scott, A. 1994. Seasonality in dynamic regression models. *Economic Journal* 104, 1324–45.

Harvey, A., Koopman, S. and Shephard, N., eds. 2004. *State Space and Unobserved Component Models: Theory and Applications*. Cambridge: Cambridge University Press.

Hood, C., Ashley, J. and Findley, D. 2004. An empirical evaluation of the performance of TRAMo/SEATS on simulated series. Technical report. Washington, DC: US Census Bureau.

Hylleberg, S. 1977. A comparative study of finite sample properties of band spectrum regression estimators. *Journal of Econometrics* 5, 167–82.

Hylleberg, S. 1986. *Seasonality in Regression.* Orlando, FL: Academic Press.

Hylleberg, S., ed. 1992. *Modelling Seasonality.* Oxford: Oxford University Press.

Hylleberg, S. 1995. Tests for seasonal unit roots: general to specific or specific to general. *Journal of Econometrics* 69, 5–25.

Hylleberg, S., Engle, R., Granger, C. and Yoo, S. 1990. Seasonal integration and cointegration. *Journal of Econometrics* 44, 215–38.

Hylleberg, S. and Pagan, A. 1997. Seasonal integration and the evolving seasonals model. *International Journal of Forecasting* 13, 329–40.

Jevons, W. 1884. *Investigations in Currency and Finances.* London: Macmillan.

Johansen, S. 1995. *Likelihood-Based Inference in Cointegrated Vector Autoregressive Models.* Oxford: Oxford University Press.

Johansen, S. and Schaumburg, E. 1999. Likelihood analysis of seasonal cointegration. *Journal of Econometrics* 88, 301–39.

Koop, G. and Dijk, H. 2000. Testing for integration using evolving trend and seasonals models: a Bayesian approach. *Journal of Econometrics* 97, 261–91.

Kunst, R. 1997. Testing for cyclical non-stationarity in autoregressive processes. *Journal of Time Series Analysis* 18, 123–35.

Kwiatkowski, D., Phillips, P., Schmidt, P. and Shin, Y. 1992. Testing the null hypothesis of stationarity against the alternative of a unit root – how sure are we that economic time series have a unit root? *Journal of Econometrics* 54, 159–78.

Lee, H. 1992. Maximum likelihood inference on cointegration and seasonal cointegration. *Journal of Econometrics* 54, 1–47.

Lovell, M. 1963. Seasonal adjustment of economic time series. *Journal of the American Statistical Association* 58, 993–1010.

Mazzi, G.L. and Savio, G. 2005. The seasonal adjustment of short time series. Technical Report KS-DT-05-002. EUROSTAT.

Mills, F. 1924. *Statistical Methods.* London: Pitman.

Miron, J. 1996. *The Economics of Seasonal Cycles.* Cambridge, MA: MIT Press.

Miron, J. and Zeldes, S. 1988. Seasonality, cost shocks and the production smoothing model of inventories. *Econometrica* 56, 877–908.

OECD. 1999. Feature article: seasonal adjustment. *Main Economic Indicators.* November. Paris: OECD.

Osborn, D. 1988. Seasonality and habit persistence in a life cycle model of consumption. *Journal of Applied Econometrics* 3, 255–66.

Osborn, D. 1991. The implications of periodically varying coefficients for seasonal time-series processes. *Journal of Econometrics* 48, 373–84.

Shiskin, J., Young, A. and Musgrave, J. 1967. The X-11 variant of the census method II seasonal adjustment program. Technical Paper No. 15. Washington, DC: US Census Bureau.

Taylor, A. 2005. Variance ratio tests of the seasonal unit root hypothesis. *Journal of Econometrics* 124, 33–54.

Vahid, F. and Engle, R. 1993. Common trends and common cycles. *Journal of Applied Econometrics* 8, 341–60.

Wallis, K. 1998. Comment. *Journal of Business and Economic Statistics* 16, 164–5.

Wells, J. 1997. Business cycles, seasonal cycles, and common trends. *Journal of Macroeconomics* 19, 443–69.

serial correlation and serial dependence

1. Introduction

Serial correlation and serial dependence have been central to time series econometrics. The existence of serial correlation complicates statistical inference of econometric models; and in time series analysis, inference of serial correlation, or more generally, serial dependence, is crucial to characterize the dynamics of time series processes. Lack of serial correlation is also an important implication of many economic theories and economic hypotheses. For example, the efficient market hypothesis implies that asset returns are an martingale difference sequence (*m.d.s.*), and so are serially uncorrelated. More generally, rational expectations theory implies that the expectational errors of the economic agent are serially uncorrelated. In this article we first discuss various tests for serial correlation, for both estimated model residuals and observed raw data, and we discuss their relationships. We then discuss serial dependence in a nonlinear time series context, introducing related measures and tests for serial dependence.

2. Testing for serial correlation

Consider a linear regression model

$$Y_t = X_t'\beta^0 + \varepsilon_t, \quad t = 1, \cdots, n, \tag{2.1}$$

where Y_t is a dependent variable, X_t is a $k \times 1$ vector of explanatory variables, β^0 is an unknown $k \times 1$ parameter vector, and ε_t is an unobservable disturbance with $E(\varepsilon_t|X_t) = 0$. Suppose X_t is strictly exogenous such that $\text{cov}(X_t, \varepsilon_s) = 0$ for all t, s. Then (2.1) is called a static regression model. If X_t contains lagged dependent variables, (2.1) is called a dynamic regression model.

For a linear dynamic regression model, serial correlation in $\{\varepsilon_t\}$ will generally render inconsistent the OLS estimator. To see this, we consider an AR(1) model

$$Y_t = \beta_0^0 + \beta_1^0 Y_{t-1} + \varepsilon_t = X_t'\beta^0 + \varepsilon_t,$$

where $X_t = (1, Y_{t-1})'$. If $\{\varepsilon_t\}$ also follows an AR(1) process, we will have $E(X_t\varepsilon_t) \neq 0$, rendering inconsistent the OLS estimator for β^0. It is therefore important to check serial correlation for estimated model residuals, which serves as a misspecification test for a linear dynamic regression model. For a static linear regression model, it is also useful to check serial correlation. In particular, if there exists no serial correlation in $\{\varepsilon_t\}$ in a static regression model, then there is no need to use a long-run variance estimator of the OLS estimator $\hat{\beta}$ (for example, Andrews, 1991; Newey and West, 1987).

2.1 Durbin–Watson test

Testing for serial correlation has been a longstanding problem in time series econometrics. The most well known test for serial correlation in regression disturbances is the Durbin–Watson test, which is the first formal procedure developed for testing first order serial correlation

$$\varepsilon_t = \rho\varepsilon_{t-1} + u_t, \qquad \{u_t\} \sim i.i.d.(0, \sigma^2)$$

using the OLS residuals $\{e_t\}_{t=1}^n$ in a static linear regression model. Durbin and Watson (1950; 1951) propose a test statistic

$$d = \frac{\sum_{t=2}^n (e_t - e_{t-1})^2}{\sum_{t=1}^n e_t^2}.$$

Durbin and Watson present tables of bounds at the 0.05, 0.025 and 0.01 significance levels of the d statistic for static regressions with an intercept. Against the one-sided alternative that $\rho > 0$, if d is less than the lower bound d_L, the null hypothesis that $\rho = 0$ is rejected; if ρ is greater than the upper bound d_U, the null hypothesis is accepted. Otherwise, the test is equivocal. Against the one-sided alternative that $\rho < 0$, $4-d$ can be used to replace d in the above procedure.

The Durbin–Watson test has been extended to test for lag 4 autocorrelation by Wallis (1972) and for autocorrelation at any lag by Vinod (1973).

2.2 Durbin's h *test*

The Durbin–Watson d test is not applicable to dynamic linear regression models, because parameter estimation uncertainty in the OLS estimator $\hat{\beta}$ will have nontrivial impact on the distribution of d. Durbin (1970) developed the so-called h test for first-order autocorrelation in $\{\varepsilon_t\}$ that takes into account parameter estimation uncertainty in $\hat{\beta}$. Consider a simple dynamic linear regression model

$$Y_t = \beta_0^0 + \beta_1^0 Y_{t-1} + \beta_2^0 X_t + \varepsilon_t,$$

where X_t is strictly exogenous. Durbin's h statistic is defined as:

$$h = \hat{\rho}\sqrt{\frac{n}{1 - n\hat{\text{v}}\text{ar}(\hat{\beta}_1)}},$$

where $\hat{\text{v}}\text{ar}(\hat{\beta}_1)$ is an estimator for the asymptotic variance of $\hat{\beta}_1$, $\hat{\rho}$ is the OLS estimator from regressing e_t on e_{t-1} (in fact, $\hat{\rho} \approx 1 - d/2$). Durbin (1970) shows that $h \xrightarrow{d} N(0, 1)$ as $n \to \infty$ under null hypothesis that $\rho = 0$.

2.3 Breusch–Godfrey test

A more convenient and generally applicable test for serial correlation is the Lagrange multiplier test developed by Breusch (1978) and Godfrey (1978). Consider an auxiliary autoregression of order p:

$$\varepsilon_t = \sum_{j=1}^{p} \alpha_j \varepsilon_{t-j} + z_t, \qquad t = p+1, \ldots, n. \tag{2.2}$$

The null hypothesis of no serial correlation implies $\alpha_j = 0$ for all $1 \leq j \leq p$. Under the null hypothesis, we have $nR^2_{uc} \xrightarrow{d} \chi^2_p$, where R^2_{uc} is the uncentred R^2 of (2.2). However, the autoregression (2.2) is infeasible because ε_t is unobservable. One can replace ε_t with the OLS residual e_t:

$$e_t = \sum_{j=1}^{p} \alpha_j e_{t-j} + v_t, \qquad t = p+1, \ldots, n.$$

Such a replacement, however, may contaminate the asymptotic distribution of the test statistic because $e_t = \varepsilon_t - (\hat{\beta} - \beta)' X_t$ contains the estimation error $(\hat{\beta} - \beta)' X_t$ where X_t may have nonzero correlation with the regressors e_{t-j} for $1 \leq j \leq p$ in dynamic regression models. This correlation affects the asymptotic distribution of nR^2_{uc} so that it will not be χ^2_p. To purge this impact of the asymptotic distribution of the test statistic, one can consider the augmented auxiliary regression

$$e_t = X_t' \gamma + \sum_{j=1}^{p} \alpha_j e_{t-j} + v_t, \qquad t = p+1, \ldots, n. \tag{2.3}$$

The inclusion of X_t will capture the impact of estimation error $(\hat{\beta} - \beta)' X_t$. As a result, the test statistic $nR^2 \xrightarrow{d} \chi^2_p$ under the null hypothesis, where, assuming that X_t contains an intercept, R^2 is the centred squared multi-correlation coefficient in (2.3). For a static linear regression model, it is not necessary to include X_t in the auxiliary regression, because $\{X_t\}$ and $\{\varepsilon_t\}$ are uncorrelated, but it does not harm the size of the test if X_t is included. Therefore, the nR^2 test is applicable to both static and dynamic regression models. We note that Durbin's h test is asymptotically equivalent to the nR^2 test of (2.3) with $p = 1$.

2.4 Box–Pierce–Ljung test

In time series ARMA modelling, Box and Pierce (1970) propose a portmanteau test as a diagnostic check for the adequacy of an ARMA model

$$Y_t = \psi_0 + \sum_{j=1}^{r} \psi_j Y_{t-j} + \sum_{j=1}^{q} \theta_j \varepsilon_{t-j} + \varepsilon_t, \qquad \{\varepsilon_t\} \sim i.i.d.(0, \sigma^2). \tag{2.4}$$

Suppose e_t is an estimated residual obtained from a maximum likelihood estimator. One can define the residual sample autocorrelation function

$$\hat{\rho}(j) = \frac{\hat{\gamma}(j)}{\hat{\gamma}(0)}, \qquad j = 0, \pm 1, \ldots, \pm(n-1),$$

where $\hat{\gamma}(j) = n^{-1} \sum_{t=|j|+1}^{n} e_t e_{t-|j|}$ is the residual sample autocovariance function.

Box and Pierce (1970) propose a portmanteau test

$$Q \equiv n \sum_{j=1}^{p} \hat{\rho}^2(j) \xrightarrow{d} \chi^2_{p-(r+q)},$$

where the asymptotic χ^2 distribution follows under the null hypothesis of no serial correlation, and the adjustment of degrees of freedom $r+q$ is due to the impact of parameter estimation uncertainty for the r autoregressive coefficients and q moving average coefficients in (2.4).

To improve small sample performance of the Q test, Ljung and Box (1978) propose a modified Q test statistic:

$$Q^* \equiv n(n+2) \sum_{j=1}^{p} (n-j)^{-1} \hat{\rho}^2(j) \xrightarrow{d} \chi^2_{p-(r+q)}.$$

The modification matches the first two moments of Q^* with those of the χ^2 distribution. This improves the size in small samples, although not the power of the test.

The Q test is applicable to test serial correlation in the OLS residuals $\{e_t\}$ of a linear static regression model, with $Q \xrightarrow{d} \chi^2_p$ under the null hypothesis. Unlike for ARMA models, there is no need to adjust the degrees of freedom for the χ^2 distribution because the estimation error $(\hat{\beta} - \beta)' X_t$ has no impact on it, due to the fact that $\text{cov}(X_t, \varepsilon_s) = 0$ for all t, s. In fact, it could be shown that the nR^2 and Q statistics are asymptotically equivalent under the null hypothesis. However, when applied to the estimated residual of a dynamic regression model which contains both endogenous and exogenous variables, the asymptotic distribution of the Q test is generally unknown (Breusch and Pagan, 1980). One solution is to modify the Q test statistic as follows:

$$\hat{Q} \equiv n\hat{\rho}'(I - \hat{\Phi})^{-1}\hat{\rho} \xrightarrow{d} \chi^2_p \quad \text{as } n \to \infty,$$

where $\hat{\rho} = [\hat{\rho}(1), \ldots, \hat{\rho}(p)]'$ and $\hat{\Phi}$ captures the impact caused by nonzero correlation between $\{X_t\}$ and $\{\varepsilon_s\}$. See Hayashi (2000, Section 2.10) for more discussion.

2.5 Spectral density-based test

Much criticism has been levelled at the possible low power of the Box–Pierce–Ljung portmanteau tests, which also applies to the nR^2 test, due to the asymptotic equivalence between the Q test and the nR^2 test for a static regression. Moreover, there is no theoretical guidance on the choice of p for these tests. A fixed lag order p will render inconsistent any test for serial correlation of unknown form.

To test serial correlation of unknown form in the estimated residuals of a linear regression model, which can be static or dynamic, Hong (1996) uses a kernel spectral density estimator

$$\hat{h}(\omega) = \frac{1}{2\pi} \sum_{j=1-n}^{n-1} k(j/p)\hat{\gamma}(j)e^{-ij\omega}, \qquad \omega \in [-\pi, \pi],$$

and compares it with the flat spectrum implied by the null hypothesis of no serial correlation:

$$\hat{h}_0(\omega) = \frac{1}{2\pi}\hat{\gamma}(0), \qquad \omega \in [-\pi, \pi].$$

Under the null hypothesis, $\hat{h}(\omega)$ and $\hat{h}_0(\omega)$ are close. If $\hat{h}(\omega)$ is significantly different from $\hat{h}_0(\omega)$ there is evidence of serial correlation. A global measure of the divergence between $\hat{h}(\omega)$ and $\hat{h}_0(\omega)$ is the quadratic form

$$\begin{aligned} L^2(\hat{h}, \hat{h}_0) &\equiv \int_{-\pi}^{\pi} \left[\hat{h}(\omega) - \hat{h}_0(\omega)\right]^2 d\omega \\ &= \sum_{j=1}^{n-1} k^2(j/p)\hat{\gamma}^2(j). \end{aligned}$$

The test statistic is a normalized version of the quadratic form:

$$M_o \equiv \left[n\sum_{j=1}^{n-1} k^2(j/p)\hat{\rho}^2(j) - \hat{C}_o(p) \right] \Big/ \sqrt{\hat{D}_o(p)} \xrightarrow{d} N(0,1)$$

where the centring and scaling factors

$$\hat{C}_o(p) = \sum_{j=1}^{n-1}(1 - j/n)k^2(j/p),$$

$$\hat{D}_o(p) = 2\sum_{j=1}^{n-2}(1 - j/n)[1 - (j+1)/n]k^4(j/p).$$

This test can be viewed as a generalized version of Box and Pierce's (1970) portmanteau test, the latter being equivalent to using the truncated kernel $k(z) = \mathbf{1}(|z| \leq 1)$, which gives equal weighting to each of the first p lags. In this case, M_o is asymptotically equivalent to

$$M_T \equiv \frac{n\sum_{j=1}^{p}\hat{\rho}^2(j) - p}{\sqrt{2p}} \xrightarrow{d} \frac{\chi_p^2 - p}{\sqrt{2p}} \sim N(0,1) \;\; as \;\; p \to \infty.$$

However, uniform weighting to different lags may not be powerful when a large number of lags is employed. For any weakly stationary process, the autocovariance function $\gamma(j)$ typically decays to 0 as lag order j increases. Thus, it is more efficient to discount higher order lags. This can be achieved by using non-uniform kernels. Most commonly used kernels, such as the Bartlett, Pazren and quadratic-spectral kernels, discount higher order lags. Hong (1996) shows that the Daniell kernel $k(z) = \sin(\pi z)/(\pi z)$, $-\infty < z < \infty$, maximizes the power of the M test over a wide class of the kernel functions when $p \to \infty$. The optimal kernel for hypothesis testing differs from the optimal kernel for spectral density estimation.

It is important to note that the spectral density test M applies to both static and dynamic regression models, and no modification is needed when applied to a dynamic regression model. Intuitively, parameter estimation uncertainty causes some adjustment of degrees of freedom, which becomes asymptotically independent when the lag order $p \to \infty$ as $n \to \infty$. This differs from the case where p is fixed.

For similar spectral density-based tests for serial correlation, see Paparoditis (2000), Chen and Deo (2004), and Fan and Zhang (2004).

2.6 Heteroskedasticity–robust tests

All the aforementioned tests assume conditional homoskedasticity or even *i.i.d.* on $\{\varepsilon_t\}$. This rules out high frequency financial time series, which have been documented to have persistent volatility clustering. Some effort has been devoted to robustifying tests for serial correlation. Wooldridge (1990; 1991) proposes a two-stage procedure to robustify the nR^2 test for serial correlation in estimated residuals $\{e_t\}$ of a linear regression model (2.1): (i) regress $(e_{t-1}, \ldots, e_{t-p})$ on X_t and save the estimated $p \times 1$ residual vector $\hat{v}_t$; (ii) regress 1 on $\hat{v}_t e_t$ and obtain SSR, the sum of squared residuals; (iii) compare the $n-SSR$ statistic with the asymptotic χ^2_p distribution. The first auxiliary regression purges the impact of parameter estimation uncertainty in the OLS estimator $\hat{\beta}$ and the second auxiliary regression delivers a test statistic robust to conditional heteroskedasticity of unknown form.

Whang (1998) also proposes a semiparametric test for serial correlation in estimated residuals of a possibly nonlinear regression model. Assuming that $\varepsilon_t = \sigma[Z_t(\alpha)]z_t$, where $\{z_t\} \sim$ i.i.d.(0, 1), $\text{var}(\varepsilon_t|I_{t-1}) = \sigma^2[Z_t(\alpha)]$ depends on a random vector with fixed dimension (for example, $Z_t(\alpha) = (\varepsilon^2_{t-1}, \ldots, \varepsilon^2_{t-K})'$ for a fixed K), but the functional form $\sigma^2(\cdot)$ is unknown. This covers a variety of conditionally heteroskedastic processes, although it rules out non-Markovian processes such as Bollerslev's (1986) GARCH model. Whang (1998) first estimates $\sigma^2[Z_t(\alpha)]$ using a kernel method, and then constructs a Box–Pierce type test for serial correlation in the estimated regression residuals standardized by the square root of the nonparametric variance estimator.

The assumption imposed on $\text{var}(\varepsilon_t|I_{t-1})$ in Whang (1998) rules out GARCH models, and both Wooldridge (1991) and Whang (1998) test serial correlation up to a fixed lag order only. Hong and Lee (2007) have recently robustified Hong's (1996) spectral density-based consistent test for serial correlation of unknown form:

$$\hat{M} \equiv \left[n^{-1} \sum_{j=1}^{n-1} k^2(j/p)\hat{\gamma}^2(j) - \hat{C}(p) \right] \Big/ \sqrt{\hat{D}(p)},$$

where the centring and scaling factors

$$\hat{C}(p) \equiv \hat{\gamma}^2(0) \sum_{j=1}^{n-1} (1 - j/n)k^2(j/p) + \sum_{j=1}^{n-1} k^2(j/p)\hat{\gamma}_{22}(j),$$

$$\hat{D}(p) \equiv 2\hat{\gamma}^4(0)\sum_{j=1}^{n-2}(1-j/n)[1-(j+1)/n]k^4(j/p) + 4\hat{\gamma}^2(0)\sum_{j=1}^{n-2}k^4(j/p)\hat{\gamma}_{22}(j)$$
$$+\,2\sum_{j=1}^{n-2}\sum_{l=1}^{n-2}k^2(j/p)k^2(l/p)\hat{C}^2(0,j,l),$$

with $\hat{\gamma}_{22}(j) \equiv n^{-1}\sum_{t=j+1}^{n-1}[e_t^2-\hat{\gamma}(0)][e_{t-j}^2-\hat{\gamma}(0)]$ and $\hat{C}(0,j,l) \equiv n^{-1}\sum_{t=\max(j,l)+1}^{n}[e_t^2-\hat{\gamma}(0)]e_{t-j}e_{t-l}$. Intuitively, the centring and scaling factors have taken into account possible volatility clustering and asymmetric features of volatility dynamics, so the $\hat{M}$ test is robust to these effects. It allows for various volatility processes, including GARCH models, Nelson's (1991) EGARCH, and Glosten, Jagannathan and Runkle's (1993) Threshold GARCH models.

Martingale tests

Several tests for serial correlation are motivated for testing the *m.d.s.* property of an observed time series $\{Y_t\}$, say asset returns, rather than estimated residuals of a regression model. We now present a unified framework to view some martingale tests for observed data.

Extending an idea of Cochrane (1988), Lo and MacKinlay (1988) first rigorously present an asymptotic theory for a variance ratio test for the *m.d.s.* hypothesis of $\{Y_t\}$. Because the *m.d.s.* hypothesis implies $\gamma(j)=0$ for all $j>0$, one has

$$\frac{\mathrm{var}(\sum_{j=1}^{p}Y_{t-j})}{p\cdot\mathrm{var}(Y_t)} = \frac{p\gamma(0)+2p\sum_{j=1}^{p}(1-j/p)\gamma(j)}{p\gamma(0)} = 1.$$

This unity property of the variance ratio can be used to test the *m.d.s.* hypothesis because any departure from unity is evidence against the *m.d.s.* hypothesis.

The variance ratio test is essentially based on the statistic

$$\mathrm{VR_o} \equiv \sqrt{n/p}\sum_{j=1}^{p}(1-j/p)\hat{\rho}(j) = \frac{\pi}{2}\sqrt{n/p}\left[\hat{f}(0)-\frac{1}{2\pi}\right],$$

where $\hat{f}(0)$ is a kernel-based normalized spectral density estimator at frequency 0, with the Bartlett kernel $k(z)=(1-|z|)\,\mathbf{1}\,(|z|\leq 1)$ and a lag order p. In other words, $\mathrm{VR_o}$ is based on a spectral density estimator of frequency 0, and because of this, it is particularly powerful against long memory processes, whose spectral density at frequency 0 is infinity (see Robinson, 1994, for an excellent survey).

Under the *m.d.s.* hypothesis with conditional homoskedasticity, Lo and MacKinlay (1988) show that for any fixed p,

$$\mathrm{VR_o} \xrightarrow{d} N[0, 2(2p-1)(p-1)/3p] \quad \text{as } n\to\infty.$$

Lo and MacKinlay (1988) also consider a heteroskedasticity-consistent variance ratio test:

$$\text{VR} \equiv \sqrt{n/p}\sum_{j=1}^{p}(1-j/p)\hat{\gamma}(j)/\sqrt{\hat{\gamma}_2(j)},$$

where $\hat{\gamma}_2(j)$ is a consistent estimator for the asymptotic variance of $\hat{\gamma}(j)$ under conditional heteroskedasticity. Lo and MacKinlay (1988) assume a fourth order cumulant condition that

$$E[(Y_t-\mu)^2(Y_{t-l}-\mu)(Y_{t-1}-\mu)]=0, \qquad j,l>0, j\neq l. \tag{2.5}$$

Intuitively, this condition ensures that the sample autocovariances at different lags are asymptotically uncorrelated; that is, $\text{cov}[\sqrt{n}\hat{\gamma}(j), \sqrt{n}\hat{\gamma}(l)] \to 0$ for all $j\neq l$. As a result, the heroskedasticity-consistent VR has the same asymptotic distribution as VR_o. However, the condition in (2.5) rules out many important volatility processes, such as EGARCH and Threshold GARCH models. Moreover, the variance ratio test only exploits the implication of the *m.d.s.* hypothesis on the spectral density at frequency 0; it does not check the spectral density at nonzero frequencies. As a result, it is not consistent against serial correlation of unknown form. See Durlauf (1991) for more discussion.

Durlauf (1991) considers testing the *m.d.s.* hypothesis for observed raw data $\{Y_t\}$, using the spectral distribution function

$$H(\lambda) \equiv 2\int_0^{\pi\lambda} h(\omega)d\omega = \gamma(0)\lambda + \sqrt{2}\sum_{j=1}^{\infty}\gamma(j)\frac{\sqrt{2}\sin(j\pi\lambda)}{j\pi}, \quad \lambda\in[0,1],$$

where $h(\omega)$ is the spectral density of $\{Y_t\}$:

$$h(\omega) = \frac{1}{2\pi}\sum_{j=-\infty}^{\infty}\gamma(j)\cos(j\omega), \quad \omega\in[-\pi,\pi].$$

Under the *m.d.s.* hypothesis, $H(\lambda)$ becomes a straight line:

$$H_0(\lambda) = \gamma(0)\lambda, \ \ \lambda\in[0,1].$$

An *m.d.s.* test can be obtained by comparing a consistent estimator for $H(\lambda)$ and $\hat{H}_0(\lambda) = \hat{\gamma}(0)\lambda$.

Although the periodogram (or sample spectral density function)

$$\hat{I}(\omega) \equiv \frac{1}{2\pi n}\left|\sum_{t=1}^{n}(Y_t - Y)e^{it\omega}\right|^2 = \frac{1}{2\pi}\sum_{j=1-n}^{n-1}\hat{\gamma}(j)e^{-ij\omega}$$

is not consistent for the spectral density $h(\omega)$, the integrated periodogram

$$\hat{H}(\lambda) \equiv 2\int_0^{\lambda\pi} \hat{I}(\omega)d\omega = \hat{\gamma}(0)\lambda + \sqrt{2}\sum_{j=1}^{n-1}\hat{\gamma}(j)\frac{\sqrt{2}\sin(j\pi\lambda)}{j\pi}$$

is consistent for $H(\lambda)$, thanks to the smoothing provided by the integration. Among other things, Durlauf (1991) proposes a Cramer–von Mises type statistic

$$CVM \equiv \frac{1}{2}n\int_0^1 [\hat{H}(\lambda)/\hat{\gamma}(0) - \lambda]^2 d\lambda = n\sum_{j=1}^{n-1}\hat{\rho}^2(j)/(j\pi)^2.$$

Under the *m.d.s.* hypothesis with conditional homoskedasticity, Durlauf (1991) shows

$$CVM \xrightarrow{d} \sum_{j=1}^{\infty}\chi_j^2(1)/(j\pi)^2,$$

where $\{\chi_j^2(1)\}_{j=1}^{\infty}$ is a sequence of *i. i. d.* χ^2 random variables with one degree of freedom. This asymptotic distribution is nonstandard, but it is distribution-free and can be easily tabulated or simulated. An appealing property of Durlauf's (1991) test is its consistency against serial correlation of unknown form, and there is no need to choose a lag order p.

Deo (2000) shows that under the *m.d.s.* hypothesis with conditional heteroskedasticity, Durlauf's (1991) test statistic can be robustified as follows:

$$CVM = \sum_{j=1}^{n-1}\frac{[\hat{\gamma}^2(j)/\hat{\gamma}_2(j)]}{(j\pi)^2} \xrightarrow{d} \sum_{j=1}^{\infty}\chi_j^2(1)/(j\pi)^2.$$

where $\hat{\gamma}_2(j)$ is a consistent estimator for the asymptotic variance of $\hat{\gamma}(j)$ and the asymptotic distribution remains unchanged. Like Lo and MacKinlay (1988), Deo (2000) also imposes the crucial fourth order joint cumulant condition in (2.5).

3. Serial dependence in nonlinear models

The autocorrelation function $\gamma(j)$, or equivalently, the power spectrum $h(\omega)$, of a time series $\{Y_t\}$, is a measure for linear association. When $\{Y_t\}$ is a stationary Gaussian process, $\gamma(j)$ or $h(\omega)$ can completely determine the full dynamics of $\{Y_t\}$.

It has been well documented, however, that most economic and financial time series, particularly high-frequency economic and financial time series, are not Gaussian. For non-Gaussian processes, $\gamma(j)$ and $h(\omega)$ may not capture the full dynamics of $\{Y_t\}$. We consider two nonlinear process examples:

- Bilinear (BL) autoregressive process:

$$Y_t = \alpha\varepsilon_{t-1}Y_{t-2} + \varepsilon_t, \{\varepsilon_t\} \sim i.i.d.(0, \sigma^2). \tag{3.1}$$

- Nonlinear moving average (NMA) process:

$$Y_t = \alpha\varepsilon_{t-1}\varepsilon_{t-2} + \varepsilon_t, \{\varepsilon_t\} \sim i.i.d.(0, \sigma^2). \tag{3.2}$$

For these two processes, there exists nonlinearity in conditional mean: $E(Y_t|I_{t-1}) = \alpha\varepsilon_{t-1}Y_{t-2}$ under (3.1) and $E(Y_t|I_{t-1}) = \alpha\varepsilon_{t-1}Y_{t-2}$ under (3.2). However, both processes are serially uncorrelated. If $\{Y_t\}$ follows either a BL process in (3.1) or a NMA process in (3.2), $\{Y_t\}$ is not *m.d.s.* but $\gamma(j)$ and $h(\omega)$ will miss it. Hong and Lee (2003a) document that indeed, for foreign currency markets, most foreign exchange changes are serially uncorrelated, but they are all not *m.d.s.* There exist predictable nonlinear components in the conditional mean of foreign exchange markets.

Serial dependence may also exist only in higher order conditional moments. An example is Engle's (1982) first order autoregressive conditional heteroskedastic (ARCH (1)) process:

$$\begin{cases} Y_t = \sigma_t \varepsilon_t, \\ \sigma_t^2 = \alpha_0 + \alpha_1 Y_{t-1}^2, \\ \{\varepsilon_t\} \sim i.i.d.(0,1). \end{cases} \tag{3.3}$$

For this process, the conditional mean $E(Y_t|I_{t-1}) = 0$; which implies $\gamma(j) = 0$ for all $j > 0$.

However, the conditional variance, $var(Y_t|I_{t-1}) = \alpha_0 + \alpha_1 Y_{t-1}^2$, depends on the previous volatility. Both $\gamma(j)$ and $h(\omega)$ will miss such higher order dependence.

In nonlinear time series modelling, it is important to measure serial dependence, that is, any departure from *i.i.d.*, rather than merely serial correlation. As Priestley (1988) points out, the main purpose of nonlinear time series analysis is to find a filter $h(\cdot)$ such that

$$h(Y_t, Y_{t-1}, \ldots) = \varepsilon_t \sim i.i.d.(0, \sigma^2).$$

In other words, the filter $h(\cdot)$ can capture all serial dependence in $\{Y_t\}$ so that the 'residual' $\{\varepsilon_t\}$ becomes an *i.i.d.* sequence. One example of $h(\cdot)$ in modelling the conditional probability distribution of Y_t given I_{t-1}, is the probability integral transform

$$Z_t(\beta) = \int_{-\infty}^{Y_t} f(y|I_{t-1}, \beta)\, dy,$$

where $f(y|I_{t-1}, \beta)$ is a conditional density model for Y_t given and I_{t-1}, and β is an unknown parameter. When $f(y|I_{t-1}, \beta)$ is correctly specified for the conditional probability density of Y_t given I_{t-1}, that is, when the true conditional density coincides with $f(y|I_{t-1}, \beta^0)$ for some β^0, the probability integral transforms becomes

$$\{Z_t(\beta^0)\} \sim i.i.d.\mathrm{U}[0,1]. \tag{3.4}$$

Thus, one can test whether $f(y|I_{t-1}, \beta)$ is correctly specified by checking the $i.i.d.$U[0,1] for the probability integral transform series.

3.1 Bispectrum and higher-order spectra

Because the autocorrelation function $\gamma(j)$ and the spectral density $h(\omega)$ are rather limited in nonlinear time series analysis, various alternative tools have been proposed to capture nonlinear serial dependence (for example, Granger and Terasvirta, 1993; Tjøstheim, 1996). For example, one often uses the third-order cumulant function

$$C(j,k) \equiv E[(Y_t - \mu)(Y_{t-j} - \mu)(Y_{t-k} - \mu)], \qquad j,k = 0, \pm 1, \ldots.$$

This is also called the biautocovariance function of $\{Y_t\}$. It can capture certain nonlinear time series, particularly those displaying asymmetric behaviours such as skewness. Hsieh (1989) proposes a test based on $C(j, k)$ for a given pair of (j, k) which can detect some predictable nonlinear components in asset returns.

The Fourier transform of $C(j, k)$,

$$b(\omega_1, \omega_2) \equiv \frac{1}{(2\pi)^2} \sum_{j=-\infty}^{\infty} \sum_{k=-\infty}^{\infty} C(j,k) e^{-ij\omega_1 - ik\omega_2}, \qquad \omega_1, \omega_2 \in [-\pi, \pi],$$

is called the bispectrum of $\{Y_t\}$. When $\{Y_t\}$ is *i.i.d.*, $b(\omega_1, \omega_2)$ becomes a flat bispectral surface:

$$b_0(\omega_1, \omega_2) \equiv \frac{E(Y_t^3)}{(2\pi)^2}, \qquad \omega_1, \omega_2 \in [-\pi, \pi].$$

Any deviation from a flat bispectral surface will indicate the existence of serial dependence in $\{Y_t\}$. Moreover, $b(\omega_1, \omega_2)$ can be used to distinguish some linear time series processes from nonlinear time series processes. When $\{Y_t\}$ is a linear process with *i.i.d.* innovations, that is, when

$$Y_t = \alpha_0 + \sum_{j=1}^{\infty} \alpha_j \varepsilon_{t-j} + \varepsilon_t, \qquad \{\varepsilon_t\} \sim i.i.d.(0, \sigma^2),$$

the normalized bispectrum

$$|\tilde{b}(\omega_1, \omega_2)|^2 \equiv \frac{|b(\omega_1, \omega_2)|^2}{h(\omega_1)h(\omega_2)h(\omega_1 + \omega_2)} = \frac{[E(\varepsilon_t^3)]^2}{2\pi\sigma^6}$$

is a flat surface. Any departure from a flat normalized bispectral surface will indicate that $\{Y_t\}$ is not a linear time series with *i.i.d.* innovations.

The bispectrum $b(\omega_1, \omega_2)$ can capture the BL and NMA processes in (3.1) and (3.2), because the third order cumulant $C(j, k)$ can distinguish them from an *i.i.d* process. However, it may still miss some important alternatives. For example, it will easily miss ARCH (1) with *i.i.d.* N(0,1) innovation $\{\varepsilon_t\}$. In this case, $b(\omega_1, \omega_2)$ becomes a flat bispectrum and cannot distinguish ARCH (1) from an *i.i.d.* sequence. One could use higher order spectra or polyspectra (Brillinger and Rosenblatt, 1967a; 1967b), which are the Fourier transforms of higher order cumulants. However, higher-order spectra have met with some difficulty in practice: Their spectral shapes are difficult to interpret, and their estimation is not stable in finite samples, due to the assumption of the existence of higher order moments. Indeed, it is often a question whether economic and financial data, particularly high-frequency data, have finite higher order moments.

3.2 Nonparametric measures of serial dependence

Nonparametric measures for serial dependence have been proposed in the literature, which avoid assuming the existence of moments. Granger and Lin (1994) propose a nonparametric entropy measure for serial dependence to identify significant lags in nonlinear time series. Define the Kullback–Leibler information criterion

$$I(j) = \int ln[\frac{f_j(x,y)}{g(x)g(y)}]f_j(x,y)dxdy, \qquad j = 1,2,\ldots.$$

where $f_j(x, y)$ is the joint probability density of Y_t and Y_{t-j}, and $g(x)$ is the marginal probability density of $\{Y_t\}$. The Granger–Lin normalized entropy measure is defined as follows:

$$e^2(j) = 1 - \exp[-2I(j)],$$

which enjoys some appealing features. For example, $e(j) = 0$ if and only if Y_t and Y_{t-j} are independent, and it is invariant to any monotonic continuous transformation. Because $f_j(x, y)$ and $g(x)$ are unknown, Granger and Lin (1994) use nonparametric kernel density estimators. They establish the consistency of their entropy estimator (say $\hat{I}(j)$) but do not derive its asymptotic distribution, which is important for confidence interval estimation and hypothesis testing.

In fact, Robinson (1991) has elegantly explained the difficulty of obtaining the asymptotic distribution of $\hat{I}(j)$ for serial dependence, namely it is a degenerate statistic so that the usual root-n normalization does not deliver a well-defined asymptotic distribution. Robinson (1991) considers a modified entropy estimator

$$\hat{I}_\gamma(j) = n^{-1}\sum_{t=j+1}^{n} C_t(\gamma)\, ln\left[\frac{\hat{f}_j(Y_t, Y_{t-j})}{\hat{g}(Y_t)\hat{g}(Y_{t-j})}\right],$$

where $\hat{f}_j(\cdot,\cdot)$ and $\hat{g}(\cdot)$ are nonparametric kernel density estimators, $C_t(\gamma) = 1-\gamma$ if t is odd, $C_t(\gamma) = 1+\gamma$ if t is even, and γ is a pre-specified parameter. The weighting device $C_t(\gamma)$ does not affect the consistency of $\hat{I}_\gamma(j)$ to $I(j)$ and affords a well-defined asymptotic N(0,1) distribution under the *i.i.d.* hypothesis.

Skaug and Tjøstheim (1993a; 1996) use a different weighting function to avoid the degeneracy of the entropy estimator for serial dependence:

$$\hat{I}_w(j) = n^{-1}\sum_{t=1}^{n} w(Y_t,\ Y_{t-j})\, ln\left[\frac{\hat{f}_j(Y_t, Y_{t-j})}{\hat{g}(Y_t)\hat{g}(Y_{t-j})}\right],$$

where $w(Y_t,\ Y_{t-j})$ is a weighting function of observations X_t and X_{t-j}. Unlike using Robinson's (1991) weighting device, $\hat{I}_w(j)$ is not consistent for the population entropy $I(j)$, but it also delivers a well-defined asymptotic N(0, 1) distribution after a root-n normalization.

Intuitively, the use of weighting devices slows down the convergence rate of the entropy estimators, giving a well-defined asymptotic N(0,1) distribution after the usual root-n normalization. However, this is achieved at the cost of an efficiency loss, due to the slower convergence rate. Moreover, this approach breaks down when $\{Y_t\}$ is uniformly distributed, as in the case of the probability integral transforms of the conditional density in (3.4). Instead of using a weighting device, Hong and White (2005) exploit the degeneracy of $\hat{I}(j)$ and use a degenerate U-statistic theory to establish its asymptotic normality. Specifically, Hong and White (2005) show

$$nh\hat{I}(j) + hd_n^0 \xrightarrow{d} N(0, V),$$

where $h = h(n)$ is the bandwidth, and d_n^0 and V are nonstochastic factors. This approach preserves the convergence rate of the unweighted entropy estimator, giving sharper confidence interval estimation and more powerful hypothesis tests. It is applicable when $\{Y_t\}$ is uniformly distributed.

Skaug and Tjøstheim (1993b) also use an Hoeffding measure to test serial dependence (see also Delgado, 1996; Hong, 1998; 2000). The empirical Hoeffding measures are based on the empirical distribution functions, which avoid smoothed nonparametric density estimation.

3.3 Generalized spectrum

Without assuming the existence of higher order moments, Hong (1999) proposes a generalized spectrum as an alternative analytic tool to the power spectrum and higher order spectra. The basic idea is to transform $\{Y_t\}$ via a complex-valued exponential function

$$Y_t \rightarrow \exp(iuY_t), \quad u \in (-\infty, \infty),$$

and then consider the spectrum of the transformed series. Let $\psi(u) \equiv E(e^{iuY_t})$ be the marginal characteristic function of $\{Y_t\}$ and let $\psi_j(u, v) \equiv E[e^{i(uY_t + vY_{t-|j|})}]$, $j = 0, \pm 1, \ldots,$ be the pairwise joint characteristic function of. $(Y_t, Y_{t-|j|})$. Define the covariance function between transformed variables e^{iuY_t} and $e^{ivY_{t-|j|}}$:

$$\sigma_j(u, v) \equiv \text{cov}(e^{iuY_t}, e^{ivY_{t-|j|}}), \quad j = 0, \pm 1, \ldots.$$

Straightforward algebra yields $\sigma_j(u, v) = \psi_j(u, v) - \psi(u)\,\psi(v)$, which is zero for all u, v if and only if Y_t and $Y_{t-|j|}$ are independent. Thus $\sigma_j(u, v)$ can capture any type of pairwise serial dependence over various lags, including those with zero autocorrelation. For example, $\sigma_j(u, v)$ can capture the BL, NMA and ARCH (1) processes in (3.1)–(3.3), all of which are serially uncorrelated.

The Fourier transform of the generalized covariance $\sigma_j(u, v)$:

$$f(\omega, u, v) \equiv \frac{1}{2\pi} \sum_{j=-\infty}^{\infty} \sigma_j(u, v) e^{-ij\omega}, \quad \omega \in [-\pi, \pi],$$

is called the 'generalized spectral density' of $\{Y_t\}$. Like $\sigma_j(u, v)$, $f(\omega, u, v)$ can capture any type of pairwise serial dependencies in $\{Y_t\}$ over various lags. Unlike the power spectrum and higher order spectra, $f(\omega, u, v)$ does not require any moment condition on $\{Y_t\}$. When var(Y_t) exists, the power spectrum of $\{Y_t\}$ can be obtained by differentiating $f(\omega, u, v)$ with respect to $(u\ v)$ at $(0, 0)$:

$$h(\omega) \equiv \frac{1}{2\pi}\sum_{j=-\infty}^{\infty}\gamma(j)e^{-ij\omega}$$
$$= -\frac{\partial^2}{\partial u \partial v} f(\omega, u, v)|_{(u,v)=(0,0)}, \omega \in [-\pi, \pi].$$

This is the reason why $f(\omega, u, v)$ is called the 'generalized spectral density' of $\{Y_t\}$.

When $\{Y_t\}$ is *i.i.d.*, $f(\omega, u, v)$ becomes a flat generalized spectrum as a function of ω:

$$f_0(\omega, u, v) = \frac{1}{2\pi}\sigma_0(u, v), \quad \omega \in [-\pi, \pi].$$

Any deviation of $f(\omega, u, v)$ from the flat generalized spectrum $f_0(\omega, u, v)$ is evidence of serial dependence. Thus, $f(\omega, u, v)$ is suitable to capture any departures from *i.i.d.* Hong and Lee (2003b) use the generalized spectrum to develop a test for the adequacy of nonlinear time series models by checking whether the standardized model residuals are *i.i.d.* Tests for *i.i.d.* are more suitable than tests for serial correlation in nonlinear contexts. Indeed, Hong and Lee (2003b) find that some popular EGARCH models are inadequate in capturing the full dynamics of stock returns, although the standardized model residuals are serially uncorrelated.

Insight into the ability of $f(\omega, u, v)$ can be gained by considering a Taylor series expansion

$$f(\omega, u, v) = \sum_{m=0}^{\infty}\sum_{l=0}^{\infty}\frac{(iu)^m(iv)^l}{m!l!}\left[\frac{1}{2\pi}\sum_{j=-\infty}^{\infty}\text{cov}(X_t^m, X_{t-|j|}^l)e^{-ij\omega}\right].$$

Although $f(\omega, u, v)$ has no physical interpretation, it can be used to characterize cyclical movements caused by linear and nonlinear serial dependence. Examples of nonlinear cyclical movements include cyclical volatility clustering, and cyclical distributional tail clustering (for example, Engle and Manganelli's (2004) CAVaR model). Intuitively, the supremum function

$$s(\omega) = \sup_{-\infty < u,v < \infty} |f(\omega, u, v)|, \quad \omega \in [-\pi, \pi],$$

can measure the maximum dependence at frequency ω of $\{Y_t\}$. It can be viewed as an operational frequency domain analogue of Granger and Terasvirta's (1993) maximum correlation measure

$$mm\rho(j) = \max_{g(\cdot),h(\cdot)} |corr[g(Y_t), h(X_{t-j})]|.$$

Once generic serial dependence is detected using $f(\omega, u, v)$ or any other dependence measure, one may like to explore the nature and pattern of serial dependence. For example, one may be interested in the following questions:

- Is serial dependence operative primarily through the conditional mean or through conditional higher order moments?
- If serial dependence exists in conditional mean, is it linear or nonlinear?
- If serial dependence exists in conditional variance, does there exist linear or nonlinear and asymmetric ARCH?

Different types of serial dependence have different economic implications. For example, the efficient market hypothesis fails if and only if there is no serial dependence in conditional mean.

Just as the characteristic function can be differentiated to generate various moments, generalized spectral derivatives, when they exist, can capture various specific aspects of serial dependence, thus providing information on possible types of serial dependence. Suppose $E[(Y_t)^{2\max(m,l)}] < \infty$ for some nonnegative integers m, l. Then the following generalized spectral derivative exists:

$$f^{(0,m,l)}(\omega, u, v) = \frac{\partial^{m+l}}{\partial u^m \partial v^l} f(\omega, u, v) = \frac{1}{2\pi} \sum_{j=-\infty}^{\infty} \sigma_j^{(m,l)}(u, v) e^{-ij\omega},$$

where $\sigma_j^{(m,l)}(u, v) \equiv \partial^{m+l} \sigma_j(u, v) / \partial u^m \partial v^l$. As an illustrative example, we consider the generalized spectral derivative of order $(m, l)=(1, 0)$:

$$f^{(0,1,0)}(\omega, u, v) = \frac{1}{2\pi} \sum_{j=-\infty}^{\infty} \sigma_j^{(1,0)}(u, v) e^{-ij\omega}.$$

Observe $\sigma_j^{(1,0)}(0, v) \equiv \mathrm{cov}(iY_t, e^{ivY_{t-|j|}}) = 0$ for all $v \in (-\infty, \infty)$ if and only if $E(Y_t|Y_{t-|j|}) = E(Y_t)$ a.s. The function $E(Y_t|Y_{t-|j|})$ is called the autoregression function of $\{Y_t\}$ at lag j. It can capture a variety of linear and nonlinear dependencies in conditional mean, including the BL and NMA processes in (3.1) and (3.2). (The use of $\sigma_j^{(1,0)}(0, v)$, which can be easily estimated by a sample average, avoids smoothed nonparametric estimation of $E(Y_t|Y_{t-|j|})$.) Thus, the generalized spectral derivative $f^{(0,1,0)}(\omega, u, v)$ can be used to capture a wide range of serial dependence in conditional mean. In particular, the function

$$s(\omega) = \sup_{-\infty < v < +\infty} |f^{(0,1,0)}(\omega, 0, v))|$$

can be viewed as an operational frequency domain analogue of Granger and Terasvirta's (1993) maximum mean correlation measure

$$mm(j) = \max_{h(\cdot)} |\mathrm{corr}(Y_t, h(Y_{t-j}))|.$$

See Hong and Lee (2005) for more discussion.

Suppose one has found evidence of serial dependence in conditional mean using $f^{(0,\,1,\,0)}(\omega, u, v)$ or any other suitable measure, one can go further to explore whether there exists linear serial dependence in mean. This can be done by using the (1, 1)-th order generalized derivative

$$f^{(0,1,1)}(\omega, 0, 0) = -h(\omega),$$

which checks serial correlation. Moreover, one can further use $f^{(0,1,l)}(\omega, u, v)$ for $l \geq 2$ to reveal nonlinear serial dependence in mean. In particular, these higher-order derivatives can suggest that there exist: (i) an ARCH-in-mean effect (for example, Engle, Lilien and Robins, 1987) if $\text{cov}(Y_t, Y_{t-j}^2) \neq 0$, (ii) a skewness-in-mean effect (for example, Harvey and Siddique, 2000) if $\text{cov}(Y_t, Y_{t-j}^3) \neq 0$, and (iii) kurtosis-in-mean effect (for example, Brooks, Burke and Persand, 2005) if $\text{cov}(Y_t, Y_{t-j}^4) \neq 0$. These effects may arise from the existence of a time-varying risk premium, asymmetry of market behaviours, and inadequate account for large losses, respectively.

YONGMIAO HONG

See also **spectral analysis.**

I thank Steven Durlauf (editor) for suggesting this topic and comments on an earlier version, and Jing Liu for excellent research assistance and references. This research is supported by the Cheung Kong Scholarship of the Chinese Ministry of Education and Xiamen University. All remaining errors are solely mine.

Bibliography

Andrews, D.W.K. 1991. Heteroskedasticity and autocorrelation consistent covariance matrix estimation. *Econometrica* 59, 817–58.

Bollerslev, T. 1986. Generalized autoregressive conditional heteroskedastcity. *Journal of Econometrics* 31, 307–27.

Box, G.E.P. and Pierce, D.A. 1970. Distribution of residual autocorrelations in autoregressive moving average time series models. *Journal of the American Statistical Association* 65, 1509–26.

Breusch, T.S. 1978. Testing for autocorrelation in dynamic linear models. *Australian Economic Papers* 17, 334–55.

Breusch, T.S. and Pagan, A. 1980. The Lagrange multiplier test and its applications to model specification in econometrics. *Review of Economic Studies* 47, 239–53.

Brillinger, D.R. and Rosenblatt, M. 1967a. Asymptotic theory of estimates of Kth order spectra. In *Spectral Analysis of Time Series*, ed. B. Harris. New York: Wiley.

Brillinger, D.R. and Rosenblatt, M. 1967b. Asymptotic theory of estimates of Kth order spectra. In *Spectral Analysis of Time Series*, ed. B. Harris. New York: Wiley.

Brooks, C., Burke, S. and Persand, G. 2005. Autoregressive conditional kurtosis. *Journal of Financial Econometrics* 3, 399–421.

Campbell, J.Y., Lo, A.W. and MacKinlay, A.C. 1997. *The Econometrics of Financial Markets.* Princeton, NJ: Princeton University Press.

Chen, W. and Deo, R. 2004. A generalized portmanteau goodness-of-fit test for time series models. *Econometric Theory* 20, 382–416.

Cochrane, J.H. 1988. How big is the random walk in GNP? *Journal of Political Economy* 96, 893–920.

Delgado, M.A. 1996. Testing serial independence using the sample distribution function. *Journal of Time Series Analysis* 17, 271–85.

Deo, R.S. 2000. Spectral tests of the martingale hypothesis under conditional heteroscedasticity. *Journal of Econometrics* 99, 291–315.

Durbin, J. 1970. Testing for serial correlation in least squares regression when some of the regressors are lagged dependent variables. *Econometrica* 38, 422–1.

Durbin, J. and Watson, G.S. 1950. Testing for serial correlation in least squares regression: I. *Biometrika* 37, 409–28.

Durbin, J. and Watson, G.S. 1951. Testing for serial correlation in least squares regression: II. *Biometrika* 38, 159–78.

Durlauf, S.N. 1991. Spectral based testing of the martingale hypothesis. *Journal of Econometrics* 50, 355–76.

Engle, R. 1982. Autoregressive conditional hetersokedasticity with estimates of the variance of United Kingdom inflation. *Econometrica* 50, 987–1008.

Engle, R., Lilien, D. and Robins, R.P. 1987. Estimating time varying risk premia in the term structure: the ARCH-M model. *Econometrica* 55, 391–407.

Engle, R. and Manganelli, S. 2004. CARViaR: conditional autoregressive value at risk by regression quantiles. *Journal of Business and Economic Statistics* 22, 367–91.

Fan, J. and Zhang, W. 2004. Generalized likelihood ratio tests for spectral density. *Biometrika* 91, 195–209.

Glosten, R., Jagannathan, R. and Runkle, D. 1993. On the relation between the expected value and the volatility of the nominal excess return on stocks. *Journal of Finance* 48, 1779–801.

Godfrey, L.G. 1978. Testing against general autoregressive and moving average error models when the regressors include lagged dependent variables. *Econometrica* 46, 1293–301.

Granger, C.W.J. and Lin, J.L. 1994. Using the mutual information coefficient to identify lags in nonlinear models. *Journal of Time Series Analysis* 15, 371–84.

Granger, C.J.W. and Terasvirta, T. 1993. *Modeling Nonlinear Economic Relationships*. Oxford: Oxford University Press.

Harvey, C.R. and Siddique, A. 2000. Conditional skewness in asset pricing tests. *Journal of Finance* 51, 1263–95.

Hayashi, F. 2000. *Econometrics*. Princeton: Princeton University Press.

Hong, Y. 1996. Consistent testing for serial correlation of unknown form. *Econometrica* 64, 837–64.

Hong, Y. 1998. Testing for pairwise serial independence via the empirical distribution function. *Journal of the Royal Statistical Society, Series B* 60, 429–53.

Hong, Y. 1999. Hypothesis testing in time series via the empirical characteristic function: a generalized spectral density approach. *Journal of the American Statistical Association* 94, 1201–20.

Hong, Y. 2000. Generalized spectral tests for serial dependence. *Journal of the Royal Statistical Society, Series B* 62, 557–74.

Hong, Y. and Lee, T.H. 2003a. Inference on predictability of foreign exchange rates via generalized spectrum and nonlinear time series models. *Review of Economics and Statistics* 85, 1048–62.

Hong, Y. and Lee, T.H. 2003b. Diagnostic checking for the adequacy of nonlinear time series models. *Econometric Theory* 19, 1065–121.

Hong, Y. and Lee, Y.J. 2005. Generalized spectral testing for conditional mean models in time series with conditional heteroskedasticity of unknown form. *Review of Economic Studies* 72, 499–51.

Hong, Y. and Lee, Y.J. 2007. Consistent testing for serial correlation of unknown form under general conditional heteroskedasticity. Working paper, Department of Economics, Cornell University, and Department of Economics, Indiana University.

Hong, Y. and White, H. 2005. Asymptotic distribution theory for nonparametric entropy measures of serial dependence. *Econometrica* 73, 837–901.

Hsieh, D.A. 1989. Testing for nonlinear dependence in daily foreign exchange rates. *Journal of Business* 62, 339–68.

Ljung, G.M. and Box, G.E.P. 1978. On a measure of lack of fit in time series models. *Biometrika* 65, 297–303.
Lo, A.W. and MacKinlay, A.C. 1988. Stock market prices do not follow random walks: evidence from a simple specification test. *Review of Financial Studies* 1, 41–66.
Nelson, D. 1991. Conditional heteroskedasticity in asset returns: a new approach. *Econometrica* 59, 347–70.
Newey, W.K. and West, K.D. 1987. A simple, positive semi-definite, heteroscedasticity and autocorrelation consistent covariance matrix. *Econometrica* 55, 703–8.
Paparoditis, E. 2000. Spectral density based goodness-of-fit tests for time series models. *Scandinavian Journal of Statistics* 27, 143–76.
Priestley, M.B. 1988. *Non-Linear and Non-Stationary Time Series Analysis.* London: Academic Press.
Robinson, P.M. 1991. Consistent nonparametric entropy-based testing. *Review of Economic Studies* 58, 437–53.
Robinson, P.M. 1994. Time series with strong dependence. In *Advances in Econometrics, Sixth World Congress*, vol. 1, ed. C. Sims. Cambridge: Cambridge University Press.
Skaug, H.J. and Tjøstheim, D. 1993a. Nonparametric tests of serial independence. In *Developments in Time Series Analysis*, ed. S. Rao. London: Chapman and Hall.
Skaug, H.J. and Tjøstheim, D. 1993b. A nonparametric test of serial independence based on the empirical distribution function. *Biometrika* 80, 591–602.
Skaug, H.J. and Tjøstheim, D. 1996. Measures of distance between densities with application to testing for serial independence. In *Time Series Analysis in Memory of E.J. Hannan*, ed. P. Robinson and M. Rosenblatt. New York: Springer.
Tjøstheim, D. 1996. Measures and tests of independence: a survey. *Statistics* 28, 249–84.
Vinod, H.D. 1973. Generalization of the Durbin–Watson statistic for higher order autoregressive processes. *Communications in Statistics* 2, 115–44.
Wallis, K.F. 1972. Testing for fourth order autocorrelation in quarterly regression equations. *Econometrica* 40, 617–36.
Whang, Y.J. 1998. A test of autocorrelation in the presence of heteroskedasticity of unknown form. *Econometric Theory* 14, 87–122.
Wooldridge, J.M. 1990. An encompassing approach to conditional mean tests with applications to testing nonnested hypotheses. *Journal of Econometrics* 45, 331–50.
Wooldridge, J.M. 1991. On the application of robust, regression-based diagnostics to models of conditional means and conditional variances. *Journal of Econometrics* 47, 5–46.

SNP: nonparametric time series analysis

SNP is a method of multivariate nonparametric time series analysis. SNP is an abbreviation of 'semi-nonparametric' which was introduced by Gallant and Nychka (1987) to suggest the notion of a statistical inference methodology that lies halfway between parametric and nonparametric inference. The method employs an expansion in Hermite functions to approximate the conditional density of a multivariate process.

The leading term of this expansion can be chosen through selection of model parameters to be a Gaussian vector autoregression (VAR) model, a semi-parametric VAR model, a Gaussian ARCH model (Engle, 1982), a semiparametric ARCH model, a Gaussian GARCH model (Bollerslev, 1986), or a semiparametric GARCH model, either univariate or multivariate in each case. The unrestricted SNP expansion is more general than that of any of these models. The SNP model is fitted using maximum likelihood together with a model selection strategy that determines the appropriate order of expansion. Because the SNP model possesses a score, it is an ideal candidate for the auxiliary model in connection with efficient method of moment estimation (Gallant and Tauchen, 1996). Due to its leading term, the SNP approach does not suffer from the curse of dimensionality to the same extent as kernels and splines. In regions where data are sparse, the leading term helps to fill in smoothly between data points. Where data are plentiful, the higher-order terms accommodate deviations from the leading term. The method was first proposed by Gallant and Tauchen (1989) in connection with an asset pricing application. A C+ + implementation of SNP is at http://econ.duke.edu/webfiles/arg/snp/, together with a User's Guide, which is an excellent tutorial introduction to the method.

Important adjuncts to SNP estimation are a rejection method for simulating from the SNP density developed in Gallant and Tauchen (1992), which can be used, for example, to set bootstrapped confidence intervals as in Gallant, Rossi and Tauchen (1992); nonlinear error shock analysis as described in Gallant, Rossi and Tauchen (1993), which develops the nonlinear analog of conventional error shock analysis for linear VAR models; and re-projection, which is a form of nonlinear Kalman filtering that can be used to forecast the unobservables of nonlinear latent variables models (Gallant and Tauchen, 1998).

As stated above, the SNP method is based on the notion that a Hermite expansion can be used as a general purpose approximation to a density function. Letting z denote an M–vector, we can write the Hermite density as $h(z) \propto [\mathscr{P}(z)]^2 \phi(z)$ where $\mathscr{P}(z)$ denotes a multivariate polynomial of degree K_z and $\phi(z)$ denotes the density function of the (multivariate) Gaussian distribution with mean zero and variance the identity matrix. Denote the coefficients of $\mathscr{P}(z)$ by a, which is a vector whose length depends on K_z and M. When we wish to call attention to the coefficients, we write $\mathscr{P}(z|a)$.

The constant of proportionality is $1/\int[\mathscr{P}(s)]^2\phi(s)ds$ which makes $h(z)$ integrate to one. As seen from the expression that results, namely

$$h(z) = \frac{[\mathscr{P}(z)]^2\phi(z)}{\int[\mathscr{P}(s)]^2\phi(s)ds},$$

we are effectively expanding the square root of the density in Hermite functions of the form $\mathscr{P}(z)\sqrt{\phi(z)}$. Because the square root of a density is always square integrable and because the Hermite functions of the form $\mathscr{P}(z)\sqrt{\phi(z)}$ are dense for the collection of square integrable functions (Fenton and Gallant, 1996), every density has such an expansion. Because $[\mathscr{P}(z)]^2/\int[\mathscr{P}(s)]^2\phi(s)ds$ is a homogeneous function of the coefficients of the polynomial $\mathscr{P}(z)$, the coefficients can only be determined to within a scalar multiple. To achieve a unique representation, the constant term of the polynomial part is put to 1. Customarily the Hermite density is written with its terms orthogonalized and the C++ code is written in the orthogonalized form for numerical efficiency. But reflecting that here would lead to cluttered notation and add nothing to the ideas.

A change of variables using the location-scale transformation $y = Rz + \mu$, where R is an upper triangular matrix and μ is an M-vector, gives

$$f(y|\theta) \propto \{\mathscr{P}[R^{-1}(y-\mu)]\}^2 \times \{\phi[R^{-1}(y-\mu)]/|\det(R)|\}$$

The constant of proportionality is the same as above, $1/\int[\mathscr{P}(s)]^2\phi(s)ds$. Because $\{\phi[R^{-1}(y-\mu)]/|\det(R)|\}$ is the density function of the M-dimensional, multivariate, Gaussian distribution with mean μ and variance-covariance matrix $\Sigma = RR'$, and because the leading term of the polynomial part is 1, the leading term of the entire expansion is proportional to the multivariate, Gaussian density function. Denote the Gaussian density of dimension M with mean vector μ and variance matrix Σ by $n_M(y|\mu, \Sigma)$ and write

$$f(y|\theta) \propto [\mathscr{P}(z)]^2 n_M(y|\mu, \Sigma)$$

where $z = R^{-1}(y-\mu)$ for the density above.

When K_z is put to zero, one gets $f(y|\theta) = n_M(y|\mu, \Sigma)$ exactly. When K_z is positive, one gets a Gaussian density whose shape is modified due to multiplication by a polynomial in $z = R^{-1}(y-\mu)$. The shape modifications thus achieved are rich enough to accurately approximate densities from a large class that includes densities with fat, t-like tails, densities with tails that are thinner than Gaussian, and skewed densities (Gallant and Nychka, 1987).

The parameters θ of $f(y|\theta)$ are made up of the coefficients a of the polynomial $\mathscr{P}(z)$ plus μ and R and are estimated by maximum likelihood which is accomplished by minimizing $s_n(\theta) = (-1/n)\sum_{t=1}^{n}\log[f(y_t|\theta)]$. As mentioned above, if the number of parameters p_θ grows with the sample size n, the true density and various features of it such as derivatives and moments are estimated consistently (Gallant and Nychka, 1987).

This basic approach can be adapted to the estimation of the conditional density of a multiple time series $\{y_t\}$ that has a Markovian structure. Here, the term 'Markovian structure' is taken to mean that the conditional density of the M–vector y_t given the entire past $y_{t-1}, y_{t-2}, \ldots$ depends only on L lags from the past. For convenience, we will presume that the data are from a process with a Markovian structure, but one should be aware that, if L is sufficiently large, then non-Markovian data can be well approximated by an SNP density (Gallant and Long, 1997). Collect these lags together as $x_{t-1} = (y_{t-1}, y_{t-2}, \ldots, y_{t-L})$, where L exceeds all lags in the following discussion.

To approximate the conditional density of $\{y_t\}$ using the ideas above, begin with a sequence of innovations $\{z_t\}$. First consider the case of homogeneous innovations; that is, the distribution of z_t does not depend on x_{t-1}. Then, as above, the density of z_t can be approximated by $h(z) \propto [\mathscr{P}(z)]^2\phi(z)$ where $\mathscr{P}(z)$ is a polynomial of degree K_z. Follow with the location-scale transformation $y_t = Rz_t + \mu_x$ where μ_x is a linear function that depends on L_u lags

$$\mu_x = b_0 + Bx_{t-1}.$$

(If $L_u < L$, then some elements of B are zero.) The density that results is

$$f(y|x, \theta) \propto [\mathscr{P}(z)]^2 n_M(y|\mu_x, \Sigma)$$

where $z = R^{-1}(y - \mu_x)$. The constant of proportionality is as above, $1/\int[\mathscr{P}(s)]^2\phi(s)ds$. The leading term of the expansion is $n_M(y|\mu_x, \Sigma)$ which is a Gaussian vector autoregression or Gaussian VAR. When K_z is put to zero, one gets $n_M(y|\mu_x, \Sigma)$ exactly. When Kz is positive, one gets a semiparametric VAR density.

To approximate conditionally heterogeneous processes, proceed as above but let each coefficient of the polynomial $\mathscr{P}(z)$ be a polynomial of degree K_x in x. A polynomial in z of degree K_z whose coefficients are polynomials of degree K_x in x is, of course, a polynomial in (z, x) of degree $K_z + K_x$. Denote this polynomial by $\mathscr{P}(z, x)$. Denote the mapping from x to the coefficients a of $\mathscr{P}(z)$ such that $\mathscr{P}(z|a_x) = \mathscr{P}(z, x)$ by a_x and the number of lags on which it depends by L_p. The form of the density with this modification is

$$f(y|x, \theta) \propto [\mathscr{P}(z, x)]^2 n_M(y|\mu_x, \Sigma)$$

where $z = R^{-1}(y - \mu_x)$. The constant of proportionality is $1/\int[\mathscr{P}(s, x)]^2\phi(s)ds$. When K_x is zero, the density reverts to the density above. When K_x is positive, the shape of the density will depend upon x. Thus, all moments can depend upon x and the density can, in principal, approximate any form of conditional heterogeneity (Gallant and Tauchen, 1989).

In practice the second moment can exhibit marked dependence upon x. In an attempt to track the second moment, K_x can get quite large. To keep K_x small when data are markedly conditionally heteroskedastic, the leading term $n_M(y|\mu_x, \Sigma)$ of the expansion can be put to a Gaussian GARCH rather than a Gaussian VAR. SNP uses a modified BEKK expression as described in Engle and Kroner (1995); the modifications are to add leverage and level effects.

$$\Sigma_{x_{t-1}} = R_0 R_0' + \sum_{i=1}^{L_g} Q_i \Sigma_{x_{t-1-i}} Q_i' + \sum_{i=1}^{L_r} P_i (y_{t-i} - \mu_{x_{t-1-i}})(y_{t-i} - \mu_{x_{t-1-i}})' P_i'$$
$$+ \sum_{i=1}^{L_v} \max[0, V_i(y_{t-i} - \mu_{x_{t-1-i}})] \max[0, V_i(y_{t-i} - \mu_{x_{t-1-i}})]'$$
$$+ \sum_{i=1}^{L_w} W_i x_{(1),t-1} x_{(1),t-i}' W_i'.$$

Above, R_0 is an upper triangular matrix. The matrices P_i, Q_i, V_i, and W_i can be scalar, diagonal, or full M by M matrices. The notation $x_{(1),t-i}$ indicates that only the first column of x_{t-i} enters the computation. The max(0, x) function is applied elementwise. Because $\Sigma_{x_{t-1}}$ must be differentiable with respect to the parameters of $\mu_{x_{t-2-i}}$, the max(0,x) function is approximated by a twice continuously differentiable cubic spline. Defining $R_{x_{t-1}}$ by the factorization $\Sigma_{x_{t-1}} = R_{x_{t-1}} R_{x_{t-1}}'$ and writing x for x_{t-1}, the SNP density becomes

$$f(y|x,\theta) \propto [\mathscr{P}(z,x)]^2 n_M(y|\mu_x, \Sigma_x)$$

where $z = R_x^{-1}(y - \mu_x)$. The constant of proportionality is $1/\int [\mathscr{P}(s,x)]^2 \phi(s) ds$. The leading term $n_M(y|\mu_x, \Sigma_x)$ is Gaussian ARCH if $L_g = 0$ and $L_r > 0$ and Gaussian GARCH if both $L_g > 0$ and $L_r > 0$ (leaving aside the implications of L_v and L_w).

A. RONALD GALLANT

See also **ARCH models; impulse response function; nonlinear time series analysis.**

Research for this article was supported by the National Science Foundation.

Bibliography

Bollerslev, T. 1986. Generalized autoregressive conditional heteroskedasticity. *Journal of Econometrics* 31, 307–27.

Engle, R.F. 1982. Autoregressive conditional heteroskedasticity with estimates of the variance of United Kingdom inflation. *Econometrica* 50, 987–1007.

Engle, R.F. and Kroner, K.F. 1995. Multivariate simultaneous generalized ARCH. *Econometric Theory* 11, 122–50.

Fenton, V.M. and Gallant, A.R. 1996. Qualitative and asymptotic performance of SNP density estimators. *Journal of Econometrics* 74, 77–118.

Gallant, A.R., Hsieh, D. and Tauchen, G. 1997. Estimation of stochastic volatility models with diagnostics. *Journal of Econometrics* 81, 159–92.

Gallant, A.R. and Long, J.R. 1997. Estimating stochastic differential equations efficiently by minimum chi-square. *Biometrika* 84, 125–41.

Gallant, A.R. and Nychka, D.W. 1987. Seminonparametric maximum likelihood estimation. *Econometrica* 55, 363–90.

Gallant, A.R., Rossi, P.E and Tauchen, G. 1992. Stock prices and volume. *Review of Financial Studies* 5, 199–242.

Gallant, A.R., Rossi, P.E. and Tauchen, G. 1993. Nonlinear dynamic structures. *Econometrica* 61, 871–907.

Gallant, A.R. and Tauchen, G. 1989. Seminonparametric estimation of conditionally constrained heterogeneous processes: asset pricing applications. *Econometrica* 57, 1091–120.

Gallant, A.R. and Tauchen, G. 1992. A nonparametric approach to nonlinear time series analysis: estimation and simulation. In *New Directions in Time Series Analysis, Part II.*, ed. D. Brillinger et al. New York: Springer-Verlag.

Gallant, A.R. and Tauchen, G. 1996. Which moments to match? *Econometric Theory* 12, 657–81.

Gallant, A.R. and Tauchen, G. 1998. Reprojecting partially observed systems with application to interest rate diffusions. *Journal of the American Statistical Association* 93, 10–24.

spectral analysis

Spectral analysis is a statistical approach for analysing stationary time series data in which the series is decomposed into cyclical or periodic components indexed by the frequency of repetition. Spectral analysis falls within the frequency domain approach to time series analysis. This is in contrast with the time domain approach in which a time series is characterized by its correlation structure over time. While spectral analysis provides a different interpretation of a time series from time domain approaches, the two approaches are directly linked to each other.

Statistical spectral analysis tools were first developed in the middle of the 20th century in the mathematical statistics and engineering literatures, and many of the important early contributions are discussed in the classic textbook treatment by Priestley (1981). The label 'spectral' has been adopted because of the close link to the physics of light. While the analogy with the physics of light is fairly useless in economics, economists recognized by the 1960s that spectral analysis is a useful empirical tool for understanding the cyclical nature of many time series, and it provides a powerful theoretical framework for developing econometric methodology: for example, the theoretical underpinnings of Granger causality (Granger, 1969) are based in spectral analysis. Since the 1960s, spectral analysis tools have become standard parts of the time series econometrics toolkit, and have influenced a broad range of areas within econometrics. A comprehensive list of references would be long but some notable examples are: band spectral regression (Engle, 1974), generalized method of moments (GMM) (Hansen, 1982), heteroskedasticity autocorrelation (HAC) covariance matrix estimation and inference (Newey and West, 1987; Andrews, 1991; Kiefer and Vogelsang, 2005), unit root testing (Phillips and Perron, 1988), cointegration (Stock and Watson, 1988; Phillips and Hansen, 1990), semiparametric methods (Robinson, 1991), structural identification of empirical macroeconomics models (Blanchard and Quah, 1989; King et al., 1991), testing for serial correlation (Hong, 1996), measures of persistence (Cochrane, 1988), measures of fit for calibrated macro models (Watson, 1993), estimation of long memory models (Geweke and Porter-Hudak, 1983).

Let y_t, $t = 1, 2, \ldots$ denote a second-order stationary time series with mean $\mu = E(y_t)$ and autocovariance function $\gamma_j = cov(y_t, y_{t-j})$. Most empirical economists find it natural to characterize relationships between random variables in terms of correlation structure, and γ_j conveniently summarizes the statistical structure of y_t. Autocovariances are fundamental population moments of a time series not directly connected to any specific modelling choice. In contrast, the idea of decomposing y_t into cyclical components may appear to impose restrictions on y_t; but an important result, the spectral representation theorem, indicates that nearly any stationary time series can be represented in terms of cyclical components. By using notation from Hamilton

(1994), nearly any stationary (discrete-time) time series can be represented as

$$y_t = \mu + \int_0^{\pi} [\alpha(\omega)\cos(\omega t) + \delta(\omega)\sin(\omega t)]\, d\omega, \tag{1}$$

where ω denotes frequency and $\alpha(\omega)$ and $\delta(\omega)$ are mean zero random processes such that for any frequencies $0<\omega_1<\omega_2<\omega_3<\omega_4<\pi$,

$$cov\left(\int_{\omega_1}^{\omega_2} \alpha(\omega)\, d\omega, \int_{\omega_3}^{\omega_4} \alpha(\omega)\, d\omega\right) = 0, \quad cov\left(\int_{\omega_1}^{\omega_2} \delta(\omega)\, d\omega, \int_{\omega_3}^{\omega_4} \delta(\omega)\, d\omega\right) = 0,$$

and for any frequencies $0<\omega_1<\omega_2<\pi$ and $0<\omega_3<\omega_4<\pi$,

$$cov\left(\int_{\omega_1}^{\omega_2} \alpha(\omega)\, d\omega, \int_{\omega_3}^{\omega_4} \delta(\omega)\, d\omega\right) = 0.$$

It is fundamental that a stationary time series can be decomposed into (random) cyclical (cosine and sine) components.

A useful way of interpreting (1) is to measure how the cyclical components contribute to the variation in y_t. Similar to the way in which the area under a density of a random variable determines the probability of a range of values, the area under the spectral density of y_t measures the contribution to the variance of y_t from the cyclical components for a range of frequencies.

Let $f(\omega)$ denote the spectral density of y_t where $\omega \in [-\pi, \pi]$. It can be shown that $f(\omega) \geq 0$ and that $f(-\omega) = f(\omega)$. A fundamental property of $f(\omega)$ is that $var(y_t) \equiv \gamma_0 = \int_{-\pi}^{\pi} f(\omega) d\omega$ and the contribution to the variance of y_t from components with frequencies $\omega \in (\omega_1,\omega_2)$ where $-\pi<\omega_1<\omega_2<\pi$ is given by $\int_{\omega_1}^{\omega_2} f(\omega)\, d\omega$. Therefore, loosely speaking, frequencies for which $f(\omega)$ takes on large values correspond to cyclical components that make relatively large contributions to the variation in y_t. For those more comfortable thinking in terms of the time domain cycle length, it is easy to convert frequency to cycle length. Suppose $f(\omega_1)$ corresponds to a peak (global or local) of $f(\omega)$; then components with frequencies close to ω_1 make important contributions to the variation in y_t. Consider $\cos(\omega_1 t)$ and $\sin(\omega_1 t)$ and rewrite them as $\cos(2\pi\frac{\omega_1 t}{2\pi})$ and $\sin(2\pi\frac{\omega_1 t}{2\pi})$. Recall that the cosine and sine functions are periodic with period 2π in their argument. Therefore, $\cos(2\pi\frac{\omega_1 t}{2\pi})$ and $\sin(2\pi\frac{\omega_1 t}{2\pi})$ repeat whenever $\frac{\omega_1 t}{2\pi}$ is an integer. Setting $\frac{\omega_1 t}{2\pi} = 1$ indicates that the functions repeat every $t = \frac{2\pi}{\omega_1}$ time periods. The quantity $\frac{2\pi}{\omega_1}$ is called the period corresponding to frequency ω_1.

For a concrete example consider a monthly time series where $f(\omega)$ has a peak at $\omega = \frac{\pi}{6}$. Thus cycles with period $2\pi/(\pi/6) = 12$ months (annual cycles) are important for variation of y_t. Suppose quarterly cycles (period is three months) are also important, then $f(\omega)$ will have a peak at $\omega = \frac{2\pi}{3}$. The highest frequency for which we can learn about y_t is $\omega = \pi$ or two-period cycles because cycles that last fewer than two periods do not have data observed within the cycle. This practical limitation on the frequency is called the 'alias effect'. The length of the sample size, T, also limits what can be learned about long cycles. For cycles that last T time periods, that is, frequency $\omega = \frac{2\pi}{T}$, we observe data for exactly one cycle. For cycles longer than T time periods the data does contain information about those cycles, but the information is incomplete because only part of the cycle is observed. It is difficult to learn about very low-frequency components from the data, and in particular it is difficult to learn about $f(0)$.

What does $f(\omega)$ look like? The spectral representation theorem implicitly defines the integral of $f(\omega)$ but not $f(\omega)$ itself. Because the variance of y_t is the area under $f(\omega)$, there is a direct link between $f(\omega)$ and γ_j. Suppose $\sum_{j=-\infty}^{\infty}|\gamma_j|<\infty$, then the spectral density can be expressed as

$$f(\omega) = \frac{1}{2\pi}\sum_{j=-\infty}^{\infty}\cos(\omega j)\gamma_j = \frac{1}{2\pi}[\gamma_0 + 2\sum_{j=1}^{\infty}\cos(\omega j)\gamma_j] \tag{2}$$

where the last expression uses $\cos(0)=1$, $\cos(-\omega j)=\cos(\omega j)$ and $\gamma_{-j}=\gamma_j$. It straightforward to show that a converse relationship holds

$$\gamma_j = \int_{-\pi}^{\pi} f(\omega)\cos(\omega j)\, d\omega.$$

This dual relationship between $f(\omega)$ and γ_j makes spectral analysis a powerful analytical tool beyond the direct interpretation of the spectral density in assessing importance of cyclical components.

For the class of stationary autoregressive moving average (ARMA) models $f(\omega)$ takes on a simple form. Let L denote the lag operator, $Ly_t=y_{t-1}$, and define lag polynomials $\phi(L) = 1 - \phi_1 L - \phi_2 L^2 - \ldots - \phi_p L^p$ and $\theta(L) = 1 + \theta_1 L + \theta_2 L^2 + \ldots + \theta_q L^q$. Suppose y_t is a stationary $ARMA(p, q)$ process given by $\phi(L)(y_t - \mu) = \theta(L)\varepsilon_t$ where ε_t is a mean zero uncorrelated time series (white noise process) with $var(\varepsilon_t) = \sigma_\varepsilon^2$. Then

$$f(\omega) = \frac{\sigma_\varepsilon^2\theta(e^{i\omega})\theta(e^{-i\omega})}{2\pi\phi(e^{i\omega})\phi(e^{-i\omega})} \tag{3}$$

where $i = \sqrt{-1}$. If the lag polynomials can be factored as $\phi(L) = (1-\lambda_1 L)(1-\lambda_2 L)\ldots(1-\lambda_p L)$ and $\theta(L) = (1-\delta_1 L)(1-\delta_2 L)\cdots(1-\delta_q L)$, then $f(\omega)$ can be written as

$$f(\omega) = \frac{\sigma_\varepsilon^2 \prod_{j=1}^{q}(1+\delta_j^2 - 2\delta_j\cos(\omega))}{2\pi\prod_{j=1}^{p}(1+\lambda_j^2 - 2\lambda_j\cos(\omega))}.$$

In the $AR(1)$ case $f(\omega) = \frac{\sigma_\varepsilon^2}{2\pi}(1+\phi_1^2 - 2\phi_1\cos(\omega))^{-1}$. If $\phi_1>0$ (typical for many macroeconomic and finance time series), $f(\omega)$ has a single peak at $\omega=0$. As ω increases, $f(\omega)$ steadily declines. As ϕ_1 approaches one, the peak at $\omega=0$ increases and sharpens. Thus, variation of autoregressive time series with strong persistence is driven primarily by short frequency/long cycle components. At the other extreme, when the time series is uncorrelated ($\phi_1 = 0$), the spectral density is constant/flat for all ω, so cyclical components contribute equally at all frequencies to the variation in y_t. An uncorrelated series is called a white noise process because of the analogue to white light which is comprised equally of all visible frequencies of light (all colours).

The special case of $f(0)$ is important for inference in time series models because the asymptotic variance of many time series estimators depends on $f(0)$. For example, consider $\bar{y} = T^{-1}\sum_{t=1}^{T} y_t$, the natural estimator of μ. A simple calculation gives

$$var(\bar{y}) = T^{-1}(\gamma_0 + 2\sum_{j=1}^{T-1}(1 - \frac{j}{T})\gamma_j).$$

If $\sum_{j=-\infty}^{\infty}|\gamma_j| < \infty$, then $\lim_{T\to\infty} Tvar(\bar{y}) = \gamma_0 + 2\sum_{j=1}^{\infty}\gamma_j = 2\pi f(0)$. Therefore, the asymptotic variance of a sample average, often called the long-run variance, is proportional to the spectral density at frequency zero. Inference about the population mean would require a standard error, that is, an estimate of $f(0)$. The link between asymptotic variances and zero frequency spectral densities extends to estimation of linear regression parameters and nonlinear estimation obtained using GMM. The estimation of asymptotic variance matrices that are proportional to a zero frequency spectral density is commonly known as HAC covariance matrix estimation.

Estimates of the spectral density can be obtained either parametrically or non-parametrically. For the case of ARMA models, parametric estimators are straightforward in principle and involve replacing the lag polynomial coefficients in (3) with estimators. Although estimation methods for ARMA models are well established, there are numerical and identification issues that can complicate matters when an MA component is included, especially in the case of vector time series. In contrast, pure AR models are easy to estimate (including the vector case) and, in principle, AR models can well approximate a stationary time series with suitable choice of the lag order. For these reasons, autoregressive spectral density estimators are the most widely used parametric estimators (see, for example, Berk, 1974; Perron and Ng, 1998; den Haan and Levin, 1997). One important practical challenge of implementing autoregressive spectral density estimators is the choice of autoregressive lag order. Advice in the literature on choice of lag order often depends on the intended use of the spectral density estimator.

Nonparametric estimators of the spectral density are appealing at the conceptual level because they do not depend on specific parameterization of the model. In principle nonparametric estimators are flexible enough to provide good estimators for a very wide range of stationary time series. In practice, though, implementation of nonparametric estimators can be a delicate matter, and large sample sizes are required for accuracy. Notwithstanding the practical challenges, nonparametric spectral density estimators are widely used in econometrics primarily because of the central role they play in HAC covariance matrix literature due to the influential contributions by Newey and West (1987) and Andrews (1991). These so-called Newey–West or Newey–West–Andrews standard errors are routinely used in practice; yet many empirical researchers are unaware of the direct link to nonparametric spectral density estimation.

Nonparametric estimators are obtained by estimation of (2) using sample autocovariances

$$\hat{\gamma}_j = T^{-1} \sum_{t=j+1}^{T} (y_t - \bar{y})(y_{t-j} - \bar{y}).$$

The challenge is that $f(\omega)$ depends on an infinite number of autocovariances of which only a finite number can be estimated. The highest-order autocovariance that be estimated is γ_{T-1}, but it is estimated badly because there is only one observation, $(y_T - \bar{y})(y_1 - \bar{y})$. Plugging the $\hat{\gamma}_j$ into (2) gives the estimator

$$I_T(\omega) = \frac{1}{2\pi}\left[\hat{\gamma}_0 + 2\sum_{j=1}^{T-1} \cos(\omega j)\hat{\gamma}_j\right],$$

which is the periodogram. Like $f(\omega)$ the periodogram is non-negative. For $\omega \neq 0$ the periodogram is asymptotically unbiased but its variance does not shrink as the sample size grows, so it is not a consistent estimator. At frequency zero, the situation is even more problematic because simple algebra can used to show that $I_T(0) = 0$. This result holds as long as $\hat{\gamma}_j$ is computed using a quantity that sums to zero (like $y_t - \bar{y}$). Therefore, $I_T(0)$ is useless for estimating $f(0)$. Fortunately, the periodogram can be modified to obtain better estimates of $f(\omega)$. Consider

$$\hat{f}(\omega) = \frac{1}{2\pi}\left[\hat{\gamma}_0 + 2\sum_{j=1}^{T-1} k\left(\frac{j}{M}\right)\cos(\omega j)\hat{\gamma}_j\right]$$

where $k(x)$ is a weighting function or kernel such that $k(0) = 1$, $k(-x) = k(x)$, $\int_{-\infty}^{\infty} k(x)^2 dx < \infty$ and $|k(x)| \leq 1$. The number M is called the bandwidth or, for some $k(x)$ functions, the truncation lag. The kernel downweights the higher order $\hat{\gamma}_j$, and the bandwidth controls the speed at which downweighting occurs. A recent paper by Phillips, Sun and Jin (2006) achieves downweighting by exponentiating the kernel, for example by using $k(j/T)^{\rho}$, where ρ controls the degree of downweighting.

While a large number of kernel functions have been proposed and analysed since the 1940s, two have become widely used in econometrics: the Bartlett kernel and the quadratic spectral (QS) or Bartlett–Priestley kernel. These kernels are in the class of kernels that guarantee $\hat{f}(\omega) \geq 0$. The Bartlett kernel is

$$k(x) = \begin{cases} 1 - |x| & \text{for} |x| \leq 1 \\ 0 & \text{for} |x| > 1 \end{cases},$$

and it puts linearly declining weights on $\hat{\gamma}_j$ up to lag $M-1$ and weight zero on higher lags so that M plays the role of a truncation lag. Consistency of zero frequency Bartlett kernel estimators was established by Newey and West (1987) in a very general setting.

The QS kernel is

$$k(x) = \frac{25}{12\pi^2 x^2}\left(\frac{\sin(6\pi x/5)}{6\pi x/5} - \cos(6\pi x/5)\right),$$

and it does not truncate; weight is placed on all $\hat{\gamma}_j$. The weights decline in magnitude as j increases but some weights can be negative. Andrews (1991) showed in a general setting that the QS kernel minimizes the approximate mean square error of $\hat{f}(0)$ for a particular class of kernels.

The idea of downweighting $\hat{\gamma}_j$ is natural and is not merely a technical trick. For any stationary time series, $\lim_{j\to\infty}|\gamma_j| = 0$, therefore downweighting $\hat{\gamma}_j$, or replacing it with zero when j is large, is similar to replacing an unbiased, high variance estimator with a biased, small variance estimator. If γ_j shrinks quickly as j increases, the bias induced by downweighting is small. For $\hat{f}(\omega)$ to be a consistent estimator $M\to\infty$ and $M/T\to 0$ as $T\to\infty$. These conditions suggest that while downweighting is required it cannot be too severe. Unfortunately, these conditions do not restrict the value of M that can be used for a given sample because, given T, any value of M can be embedded in a rule that satisfies these conditions. For example, suppose $T=100$. Then the bandwidth rules $M = 10\sqrt{T}$ and $M = 0.2\sqrt{T}$ satisfy the conditions for consistency but yield very different bandwidths of $M=100$ and $M=2$.

The finite sample properties of $\hat{f}(\omega)$ are complicated and depend on M and $k(x)$. Formulas for the exact bias and variance of $\hat{f}(\omega)$ have been worked out by Neave (1971) when $\omega\neq 0$, and by Ng and Perron (1996) for $\omega=0$. Because the exact formulas are complicated, approximations are often used. The variance can be approximated by

$$\frac{T}{M} var(\hat{f}(\omega)) \approx \begin{cases} V & \text{for} \quad 0<\omega<\pi \\ 2V & \text{for} \quad \omega = 0, \pi \end{cases} \qquad (4)$$

where $V = f(\omega)^2 \int_{-\infty}^{\infty} k(x)^2 dx$. An approximation for the bias was derived by Parzen (1957) and it depends on the behaviour of $k(x)$ around $x=0$. For the Bartlett and QS kernels the approximate bias formulas are $-\frac{1}{M}\sum_{j=-\infty}^{\infty}|j|\gamma_j$ and $-\frac{18}{125}\pi^2\frac{1}{M^2}\sum_{j=-\infty}^{\infty}j^2\gamma_j$ respectively. Under suitable regularity conditions, an asymptotic normality result holds:

$$\sqrt{\tfrac{T}{M}}(\hat{f}(\omega) - E(\hat{f}(\omega)))\to^d N(0, V).$$

According to these approximations, the variance is proportional to $f(\omega)^2$ and increases as M increases whereas the bias depends on additional nuisance parameters but decreases as M increases. These well known results are discussed at length in Priestley (1981) and are the source of commonly held intuition that says that, as M increases, bias decreases but variance increases. This intuition is usually valid but only holds for $\hat{f}(0)$ when M is small. As M increases the relationship between bias/variance and M is more complicated, as discussed by Ng and Perron (1996). Recall that if no downweighting of $\hat{\gamma}_j$ is used, then $\hat{f}(0)$ becomes $I_T(0) = 0$. Obviously, this estimator has a large bias and zero variance. As M increases, less downweighting is used and, once M is large enough, the bias/variance relationship flips with bias increasing and variance decreasing in M.

An asymptotic approximation that can capture this more complex bias/variance bandwidth relationship for $\hat{f}(0)$ can be obtained using fixed-b asymptotics. Suppose $\hat{f}(0)$ is embedded into a sequence of random variables under the assumption that $b = M/T$ is a fixed constant with $b \in (0,1)$. Neave (1970) first used this approach to derive an alternate asymptotic variance formula for $\hat{f}(\omega)$. Let $B(r)$ denote a standard Brownian bridge, that is, $B(r) = W(r) - rW(1)$ where $W(r)$ is a standard Wiener process, and let $\Rightarrow$ denote weak convergence. Under suitable regularity conditions Kiefer and Vogelsang (2005) show that $\hat{f}(0) \Rightarrow f(0)Q(b)$, where

$$Q(b) = \frac{2}{b}\left(\int_0^1 B(r)^2 dr - \int_0^{1-b} B(r+b)B(r)\, dr\right)$$

for the Bartlett kernel and

$$Q(b) = -\frac{1}{b^2}\int_0^1 \int_0^1 k''\left(\frac{r-s}{b}\right)B(r)B(s)\, drds$$

for the QS kernel with analogous results for $\omega \neq 0$ obtained by Hashimzade and Vogelsang (2007). Phillips, Sun and Jin (2006) obtain similar results for exponentiated kernels. The fixed-b asymptotic result approximates $\hat{f}(0)$ by the random variable $Q(b)$ which is similar to a chi-square random variable. When $\hat{f}(0)$ is used to construct standard errors of an estimator like $\bar{y}$, fixed-b asymptotics provides an approximation for $t = (\bar{y} - \mu)/\sqrt{2\pi\hat{f}(0)/T}$ of the form $t \Rightarrow W(1)/\sqrt{Q(b)}$. This limiting random variable is invariant to $f(0)$ but depends on the random variable $Q(b)$. Because $Q(b)$ depends on M (through $b = M/T$) and $k(x)$, the fixed-b approximation captures much (but not all) of the randomness in $\hat{f}(0)$. In contrast, the standard approach appeals to a consistency result for $\hat{f}(0)$ to justify approximating $\hat{f}(0)$ by $f(0)$ and t is approximated by a $N(0,1)$ random variable that does not depend on M or $k(x)$. Theoretical work by Jansson (2004) and Phillips, Sun and Jin (2005) has established that the fixed-b approximation for t in the case of Gaussian data is more accurate than the standard normal approximation. Some results for the non-Gaussian case have been obtained by Goncalves and Vogelsang (2006).

The fixed-b approximation provides approximations of the bias and variance of $\hat{f}(0)$ that are polynomials in b. For the Bartlett kernel Hashimzade, Kiefer and Vogelsang (2005) show that $E(\hat{f}(0) - f(0)) \approx f(0)(-b + \frac{1}{3}b^2)$ and

$$\begin{aligned} var(\hat{f}(0)) &\approx var(f(0)Q(b)) \\ &= f(0)^2\left(\frac{4}{3}b - \frac{7}{3}b^2 + \frac{14}{15}b^3 + \frac{2}{9}b^4 - \frac{1}{15b^2}(2b-1)^5 1\left(b > \frac{1}{2}\right)\right), \end{aligned} \tag{5}$$

where $1(b > \frac{1}{2}) = 1$ if $b > \frac{1}{2}$ and 0 otherwise. Because $b = M/T$, the bias and variance of $\hat{f}(0)$ are approximated by high order polynomials in M/T. The leading term in the variance exactly matches the standard variance formula (4). Because of the higher order terms, the fixed-b variance is more closely related to the exact variance. A plot of the variance polynomial would show that as M increases, variance is initially increasing but once M becomes large enough, variance decreases in M. The fixed-b bias can be combined with the Parzen bias to give

$$E(\hat{f}(0) - f(0)) \approx -\frac{1}{M}\sum_{j=-\infty}^{\infty} |j|\gamma_j + f(0)\left(-\frac{M}{T} + \frac{1}{3}\left(\frac{M}{T}\right)^2\right). \tag{6}$$

This combined formula better approximates the behaviour of the exact bias. As M increases, the first term shrinks, but the second and third terms increase in magnitude. Depending on the relative magnitudes of $\sum_{j=-\infty}^{\infty}|j|\,\gamma_j$ and $f(0)$, bias will be decreasing in M when M is small, but as M increases further bias becomes increasing in M. It is interesting to note that $1/M$ and M/T terms in (6) match the terms in the type of bias approximations used by Velasco and Robinson (2001) in third order Edgeworth calculations.

Bandwidth choices that minimize approximate mean square error (MSE) of $\hat{f}(0)$ were used by Andrews (1991) and Newey and West (1994) where the bias and variance were approximated using only the leading terms in (5) and (6). A simple, closed form, solution is obtained for M that depends on $\sum_{j=-\infty}^{\infty}|j|\,\gamma_j$ and $f(0)$. Andrews (1991) recommends plugging in parametric estimators of these unknown quantities, whereas Newey and West (1994) recommend using nonparametric estimators. Including the higher order terms provided by (5) and (6) would allow a higher order approximation to the MSE. Given the polynomial structure of (5) and (6) with respect to M, the first order condition to this optimization problem is a high order polynomial in M with coefficients that depend $\sum_{j=-\infty}^{\infty}|j|\,\gamma_j$ and $f(0)$. Given plug-in estimates, obtaining the value of M that minimizes the approximate MSE would amount to numerically finding the root of a polynomial, which is not difficult. Such an analysis does not appear to exist in the econometrics literature.

While the focus of this article has been on the spectral analysis of a univariate time series, extending the concepts, notation, and estimation methods to the case of a vector of time series is straightforward. A vector of time series can be characterized by what is called the spectral density matrix. The diagonal elements of this matrix are the individual spectral densities. The off-diagonal elements are called the cross-spectral densities. The cross-spectral densities in general can be complex valued functions even when the data is real valued. The cross-spectral densities capture correlation between series and co-movements of series can be characterized in terms of cross-amplitude, phase and coherency, which are real valued functions. Many of the ideas and concepts in the original Granger (1969) causality paper were expressed in terms of cross-spectral densities.

TIMOTHY J. VOGELSANG

See also **generalized method of moments estimation; heteroskedasticity and autocorrelation corrections.**

Bibliography

Andrews, D.W.K. 1991. Heteroskedasticity and autocorrelation consistent covariance matrix estimation. *Econometrica* 59, 817–54.

Berk, K. 1974. Consistent autoregressive spectral estimates. *Annals of Statistics* 2, 489–502.

Blanchard, O. and Quah, D. 1989. The dynamic effects of aggregate demand and supply disturbances. *American Economic Review* 79, 655–73.

Cochrane, J.H. 1988. How big is the random walk in GNP? *Journal of Political Economy* 96, 893–920.

den Haan, W.J. and Levin, A. 1997. A practictioner's guide to robust covariance matrix estimation. In *Handbook of Statistics: Robust Inference*, vol. 15, ed. G. Maddala and C. Rao. New York: Elsevier.

Engle, R. 1974. Band spectrum regression. *International Economic Review* 15, 1–11.

Geweke, J. and Porter-Hudak, S. 1983. The estimation and application of long memory time series. *Journal of Time Series Analysis* 4, 221–38.

Goncalves, S. and Vogelsang, T.J. 2006. Block bootstrap puzzles in HAC robust testing: the sophistication of the naive bootstrap. Working paper, Department of Economics, Michigan State University.

Granger, C. 1969. Investigating causal relations by econometric models and cross-spectral methods. *Econometrica* 37, 424–38.

Hamilton, J.D. 1994. *Time Series Analysis.* Princeton, NJ: Princeton University Press.

Hansen, L.P. 1982. Large sample properties of generalized method of moments estimators. *Econometrica* 50, 1029–54.

Hashimzade, N., Kiefer, N.M. and Vogelsang, T.J. 2005. Moments of HAC robust covariance matrix estimators under fixed-*b* asymptotics. Working paper, Department of Economics, Cornell University.

Hashimzade, N. and Vogelsang, T.J. 2007. Fixed-*b* asymptotic approximation of the sampling behavior of nonparametric spectral density estimators. *Journal of Time Series Analysis.*

Hong, Y. 1996. Consistent testing for serial correlation of unknown form. *Econometrica* 64, 837–64.

Jansson, M. 2004. The error rejection probability of simple autocorrelation robust tests. *Econometrica* 72, 937–46.

Kiefer, N.M. and Vogelsang, T.J. 2005. A new asymptotic theory for heteroskedasticity autocorrelation robust tests. *Econometric Theory* 21, 1130–64.

King, R., Plosser, C., Stock, J. and Watson, M. 1991. Stochastic trends and economic fluctuations. *American Economic Review* 81, 819–40.

Neave, H.R. 1970. An improved formula for the asymptotic variance of spectrum estimates. *Annals of Mathematical Statistics* 41, 70–7.

Neave, H.R. 1971. The exact error in spectrum estimates. *Annals of Mathematical Statistics* 42, 901–75.

Newey, W.K. and West, K.D. 1987. A simple, positive semi-definite, heteroskedasticity and autocorrelation consistent covariance matrix. *Econometrica* 55, 703–8.

Newey, W.K. and West, K.D. 1994. Automatic lag selection in covariance estimation. *Review of Economic Studies* 61, 631–54.

Ng, S. and Perron, P. 1996. The exact error in estimating the spectral density at the origin. *Journal of Time Series Analysis* 17, 379–408.

Parzen, E. 1957. On consistent estimates of the spectrum of a stationary time series. *Annals of Mathematical Statistics* 28, 329–48.

Perron, P. and Ng, S. 1998. An autoregressive spectral density estimator at frequency zero for nonstationarity tests. *Econometric Theory* 14, 560–603.

Phillips, P.C.B. and Hansen, B.E. 1990. Statistical inference in instrumental variables regression with i(1) processes. *Review of Economic Studies* 57, 99–125.

Phillips, P.C.B. and Perron, P. 1988. Testing for a unit root in time series regression. *Biometrika* 75, 335–346.

Phillips, P.C.B., Sun, Y. and Jin, S. 2005. Optimal bandwidth selection in heteroskedasticity-autocorrelation robust testing. Working Paper No. 2005-12, Department of Economics, UCSD.

Phillips, P.C.B., Sun, Y. and Jin, S. 2006. Spectral density estimation and robust hypothesis testing using steep origin kernels without truncation. *International Economic Review* 47, 837–94.

Priestley, M.B. 1981. *Spectral Analysis and Time Series*, vol. 1. New York: Academic Press.

Robinson, P. 1991. Automatic frequency domain inference on semiparametric and nonparametric models. *Econometrica* 59, 1329–63.

Stock, J. and Watson, M. 1988. Testing for common trends. *Journal of the American Statistical Association* 83, 1097–107.

Velasco, C. and Robinson, P.M. 2001. Edgeworth expansions for spectral density estimates and studentized sample mean. *Econometric Theory* 17, 497–539.

Watson, M.W. 1993. Measures of fit for calibrated models. *Journal of Political Economy* 101, 1011–41.

spline functions

In the everyday use of the word, a 'spline' is a flexible strip of material used by draftsmen in the same manner as French curves to draw a smooth curve between specified points. The mathematical *spline function* is similar to the draftsman's spline. It has roots in the aircraft, automobile and shipbuilding industries. Formally, a spline function is a piecewise continuous function with a specified degree of continuity imposed on its derivatives. Usually the pieces are polynomials. The abscissa values, which define the segments, are referred to as 'knots', and the set of knots is referred to as the 'mesh'.

The terminology and impetus for most contemporary work on spline functions can be traced to the seminal work of I.J. Schoenberg (1946), although the basic idea can be found in the writings of E.T. Whittaker (1923) and, in Schoenberg's (1946, p. 68) own modest opinion, in the earlier work of Laplace. Today the literature on spline functions comprises an integral part of modern approximation theory. Useful monographs covering splines are De Boor (2001), Eubank (1988), Green and Silverman (1994), Poirier (1976), Schumaker (1981), and Wahba (1990). The many important contributions of Grace Wahba in the 1970s and 1980s (for example, Kimeldorf and Wahba, 1970; Wahba, 1978; 1983) united the approximation theory and the emerging statistics literatures involving spline functions.

Given a degree d and a knot vector $t = [t_1, t_2, \ldots, t_K]'$, where $t_1 < t_2 < \cdots < t_K$, the collection of polynomial splines having s continuous derivatives forms a linear space. For example, the collection of *linear* splines with knot sequence t is spanned by the functions

$$1, x, (x - t_1)_+, \ldots, (x - t_K)_+.$$

where $(\cdot)_+ = \max(\cdot, 0)$. This set is called the *truncated power basis* of the space. In general, the basis for a spline space of degree d and smoothness s is made up of monomials up to degree d together with terms of the form $(x - t_k)_+^{s+j}$, where $1 \leq j \leq d - s$. For example, cubic splines have $d = 3$ and $s = 2$ so that the basis has elements

$$1, x, x^2, x^3, (x - t_1)_+^3, \ldots, (x - t_K)_+^3.$$

Unfortunately, these truncated power functions have poor numerical properties. For example, in linear regression problems the condition of the design matrix deteriorates rapidly as the number of knots increases. A popular alternative representation is the so-called *B-spline basis* (see De Boor. 2001). These functions are constructed to have support only on a few neighbouring intervals defined by the knots.

The importance of spline functions in approximation theory is explained by the following *best approximation property*. Consider the data points $(x_i, y_i)(i = 1, 2, \ldots, n)$ and suppose without loss of generality that $0 < x_1 < x_2 < x_3 < \cdots < x_n < 1$. Given $\lambda > 0$, consider the optimization problem

$$\min_{f(\cdot)} \sum_{i=1}^{n} [y_i - f(x_i)]^2 + \lambda \int_0^1 [D^m f(x)]^2 dx, \tag{1}$$

where D^m denotes the differentiation operator of degree m, $f(\cdot)$ is a function defined on [0, 1] such that $D^j f$, $j \leq m - 1$, is absolutely continuous, and $D^m f$ is in the set of measurable square integrable functions on [0, 1]. The first term in (1) comprises the familiar least squares measure of fit and the second term comprises a measure of the smoothness in $f(\cdot)$. The parameter λ measures the trade-off between fit and smoothness. The solution to (1) is a polynomial *smoothing spline* of degree $2m - 1$ with knots at all the abscissa data points. As $\lambda \rightarrow 0$, the solution is referred to as an *interpolating spline* and it fits the data exactly. The choice of λ is crucial and the method of cross-validation is a popular method for choosing λ (see for example Craven and Wahba, 1979, or Green and Silverman, 1994, pp. 30–8). The most popular choice for m is $m = 2$ yielding a *natural cubic spline* as the solution to (1). The adjective 'natural' implies that the second derivative equals zero at the endpoints.

To interpret the first term in (1) as the log-likelihood of normal linear regression model, smoothing splines can be viewed as the outcome of penalized (reflected in the second term in (1)) maximum likelihood estimation. A Bayesian interpretation of smoothing splines, provided by Kimeldorf and Wahba (1970) and expanded by Silverman (1985) and Wahba (1978; 1983), views (1) as a log-posterior density with $\exp[-\frac{1}{2}\lambda \int [D^m f(x)]^2 dx]$ serving as a prior density over the space of all smooth functions.

In econometrics spline functions are most often employed to parametrize a regression function. For example, splines were the functional form chosen to paremetrize the treatment in the first major social experiment in economics: the New Jersey Income-Maintenance Experiment. Such *regression splines* usually include only a few knots and not necessarily at the design points. This usage may simply reflect the flexibility and good approximation properties of splines, or the attempt to capture *structural change*. For example, a researcher may believe the relationship between two variables y and x is locally a polynomial, but that at precise points in terms of x the relationship 'changes', not in a discontinuous fashion in level but rather continuously in derivative of order $2m - 1$. Common choices for such x variables are time, age, education or income, to name a few, with a nearly unlimited number of choices of candidates for y variables.

In statistics spline functions are used in isotonic regression, histogram smoothing, density estimation, interpolation of distribution functions for which there is no closed-form analytic representation, and nonparametric regression. In the latter case spline smoothing corresponds approximately to smoothing by a kernel method with bandwidth depending on the local density of design points.

While spline functions have proved to be valuable approximation tools, they also arise naturally in their own right in economics. Income tax functions with increasing marginal tax rates constitute a *linear spline*, as do familiar 'kinked' demand curves and 'kinked' budget sets. *Quadratic splines* serve as useful ways of generating asymmetric loss functions for use in decision theory. In distributed lag analysis, spline functions have been used as natural generalizations of Almon polynomial lags. *Periodic cubic splines* have proved useful in seasonal adjustment and in analysis of electricity load curves. Spline functions in these applications are attractive partly because, given the knots, the spline can be expressed as linear functions of unknown parameters, hence facilitating statistical estimation.

Knots play different roles in the approximation theory and structural change literatures. In the former they are largely nuisance parameters, and, apart from parsimony considerations, the number and location of the knots are of no particular importance other than that they serve to define a smooth best-fitting curve. When viewed as *change points*, however, the knots become parameters of interest. In applications involving structural change, the number of potential knots is small, and their location reflects subject-matter considerations. For example, when fitting a time trend with a regression spline, the knots may reflect the effect on the dependent variable of a specific event of interest – for example, a war. A prior distribution can then be specified over the interval bounded by the start and end of the war.

Estimation of the number and location of the knots is hindered by numerical and statistic complications. The knots enter spline functions nonlinearly, and there are typically numerous local minima in the residual sum-of-squares surface. Many of these local minima correspond to knot vectors with *replicate knots*, that is, knots which pile up on top of each other, signalling that further discontinuities in the derivatives of the function are required. When knot locations are set free, knots move to areas where the function is less smooth. If, in addition, the *number* of knots is unknown, the difficulties multiply. For example, under the null hypothesis that adjacent intervals are identical, the location of the unnecessary knot is unidentified.

Different solutions have emerged to the problem of unknown location and number of knots. Some introduce a large number of potential knots from which a subset is to be selected (for example, Halpern, 1973; Friedman and Silverman, 1989; Smith and Kohn, 1996). The problem then becomes one of variable selection where each knot corresponds to a column of a design matrix from which a 'significant' subset is to be determined. In some Bayesian nonparametric regression studies (for example, DiMatteo, Genovese and Kass, 2001; Smith, Wong and Kohn, 1998; Denison, Mallick and Smith, 1998), knot locations are treated as parameters and given prior distributions. Additional constraints are usually imposed to keep knots some minimum distance apart. The definitive treatment of the problem of unknown location and number of knots has not yet emerged.

Early applications of splines to *multivariate* problems (see Green and Silverman, 1994, Ch. 7) involved tensor product spaces that of necessity depended on the choice of coordinate system. An example is the two-dimensional *thin plate spline* of Wahba (1990) which simulates how a thin metal plate would behave if forced through some control points. This is similar to the one-dimensional draftsman's spline. The tensor product structure of these spaces implicitly defines the domain of an unknown function to be a hyperrectangle, and this can restrict the ability to capture important features in the data that are not oriented along one of the major axes. There is a considerable literature on constructing and representing smooth, piecewise polynomial surfaces over meshes in many variables. In particular, much has been written about the case in which the underlying partition consists of triangles or high-dimensional simplicies. Because of their invariance to affine transformations, *barycentric coordinates* (that is, coordinates expressed as weighted combinations of the vertices of the triangle) are used to construct spline spaces over such meshes. The *triogram* methodology of Hansen, Kooperberg and Sardy (1998) employs continuous, piecewise linear (planar) bivariate splines defined over adaptively selected triangulations in the plane. Analogous to stepwise knot addition and deletion in a univariate spline space, the underlying triangulation is constructed adaptively by adding and deleting vertices.

DALE J. POIRIER

Bibliography

Craven, P. and Wahba, G. 1979. Smoothing noisy data with spline functions: estimating the correct degree of smoothing by the method of cross-validation. *Numerische Mathematik* 31, 377–403.

De Boor, C. 2001. *A Practical Guide to Splines*. New York: Springer-Verlag.

Denison, D., Mallick, B. and Smith, A. 1998. Automatic Bayesian curve fitting. *Journal of the Royal Statistical Society: Series B* 60, 333–50.

DiMatteo, I., Genovese, C. and Kass, R. 2001. Bayesian curve-fitting with free knot splines. *Biometrika* 88, 1055–71.

Eubank, R. 1988. *Spline Smoothing and Nonparametric Regression*. New York: Marcel-Dekker.

Friedman, J. and Silverman, B. 1989. Flexible parsimonious smoothing and additive modeling. *Technometrics* 31, 3–21.

Green, P. and Silverman, B. 1994. *Nonparametric Regression and Generalized Linear Models*. London: Chapman and Hall.

Halpern, E. 1973. Bayesian spline regression when the number of knots is unknown. *Journal of the Royal Statistical Society: Series B* 35, 347–60.

Hansen, M., Kooperberg, C. and Sardy, S. 1998. Triogram models. *Journal of the American Statistical Association* 93, 101–19.

Kimeldorf, G. and Wahba, G. 1970. A correspondence between Bayesian estimation on stochastic processes and smoothing by splines. *Annals of Mathematical Statistics* 41, 495–502.

Poirier, D. 1976. *The Econometrics of Structural Change with Special Emphasis on Spline Functions*. Amsterdam: North-Holland.

Schoenberg, I. 1946. Contributions to the problem of approximation of equidistant data by analytic functions: Parts I and II. *Quarterly Journal of Applied Mathematics* 4, 45–99; 112–41.

Schumaker, L. 1981. *Spline Functions: Basic Theory*. New York: Wiley.

Silverman, B. 1985. Some aspects of the spline smoothing approach to nonparametric regression curve fitting (with discussion). *Journal of the Royal Statistical Society: Series B* 47, 1–52.

Smith, M. and Kohn, R. 1996. Nonparametric regression using Bayesian variable selection. *Journal of Econometrics* 75, 317–43.

Smith, M., Wong, C. and Kohn, R. 1998. Additive nonparametric regression with autocorrelated errors. *Journal of the Royal Statistical Society: Series B* 60, 311–31.

Wahba, G. 1978. Improper priors, spline smoothing and the problem of guarding against model errors in regression. *Journal of the Royal Statistical Society: Series B* 40, 364–72.

Wahba, G. 1983. Bayesian 'confidence intervals' for the cross-validated smoothing spline. *Journal of the Royal Statistical Society: Series B* 45, 133–50.

Wahba, G. 1990. *Spline Models for Observational Data*. Philadelphia: Society for Industrial and Applied Mathematics.

Whittaker, E. 1923. On a new method of graduation. *Proceedings of the Edinburgh Mathematical Society* 41, 63–75.

spurious regressions

For the first three-quarters of the 20th century the main workhorse of applied econometrics was the basic regression

$$Y_t = a + bX_t + e_t. \quad (1)$$

Here the variables are indicated as being measured over time, but could be over a cross-section; and the equation was estimated by ordinary least squares (OLS). In practice more than one explanatory variable x would be likely to be used, but the form (1) is sufficient for this discussion. Various statistics can be used to describe the quality of the regression, including R^2, t-statistics for β, and Durbin–Watson statistic d which relates to any autocorrelation in the residuals. A good fitting model should have $|t|$ near 2, R^2 quite near one, and d near 2.

In standard situations the regression using OLS works well, and researchers used it with confidence. But there were several indications that in special cases the method could produce misleading results. In particular, when the individual series have strong autocorrelations, it had been realized by the early 1970s by time series analysis that the situation may not be so simple; that apparent relationships may often be observed by using standard interpretations of such regressions. Because a relationship appears to be found between independent series, they have been called 'spurious'. Note that, if $b = 0$, then e_t must have the same time series properties as Y_t, that is, it will be strongly autocorrelated, and so the assumptions of the classical OLS regression will not be obeyed. The possibility of getting incorrect results from regressions was originally pointed out by Yule (1926) in a much cited paper that discussed 'nonsense correlations'. Kendall (1954) also pointed out that a pair of independent autoregressive series of order one could have a high apparent correlation between them; and so if they were put into a regression a spurious relationship could be obtained.

The magnitude of the problem was found from a number of simulations. The first simulation on the topic was by Granger and Newbold (1974), who generated pairs of independent random walks, from (1) with $a = b = 1$. Each series had 50 terms and 100 repetitions were used. If the regression is run, using series that are temporarily uncorrelated, one would expect that roughly 95 per cent of values of $|t|$ on b would be less than 2. This original simulation using random walks found $|t| \leq 2$ on only 23 occasions; out of the 100, $|t|$ was between 2 and 4 on 24 occasions, between 4 and 7 on 34 occasions, and over 7 on the other 19 occasions.

The reaction to these results was to reassess many of the previously obtained empirical results in applied time series econometrics, which undoubtedly involved highly autocorrelated series but had not previously been concerned by this fact. Just having a high R^2 value and an apparently significant value of b was no longer sufficient for a regression to be satisfactory or its interpretations relevant. The immediate questions were how one could easily detect a spurious regression and then correct for it. Granger and Newbold (1974) concentrated on the value of the Durbin–Watson statistic: if the value is too low, it suggests that the regressions results cannot be trusted. Remedial methods such as using a Cochrane–Orcutt technique to correct autocorrelations in the residuals, or differencing the series used in a regression, were inclined to introduce further difficulties and could not be recommended. The problem arises because the equation is mis-specified; the proper reaction to having a possible spurious relationship is to add lagged dependent and independent variables until the errors appear to be white noise, according to the Durbin–Watson statistic. A random walk is an example of a I(1) process, that is, a process that needs to be differenced to become stationary. Such processes seem to be common in parts of econometrics, especially in macroeconomics and finance. One approach that is widely recommended is to test whether X_t, Y_t are I(1) and, if so, to difference before one performs the regression. There are many tests available; a popular one is due to Dickey and Fuller (1979).

A theoretical investigation of the basic unit root, ordinary least squares, spurious regression case was undertaken by Phillips (1986). He considered the asymptotic properties of the coefficients and statistics of eq. (1), $\hat{a}, \hat{b}$, the t-statistic for b, R^2 and the Durbin–Watson statistics $\hat{\rho}$. To do this he introduced the link between normed sums of functions of unit root processes and integrals of Weiner processes. For example, if a sample X_t of size T is generated from a driftless random walk, then

$$T^{-2}\sum_{1}^{T} X_t^2 \rightarrow \sigma_\varepsilon^2 \int_0^1 W^2(t)\, dt,$$

where σ_ε^2 is the variance of the shock, and $W(t)$ is a Weiner process. As a Weiner process is a continuous time random process on the real line [0,1], the various sums involved are converging and can thus be replaced by integrals of a stochastic process. This transformation makes the mathematics of the investigation much easier, once one becomes familiar with the new tools. Phillips is able to show that

- the distributions of the t-statistics for $\hat{a}$ and $\hat{b}$ from (1) diverge as t becomes large, so there is no asymptotically correct critical values for these conventional tests;
- $\hat{b}$ converges to some random variable whose value changes from sample to sample;
- Durbin–Watson statistics tend to zero; and
- R^2 does not tend to zero but to some random variable.

What is particularly interesting is not only that do these theoretical results completely explain the simulations but also that the theory deals with asymptotics, $T \rightarrow \infty$, whereas the original simulations had only $T = 50$. It seems that spurious regression occurs at all sample sizes.

Haldrup (1994) has extended Phillips's result to the case for two independent I(2) variables and obtained similar results. (An I(2) variable is one that needs differencing twice to get to stationarity, or, here, difference once to get to random walks.) Marmol (1998) has further extended these results to fractionally integrated I(d) processes. Durlauf and Phillips (1988) regress I(1) process on deterministic polynomials in time, thus polynomial trends, and found spurious relationships.

Although spurious regressions in econometrics are usually associated with I(1) processes, which were explored in Phillips's well-known theory and in the best known simulations, what is less appreciated is that the problem can also occur, although less clearly, with stationary processes.

Table 1 shows simulation results from independent series generated by two first order autoregressive models with coefficients a_1 and a_2 where $0 \leqslant a_1 = a_2 \leqslant 1$ and with inputs e_{xt}, e_{yt} both Gaussian white noise series, using regression 1 estimated using OLS with sample sizes varying between 100 and 10,000.

It is seen that sample size has little impact on the percentage of spurious regressions found (apparent significance of the b coefficient in (1)). Fluctuations down columns do not change significantly with the number of iterations used. Thus, the spurious regression problem is not a small sample property. It is also seen to be a serious problem with pairs of autoregressive series which are not unit root processes. If $a = 0.75$, for example, then 30 per cent of regressions will give spurious implications. Further results are available in the original paper but will not be reported in detail. The Gaussian error assumption can be replaced by other distributions with little or no change in the simulation results, except for an exceptional distribution such as the Cauchy. Spurious regressions also occur if $a_1 \neq a_2$, although less frequently, and particularly if the smaller of the two a values is at least 0.5 in magnitude.

The obvious implications of these results is that applied econometricians should not worry about spurious regressions *only* when dealing with I(1), unit root, processes. Thus, a strategy of first testing whether a series contains a unit root before entering into a regression is not relevant. The results suggest that many more simple regressions need to be interpreted with care when the series involved are strongly serially correlated. Again, the correct response is to move to a better specification, using lags of all variables.

Concerns about spurious regressions produced interest in tests for unit roots, of which there are now many; and empirical works with time series will usually test between I(1) or I(0), or may sometimes consider more complicated alternatives. If

Table 1 *Regression between independent AR(1) series*

Sample series	$a = 0$	$a = 0.25$	$a = 0.5$	$a = 0.75$	$a = 0.9$	$a = 1.0$
100	4.9	6.8	13.0	29.9	51.9	89.1
500	5.5	7.5	16.1	31.6	51.1	93.7
2,000	5.6	7.1	13.6	29.1	52.9	96.2
10,000	4.1	6.4	12.3	30.5	52.0	98.3

$a_1 = a_2 = a$ percentage of $|t| > 2$.
Source: Granger, Hyung and Jeon (2001).

series are found to be I(1), simple regressions have been replaced with considerations of cointegration and construction of error-correction models.

A recent survey of studies of spurious relationships is Pilatowska (2004).

CLIVE W.J. GRANGER

Bibliography

Baltagi, B. 2001. *A Companion to Theoretical Econometrics.* Oxford: Blackwell.

Dickey, D. and Fuller, W. 1979. Distribution of the estimates for autoregressive time series with a unit root. *Journal of the American Statistical Association* 74, 427–31.

Durlauf, S. and Phillips, P. 1988. Trends versus random walks in time series analysis. *Econometrica* 56, 1333–54.

Granger, C. 2001. Spurious regression in econometrics. In *A Companion to Theoretical Econometrics*, ed. B. Baltagi. Oxford: Blackwell.

Granger, C., Hyung, N. and Jeon, Y. 2001. Spurious regression with stationary series. *Applied Economics* 33, 899–904.

Granger, C. and Newbold, P. 1974. Spurious regressions in econometrics. *Journal of Econometrics* 2, 111–20.

Haldrup, N. 1994. The asymptotics of single-equation cointegration regressions with I1 and I2 variables. *Journal of Econometrics* 63, 153–81.

Kendall, M. 1954. *Exercises in Theoretical Statistics.* London: Griffin.

Marmol, F. 1998. Spurious regression theory with non-stationary fractionally integrated processes. *Journal of Econometrics* 84, 233–50.

Phillips, P. 1986. Understanding spurious regressions in econometrics. *Journal of Econometrics* 33, 311–40.

Phillips, P. 1988. New tools for understanding spurious regressions. *Econometrica* 66, 1299–325.

Pilatowska, M. 2004. Realization of the congruence postulate as a method of avoiding the effects of a spurious relationship. In *Dynamic Economic Models*, vol. 6, ed. Z. Zieliński. Torun: Nicolaus Copernicus University.

Yule, G. 1926. Why do we sometimes get nonsense correlations between time-series? A study in sampling and the nature of time-series. *Journal of the Royal Statistical Society* 89, 1–64.

state space models

State space models is a rather loose term given to time series models, usually formulated in terms of unobserved components, that make use of the state space form for their statistical treatment.

At the simplest level, structural time series models (STMs) are set up in terms of components such as trends and cycles that have a direct interpretation. Signal extraction, or smoothing, provides a description of these features that is model-based and hence avoids the ad hoc nature of procedures such as moving averages and the Hodrick–Prescott filter. While smoothing uses all the observations in the sample, filtering yields estimates at a given point in time that are constructed only from observations available at that time. For key time series, filtered estimates form the basis for 'nowcasting' in that they give an indication of the current state of the economy and the direction in which it is moving. They also provide the starting point for forecasts of future observations and components.

Local level model

A simple model with permanent and transitory components illustrates the basic ideas of filtering and smoothing. Suppose that the observations consist of a random walk component plus a random irregular term, that is

$$y_t = \mu_t + \varepsilon_t, \qquad \varepsilon_t \sim NID(0, \sigma_\varepsilon^2), \quad t = 1, \ldots, T \tag{1}$$

$$\mu_t = \mu_{t-1} + \eta_t, \qquad \eta_t \sim NID(0, \sigma_\eta^2), \tag{2}$$

where the irregular and level disturbances, ε_t and η_t respectively, are mutually independent and the notation $NID(0, \sigma^2)$ denotes normally and independently distributed with mean zero and variance σ^2. The signal–noise ratio, $q = \sigma_\eta^2/\sigma_\varepsilon^2$, plays the key role in determining how observations should be weighted for prediction and signal extraction. In a large sample, filtering is equivalent to a simple exponentially weighted moving average; the higher q is, the more past observations are discounted. When q is zero, the level is constant and all observations have the same weight. The reduced form of the model has the first differences following a first-order moving average process, that is

$$\Delta y_t = \xi_t + \theta\xi_{t-1}, \; \xi_t \sim NID(0, \sigma_\xi^2)$$

where $\theta = [(q^2 + 4q)^{1/2} - 2 - q]/2$. This produces the same forecasts. However, the structural form in (1) also yields nowcasts and smoothed estimates of the level, μ_t, throughout the series. In the middle of a large sample, the smoothed estimates are approximately equal to

$$\frac{1+\theta}{1-\theta}\sum_j (-\theta)^{|j|} y_{t+j}, \; q > 0, \tag{3}$$

while the filtered estimates are

$$(1+\theta)\sum_{j\geq 0}(-\theta)^j y_{t-j},\ q>0. \tag{4}$$

At the end of the sample this estimate also yields the forecast of future levels and future observations. Note that, although the above expressions are useful for displaying the weighting of the observations, finite sample computation is best done by a simple forward recursion for filtering and a subsequent backward one for smoothing.

State space form

The state space form (SSF) is a simple device whereby a dynamic model is written in terms of just two equations. The model in (1) and (2) is a special case. The general linear SSF applies to a multivariate time series, $\mathbf{y}_t$, containing N elements. These observable variables are related to an $m \times 1$ vector, α_t, known as the *state vector*, through a *measurement equation*

$$\mathbf{y}_t = \mathbf{Z}_t\boldsymbol{\alpha}_t + \mathbf{d}_t + \boldsymbol{\varepsilon}_t, \qquad t = 1, \ldots, T \tag{5}$$

where $\mathbf{Z}_t$ is an $N \times m$ matrix, $\mathbf{d}_t$ is an $N \times 1$ vector and $\boldsymbol{\varepsilon}_t$ is an $N \times 1$ vector of serially uncorrelated disturbances with mean zero and covariance matrix $\mathbf{H}_t$. In general the elements of $\boldsymbol{\alpha}_t$ are not observable. However, they are assumed to be generated by a first-order Markov process,

$$\boldsymbol{\alpha}_t = \mathbf{T}_t\boldsymbol{\alpha}_{t-1} + \mathbf{c}_t + \mathbf{R}_t\boldsymbol{\eta}_t, \qquad t = 1, \ldots, T \tag{6}$$

where $\mathbf{T}_t$ is an $m \times m$ matrix, $\mathbf{c}_t$ is an $m \times 1$ vector, $\mathbf{R}_t$ is an $m \times g$ matrix and $\boldsymbol{\eta}_t$ is a $g \times 1$ vector of serially uncorrelated disturbances with mean zero and covariance matrix, $\mathbf{Q}_t$. Equation (6) is the *transition equation*. The specification of the system is completed by assuming that the initial state vector, $\boldsymbol{\alpha}_0$, has a mean of $\mathbf{a}_0$ and a covariance matrix $\mathbf{P}_0$, and that the disturbances $\boldsymbol{\varepsilon}_t$ and $\boldsymbol{\eta}_t$ are uncorrelated with the initial state. The disturbances are often assumed to be uncorrelated with each other in all time periods, though this assumption may be relaxed to allow contemporaneous correlation, the consequence being a slight complication in some of the filtering formulae.

The definition of $\boldsymbol{\alpha}_t$ for any particular statistical model is determined by construction. Its elements may or may not be identifiable with components that have a substantive interpretation, for example as a trend or a seasonal. From the technical point of view, the aim of the state space formulation is to set up $\boldsymbol{\alpha}_t$ in such a way that it contains all the relevant information on the system at time t and that it does so by having as small a number of elements as possible. The SSF is not, in general, unique.

The Kalman filter (KF) is a recursive procedure for computing the optimal estimator of the state vector at time t, based on the observations up to and including $\mathbf{y}_t$. In a Gaussian model, the disturbances $\boldsymbol{\varepsilon}_t$ and $\boldsymbol{\eta}_t$, and the initial state, are all normally distributed. Because a normal distribution is characterized by its first two moments, the

Kalman filter can be interpreted as updating the mean and covariance matrix of the conditional distribution of the state vector as new observations become available. The conditional mean minimizes the mean square error and when viewed as a rule for all realizations it is the minimum mean square error estimator (MMSE). Since the conditional covariance matrix does not depend on the observations, it is the unconditional MSE matrix of the MMSE. When the normality assumption is dropped, the KF is still optimal in the sense that it minimizes the mean square error within the class of all linear estimators. Given initial conditions, $\mathbf{a}_0$ and $\mathbf{P}_0$, the Kalman filter delivers the optimal estimator of the state vector as each new observation becomes available. When all T observations have been processed, it yields the optimal estimator of the current state vector based on the full information set. When the initial conditions cannot be specified a diffuse prior is often placed on the initial state. This amounts to setting $\mathbf{P}_0 = \boldsymbol{\kappa}\mathbf{I}$, and letting the scalar κ go to infinity. Stable algorithms for handling diffuse priors are set out in Durbin and Koopman (2001).

Prediction is carried out straightforwardly by running the KF without updating. Mean square errors of the forecasts are produced at the same time. Smoothing is carried out by a backward filter initialized with the estimates delivered by the KF at time T. The aim is to compute the optimal estimator of the state vector at time t using information made available after time t as well as before. Efficient smoothing algorithms are described in Durbin and Koopman (2001, pp. 70–3). The weights are implicit, but Koopman and Harvey (2003) give an algorithm for computing and displaying them at any point in time. The state space smoother is far more general than the classic Wiener–Kolomogorov (WK) filter. The WK filter computes weights explicitly and for simple models it is possible to obtain expressions for the estimator in the middle of a doubly infinite sample without too much difficulty. Formula (3) is a case in point. However, the WK filter is limited to time-invariant models and even here it has no computational advantages over the state space fixed-interval smoothing algorithm. In the second edition of his celebrated text describing the WK filter, Whittle (1984, p. xi) writes 'In its preoccupation with the stationary case and generating function methods, the 1963 text essentially missed the fruitful concept of state structure. This … has now come to dominate the subject.'

The system matrices $\mathbf{Z}_t, \mathbf{H}_t, \mathbf{T}_t, \mathbf{R}_t$ and $\mathbf{Q}_t$ may depend on a set of unknown parameters, and one of the main statistical tasks will often be the estimation of these parameters. Thus in the random walk plus noise model, (1), the parameters σ_η^2 and σ_ε^2 will usually be unknown. As a by-product, the KF produces a vector of prediction errors or *innovations* and in a Gaussian model these can be used to construct a likelihood function that can be maximized numerically with respect to the unknown parameters.

Since the state vector is a vector of random variables, a Bayesian interpretation of the Kalman filter as a way of updating a Gaussian prior distribution on the state to give a posterior is quite natural. The mechanics of filtering, smoothing and prediction are the same irrespective of whether the overall framework is Bayesian or classical. Smoothing gives the mean and variance of the state, conditional on all the observations. For the classical statistician, the conditional mean is the MMSE, while for the Bayesian it minimizes the expected loss for a symmetric loss function. With a quadratic loss function, the expected loss is given by the conditional variance. The real differences between classical and Bayesian treatments arise when the parameters are unknown. In a Bayesian framework, the hyperparameters, as they are often called, are random variables. The development of simulation techniques based on Markov chain Monte Carlo (MCMC) has now made a full Bayesian treatment a feasible proposition. This means that it is possible to simulate a distribution for the state that takes account of hyperparameter uncertainty.

Applications

The use of unobserved components opens up a new range of possibilities for economic modelling. Furthermore, it provides insights and a unified approach to many other problems. The examples below give a flavour.

The local linear trend model generalizes (1) by the introduction of a stochastic slope, β_t, which itself follows a random walk. Thus

$$\begin{aligned} \mu_t &= \mu_{t-1} + \beta_{t-1} + \eta_t, \qquad && \eta_t \sim NID(0, \sigma_\eta^2), \\ \beta_t &= \beta_{t-1} + \zeta_t, && \zeta_t \sim NID(0, \sigma_\zeta^2), \end{aligned} \tag{7}$$

where the irregular, level and slope disturbances, ε_t, η_t and ζ_t, respectively, are mutually independent. If both variances σ_η^2 and σ_ζ^2 are zero, the trend is deterministic. When only σ_ζ^2 is zero, the slope is fixed and the trend reduces to a random walk with drift. Allowing σ_ζ^2 to be positive but setting σ_η^2 to zero gives an integrated random walk trend, which when estimated tends to be relatively smooth. Signal extraction of the trend by setting the signal–noise ratio, $q = \sigma_\zeta^2/\sigma_\varepsilon^2$, to 1/1600 gives the Hodrick–Prescott filter for quarterly data. Adding a cyclical component to (1) provides a vehicle for detrending based on a model the parameters of which can be estimated from the data.

Orphanides and van Norden (2002) have recently stressed the importance of tracking the output gap in *real time*. Given the parameter estimates, real time estimation of components such as the output gap is just an exercise in filtering. However, as new observations become available the estimate of the gap at a particular point in time can be improved by smoothing. Harvey, Trimbur and van Dijk (2007) adopt a Bayesian approach which has the advantage of giving the full distribution of the output gap. Statistics such as the probability that the output gap is increasing are readily calculated. Following Kuttner (1994), Harvey, Trimbur and van Dijk (2007) also construct an unobserved components model relating the output gap to inflation in what is effectively a Phillips curve relationship.

A number of authors, beginning with Sargent (1989), have estimated the structural parameters of dynamic stochastic general equilibrium (DSGE) models using state space methods. The linear rational expectations model is first solved for the reduced-form state equation in its predetermined variables. Once this has been done, the model is put in state space form and the parameters are estimated by maximum likelihood. Alternatively a Bayesian approach can be adopted; see Smets and Wouter (2003, p. 1138).

Data irregularities

Some of the most striking benefits of the structural approach to time series modelling become apparent only when we start to consider more complex problems. In particular, the SSF offers considerable flexibility with regard to dealing with data irregularities, such as missing observations and observations at mixed frequencies. Missing observations are easily handled in the SSF simply by omitting the updating equations while retaining the prediction equations. Filtering and smoothing then go through automatically and the likelihood function is constructed using prediction errors corresponding to actual observations. With flow variables, such as income, the issue is one of temporal aggregation. This may be dealt with by the introduction of a cumulator variable into the state. The study by Harvey and Chung (2000) on the measurement of British unemployment provides an illustration of how mixed frequencies are handled and how using an auxiliary series can improve the efficiency of nowcasting and forecasting a target series. The challenge was how to obtain timely estimates of the underlying change in unemployment. Estimates of the numbers of unemployed according to the International Labor Organization (ILO) definition are given by the Labour Force Survey (LFS), which consists of a rotating sample of approximately 60,000 households. These estimates have been published on a quarterly basis since the spring of 1992, but from 1984 to 1991 estimates were available for the spring quarter only. Another measure of unemployment, based on administrative sources, is the number of people claiming unemployment benefit. This measure, known as the claimant count, is available monthly, with very little delay and is an exact figure. It does not provide a figure corresponding to the ILO definition, but it moves roughly in the same way as the LFS figure. The first problem is how to extract the best estimate of the underlying monthly change in a series which is subject to sampling error and which may not have been recorded every month. The second is how to use a related series to improve this estimate. These two issues are of general importance, for example in the measurement of the underlying rate of inflation or the way in which monthly figures on industrial production might be used to produce more timely estimates of national income. State space methods deal with the mixed frequencies in the target series, with the rather complicated error structure coming from the rotating sample (see Pfeffermann, 1991) and with the different frequency of the auxiliary series.

Continuous time

Continuous time STMs observed at discrete intervals can easily be put in SSF (see Harvey, 1989, Ch. 9). An important case is the continuous time version of (7) where the smoothed trend is a cubic spline. Setting up such a model for a cubic spline enables the smoothness parameter to be estimated by maximum likelihood and the fact that irregularly spaced data may be handled means that it can be used to fit a nonlinear function to cross-sectional data. The model can easily be extended, for example to include other components, and it can be compared with alternative models using standard statistical criteria (see Kohn, Ansley and Wong, 1992).

Nonlinear and non-Gaussian models

Some of the most exciting recent developments in time series have been in nonlinear and non-Gaussian models. For example, it is possible to fit STMs with heavy-tailed distributions on the disturbances, thereby making them robust with respect to outliers and structural breaks. Similarly, non-Gaussian models, designed to deal with count data and qualitative observations, can be set up with stochastic components. In the general formulation of a state space model, the distribution of the observations is specified conditional on the current state and past observations, that is

$$p(\mathbf{y}_t|\alpha_t, \mathbf{Y}_{t-1}), \qquad t = 1, \ldots, T \tag{8}$$

where $\mathbf{Y}_{t-1} = \{\mathbf{y}_{t-1}, \mathbf{y}_{t-2}, \ldots\}$. Similarly the distribution of the current state is specified conditional on the previous state and observations so that

$$p(\boldsymbol{\alpha}_t|\boldsymbol{\alpha}_{t-1}, \mathbf{Y}_{t-1}). \tag{9}$$

The initial distribution of the state is given as $p(\boldsymbol{\alpha}_0)$. In a linear Gaussian model the conditional distribution in (8) and (9) are characterized by their first two moments and so they are specified by the measurement and transition equations. The Kalman filter updates the mean and covariance matrix of the state. In more general models, computer-intensive methods, using techniques such as importance sampling, have to be applied (see Durbin and Koopman, 2001). Within a Bayesian framework, methods are normally based on MCMC. Particle filtering is often used for signal extraction; see the review in Harvey and de Rossi (2006).

The use of state space methods highlights a fundamental distinction in time series models between those motivated by description and those set up to deal directly with forecasting. This is epitomized by the contrast between STMs on the one hand and autoregressions and autoregressive-integrated-moving average (ARIMA) models on the other. In a linear Gaussian world, the reduced form of an STM is an ARIMA model and questions regarding the merits of STMs for forecasting revolve round the gains, or losses, from the implied restrictions on the reduced form and the guidance, or lack of it, given to the selection of a suitable model (see the discussion in Harvey, 2006). Once nonlinearity and non-Gaussianity enter the picture, the two approaches can be very different. Models motivated solely by forecasting tend to be set up in terms of a distribution for the current observations conditional on past observations rather than in terms of components. For example, changing variance can be captured by a model from the generalized autoregressive conditional heteroscedasticity (GARCH) class, where conditional variance is a function of past observations, as opposed to a stochastic volatility (SV) model in which the variance is a dynamic unobserved component. The readings in Shephard (2005) describe SV models and discuss the use of computationally intensive methods for estimating them.

The realization that the statistical treatment of a wide range of dynamic models can be dealt with directly in a unified framework is important. For engineers, using state space methods is a natural way to proceed. For many economists, brought up with regression and autoregression, state space is an alien concept. This is changing. State space methods are now becoming an important part of the toolkit of econometricians and economists.

ANDREW HARVEY

See also **data filters; Kalman and particle filtering; prediction formulas.**

Bibliography

Durbin, J. and Koopman, S. 2001. *Time Series Analysis by State Space Methods.* Oxford: Oxford University Press.

Harvey, A. 1989. *Forecasting, Structural Time Series Models and Kalman Filter.* Cambridge: Cambridge University Press.

Harvey, A. 2006. Forecasting with unobserved components time series models. In *Handbook of Economic Forecasting*, vol. 1, ed. G. Elliot, C. Granger and A. Timmermann. Amsterdam: North-Holland.

Harvey, A. and Chung, C.-H. 2000. Estimating the underlying change in unemployment in the UK (with discussion). *Journal of the Royal Statistical Society, Series A* 163, 303–39.

Harvey, A. and de Rossi, G. 2006. Signal extraction. In *Palgrave Handbook of Econometrics*, vol. 1, ed. K. Patterson and T. Mills. Basingstoke: Palgrave Macmillan.

Harvey, A., Trimbur, T. and van Dijk, H. 2007. Trends and cycles in economic time series: a Bayesian approach. *Journal of Econometrics* 140(2), 618–49.

Kohn, R., Ansley, C. and Wong, C.-H. 1992. Nonparametric spline regression with autoregressive moving average errors. *Biometrika* 79, 335–46.

Koopman, S. and Harvey, A. 2003. Computing observation weights for signal extraction and filtering. *Journal of Economic Dynamics and Control* 27, 1317–33.

Kuttner, K. 1994. Estimating potential output as a latent variable. *Journal of Business and Economic Statistics* 12, 361–8.

Orphanides, A. and van Norden, S. 2002. The unreliability of output gap estimates in real-time. *Review of Economics and Statistics* 84, 569–83.

Pfeffermann, D. 1991. Estimation and seasonal adjustment of population means using data from repeated surveys. *Journal of Business and Economic Statistics* 9, 163–75.

Sargent, T. 1989. Two models of measurements and the investment accelerator. *Journal of Political Economy* 97, 251–87.

Smets, F. and Wouter, R. 2003. An estimated dynamic stochastic general equilibrium model of the euro area. *Journal of the European Economic Association* 1, 1123–75.

Shephard, N. 2005. *Stochastic Volatility.* Oxford: Oxford University Press.

Whittle, P. 1984. *Prediction and Regulation*, 2nd edn. Blackwell: Oxford.

stochastic volatility models

Stochastic volatility (SV) is the main concept used in the fields of financial economics and mathematical finance to deal with the endemic time-varying volatility and codependence found in financial markets. Such dependence has been known for a long time; early commentators include Mandelbrot (1963) and Officer (1973). It was also clear to the founding fathers of modern continuous time finance that homogeneity was an unrealistic if convenient simplification; for example, Black and Scholes (1972, p. 416) wrote, '… there is evidence of non-stationarity in the variance. More work must be done to predict variances using the information available.' Heterogeneity has deep implications for the theory and practice of financial economics and econometrics. In particular, asset pricing theory is dominated by the idea that higher rewards may be expected when we face higher risks, but these risks change through time in complicated ways. Some of the changes in the level of risk can be modelled stochastically, where the level of volatility and degree of codependence between assets is allowed to change over time. Such models allow us to explain, for example, empirically observed departures from Black–Scholes–Merton prices for options and understand why we should expect to see occasional dramatic moves in financial markets.

The outline of this article is as follows. In the first section I trace the origins of SV and provide links with the basic models used today in the literature. In the second section I briefly discuss some of the innovations in the second generation of SV models. In the third section I briefly discuss the literature on conducting inference for SV models. In the fourth section I talk about the use of SV to price options. In the fifth section I consider the connection of SV with realized volatility. An extensive review of this literature is given in Shephard (2005).

The origin of SV models

The origins of SV are messy. I will give five accounts, which attribute the subject to different sets of people. Clark (1973) introduced Bochner's (1949) time-changed Brownian motion (BM) into financial economics. He wrote down a model for the log-price M as

$$M_t = W_{\tau_t}, \quad t \geq 0, \tag{1}$$

where W is Brownian motion (BM), t is continuous time, τ is a time change and $W \perp\!\!\!\perp \tau$, where $\perp\!\!\!\perp$ denotes independence. The definition of a time-change is a non-negative process with non-decreasing sample paths, although Clark also assumed τ has independent increments. Then $M_t|\tau_t \sim N(0, \tau_t)$. Further, so long (for each t) as $E\sqrt{\tau_t} < \infty$, then M is a martingale (written $M \in \mathcal{M}$) for this is necessary and sufficient to ensure that $E|M_t| < \infty$. More generally, if (for each t) $\tau_t < \infty$, then M is a local martingale (written $M \in \mathcal{M}_{loc}$). Hence Clark was solely modelling the instantly risky component of the log of an asset price, written Y, which in modern semimartingale (written $Y \in \mathcal{SM}$) notation we would write as

$$Y = A + M.$$

The increments of A can be thought of as the instantly available reward component of the asset price, which compensates the investor for being exposed to the risky increments of M. The A process is assumed to be of finite variation (written $A \in \mathcal{FV}$).

To the best of my understanding, the first published direct volatility clustering SV paper is that by Taylor (1982). His discrete time model of daily returns, computed as the difference of log-prices

$$y_i = Y_i - Y_{i-1}, \quad i = 1, 2, \ldots,$$

where I have assumed that $t = 1$, represents one day to simplify the exposition. He modelled the risky part of returns, $m_i = M_i - M_{i-1}$, as a product process

$$m_i = \sigma_i \varepsilon_i. \tag{2}$$

Taylor assumed ε has a mean of zero and unit variance, while σ is some non-negative process, finishing the model by assuming $\varepsilon \perp\!\!\!\perp \sigma$. Taylor modelled ε as an autoregression and

$$\sigma_i = \exp(h_i/2),$$

where h is a non-zero mean Gaussian linear process. The leading example of this is the first order autoregression

$$h_{i+1} = \mu + \phi(h_i - \mu) + \eta_i, \quad \eta_i \sim NID(0, \sigma^2). \tag{3}$$

In the modern SV literature the model for ε is typically simplified to an i.i.d. process, for we deal with the predictability of asset prices through the A process rather than via M. This is now often called the log-normal SV model in the case where ε is also assumed to be Gaussian. In general, M is always a local martingale.

A key feature of SV, which is not discussed by Taylor, is that it can deal with leverage effects. Leverage effects are associated with the work of Black (1976) and Nelson (1991), and can be implemented in discrete time SV models by negatively correlating the Gaussian ε_i and η_i. This still implies that $M \in \mathcal{M}_{loc}$, but allows the direction of returns to influence future movements in the volatility process, with falls in prices associated with rises in subsequent volatility.

Taylor's discussion of the product process was pre-dated by a decade in the (until recently) unpublished Rosenberg (1972). Rosenberg introduces product processes, empirically demonstrating that time-varying volatility is partially forecastable and so breaks with the earlier work by Clark. He suggests an understanding of aggregational Gaussianity of returns over increasing time intervals and pre-dates a variety of econometric methods for analysing heteroskedasticity.

In continuous time the product process is the standard SV model

$$M_t = \int_0^t \sigma_s dW_s, \tag{4}$$

where the non-negative spot volatility σ is assumed to have càdlàg sample paths (which means it can possess jumps). The squared volatility process is often called the spot variance.

The first use of continuous-time SV models in financial economics was, to my knowledge, by Johnson (1979), who studied the pricing of options using time-changing volatility models in continuous time (see also Johnson and Shanno, 1987; Wiggins, 1987). The best-known paper in this area is Hull and White (1987). Each of these authors desired to generalize the Black and Scholes (1973) approach to option pricing models to deal with volatility clustering. In the Hull and White approach, σ^2 follows the solution to the univariate SDE

$$d\sigma^2 = \alpha(\sigma^2)dt + \omega(\sigma^2)dB,$$

where B is a second Brownian motion and $\omega(.)$ is a non-negative deterministic function.

The probability literature has demonstrated that SV models and their time-changed BM relatives are fundamental. This theoretical development will be the fifth strand of literature that I think of as representing the origins of modern stochastic volatility research. Suppose we simply assume that $M \in \mathcal{M}^c_{loc}$, a process with continuous local martingale sample paths. Then the celebrated Dambis–Dubins–Schwartz theorem shows that M can be written as a time-changed Brownian motion. Further, the time-change is the quadratic variation (QV) process

$$[M]_t = \text{p} - \lim_{n\to\infty} \sum_{j=1}^{n} (M_{t_j} - M_{t_j-1})(M_{t_j} - M_{t_j-1})', \tag{5}$$

for any sequence of partitions $t_0 = 0 < t_1 < \ldots < t_n = t$ with $\sup_j\{t_j - t_{j-1}\} \to 0$ for $n \to \infty$. What is more, as M has continuous sample paths, so must $[M]$. Under the stronger condition that $[M]$ is absolutely continuous, then M can be written as a stochastic volatility process. This latter result, which is called the martingale representation theorem, is due to Doob (1953). Taken together, this implies that time-changed BMs are canonical in continuous sample path price processes, and SV models are special cases of this class. A consequence of the fact that, for continuous sample path time-change BM, $[M] = \tau$, is that in the SV case

$$[M]_t = \int_0^t \sigma_s^2 ds.$$

The SV framework has an elegant multivariate generalization. In particular, write a p-dimensional price process M as (4) but where σ is a matrix process whose elements are all càdlàg, W is a multivariate BM process. Further $[M]_t = \int_0^t \sigma_s \sigma_s' ds$.

Second-generation model building

Univariate models

General observations

In initial diffusion-based models the volatility was Markovian with continuous sample paths. Research in the late 1990s and early 2000s has shown that more complicated volatility dynamics are needed to model either options data or high frequency return data. Leading extensions to the model are to allow jumps into the volatility SDE (for example, Barndorff-Nielsen and Shephard, 2001; Eraker, Johannes and Polson, 2003) or to model the volatility process as a function of a number of separate stochastic processes or factors (for example, Chernov et al., 2003; Barndorff-Nielsen and Shephard, 2001).

Long memory

In the SV literature considerable progress has been made on working with both discrete and continuous time long-memory SV. This involves specifying a long-memory model for σ in discrete or continuous time.

Breidt, Crato and de Lima (1998) and Harvey (1998) looked at discrete time models where the log of the volatility was modelled as a fractionally integrated process. In continuous time there is work on modelling the log of volatility as fractionally integrated Brownian motion by Comte and Renault (1998). More recent work, which is econometrically easier to deal with, is the square-root model driven by fractionally integrated BM introduced in an influential paper by Comte, Coutin and Renault (2003) and the infinite superposition of non-negative OU processes introduced by Barndorff-Nielsen (2001).

Jumps

In detailed empirical work a number of researchers have supplemented standard SV models by adding jumps to the price process or to the volatility dynamics. Bates (1996) was particularly important as it showed the need to include jumps in addition to SV, at least when volatility is Markovian. Eraker, Johannes and Polson (2003) deals with the efficient inference of these types of models. A radical departure in SV models was put forward by Barndorff-Nielsen and Shephard (2001), who suggested building volatility models out of pure jump processes called non-Gaussian OU processes. Closed form option pricing based on this structure is studied briefly in Barndorff-Nielsen and Shephard (2001) and in detail by Nicolato and Venardos (2003). All these non-Gaussian OU processes are special cases of the affine class advocated by Duffie, Pan and Singleton (2000) and Duffie, Filipovic and Schachermayer (2003).

Multivariate models

Diebold and Nerlove (1989) introduced volatility clustering into traditional factor models, which are used in many areas of asset pricing. In continuous time their type of model has the interpretation

$$M_t = \sum_{j=1}^{J} \int \beta_{(j)s} dF_{(j)s} + G_t,$$

where the factors $F_{(1)}, F_{(2)}, \ldots, F_{(J)}$ are independent univariate SV models and G is correlated multivariate BM. Some of the related papers on the econometrics of this topic include King, Sentana and Wadhwani (1994) and Fiorentini, Sentana and Shephard (2004), who all fit this kind of model. These papers assume that the factor loading vectors are constant through time.

A more limited multivariate discrete time model was put forward by Harvey, Ruiz and Shephard (1994) who allowed $M_t = C\int_0^t \sigma_s dW_s$, where σ is a diagonal matrix process and C is a fixed matrix of constants with a unit leading diagonal. This means that the risky part of prices is simply a rotation of a p-dimensional vector of independent univariate SV processes.

Inference based on return data

Moment-based inference

The task is to carry out inference on $\theta = (\theta_1, \ldots, \theta_K)'$, the parameters of the SV model based on a sequence of returns $y = (y_1, \ldots, y_T)'$. Taylor (1982) and Melino and Turnbull (1990) calibrated their models using the method of moments. Systematic studies, using a GMM approach, of which moments to heavily weight in SV models was given in Andersen and Sørensen (1996), Genon-Catalot, Jeantheau and Larédo (2000), Sørensen (2000) and Hoffmann (2002).

A difficulty with using moment-based estimators for continuous time SV models is that it is not straightforward to compute the moments y. In the case of no leverage, general results for the second order properties of y and their squares were given in Barndorff-Nielsen and Shephard (2001). Some quite general results under leverage are also given in Meddahi (2001).

In the discrete time log-normal SV models the approach advocated by Harvey, Ruiz and Shephard (1994) has been influential. Their approach was to remove the predictable part of the returns, so we think of $Y = M$ again, and work with $\log y_i^2 = h_i + \log \varepsilon_i^2$. If the volatility has short memory, then this form of the model can be handled using the Kalman filter, while long-memory models are often dealt with in the frequency domain. Either way, this delivers a Gaussian quasi-likelihood which can be used to estimate the parameters of the model. The linearized model is non-Gaussian due to the long left-hand tail of $log\, \varepsilon_i^2$ which generates outliers when ε_i is small.

Simulation-based inference

In the 1990s a number of econometricians started to use simulation-based inference to tackle SV models. To discuss these methods it will be convenient to focus on the simplest discrete time log-normal SV model given by (2) and (3).

MCMC allows us to simulate from $\theta, h|y$, where $h = (h_1, \ldots, h_T)'$. Discarding the h draws yields samples from $\theta|y$. Summarizing yields fully efficient parametric inference. In an influential paper, Jacquier, Polson and Rossi (1994) implemented an MCMC algorithm for this problem. A subsequent paper by Kim, Shephard and Chib (1998) gave quite an extensive discussion of various MCMC algorithms. This is a subtle issue and makes a very large difference to the computational efficiency of the methods (see, for example, Jacquier, Polson and Rossi, 2004; Yu, 2005).

Kim, Shephard and Chib (1998) introduced the first filter using a so-called particle filter. As well as being of substantial scientific interest for decision making, the advantage of a filtering method is that it allows us to compute marginal likelihoods for model comparison and one-step-ahead predictions for model testing.

Although MCMC-based papers are mostly couched in discrete time, a key advantage of the general approach is that it can be adapted to deal with continuous time models by the idea of augmentation. This was fully worked out in Elerian, Chib and Shephard (2001), Eraker (2001) and Roberts and Stramer (2001).

A more novel non-likelihood approach was introduced by Smith (1993) and later developed by Gourieroux, Monfort and Renault (1993) and Gallant and Tauchen (1996) into what is now called indirect inference or the efficient method of moments. Here I briefly give a stylized version of this approach.

Suppose there is an auxiliary model for the returns (for example, GARCH) whose density, $g(y;\psi)$, is easy to compute and, for simplicity of exposition, has $\dim(\psi) = \dim(\theta)$. Then compute its MLE, which we write as $\hat{\psi}$. We assume this is a regular problem so that $\partial \log g(y;\hat{\psi})/\partial\psi = 0$ recalling that y is the observed return vector. Simulate a very long process from the SV model using parameters θ, which we denote by y^{+}, and evaluate the score using not the data but this simulation. This produces

$$\frac{\partial \log g(y^{+};\psi)}{\partial\psi}\Big|_{\psi=\hat{\psi}}, y^{+} \sim f(y;\theta).$$

Then move θ around until the score is again zero, but now under the simulation. Write the point where this happens as $\widetilde{\theta}$. It is called the indirect inference estimator.

Options

Models

SV models provide a basis for realistic modelling of option prices. We recall the central role played by Johnson and Shanno (1987) and Wiggins (1987). The best-known paper in this area is by Hull and White (1987), who looked at a diffusion volatility model with leverage effects. They assumed that volatility risk was unrewarded and priced their options either by approximation or by simulation. Hull and White (1987) indicated that SV models could produce smiles and skews in option prices, which are frequency observed in market data. The skew is particularly important in practice, and Renault and Touzi (1996) proved that can be achieved in SV models via leverage effects.

The first analytic option pricing formulae were developed by Stein and Stein (1991) and Heston (1993). The only other closed form solution I know of is the one based on the Barndorff-Nielsen and Shephard (2001) class of non-Gaussian OU SV models. Nicolato and Venardos (2003) provided a detailed study of such option pricing solutions; see also the textbook exposition in Cont and Tankov (2004, ch. 15). Slightly harder computationally to deal with is the more general affine class of models highlighted by Duffie, Filipovic and Schachermayer (2003).

Econometrics of SV option pricing

In theory, option prices themselves should provide rich information for estimating and testing volatility models. I discuss the econometrics of options in the context of the stochastic discount factor (SDF) approach, which has a long history in financial economics and is emphasized in, for example, Cochrane (2001) and Garcia, Ghysels and Renault (2006). For simplicity I assume interest rates are constant. We start with the standard Black–Scholes (BS) problem, which will take a little time to recall, before being able to rapidly deal with the SV extension. We model

$$\mathrm{d}\log Y = (r + p - \sigma^2/2)\mathrm{d}t + \sigma \mathrm{d}W, \quad \mathrm{d}\log \widetilde{M} = h\mathrm{d}t + b\mathrm{d}W,$$

where $\widetilde{M}$ is the SDF process and r the riskless short rate, and σ, h, b and p, the risk premium, are assumed constant for the moment.

We price all contingent payoffs $g(Y_T)$ as $C_t = E(\widetilde{M}_T/\widetilde{M}_t)g(Y_T)|\mathscr{F}_t$, the expected discounted value of the claim where $T > t$. For this model to make financial sense we require that $\widetilde{M}_t Y_t$ and $\widetilde{M}_t \exp(tr)$ are local martingales, which is enough to mean that adding other independent BMs to the $\log \widetilde{M}$ process makes no difference to C or Y, the observables. These two constraints imply, respectively, $p + b\sigma = 0$ and $h = -r - b^2/2$. This means that (C^{BS}, Y) is driven by a single W.

When we move to the standard SV model we can remove this degeneracy. The functional form for the SV Y process is unchanged, but we now allow

$$\mathrm{d}\log \widetilde{M} = h\mathrm{d}t + a\mathrm{d}B + b\mathrm{d}W, \qquad \mathrm{d}\sigma^2 = \alpha \mathrm{d}t + \omega \mathrm{d}B,$$

where we assume that $B \perp\!\!\!\perp W$ to simplify the exposition. The SV structure means that p will have to change through time in response to the moving σ^2. B is again redundant in the SDF (but not in the volatility) so the usual SDF conditions again imply $h = -r - \frac{1}{2}a^2$ and $p + b\sigma = 0$. This implies that the move to the SV case has little impact, except that the sample path of $\sigma^2 \perp\!\!\!\perp W$. So the generalized BS (GBS) price is

$$C_t^{GBS}(\sigma_t^2) = E\left(\frac{\widetilde{M}_T}{\widetilde{M}_t} g(Y_T)|\mathscr{F}_t\right) = E\left\{C_t^{BS}\left(\frac{1}{T-t}\int_t^T \sigma_u^2 du\right)|\sigma_t^2, Y_t\right\}.$$

Now C^{GBS} is a function of both Y_t and σ_t^2, which means that (C^{GBS}, Y) is not degenerate. From an econometric viewpoint this is an important step, meaning inference on options is just the problem of making inference on a complicated bivariate diffusion process. When we allow leverage back into the model, the analysis becomes slightly more complicated algebraically.

In some recent work econometricians have been trying to use data from underlying assets and option markets to jointly model the dynamics of (C^{GBS}, Y). The advantage of this joint estimation is that we can pool information across data types and estimate all relevant effects which influence Y, σ^2 and $\widetilde{M}$. Relevant papers include Chernov and Ghysels (2000), Pastorello, Patilea and Renault (2003), Das and Sundaram (1999) and Bates (2000).

Realized volatility

The advent of very informative high-frequency data has prompted econometricians to study estimators of the increments of the quadratic variation (QV) process and then to use this estimate to project QV into the future in order to predict future levels of volatility. The literature on this starts with independent, concurrent papers by Andersen and Bollerslev (1998), Barndorff-Nielsen and Shephard (2001) and Comte and Renault (1998). Some of this work echoes earlier important contributions from, for example, Rosenberg (1972) and Merton (1980).

A simple estimator of $[Y]$ is the *realized QV process*

$$[Y_\delta]_t = \sum_{j=1}^{\lfloor t/\delta \rfloor} (Y_{\delta j} - Y_{\delta(j-1)})(Y_{\delta j} - Y_{\delta(j-1)})',$$

thus as $\delta \downarrow 0$ so $[Y_\delta]_t \xrightarrow{p} [Y]_t$. If $A \in \mathscr{F}V^c$, then $[Y] = [M]$, while if we additionally assume that M is SV then $[Y_\delta]_t \xrightarrow{p} \int_0^t \sigma_s \sigma_s' ds$.

In practice it makes sense to look at the increments of the QV process. Suppose we are interested in analysing daily return data, but in addition have higher-frequency data measured at the time interval δ. The i-th daily realized QV is defined as

$$V(Y_\delta)_i = \sum_{j=1}^{\lfloor 1/\delta \rfloor} (Y_{i+\delta j} - Y_{i+\delta(j-1)}) \times (Y_{i+\delta j} - Y_{i+\delta(j-1)})' \xrightarrow{p} V(Y)_i = [Y]_i - [Y]_{i-1},$$

the i-th daily QV. The diagonal elements of $V(Y_\delta)_i$ are called realized variances and their square roots are called realized volatilities.

Andersen et al. (2001) have shown that to forecast the volatility of future asset returns a key input should be predictions of future daily QV. Recall, from Ito's formula, that, if $Y \in \mathscr{S}\mathscr{M}^c$ and $M \in \mathscr{M}$, then writing F_t as the filtration generated by the continuous history of Y up to time t then

$$E(y_i y_i' | \mathscr{F}_{i-1}) \simeq E(V(Y)_i | \mathscr{F}_{i-1}).$$

A review of some of this material is given by Barndorff-Nielsen and Shephard (2006a).

A difficulty with this line of argument is that the QV theory tells us only that $V(Y_\delta)_i \xrightarrow{p} V(Y)_i$; it gives no impression of the size of $V(Y_\delta)_i - V(Y)_i$. Jacod (1994) and Barndorff-Nielsen and Shephard (2002) have strengthened the consistency result to provide a univariate central limit theory

$$\frac{\delta^{-1/2}([Y_\delta]_t - [Y]_t)}{\sqrt{2\int_0^t \sigma_s^4 ds}} \xrightarrow{d} N(0,1),$$

while giving a method for consistently estimating the integrated quarticity $\int_0^t \sigma_s^4 ds$ by using high-frequency data. This analysis was generalized to the multivariate case by Barndorff-Nielsen and (Shephard 2004a).This type of analysis greatly simplifies parametric estimation of SV models, for we can now have estimates of the volatility quantities SV models directly parameterize. Barndorff-Nielsen and Shephard (2002), Bollerslev and Zhou (2002) and Phillips and Yu (2005) study this topic from different perspectives.

Recently there has been interest in studying the impact of market microstructure effects on the estimates of realized covariation. This causes the estimator of the QV to become biased. Leading papers on this topic are Zhou (1996), Fang (1996), Bandi and Russell (2003), Hansen and Lunde (2006) and Zhang, Mykland and Aït-Sahalia (2005). Further, one can estimate the QV of the continuous component of prices in the presence of jumps using the so-called realized bipower variation process. This was introduced by Barndorff-Nielsen and Shephard (2004b; 2006b).

NEIL SHEPHARD

My research is supported by the Economic and Social Science Research Council (UK) through the grant 'High frequency financial econometrics based upon power variation'.

Bibliography

Andersen, T. and Bollerslev, T. 1998. Answering the skeptics: yes, standard volatility models do provide accurate forecasts. *International Economic Review* 39, 885–905.

Andersen, T., Bollerslev, T., Diebold, F. and Labys, P. 2001. The distribution of exchange rate volatility. *Journal of the American Statistical Association* 96, 42–55. Correction published in vol. 98 (2003), p. 501.

Andersen, T. G. and Sørensen, B. 1996. GMM estimation of a stochastic volatility model: a Monte Carlo study. *Journal of Business and Economic Statistics* 14, 328–52.

Bandi, F. and Russell, J. 2003. Microstructure noise, realized volatility, and optimal sampling. Mimeo. Graduate School of Business, University of Chicago.

Barndorff-Nielsen, O. 2001. Superposition of Ornstein–Uhlenbeck type processes. *Theory of Probability and its Applications* 45, 175–94.

Barndorff-Nielsen, O. and Shephard, N. 2001. Non-Gaussian Ornstein–Uhlenbeck-based models and some of their uses in financial economics (with discussion). *Journal of the Royal Statistical Society, Series B* 63, 167–241.

Barndorff-Nielsen, O. and Shephard, N. 2002. Econometric analysis of realised volatility and its use in estimating stochastic volatility models. *Journal of the Royal Statistical Society, Series B* 64, 253–80.

Barndorff-Nielsen, O. and Shephard, N. 2004a. Econometric analysis of realised covariation: high frequency covariance, regression and correlation in financial economics. *Econometrica* 72, 885–925.

Barndorff-Nielsen, O. and Shephard, N. 2004b. Power and bipower variation with stochastic volatility and jumps (with discussion). *Journal of Financial Econometrics* 2, 1–48.

Barndorff-Nielsen, O. and Shephard, N. 2006a. Variation, jumps, and high frequency data in financial econometrics. In *Advances in Economics and Econometrics: Theory and Applications*, vol. 1, ed. R. Blundell, P. Torsten and W. Newey. Cambridge: Cambridge University Press.

Barndorff-Nielsen, O. and Shephard, N. 2006b. Econometrics of testing for jumps in financial economics using bipower variation. *Journal of Financial Econometrics* 4, 1–30.

Bates, D. 1996. Jumps and stochastic volatility: exchange rate processes implicit in deutsche mark options. *Review of Financial Studies* 9, 69–107.

Bates, D. 2000. Post-'97 crash fears in the S-&P 500 futures option market. *Journal of Econometrics* 94, 181–238.

Black, F. 1976. Studies of stock price volatility changes. *Proceedings of the Business and Economic Statistics Section, American Statistical Association*, 177–81.

Black, F. and Scholes, M. 1972. The valuation of option contracts and a test of market efficiency. *Journal of Finance* 27, 399–418.

Black, F. and Scholes, M. 1973. The pricing of options and corporate liabilities. *Journal of Political Economy* 81, 637–54.

Bochner, S. 1949. Diffusion equation and stochastic processes. *Proceedings of the National Academy of Science of the United States of America* 85, 369–70.

Bollerslev, T. and Zhou, H. 2002. Estimating stochastic volatility diffusion using conditional moments of integrated volatility. *Journal of Econometrics* 109, 33–65.

Breidt, F., Crato, N. and de Lima, P. 1998. On the detection and estimation of long memory in stochastic volatility. *Journal of Econometrics* 83, 325–48.

Chernov, M. and Ghysels, E. 2000. A study towards a unified approach to the joint estimation of objective and risk neutral measures for the purpose of options valuation. *Journal of Financial Economics* 56, 407–58.

Chernov, M., Gallant, A., Ghysels, E. and Tauchen, G. 2003. Alternative models of stock price dynamics. *Journal of Econometrics* 116, 225–57.

Clark, P. 1973. A subordinated stochastic process model with fixed variance for speculative prices. *Econometrica* 41, 135–56.

Cochrane, J. 2001. *Asset Pricing*. Princeton: Princeton University Press.

Comte, F., Coutin, L. and Renault, E. 2003. Affine fractional stochastic volatility models. Mimeo. University of Montreal.

Comte, F. and Renault, E. 1998. Long memory in continuous-time stochastic volatility models. *Mathematical Finance* 8, 291–323.

Cont, R. and Tankov, P. 2004. *Financial Modelling with Jump Processes*. London: Chapman and Hall.

Das, S. and Sundaram, R. 1999. Of smiles and smirks: a term structure perspective. *Journal of Financial and Quantitative Analysis* 34, 211–40.

Diebold, F. and Nerlove, M. 1989. The dynamics of exchange rate volatility: a multivariate latent factor ARCH model. *Journal of Applied Econometrics* 4, 1–21.

Doob, J. 1953. *Stochastic Processes*. New York: John Wiley and Sons.

Duffie, D., Filipovic, D. and Schachermayer, W. 2003. Affine processes and applications in finance. *Annals of Applied Probability* 13, 984–1053.

Duffie, D., Pan, J. and Singleton, K. 2000. Transform analysis and asset pricing for affine jump-diffusions. *Econometrica* 68, 1343–76.

Elerian, O., Chib, S. and Shephard, N. 2001. Likelihood inference for discretely observed non-linear diffusions. *Econometrica* 69, 959–93.

Eraker, B. 2001. Markov chain Monte Carlo analysis of diffusion models with application to finance. *Journal of Business and Economic Statistics* 19, 177–91.

Eraker, B., Johannes, M. and Polson, N. 2003. The impact of jumps in returns and volatility. *Journal of Finance* 53, 1269–300.

Fang, Y. 1996. Volatility modeling and estimation of high-frequency data with Gaussian noise. Ph.D. thesis. Sloan School of Management, MIT.

Fiorentini, G., Sentana, E. and Shephard, N. 2004. Likelihood-based estimation of latent generalised ARCH structures. *Econometrica* 12, 1481–517.

Gallant, A. and Tauchen, G. 1996. Which moments to match. *Econometric Theory* 12, 657–81.

Garcia, R., Ghysels, E. and Renault, E. 2006. The econometrics of option pricing. In *Handbook of Financial Econometrics*, ed. Y. Aït-Sahalia and L. Hansen. Amsterdam: North-Holland.

Genon-Catalot, V., Jeantheau, T. and Larédo, C. 2000. Stochastic volatility as hidden Markov models and statistical applications. *Bernoulli* 6, 1051–79.

Gourieroux, C., Monfort, A. and Renault, E. 1993. Indirect inference. *Journal of Applied Econometrics* 6, S85–S118.

Hansen, P. and Lunde, A. 2006. Realized variance and market microstructure noise (with discussion). *Journal of Business and Economic Statistics* 24, 127–61.

Harvey, A. 1998. Long memory in stochastic volatility. In *Forecasting Volatility in Financial Markets*, ed. J. Knight and S. Satchell. Oxford: Butterworth-Heinemann.

Harvey, A., Ruiz, E. and Shephard, N. 1994. Multivariate stochastic variance models. *Review of Economic Studies* 61, 247–64.

Heston, S. 1993. A closed-form solution for options with stochastic volatility, with applications to bond and currency options. *Review of Financial Studies* 6, 327–43.

Hoffmann, M. 2002. Rate of convergence for parametric estimation in stochastic volatility models. *Stochastic Processes and their Application* 97, 147–70.

Hull, J. and White, A. 1987. The pricing of options on assets with stochastic volatilities. *Journal of Finance* 42, 281–300.

Jacod, J. 1994. Limit of random measures associated with the increments of a Brownian semimartingale. Preprint No. 120. Laboratoire de Probabilitiés, Université Pierre et Marie Curie, Paris.

Jacquier, E., Polson, N. and Rossi, P. 1994. Bayesian analysis of stochastic volatility models (with discussion). *Journal of Business and Economic Statistics* 12, 371–417.

Jacquier, E., Polson, N. and Rossi, P. 2004. Stochastic volatility models: univariate and multivariate extensions. *Journal of Econometrics* 122, 185–212.

Johnson, H. 1979. Option pricing when the variance rate is changing. Working paper. University of California, Los Angeles.

Johnson, H. and Shanno, D. 1987. Option pricing when the variance is changing. *Journal of Financial and Quantitative Analysis* 22, 143–51.

Kim, S., Shephard, N. and Chib, S. 1998. Stochastic volatility: likelihood inference and comparison with ARCH models. *Review of Economic Studies* 65, 361–93.

King, M., Sentana, E. and Wadhwani, S. 1994. Volatility and links between national stock markets. *Econometrica* 62, 901–33.

Mandelbrot, B. 1963. The variation of certain speculative prices. *Journal of Business* 36, 394–419.

Meddahi, N. 2001. An eigenfunction approach for volatility modeling. Cahiers de recherche No. 2001–29. Department of Economics, University of Montreal.

Melino, A. and Turnbull, S. 1990. Pricing foreign currency options with stochastic volatility. *Journal of Econometrics* 45, 239–65.

Merton, R. 1980. On estimating the expected return on the market: an exploratory investigation. *Journal of Financial Economics* 8, 323–61.
Nelson, D. 1991. Conditional heteroskedasticity in asset pricing: a new approach. *Econometrica* 59, 347–70.
Nicolato, E. and Venardos, E. 2003. Option pricing in stochastic volatility models of the Ornstein–Uhlenbeck type. *Mathematical Finance* 13, 445–66.
Officer, R. 1973. The variability of the market factor of the New York stock exchange. *Journal of Business* 46, 434–53.
Pastorello, S., Patilea, V. and Renault, E. 2003. Iterative and recursive estimation in structural non-adaptive models. *Journal of Business and Economic Statistics* 21, 449–509.
Phillips, P. and Yu, J. 2005. A two-stage realized volatility approach to the estimation for diffusion processes from discrete observations. Discussion Paper No. 1523. Cowles Foundation, Yale University.
Renault, E. and Touzi, N. 1996. Option hedging and implied volatilities in a stochastic volatility model. *Mathematical Finance* 6, 279–302.
Roberts, G. and Stramer, O. 2001. On inference for nonlinear diffusion models using the Hastings–Metropolis algorithms. *Biometrika* 88, 603–21.
Rosenberg, B. 1972. The behaviour of random variables with nonstationary variance and the distribution of security prices. Working paper 11, Graduate School of Business Administration, University of California, Berkeley. Reprinted in N. Shephard (2005).
Shephard, N. 2005. *Stochastic Volatility: Selected Readings*. Oxford: Oxford University Press.
Smith, A. 1993. Estimating nonlinear time series models using simulated vector autoregressions. *Journal of Applied Econometrics* 8, S63–S84.
Sørensen, M. 2000. Prediction based estimating equations. *Econometrics Journal* 3, 123–47.
Stein, E. and Stein, J. 1991. Stock price distributions with stochastic volatility: an analytic approach. *Review of Financial Studies* 4, 727–52.
Taylor, S. 1982. Financial returns modelled by the product of two stochastic processes – a study of daily sugar prices 1961–79. In *Time Series Analysis: Theory and Practice*, vol. 1, ed. O. Anderson. Amsterdam: North-Holland.
Wiggins, J. 1987. Option values under stochastic volatilities. *Journal of Financial economics* 19, 351–72.
Yu, J. 2005. On leverage in a stochastic volatility model. *Journal of Econometrics* 127, 165–78.
Zhang, L., Mykland, P. and Aït-Sahalia, Y. 2005. A tale of two time scales: determining integrated volatility with noisy high-frequency data. *Journal of the American Statistical Association* 100, 1394–411.
Zhou, B. 1996. High-frequency data and volatility in foreign-exchange rates. *Journal of Business and Economic Statistics* 14, 45–52.

structural change, econometrics of

This article covers methodological issues related to estimation, testing and computation for models involving structural changes. The amount of work on this subject since the 1950s is truly voluminous in both the statistics and econometrics literature. Accordingly, any survey is bound to focus on specific aspects. Our aim is to review developments as they relate to econometric applications based on linear models. Recently, substantial advances have been made to cover models at a level of generality that allows a host of interesting practical applications. These include models with general stationary regressors and errors that can exhibit temporal dependence and heteroskedasticity, models with trending variables and possible unit roots and cointegrated models, among others. Advances have been made pertaining to computational aspects of constructing estimates, their limit distributions, tests for structural changes, and methods to determine the number of changes present. For a more extensive review the reader is referred to Perron (2006).

We consider the following multiple linear regression with m breaks (or $m+1$ regimes):

$$y_t = x_t'\beta + z_t'\delta_j + u_t, \qquad t = T_{j-1}+1, \ldots, T_j, \tag{1}$$

for $j=1,\ldots, m+1$. In this model, y_t is the observed dependent variable; both x_t ($p \times 1$) and z_t ($q \times 1$) are vectors of covariates and β and δ_j ($j=1,\ldots,m+1$) are the corresponding vectors of coefficients; u_t is the disturbance. The break dates ($T_1,\ldots, T_m$) are explicitly treated as unknown (the convention that $T_0=0$ and $T_{m+1}=T$ is used). The purpose is to estimate the unknown regression coefficients together with the break points when T observations on (y_t, x_t, z_t) are available. This is a partial structural change model since the parameter vector β is not subject to shifts. When $p=0$, we obtain a pure structural change model where all coefficients are subject to change. Note that using a partial structural change model can be beneficial in terms of obtaining more precise estimates and having more powerful tests.

The estimates are obtained by minimizing the overall sum of squared residuals

$$\sum_{i=1}^{m+1} \sum_{t=T_{i-1}+1}^{T_i} [y_t - x_t'\beta - z_t'\delta_i]^2.$$

Let $\hat{\beta}(\{T_j\})$ and $\hat{\delta}(\{T_j\})$ denote the estimates based on the given m-partition ($T_1,\ldots, T_m$) denoted $\{T_j\}$. Substituting these in the objective function and denoting the resulting sum of squared residuals as $S_T(T_1,\ldots, T_m)$, the estimated break points ($\hat{T}_1, \ldots, \hat{T}_m$) are such that

$$(\hat{T}_1, \ldots, \hat{T}_m) = \operatorname{argmin}_{(T_1,\ldots,T_m)} S_T(T_1, \ldots, T_m), \tag{2}$$

with the minimization taken over a set of admissible partitions (see below). The parameter estimates are those associated with the partition $\{\hat{T}_j\}$, that is

$$\hat{\beta} = \hat{\beta}(\{\hat{T}_j\}), \quad \hat{\delta} = \hat{\delta}(\{\hat{T}_j\}).$$

This framework includes many contributions as special cases depending on the assumptions imposed; for example, single change, changes in the mean of a stationary process, and so on. However, since estimation is based on the least-squares principle, even if changes in the variance of u_t are allowed at the same dates as the breaks in the parameters of the regression they are not exploited to increase the precision of the break date estimators unless a quasi-likelihood framework is adopted (see below).

The assumptions and their relevance

To obtain theoretical results about the consistency and limit distribution of estimates of the break dates, some conditions need to be imposed on the regressors, the errors, the set of admissible partitions and the break dates. To our knowledge, the most general set of assumptions in the case of weakly stationary regressors are those in Perron and Qu (2006). Some are simply technical (for example, invertibility requirements), while others restrict the potential applicability of the results.

The assumptions on the regressors specifies that for $w_t = (x_t', z_t')'$, $(1/l_i)\sum_{t=T_i^0+1}^{T_i^0+[l_i v]} w_t w_t' \rightarrow_p Q_i(v)$ a non-random positive definite matrix uniformly in $v \in [0,1]$. It allows their distribution to vary across regimes. It, however, requires the data to be weakly stationary stochastic processes. This can be relaxed on a case-by-case basis though the proofs then depend on the nature of the relaxation. For instance the scaling used forbids trending regressors, unless they are of the form $\{1, (t/T), \ldots, (t/T)^p\}$, say, for a polynomial trend of order p. Casting trend functions in this form can deliver useful results in many cases. However, there are instances where specifying trends in unscaled form, that is, $\{1, t, \ldots, t^p\}$, can deliver much better results, especially if level and trend slope changes occur jointly. Results using unscaled trends with $p = 1$ are presented in Perron and Zhu (2005). A comparison of their results with other trend specifications is presented in Deng and Perron (2006).

Another important restriction is implied by the requirement that the limit be a fixed, as opposed to stochastic, matrix. This, along with the scaling, precludes integrated processes as regressors (that is, unit roots). In the single break case, this has been relaxed by Bai, Lumsdaine and Stock (1998) who considered structural changes in cointegrated relationships in system of equations. Kejriwal and Perron (2006a) provide general results for multiple structural changes in a single cointegrating vector. Consistency still applies but the rate of convergence and limit distributions of the estimates are different.

The assumptions on u_t and $\{w_t u_t\}$ impose mild restrictions on the vector $w_t u_t$ and permits a wide class of potential correlation and heterogeneity (including conditional heteroskedasticity) and lagged dependent variables. It rules out errors that have unit roots. However, unit root errors can be of interest, for example when testing for a change in the deterministic component of the trend function for an integrated series, in which case the estimates are consistent (see Perron and Zhu, 2005). The set of conditions is not the weakest possible. For example, Lavielle and Moulines (2000) allow the errors to be strongly dependent (long memory processes) but consider only the case of multiple changes in the mean.

It is also assumed that the minimization problem defined by (2) is taken over all partitions such that $T_i - T_{i-1} \geq \varepsilon T$ for some $\varepsilon > 0$. This is not restrictive in practice since ε can be small. Another assumption specifies that the break dates are asymptotically distinct, that is, we have $T_i^0 = [T\lambda_i^0]$, where $0 < \lambda_1^0 < \ldots < \lambda_m^0 < 1$. It dictates the asymptotic framework adopted, whereby all segments increase in length in the same proportions as T increases.

Under these conditions, the break fractions λ_i^0 are consistently estimated, that is, $\hat{\lambda}_i \equiv (\hat{T}_i/T) \rightarrow_p \lambda_i^0$ and that the rate of convergence is T. Note that the estimates of the break dates are not consistent themselves, but the differences between the estimates and the true values are bounded by some constant, in probability. Also, this implies that the estimates of the other parameters have the same distribution as would prevail if the break dates were known. Kejriwal and Perron (2006a) obtain similar results with $I(1)$ regressors for a cointegrated model subject to multiple changes, using the static regression or a dynamic regression augmented with leads and lags of the first differences of the $I(1)$ regressors.

Allowing for restrictions on the parameters

Perron and Qu (2006) consider the issues in a broader framework whereby arbitrary linear restrictions on the parameters of the conditional mean can be imposed in the estimation. The class of models considered is

$$y_t = z_t'\delta_j + u_t, \quad t = T_{j-1} + 1, \ldots, T_j,$$

where $R\delta = r$, with R a k by $(m+1)q$ matrix with rank k and r, a k dimensional vector of constants. The assumptions are the same as discussed above. There is no need for a distinction between variables whose coefficients are allowed to change and those whose coefficients are not allowed to change. A partial structural change model is obtained specifying restrictions that impose some coefficients to be identical across all regimes. This is a useful generalization since it permits a wider class of models of practical interests – for example, a model with a specific number of states less than the number of regimes, or one where a subset of coefficients may be allowed to change over only a limited number of regimes. Perron and Qu (2006) show that the same consistency and rate of convergence results hold. Moreover, the limit distribution of the estimates of the break dates are unaffected by the imposition of valid restrictions, but improvements can be obtained in finite samples. The main advantages of imposing restrictions are that more powerful tests and more precise estimates are obtained.

Method to compute global minimizers

To estimate the model, we need global minimizers of the objective function (2). A standard grid search requires least squares operations of order $O(T^m)$ and becomes prohibitive when the number of breaks is greater than two. Bai and Perron (2003a) discuss a method based on a dynamic programming algorithm that is very efficient (see also Hawkins, 1976). Indeed, the additional computing time needed to estimate more than two break dates is marginal compared with the time needed to estimate a two break model. Consider the case of a pure structural change model. The basic idea is that the total number of possible segments is at most $T(T+1)/2$ and is therefore of order $O(T^2)$. One then needs a method to select which combination of segments yields a minimal value of the objective function. This is achieved efficiently using a dynamic programming algorithm. For models with restrictions (including the partial structural change model), an iterative procedure is available, which in most cases requires very few iterations (see also Perron and Qu, 2006). Hence, even with large samples, the computing cost is small.

The limit distribution of the estimates of the break dates

With the assumptions on the regressors, the errors and given the asymptotic framework adopted, the limit distributions of the estimates of the break dates are independent of each other. Hence, for each break date, the analysis is the same as that of a single break. This holds because the distance between each break increases at rate T, and the mixing conditions on the regressors and errors impose a short memory property so that events that occur a long time apart are independent. This independence property does not hold with integrated data (see below).

The limit distribution of the estimates of the break dates depends on: (*a*) the magnitude of the change in coefficients (with larger changes leading to higher precision); (*b*) the (limit) sample moment matrices of the regressors for the segments pre and post break (allowed to be different); (*c*) the so-called 'long-run' variance of $\{w_t u_t\}$, which accounts for serial correlation in the errors (also allowed to be different pre and post break); (*d*) whether the regressors are trending or not. In all cases, the nuisance parameters can be consistently estimated and appropriate confidence intervals constructed, which need not be symmetric given that the data and errors can have different properties before and after the break.

A feature of the limit distribution is that, for given fixed magnitude of change, it depends on the finite sample distribution of the errors. To get rid of this dependence, the asymptotic framework is modified with the change in parameters getting smaller as T increases, but slowly enough for the estimated break fraction to remain consistent. The limit distribution obtained in Bai (1997a) and Bai and Perron (1998) applies to the case with no trending regressors. With trending regressors, a similar result is still possible (on the assumption of trends of the form (t/T)) and the reader is referred to Bai (1997a) for the case where z_t is a polynomial time trend. For an unscaled linear trend, see Perron and Zhu (2005).

The simulations in Bai and Perron (2006) show that the shrinking shifts asymptotic framework provides useful approximations to the finite sample distributions. The coverage rates are adequate, in general, unless the shifts are quite small, in which case the confidence interval is too narrow. But in such cases, the breaks are unlikely to be detected by test procedures. On the other hand, Deng and Perron (2006) show that the shrinking shift asymptotic framework leads to a poor approximation in the context of a change in a linear trend and that the limit distribution based on a fixed magnitude of shift is preferable.

In a cointegrating regression with I(1) variables, Kejriwal and Perron (2006a) show that, if the coefficients of the integrated regressors are allowed to change, the estimated break fractions are asymptotically dependent so that confidence intervals need to be constructed jointly. Methods to construct such confidence intervals are discussed. If, however, only the intercept and/or the coefficients of the stationary regressors are allowed to change, the estimates of the break dates are asymptotically independent.

Estimating breaks one at a time

Bai (1997b) and Bai and Perron (1998) showed that it is possible to consistently estimate all break fractions sequentially, that is, one at a time. When estimating a single break model in the presence of multiple breaks, the estimate of the break fraction will converge to one of the true break fractions, the one that is dominant in the sense that taking it into account allows the greatest reduction in the sum of squared residuals. Then, allowing for a break at the estimated value, a one-break model can be applied to each segment which will consistently estimate the second dominating break, and so on.

Bai (1997b) considers the limit distribution of the estimates and shows that they are not the same as those obtained when estimating all break dates simultaneously. Except for the last estimated break date, the limit distributions depend on the parameters in all segments of the sample. To remedy this problem, he suggested a repartition procedure, which re-estimates each break date conditional on the adjacent break dates. The limit distribution is then the same as when the break dates are estimated simultaneously.

Estimation in a system of regressions

Estimating structural changes in a system of regressions is relatively recent. Bai, Lumsdaine and Stock (1998) consider estimating a single break date in multivariate time series allowing stationary or integrated regressors as well as trends. They show that the width of the confidence interval decreases when series having a common break are treated as a group and estimation is carried using quasi-maximum likelihood (QML). Bai (2000) considers a segmented stationary vector autoregression (VAR) model when the breaks can occur in the parameters of the conditional mean, the covariance matrix of the error term, or both. The most general framework is that of Qu and Perron (2007), who consider models of the form

$$y_t = (I \otimes z_t')S\beta_j + u_t$$

for $T_{j-1}+1 \leq t \leq T_j$ $(j=1, \ldots, m+1)$, where y_t is an n-vector of dependent variables and z_t is a q-vector that includes the regressors from all equations, and $u_t \sim (0, \Sigma_j)$. The matrix S is of dimension nq by p with full column rank (usually a selection matrix that specifies which regressors appear in each equation). The set of basic parameters in regime j consists of the p vector β_j and of Σ_j. Qu and Perron (2007) also allow for the imposition of a set of r restrictions of the form $g(\beta, vec(\Sigma))=0$, where $\beta=(\beta_1', \ldots, \beta_{m+1}')'$, $\Sigma=(\Sigma_1, \ldots, \Sigma_{m+1})$ and $g(\cdot)$ is an r dimensional vector. Both within- and cross-equation restrictions are allowed, and in each case within or across regimes. The assumptions on the regressors z_t and the errors u_t are similar to those discussed above. Hence, the framework permits a wide class of models including VAR, SUR, linear panel data, change in means of a vector of stationary processes, and so on. Models with integrated regressors (that is, models with cointegration) are not permitted.

Allowing for general restrictions on the parameters β_j and Σ_j permits a very wide range of special cases that are of practical interest: (*a*) partial structural change models, (*b*) block partial structural change models where only a subset of the equations is subject to change; (*c*) changes in only some elements of the covariance matrix Σ_j; (*d*) changes in only the covariance matrix Σ_j, while β_j is the same for all segments; (*e*) models where the breaks occur in a particular order across subsets of equations; and so on.

The method of estimation is again QML (based on normal errors) subject to the restrictions. Qu and Perron (2007) derive the consistency, rate of convergence and limit distribution of the estimated break dates. They obtain a general result stating that, in large samples, the restricted likelihood function can be separated in two parts: one involving only the break dates and the true values of the coefficients, so that the estimates of the break dates are not affected by the restrictions; the other involving the parameters of the model, the true values of the break dates and the restrictions, showing that the limiting distributions of these estimates are influenced by the restrictions but not by the estimation of the break dates. The limit distributions for the estimates of the break dates are qualitatively similar to those discussed above. Though only root-T consistent estimates of (β, Σ) are needed to construct asymptotically valid confidence intervals, it is likely that more precise estimates of these parameters will lead to better finite sample coverage rates. Hence, it is recommended to use the estimates obtained imposing the restrictions even though imposing restrictions does not have a first-order effect on the limiting distributions of the estimates of the break dates. To make estimation possible in practice, Qu and Perron (2007) present an algorithm which extends the one discussed in Bai and Perron (2003a) using, in particular, an iterative generalized least squares (GLS) procedure to construct the likelihood function for all possible segments. The theoretical analysis shows that substantial efficiency gains can be obtained by casting the analysis in a system of regressions.

Qu and Perron (2007) also consider a novel aspect to the problem of multiple structural changes labelled 'locally ordered breaks'. This applies when the breaks across two equations are 'ordered' in the sense that we have the prior knowledge that the break in one equation occurs after the break in the other. The breaks are 'local' in the sense that the time span between their occurrence is expected to be short. Hence, the breaks cannot be viewed as occurring simultaneously, nor can the break fractions be viewed as asymptotically distinct. An algorithm to estimate such models is presented. Also, a framework to analyse the limit distribution of the estimates is introduced. Unlike the case with asymptotically distinct breaks, the distributions of the estimates of the break dates need to be considered jointly.

Tests that allow for a single break

To test for a structural change at an unknown date, Quandt (1960) suggested the likelihood ratio test evaluated at the break date that maximizes it. This is a non-standard problem since one parameter is only identified under the alternative hypothesis. This problem was treated under various degrees of specificity that culminated in the general treatment by Andrews (1993). The basic method is to use the maximum of the likelihood ratio test over all possible values of the parameter in some pre-specified set. In the case of a single change, this translates into the statistic $\sup_{\lambda_1 \in \Lambda_\varepsilon} LR_T(\lambda_1)$, where $LR_T(\lambda_1)$ denotes the likelihood ratio evaluated at some $T_1 = [T\lambda_1]$ and the maximization is restricted over break fractions that are in the set $\Lambda_\varepsilon = [\varepsilon_1, 1-\varepsilon_2]$. The limit distribution is given by

$$\sup_{\lambda_1 \in \Lambda_\varepsilon} LR_T(\lambda_1) \Rightarrow \sup_{\lambda_1 \in \Lambda_\varepsilon} \frac{[\lambda_1 W_q(1) - W_q(\lambda_1)]'[\lambda_1 W_q(1) - W_q(\lambda_1)]}{\lambda_1(1-\lambda_1)} \tag{3}$$

with $W_q(\lambda)$ a vector of independent Wiener processes of dimension q, the number of coefficients that are allowed to change (this result holds with non-trending data). The limit distribution depends on Λ_ε. If $\varepsilon_1 = \varepsilon_2 = 0$, the test diverges under the null hypothesis, and critical values grow and the power of the test decreases as ε_1 and ε_2 get smaller. Hence, the range over which we search for a maximum must be small enough for the critical values not to be too large and for the test to retain decent power, yet large enough to include break dates that are potential candidates. In the single break case, a popular choice is $\varepsilon_1 = \varepsilon_2 = .15$.

Andrews (1993) also considered tests based on the maximal value of the Wald and Lagrange multiplier (LM) tests and shows that they are asymptotically equivalent, that is, they have the same limit distribution under the null hypothesis and under a sequence of local alternatives. All tests are also consistent and have non-trivial local asymptotic power against a wide range of alternatives, namely, for which the parameters of interest are not constant over the interval specified by Λ_ε. This does not mean, however, that they all have the same behaviour in finite samples. Indeed, the simulations of Vogelsang (1999) for the special case of a change in mean with serially correlated errors, showed the sup LM_T test to be seriously affected by the problem of non-monotonic power, in the sense that, for a fixed sample size, the power of the test can rapidly decrease to zero as the change in mean increases.

For Model (1) with *i.i.d.* errors, the LR and Wald tests have similar properties, so we shall discuss the Wald test. For a single change, it is defined by (up to a scaling by q):

$$\sup_{\lambda_1 \in \Lambda_\varepsilon} W_T(\lambda_1; q) = \sup_{\lambda_1 \in \Lambda_\varepsilon} \frac{\hat{\delta}' H'(H(Z'M_X Z)^{-1} H')^{-1} H\hat{\delta}}{SSR_k/(T-2q-p)} \tag{4}$$

where $Z = diag(Z_1, \ldots, Z_{m+1})$ with $Z_i = (z_{T_{i-1}+1}, \ldots, z_{T_i})'$, H is the conventional matrix such that $(H\delta)' = (\delta_1' - \delta_2')$ and $M_X = I - X(X'X)^{-1}X'$. Here SSR_k is the sum of squared residuals under the alternative hypothesis. Note that break point that maximizes the Wald test is the same as the estimate obtained by minimizing the sum of squared residuals provided the minimization problem (2) is restricted to the set Λ_ε, that is, $\sup_{\lambda_1 \in \Lambda_\varepsilon} W_T(\lambda_1; q) = W_T(\hat{\lambda}_1; q)$. When serial correlation and/or heteroskedasticity in the errors is permitted, the Wald test must be adjusted to account for this. In this case, it is defined by

$$W_T^*(\lambda_1; q) = (T-2q-p)\hat{\delta}' H'(H\hat{V}(\hat{\delta})H')^{-1} H\hat{\delta}, \tag{5}$$

where $\hat{V}(\hat{\delta})$ is an estimate of the variance covariance matrix of $\hat{\delta}$ that is robust to serial correlation and heteroskedasticity; that is, a consistent estimate of

$$V(\hat{\delta}) = \text{plim}_{T\to\infty} T(Z'M_X Z)^{-1} Z'M_X \Omega M_X Z(Z'M_X Z)^{-1} \tag{6}$$

where Ω is the covariance matrix of the errors. Note that it can be constructed allowing identical or different distributions for the regressors and the errors across segments. This is important because if a variance shift occurs at the same time and is not taken into account, inference can be distorted (Pitarakis, 2004).

The computation of the robust version of the Wald test (5) can be involved. Since the estimate of λ_1 is T-consistent even with correlated errors, an asymptotically equivalent version is to first take the supremum of the original Wald test, as in (4), to obtain the break point, that is imposing $\Omega = \sigma^2 I$. The robust version is obtained by evaluating (5) and (6) at this estimated break date, that is, using $W_T^*(\hat{\lambda}_1; q)$ instead of $\sup_{\lambda_1 \in \Lambda_\varepsilon} W_T^*(\lambda_1; q)$, where $\hat{\lambda}_1$ is obtained by minimizing the sum of squared residuals over the set Λ_ε. This is especially convenient when testing for multiple structural changes.

Optimal tests

Andrews and Ploberger (1994) consider a class of tests that are optimal, in the sense that they maximize a weighted average of the local asymptotic power function. They are weighted functions of the standard Wald, LM or LR statistics for all permissible fixed break dates. Using either of the three basic statistics leads to tests that are asymptotically equivalent. Here, we shall proceed with the version based on the Wald test. On the assumption that equal weights are given to all break fractions in some trimmed interval $[\varepsilon_1, 1-\varepsilon_2]$, the optimal test for distant alternatives is the following so-called Exp-type test

$$Exp - W_T = \log\left(T^{-1} \sum_{T_1=[T\varepsilon_1]+1}^{T-[T\varepsilon_2]} \exp(\tfrac{1}{2} W_T(T_1/T))\right).$$

For alternatives close to the null value of no change the optimal test is the *Mean-W_T* test

$$Mean - W_T = T^{-1} \sum_{T_1=[T\varepsilon_1]+1}^{T-[T\varepsilon_2]} W_T(T_1/T)$$

Andrews and Ploberger (1994) provide critical values for both tests for a range of values for symmetric trimmings $\varepsilon_1 = \varepsilon_2$ (which can be used for some non symmetric trimmings as well). The *Mean-W_T* has highest power for small shifts and the *Exp-W_T* performs better for moderate to large shifts. None of them uniformly dominates the *Sup-W_T* test and Andrews and Ploberger (1994) recommend the *Exp-W_T* test. The *Sup-W_T* test is not a member of the class of tests that maximize some weighted version of the local asymptotic power function, though it is admissible.

Kim and Perron (2006) approach the optimality issue from a different perspective using the approximate Bahadur measure of efficiency. They show that tests based on the Mean functional are inferior to those based on the Sup and Exp (which are as efficient) when using the same base statistic. When considering tests that incorporate a correction for potential serial correlation in the errors: (*a*) for a given functional, using the LM statistic leads to tests with zero asymptotic relative efficiency compared with using the Wald statistic; (*b*) for a given statistic the Mean-type tests have zero relative efficiency compared to using the Sup and Exp versions, which are as efficient. These results are in contrast to those of Andrews and Ploberger (1994) and the practical implication is that the preferred tests should be the Sup or Exp–Wald tests. Any test based on the LM statistic should be avoided.

Non-monotonicity in power

The Sup–Wald and Exp–Wald tests have monotonic power when only one break occurs under the alternative. As shown in Vogelsang (1999), the Mean–Wald test can exhibit a non-monotonic power function, though the problem has not been shown to be severe. All of these, however, suffer from some important power problems when the alternative is one that involves two breaks (for example, Vogelsang, 1997). This suggests that a test will exhibit a non-monotonic power function if the number of breaks present under the alternative hypothesis is greater than the number of breaks explicitly accounted for in the construction of the tests. Hence, though a single break test is consistent against multiple breaks, substantial power gains can result from using tests for multiple structural changes.

Tests for multiple structural changes

The literature on tests for multiple structural changes is relatively scarce. Here, the problem with the *Mean-W_T* and *Exp-W_T* tests is practical implementation as they require the computation of the Wald test over all permissible partitions of the sample, a number of order $O(T^m)$, which is prohibitively large when $m > 2$. Consider instead the Sup-Wald test. With i.i.d. errors, maximizing the Wald statistic is equivalent to minimizing the sum of squared residuals when the search is restricted to the same possible partitions of the sample. As discussed above, this problem can be solved with a very efficient algorithm. This is the approach taken by Bai and Perron (1998). In the context of model (1) with i.i.d. errors, the Wald test for testing the null hypothesis of no change versus the alternative hypothesis of k changes is given by

$$W_T(\lambda_1, \ldots, \lambda_k; q) = \left(\frac{T-(k+1)q-p}{k} \right) \frac{\hat{\delta}' H'(H(Z'M_X Z)^{-1} H')^{-1} H\hat{\delta}}{SSR_k}$$

where H now is the matrix such that $(H\delta)' = (\delta_1' - \delta_2', \ldots, \delta_k' - \delta_{k+1}')$. The Sup–Wald test is defined by

$$\sup_{(\lambda_1,\ldots,\lambda_k)\in\Lambda_{k,\varepsilon}} W_T(\lambda_1, \ldots, \lambda_k; q) = W_T(\hat{\lambda}_1, \ldots, \hat{\lambda}_k; q),$$

where

$$\Lambda_{k,\varepsilon} = \{(\lambda_1, \ldots, \lambda_k); |\lambda_{i+1} - \lambda_i| \geq \varepsilon, \lambda_1 \geq \varepsilon, \lambda_k \leq 1 - \varepsilon\}$$

and $(\hat{\lambda}_1, \ldots, \hat{\lambda}_k) = (\hat{T}_1/T, \ldots, \hat{T}_k/T)$, with $(\hat{T}_1, \ldots, \hat{T}_k)$ the estimates of the break dates obtained by minimizing the sum of squared residuals by searching over partitions defined by the set $\Lambda_{k,\varepsilon}$. When serial correlation and/or heteroskedasticity in the residuals is allowed, the test is

$$W_T^*(\lambda_1, \ldots, \lambda_k; q) = \left(\frac{T-(k+1)q-p}{k} \right) \hat{\delta}' H'(H\hat{V}(\hat{\delta})H')^{-1} H\hat{\delta},$$

with $\hat{V}(\hat{\delta})$ as defined by (6). Again, the asymptotically equivalent version with the Wald test evaluated at the estimates $(\hat{\lambda}_1, \ldots, \hat{\lambda}_k)$ is used to make the problem tractable. The limit distribution of the tests under the null hypothesis is the same in both cases, again on the assumption of non-trending data. Critical values are presented in Bai and Perron (1998; 2003b). The importance of the choice of ε for the size and power of the test is discussed in Bai and Perron (2003a; 2006). They also discuss variations in the construction of the test that allow imposing various restrictions on the nature of the errors and regressors, which can help improve power.

Double maximum tests

Often, one may not wish to pre-specify a particular number of breaks. Then a test of the null hypothesis of no structural break against an unknown number of breaks given some upper bound M can be used. These are called 'double maximum tests'. The first is an equal-weight version defined by $UD \max W_T(M, q) = \max_{1 \leq m \leq M} W_T(\hat{\lambda}_1, \ldots, \hat{\lambda}_m; q)$. The second test applies weights to the individual tests such that the marginal p-values are equal across values of m and is denoted $WD \max F_T(M, q)$ (see Bai and Perron, 1998, for details). The choice $M = 5$ should be sufficient for most applications. In any event, the critical values vary little as M is increased beyond 5.

The double maximum tests are arguably the most useful to apply when trying to determine if structural changes are present. First, there are types of multiple structural changes that are difficult to detect with a single break test (for example, two breaks with the first and third regimes the same). Second, as discussed above, is the potential non-monotonic power problem when the number of changes is greater than specified. Third, the power of the double maximum tests is almost as high as the best power that can be achieved using the test that accounts for the correct number of breaks (for example, Bai and Perron, 2006).

Sequential tests

Bai and Perron (1998) also discuss a test of ℓ versus $\ell+1$ breaks, which can be used to estimate the number of breaks using a sequential testing procedure. For the model with ℓ breaks, the estimated break points denoted by $(\hat{T}_1, \ldots, \hat{T}_\ell)$ are obtained by a global minimization of the sum of squared residuals. The strategy proceeds by testing for the presence of an additional break in each of the $(\ell+1)$ segments obtained using the partition $\hat{T}_1, \ldots, \hat{T}_\ell$. We reject in favour of a model with $(\ell+1)$ breaks if the minimal value of the sum of squared residuals over all segments where an additional break is included is sufficiently smaller than that from the ℓ breaks model. The break date selected is the one associated with this overall minimum. The limit distribution of the test is related to that of a test for a single change. Bai (1999) considers the same problem allowing the breaks to be global minimizers of the sum of squared residuals under both the null and alternative hypotheses. The limit distribution of the test is different. A method to compute the asymptotic critical values is discussed and the results extended to the case of trending regressors.

These tests can form the basis of a sequential testing procedure by applying them successively starting from $\ell=0$, until a non-rejection occurs. The estimate of the number of breaks thus selected will be consistent provided the significance level used decreases at an appropriate rate. The simulation results of Bai and Perron (2006) show that such estimate of the number of breaks is better than those obtained using information criteria as suggested by, for example, Liu, Wu and Zidek (1997) (see also Perron, 1997). But this sequential procedure should not be applied mechanically. In several cases, it stops too early. The recommendation is to first use a double maximum test to ascertain if any break is at all present. The sequential tests can then be used starting at some value greater than 0 to determine the number of breaks.

Tests for restricted structural changes

Consider testing the null hypothesis of 0 break versus an alternative with k breaks in a model which imposes the restrictions $R\delta = r$. In this case, the limit distribution of the Sup–Wald test depends on the nature of the restrictions so that it is not possible to tabulate critical values valid in general. Perron and Qu (2006) discuss a simulation algorithm to compute the relevant critical values given some restrictions. Imposing valid restrictions results in tests with much improved power.

Tests for structural changes in multivariate systems

Bai, Lumsdaine and Stock (1998) considered a Sup–Wald test for a single change in a multivariate system. Qu and Perron (2007) extend the analysis to the context of multiple structural changes. They consider the case where only a subset of the coefficients is allowed to change, whether it be the parameters of the conditional mean, the covariance matrix of the errors, or both. The tests are based on the maximized value of the likelihood ratio over permissible partitions assuming i.i.d. errors. The tests can be corrected for serial correlation and heteroskedasticity when testing for changes in the parameters of the conditional mean assuming no change in the covariance matrix of the errors. However, when the tests involve potential changes in the covariance matrix of the errors, the limit distributions are only valid assuming a Normal distribution for these errors.

An important advantage of the general framework analysed by Qu and Perron (2007) is that it allows studying changes in the variance of the errors in the presence of simultaneous changes in the parameters of the conditional mean, thereby avoiding inference problems when changes in variance are studied in isolation. Also, it allows for the two types of changes to occur at different dates, thereby avoiding problems related to tests for changes in the parameters when a change in variance occurs at some other date.

These tests are especially important in light of Hansen's (2000) analysis. First note that the limit distribution of the tests in a single equation system has the stated limit distribution under the assumption that the regressors and the variance of the errors have distributions that are stable across the sample. Hansen shows that when the regressors are not stationary the limit distribution changes and the tests can be distorted, especially when a change in variance occurs. He proposes a fixed regressor bootstrap method to construct valid tests. But both problems of changes in the distribution of the regressors and the variance of the errors can be handled using the framework of Qu and Perron (2007). If a change in the variance of the residuals is a concern, one can perform a test for no change in some parameters of the conditional model allowing for a change in variance since the tests are based on a likelihood ratio approach. If changes in the marginal distribution of some regressors are a concern, one can use a multi-equations system with equations for these regressors.

Tests valid with $I(1)$ regressors

With $I(1)$ regressors, a case of interest is a system of cointegrated variables. For testing, Hansen (1992) considered the null hypothesis of no change in both coefficients. The tests considered are the Sup and Mean–LM tests directed against an alternative of a one time change in the coefficients. Hansen also considers a version of the LM test directed against the alternative that the coefficients are random walk processes. Kejriwal and Perron (2006b) provide a comprehensive treatment of issues related to testing for multiple structural changes at unknown dates in cointegrated regression models using the Sup–Wald test. They allow both $I(0)$ and $I(1)$ variables and derive the limiting distribution of the Sup–Wald test under the null hypothesis of no structural change against the alternative hypothesis of a given number of cointegrating regimes. They also consider the double maximum tests and provide critical values for a wide variety of models that are expected to be relevant in practice. The asymptotic results have important implications for inference. It is shown that, in models involving both $I(1)$ and $I(0)$ variables, inference is possible as long as the intercept is allowed to change across regimes. Otherwise, the limiting distributions of the tests depend on nuisance parameters. Simulation experiments show that with serially correlated errors the commonly used Sup, Mean and Exp–LM tests suffer from the problem of non-monotonic power in finite samples both with a single and multiple breaks. Kejriwal and Perron (2006b) propose a modified Sup–Wald test that has good size and power properties.

Note, however, that the Sup and Mean–Wald test will also reject when no structural change is present and the system is not cointegrated. Hence, the application of such tests should be interpreted with caution. No test are available for the null hypothesis of no change in the coefficients allowing the errors to be $I(0)$ or $I(1)$. This is because when the errors are $I(1)$, we have a spurious regression and the parameters are not identified. To be able to properly interpret the tests, they should be used in conjunction with tests for the presence or absence of cointegration allowing shifts in the coefficients (see the discussion and references in Perron, 2006). A partial solution to this problem is the following. If a spurious regression is present, the number of breaks selected will always (in large samples) be the maximum number of breaks allowed. Thus, selecting the maximum allowable number of breaks can be indicative of the presence of $I(1)$ errors (using a Sup–Wald test uncorrected for serial correlation in the errors). The same is true when information criteria are used to select the number of breaks.

Tests valid whether the errors are $I(1)$ or $I(0)$

The issue of testing for structural changes in a linear model with errors that are either $I(0)$ or $I(1)$ is interest when the regression is a polynomial time trend (for example, testing for a change in the slope of a linear trend). The problem here is to devise a procedure that has the same limit distribution in both the $I(0)$ and $I(1)$ cases. The first to provide such a solution is Vogelsang (2001). He also accounts for correlation with an autoregressive approximation so that the Wald test has a non-degenerate limit distribution in both the $I(0)$ and $I(1)$ cases. The novelty is that he weights the statistic by a unit root test scaled by some parameter. For any given significance level, a value of this scaling parameter can be chosen so that the asymptotic critical values will be the same. Vogelsang's simulations show, however, the test to have little power in the $I(1)$ case so that he resorts to advocating the joint use of that test and a normalized Wald test that has good properties in the $I(1)$ case but has otherwise very little power in the $I(0)$ case.

Perron and Yabu (2007b) builds on the work of Perron and Yabu (2007a) who analysed the problem of hypothesis testing on the slope coefficient of a linear trend model. The method is based on a feasible quasi generalized least squares approach that uses a super-efficient estimate of the sum of the autoregressive parameters α when $\alpha = 1$. The estimate of α is the OLS estimate from an autoregression applied to detrended data and is truncated to take a value 1 whenever it is in a $T^{-\delta}$

neighbourhood of 1. This makes the estimate 'super-efficient' when $\alpha = 1$ and implies that inference can be performed using the standard Normal or Chi-square distribution for all $|\alpha| \leq 1$. Theoretical arguments and simulation evidence show that $\delta = 1/2$ is the appropriate choice. Perron and Yabu (2007b) analyse the case of testing for changes in level or slope of the trend function of a univariate time series. When the break dates are known, things are similar. When the break dates are unknown, the limit distributions of the Exp, Mean and Sup functionals of the Wald test across all permissible breaks dates is no longer the same in the $I(0)$ and $I(1)$ cases. However, the limit distribution is nearly the same using the Exp functional. Hence, it is possible to have tests with nearly the same size in both cases. To improve the finite sample properties of the test, use is made of a bias-corrected version of the OLS estimate. This makes possible a testing procedure that has good size and power properties in finite samples.

Summary

There has been tremendous progress since the early 1990s in developing methods to analyse structural changes for a variety of cases that are of practical interest. Still, much remains to be done, in particular in providing tools to analyse changes in the variance of the errors without the need to assume a Normal distribution.

PIERRE PERRON

See also **cointegration; regime switching models; time series analysis; unit roots.**

Bibliography

Andrews, D.W.K. 1993. Tests for parameter instability and structural change with unknown change point. *Econometrica* 61, 821–56 (Corrigendum, 71, 395–7).

Andrews, D.W.K. and Ploberger, W. 1994. Optimal tests when a nuisance parameter is present only under the alternative. *Econometrica* 62, 1383–414.

Bai, J. 1997a. Estimation of a change point in multiple regression models. *Review of Economic and Statistics* 79, 551–63.

Bai, J. 1997b. Estimating multiple breaks one at a time. *Econometric Theory* 13, 315–52.

Bai, J. 1999. Likelihood ratio tests for multiple structural changes. *Journal of Econometrics* 91, 299–323.

Bai, J. 2000. Vector autoregressive models with structural changes in regression coefficients and in variance-covariance matrices. *Annals of Economics and Finance* 1, 303–39.

Bai, J., Lumsdaine, R.L. and Stock, J.H. 1998. Testing for and dating breaks in multivariate time series. *Review of Economic Studies* 65, 395–432.

Bai, J. and Perron, P. 1998. Estimating and testing linear models with multiple structural changes. *Econometrica* 66, 47–78.

Bai, J. and Perron, P. 2003a. Computation and analysis of multiple structural change models. *Journal of Applied Econometrics* 18, 1–22.

Bai, J. and Perron, P. 2003b. Critical values for multiple structural change tests. *Econometrics Journal* 6, 72–8.

Bai, J. and Perron, P. 2006. Multiple structural change models: a simulation analysis. In *Econometric Theory and Practice: Frontiers of Analysis and Applied Research*, ed. D. Corbea, S. Durlauf and B.E. Hansen. New York: Cambridge University Press.

Deng, A. and Perron, P. 2006. A comparison of alternative asymptotic frameworks to analyze structural change in a linear time trend. *Econometrics Journal* 9, 423–47.

Hansen, B.E. 1992. Tests for parameter instability in regressions with I(1) processes. *Journal of Business and Economic Statistics* 10, 321–35.

Hansen, B.E. 2000. Testing for structural change in conditional models. *Journal of Econometrics* 97, 93–115.

Hawkins, D.M. 1976. Point estimation of the parameters of piecewise regression models. *Applied Statistics* 25, 51–7.

Kim, D. and Perron, P. 2006. Assessing the relative power of structural break tests using a framework based on the approximate Bahadur slope. Working paper 2006-63, Boston University.

Kejriwal, M. and Perron, P. 2006a. The limit distribution of the estimates in cointegrated regression models with multiple structural changes. Working paper 2006-64, Boston University.

Kejriwal, M. and Perron, P. 2006b. Testing for multiple structural changes in cointegrated regression models. Working paper 2006-51, Boston University.

Lavielle, M. and Moulines, E. 2000. Least-squares estimation of an unknown number of shifts in a time series. *Journal of Time Series Analysis* 21, 33–59.

Liu, J., Wu, S. and Zidek, J.V. 1997. On segmented multivariate regressions. *Statistica Sinica* 7, 497–525.

Perron, P. 1997. L'estimation de modèles avec changements structurels multiples. *Actualité Économique* 73, 457–505.

Perron, P. 2006. Dealing with structural breaks. In *Palgrave Handbook of Econometrics, Vol. 1: Econometric Theory*, ed. K. Patterson and T.C. Mills. Basingstoke: Palgrave Macmillan.

Perron, P. and Qu, Z. 2006. Estimating restricted structural change models. *Journal of Econometrics* 134, 373–99.

Perron, P. and Yabu, T. 2007a. Estimating deterministic trends with an integrated or stationary noise component. Working paper 2007-21, Boston University.

Perron, P. and Yabu, T. 2007b. Testing for shifts in trend with an integrated or stationary noise component. Working paper 2007-25, Boston University.

Perron, P. and Zhu, X. 2005. Structural breaks with stochastic and deterministic trends. *Journal of Econometrics* 129, 65–119.

Pitarakis, J.-Y. 2004. Least squares estimation and tests of breaks in mean and variance under misspecification. *Econometrics Journal* 7, 32–54.

Qu, Z. and Perron, P. 2007. Estimating and testing multiple structural changes in multivariate regressions. *Econometrica* 75, 459–502.

Quandt, R.E. 1960. Tests of the hypothesis that a linear regression system obeys two separate regimes. *Journal of the American Statistical Association* 55, 324–30.

Vogelsang, T.J. 1997. Wald-type tests for detecting breaks in the trend function of a dynamic time series. *Econometric Theory* 13, 818–49.

Vogelsang, T.J. 1999. Sources of nonmonotonic power when testing for a shift in mean of a dynamic time series. *Journal of Econometrics* 88, 283–99.

Vogelsang, T.J. 2001. Testing for a shift in trend when serial correlation is of unknown form. Working paper, Cornell University.

structural vector autoregressions

Structural vector autoregressions (SVARs) are a multivariate, linear representation of a vector of observables on its own lags and (possibly) other variables as a trend or a constant. SVARs make explicit identifying assumptions to isolate estimates of policy and/or private agents' behaviour and its effects on the economy while keeping the model free of the many additional restrictive assumptions needed to give every parameter a behavioural interpretation. Introduced by Sims (1980), SVARs have been used to document the effects of money on output (Sims and Zha, 2006a), the relative importance of supply and demand shocks on business cycles (Blanchard and Quah, 1989), the effects of fiscal policy (Blanchard and Perotti, 2002), or the relation between technology shocks and worked hours (Galí, 1999), among many other applications.

Economic theory and the SVAR representation

Dynamic economic models can be viewed as restrictions on stochastic processes. From this perspective, an economic theory is a mapping between a vector of k economic shocks w_t and a vector of n observables y_t of the form $y_t = D(w^t)$, where w^t represents the whole history of shocks w_t up to period t. The economic shocks are those shocks to the fundamental elements of the theory: preferences, technology, informational sets, government policy, measurement errors, and so on. The observables are all variables that the researcher has access to. Often, y_t includes a constant to capture the mean of the process. The mapping $D(\cdot)$ is the product of the equilibrium behaviour of the agents in the model, implied by their optimal decision rules and consistency conditions like resource constraints and market clearing. The construction of the mapping $D(\cdot)$ is the sense in which economic theory tightly relates shocks and observables. Also, the mapping $D(\cdot)$ can be interpreted as the impulse response of the model to an economic shock.

Often, we restrict our attention to linear mappings of the form $y_t = D(L)w_t$, where L is the lag operator. For simplicity of exposition, w_t will be i.i.d. random variables and normally distributed, $w_t \sim N(0, \Sigma)$. More involved structures – for example, allowing for autocorrelation between the shocks – can be accommodated with additional notation.

We pick the neoclassical growth model, the workhorse of dynamic macroeconomics, to illustrate the previous paragraphs. In its basic version, the model maps productivity shocks, the w_t of the theory, into observables, y_t, like output or investment. The mapping comes from the optimal investment and labour supply decisions of the households, the resource constraint of the economy, and the law of motion for productivity. If the productivity shocks are normally distributed and we solve the model by linearizing its equilibrium conditions, we obtain a mapping of the form $y_t = D(L)w_t$ described above.

If $k = n$, that is, we have as many economic shocks as observables, and $|D(L)|$ has all its roots outside the unit circle, we can invert the mapping $D(L)$ (see Fernández-Villaverde, Rubio-Ramírez and Sargent, 2005) and obtain

$$A(L)y_t = w_t$$

where $A(L) = A_0 - \Sigma_{k=1}^{\infty} A_k L^k$ is a one-sided matrix lag polynomial that embodies all the (usually nonlinear) cross-equation restrictions derived by the equilibrium solution of the model. In general, $A(L)$ is of infinite order. This representation is known as the SVAR representation. The name comes from realizing that $A(L)y_t = w_t$ is a vector autoregression (VAR) generated by an economic model (a 'structure').

Reduced form representation, normalization, and identification

Consider now the case where a researcher does not have access to the SVAR representation. Instead, she has access to the VAR representation of y_t:

$$y_t = B_1 y_{t-1} + B_2 y_{t-2} + \cdots + a_t,$$

where $Ey_{t-j}a_t = 0$ for all j and $Ea_t a_t' = \Omega$. This representation is known as the reduced-form representation. Can the researcher recover the SVAR representation using the reduced-form representation? Fernández-Villaverde, Rubio-Ramírez, and Sargent (2005) show that, given a strictly invertible economic model, that is, $|D(L)|$ has all its roots strictly outside the unit circle, there is one and only one identification scheme to recover the SVAR from the reduced form. In addition, they show that the mapping between a_t and w_t is $a_t = A_0^{-1} w_t$. Hence, if the researcher knew A_0^{-1}, she could recover the SVAR representation from the reduced form, noting that $A_j = A_0 B_j$ for all j and $w_t = A_0 a_t$.

Hence, the recovery of w_t from y^T requires the knowledge of the dynamic economic model. Can we avoid this step? Unfortunately, the answer is, in general, 'no', because knowledge of the reduced-form matrices B_i and Ω does not imply, by itself, knowledge of the A_i and Σ, for two reasons.

The first is normalization. Reversing the signs of two rows or columns of the A_i does not matter for the B_i. Thus, without the correct normalization restrictions, statistical inference about the A_i's is essentially meaningless. Waggoner and Zha (2003) provide a general normalization rule that maintains coherent economic interpretations.

The second is identification. If we knew A_0, each equation $B_i = A_0^{-1} A_i$ would determine A_i given some B_i. But the only restrictions that the reduced-form representation imposes on the matrix A_0 comes from $\Omega = A_0 \Sigma A_0'$. In this relationship, we have $n(3n+1)/2$ unknowns (the n^2 distinct elements of A_0 and the $n(n+1)/2$ distinct elements of Σ) for n^2 knowns (the $n(n+1)/2$ distinct elements of Ω). Thus, we require n^2 identification restrictions. Since we can set the diagonal elements of A_0 equal to 1 by scaling, we are left with the need of $n(n-1)$ additional identification restrictions (alternatively, we could scale the shocks such that the diagonal of Σ is composed of ones and leave the diagonal of A_0 unrestricted). These identification restrictions are dictated by the economic theory being studied.

The literature, however, has often preferred to impose identification restrictions that are motivated by the desire to be compatible with a large class of models, instead of just one concrete model and its whole set of cross-equations restrictions. The hope is that, thanks to this generality, the inferences drawn from SVAR can be more robust and can compensate for the lack of efficiency derived from not implementing a full information method.

The most common identification restriction has been to assume that Σ is diagonal. This assumption relies on the view that economic shocks are inherently different sources of uncertainty that interact only through their effect on the decisions of the model's agents. Since this assumption imposes $n(n-1)/2$ restrictions, we still require $n(n-1)/2$ additional restrictions.

To find these additional restrictions, economists have followed two main approaches: short-run restrictions and long-run restrictions. Sims (1980) pioneered the first approach when he proposed to impose zeros on A_0. The motivation for such a scheme comes from the idea that there is a natural timing in the effect of economic shocks. For example, to place a zero on A_0, we can use the intuition that monetary policy cannot respond contemporaneously to a shock in the price level because of informational delays. Similarly, institutional constraints, like the timing of tax collections, can be exploited for identification (Blanchard and Perotti, 2002). Sims (1980) ordered variables in such a way that A_0 is lower triangular. Sims and Zha (2006a) present a non-triangular identification scheme on an eight-variable SVAR.

The long-run restrictions were popularized by Blanchard and Quah (1989). These restrictions are imposed on $A(1) = A_0 - \Sigma_{k=1}^{\infty} A_k$. Since $A^{-1}(1) = D(1)$, long-run restrictions are justified as restrictions on the long-run effects of economic shocks, usually on the first difference of an observable. For example, Blanchard and Quah (1989) assume that there are two shocks ('demand' and 'supply') affecting unemployment and output. The demand shock has no long-run effect on unemployment or output. The supply shock has no long-run effect on unemployment but may have a long-run effect on output. These differences in their long-run impacts allow Blanchard and Quah to identify the shocks and trace their impulse response function.

New identification schemes have been proposed to overcome the difficulties of the existing approaches. See, for instance, Uhlig (2005) for an identification scheme of monetary policy shocks based on sign restrictions that hold across a large class of models.

Estimation

Why is the previous discussion of the relation between the reduced and structural form of a VAR relevant? Because the reduced form can be easily estimated. An empirically implementable version of the reduced-form representation truncates the number of lags at the pth order:

$$y_t = B_1 y_{t-1} + \cdots + B_p y_{t-p} + a_t$$

where $Ea_t a_t' = \Omega$. We use hats in the matrices and the error a_t to indicate that they do not correspond exactly to the reduced form of the model but to the truncated version. The effects of the truncation on the accuracy of inference delivered by SVARs are unclear (see Chari, Kehoe and McGrattan, 2005, and Christiano, Eichenbaum and Vigfusson, 2007, for two opposite assessments). The resulting truncated VAR can be taken to the data using standard methods: GMM, maximum likelihood, or Bayesian.

The Bayesian approach is especially attractive. SVARs are proliferatively parametrized. The number of parameters in $B(L)$ grows with the square of the number of variables and the number of lags. Consequently, given the short period of data typically available to macroeconomists, classical methods become unreliable. A careful use of prior information alleviates the problem of overparametrization and improves the quality of the inference. The advent of modern simulation techniques, especially Markov chain Monte Carlo methods, has made the implementation of the Bayesian paradigm straightforward, even for sophisticated priors.

The point estimates $\hat{B}_i$ for $I = 1, \ldots, p$ and $\hat{\Omega}$ can be used to find estimates of $\hat{A}_i$ and $\hat{\Sigma}$ by solving $\hat{B}_i = \hat{A}_0^{-1}\hat{A}_i$ for $I = 1, \ldots, p$, and $\hat{\Omega} = \hat{A}_0 \hat{\Sigma} \hat{A}_0'$. With an estimate of $\hat{A}_0$ and the a_t, we can get $\hat{w}_t = \hat{A}_0 \hat{a}_t$. Thus, the reduced form plus the identifying restrictions deliver both an estimate of the economic shocks and the impulse response of the variables in the economy to those shocks. Confidence intervals for point estimates and error bands for impulse response functions can be estimated by resorting to Markov chain Monte Carlo techniques or the bootstrap.

In an important contribution, Sims and Zha (2006b) have extended the estimation of SVARs to allow for changes in equation coefficients and variances. This article opens the door for the analysis of richer dynamic models with parameter instability, arguably a more realistic description of observed aggregate variables.

Assessment of SVARs

SVARs offer an attractive approach to estimation. They promise to coax interesting patterns from the data that will prevail across a set of incompletely specified dynamic economic models with a minimum of identifying assumptions. Moreover, SVARs can be easily estimated, even with commercial software and freely available routines from the Internet. In the hands of skilful researchers, SVARs have contributed to the understanding of aggregate fluctuations, have clarified the importance of different economic shocks, and have generated fruitful debates among macroeconomists.

However, SVARs have also been criticized. We mention only three criticisms. First, it has been argued that the economic shocks recovered from an SVAR do not resemble the shocks measured by other mechanisms, such as market expectations embodied in future prices. Second, the shocks recovered from an SVAR may reflect variables omitted from the model. If these omitted variables correlate with the included variables, the estimated economic shocks will be biased. Third, the results of many SVAR exercises, even simple ones, are sensitive to the identification restrictions. Related to this criticism is the view that many of the identification schemes are the product of a specification search in which researchers look for 'reasonable' answers.

If an identification scheme matches the conventional wisdom, it is called successful; if it does not, it is called a puzzle or, even worse, a failure (Uhlig, 2005). Consequently, there is a danger that economists will get stuck in an a priori view of the data under the cloak of formal statistical inference.

JESÚS FERNÁNDEZ-VILLAVERDE AND JUAN F. RUBIO-RAMÍREZ

See also **time series analysis; vector autoregressions.**

Bibliography

Blanchard, O.J. and Quah, D. 1989. The dynamic effects of aggregate demand and supply disturbances. *American Economic Review* 79, 655–73.

Blanchard, O.J. and Perotti, R. 2002. An empirical characterization of the dynamic effects of changes in government spending and taxes on output. *Quarterly Journal of Economics* 117, 1329–68.

Chari, V.V., Kehoe, P.J. and McGrattan, E.R. 2005. A critique of structural VARs using real business cycle theory. Working Paper No. 631, Federal Reserve Bank of Minneapolis.

Christiano, L.J., Eichenbaum, M. and Vigfusson, R. 2007. Assessing structural VARs. In *NBER Macroeconomics Annual 2006*, vol. 21, ed. D. Acemoglu, K. Rogoff and M. Woodford. Cambridge, MA: MIT.

Fernández-Villaverde, J., Rubio-Ramírez, J.F. and Sargent, T.J. 2005. A, B, C's (and D's) for understanding VARs. Technical Working Paper No. 308. Cambridge, MA: NBER.

Galí, J. 1999. Technology, employment, and the business cycle: do technology shocks explain aggregate fluctuations? *American Economic Review* 89, 249–71.

Sims, C.A. 1980. Macroeconomics and reality. *Econometrica* 48, 1–48.

Sims, C.A. and Zha, T. 2006a. Does monetary policy generate recessions? *Macroeconomic Dynamics* 10, 231–72.

Sims, C.A. and Zha, T. 2006b. Were there regime switches in U.S. monetary policy? *American Economic Review* 96, 54–81.

Uhlig, H. 2005. What are the effects of monetary policy on output? Results from an agnostic identification procedure. *Journal of Monetary Economics* 52, 381–419.

Waggoner, D.F. and Zha, T. 2003. Likelihood preserving normalization in multiple equation models. *Journal of Econometrics* 114, 329–47.

threshold models

The models

Stochastic nonlinear models have been widely used in economic applications. They may arise directly from economic theory. There also exist nonlinear models that have first been suggested by statisticians, engineers and time series analysts and then found application in economics. A broad class of these models, here called threshold models, has the property that the models are either piecewise linear or may be more generally considered as linear models with time-varying parameters. This category of nonlinear models includes switching regression or threshold autoregressive models, smooth transition models, Markov-switching or hidden Markov models. Artificial neural network models may also be included in this class of nonlinear models.

The switching regression (SR) model or, in its univariate form, the threshold autoregressive (TAR) model, is defined as follows:

$$y_t = \sum_{j=1}^{r} (\alpha_j' \mathbf{z}_t + \varepsilon_{jt}) I(c_{j-1} < s_t \leq c_j) \tag{1}$$

where $\mathbf{z}_t = (\mathbf{w}_t', \mathbf{x}_t')'$ is a vector of explanatory variables, $\mathbf{w}_t = (1, y_{t-1}, \ldots, y_{t-p})'$ and $\mathbf{x}_t = (x_{1t}, \ldots, x_{kt})'$, s_t is an observable switch-variable, usually assumed to be a continuous stationary random variable, $c_0, c_1, \ldots, c_r$ are switch or threshold parameters, $c_0 = -\infty$, $c_r = M < \infty$, and $I(A)$ is an indicator function: $I(A) = 1$ when event A occurs; zero otherwise. Furthermore, $\alpha_j = (\alpha_{0j}, \alpha_{1j}, \ldots, \alpha_{mj})'$ such that $\alpha_i \neq \alpha_j$ for $i \neq j$, where $m = p + k + 1$, $\varepsilon_{jt} = \sigma_j \varepsilon_t$ with $\{\varepsilon_t\} \sim \text{iid}(0, 1)$, and $\sigma_j > 0$, $j = 1, \ldots, r$. It is seen that (1) is a piecewise linear model whose switch-points, however, are generally unknown. The most popular choice in applications is $r = 2$, that is, the model has two regimes. If $s_t = t$, eq. (1) is a linear model with $r-1$ breaks and $\alpha_j \neq \alpha_{j+1}$, $j = 1, \ldots, r - 1$. These models have recently become quite popular in econometrics and there is now a substantial literature on how to determine the number of breaks and estimate the break points. For a generalization of the threshold autoregressive regression model to the vector case, see Tsay (1998).

TAR models have been used to characterize asymmetric behaviour in GNP or unemployment rates and to consider the purchasing power parity hypothesis. They have also been applied to modelling interest rate series as well as other financial time series.

One may substitute $I(s_t = j)$ for $I(c_{j-1} < s_t \leq c_j)$ in (1), where s_t is an unobservable regime indicator with a finite set of values $\{1, \ldots, r\}$. On the assumption that s_t follows a first-order Markov chain, that is, $\Pr\{s_t = i | s_t = j\} = p_{ij}$, $i, j = 1, \ldots, r$, (1) becomes a hidden Markov or Markov-switching (MS) model (see Lindgren, 1978). Higher-order Markov chains are possible but rarely used in econometric applications.

Equation (1) with an unobservable regime indicator is not, however, the most frequently applied hidden Markov model in econometrics. Consider the univariate model

$$y_t = \mu_{s_t} + \sum_{j=1}^{p} \alpha_j (y_{t-j} - \mu_{s_{t-j}}) + \varepsilon_t \tag{2}$$

where s_t follows a first-order Markov chain as before, and $\mu_{s_t} = \mu^{(i)}$ for $s_t = i$, such that $\mu^{(i)} \neq \mu^{(j)}$, $i \neq j$. The stochastic intercept of this model, $\mu_{s_t} - \sum_{j=1}^{p} \alpha_j \mu_{s_{t-j}}$, can thus obtain r^{p+1} different values, which gives the model the desired flexibility. A comprehensive discussion of MS models can be found in Hamilton (1994, ch. 22). The model (2) has been frequently fitted, for example, to GNP series and interest rates. In the latter case, the model may be used for identifying changes in the monetary policy of the central bank.

SR and Markov-switching models contain a finite number of regimes. There is a class of models called smooth transition regression (STR) models, in which there are two 'extreme regimes', and the transition between them is smooth. A basic STR model is defined as follows:

$$y_t = \boldsymbol{\phi}' \mathbf{z}_t + \boldsymbol{\psi}' \mathbf{z}_t G(\gamma, \mathbf{c}, s_t) + \varepsilon_t \tag{3}$$

where $\boldsymbol{\phi} = (\phi_0, \phi_1, \ldots, \phi_m)'$ and $\boldsymbol{\psi} = (\psi_0, \psi_1, \ldots, \psi_m)'$ are parameter vectors, $\mathbf{c} = (c_1, \ldots, c_K)'$ is a vector of location parameters, $c_1 \leq \ldots \leq c_K$, and $\varepsilon_t \sim \text{iid}(0, \sigma^2)$. The transition function $G(\gamma, c, s_t)$ is a bounded function of s_t, continuous everywhere in the parameter space for any value of s_t. The logistic transition function has the general form

$$G(\gamma, \mathbf{c}, s_t) = \left(1 + \exp\left\{-\gamma \prod_{k=1}^{K} (s_t - c_k)\right\}\right)^{-1}, \quad \gamma > 0 \tag{4}$$

where $\gamma > 0$ is an identifying restriction. Equation (3) jointly with (4) defines the logistic STR (LSTR) model. The most common choices in practice for K are $K = 1$ and $K = 2$. For $K = 1$, the parameters $\boldsymbol{\phi} + \boldsymbol{\psi} G(\gamma, \mathbf{c}, s_t)$ change monotonically as a function of s_t from $\boldsymbol{\phi}$ to $\boldsymbol{\phi} + \boldsymbol{\psi}$. For $K = 2$, they change symmetrically around the mid-point $(c_1 + c_2)/2$ where this logistic function attains its minimum value. Slope parameter γ controls the slope and c_1 and c_2 the location of the transition function. When $K = 1$ and $\gamma \rightarrow \infty$ in (4), the model (3) becomes an SR model with $r = 2$.

The LSTR model with $K = 1$ (LSTR1 model) is capable of characterizing asymmetric behaviour. As an example, suppose that s_t measures the phase of the business cycle. Then the LSTR1 model can describe processes whose dynamic properties are different in expansions from what they are in recessions, and the transition from one extreme regime to the other is smooth. The same is true for SR and the MS models with the difference that instead of a smooth transition there is an abrupt switch. The LSTR2 model is appropriate whenever the dynamic behaviour of the process is similar at both large and small values of s_t and different in the middle.

Yet another nonlinear model that is worth mentioning because it is related to threshold models is the artificial neural network (ANN) model. The simplest single-equation case is the so-called 'single hidden-layer' ANN model. It has the following form

$$y_t = \boldsymbol{\beta}_0'\mathbf{z}_t + \sum_{j=1}^{q} \beta_j G(\gamma_j'\mathbf{z}_t) + \varepsilon_t \tag{5}$$

where y_t is the output series, $\mathbf{z}_t$ is the vector of inputs, and $\boldsymbol{\beta}_0'\mathbf{z}_t$ is a linear unit with $\boldsymbol{\beta}_0 = (\beta_{00}, \beta_{01}, \ldots, \beta_{0,p+k})'$. Furthermore, β_j, $j = 1, \ldots, q$, are parameters, called "connection strengths" in the neural network literature. Function $G(\cdot)$ is a bounded, asymptotically constant function called the 'squashing function' and $\boldsymbol{\gamma}_j$, $j = 1, \ldots, q$, are parameter vectors. They form the hidden layer which the name of the model refers to. Typical squashing functions are monotonically increasing ones such as the logistic function and the hyperbolic tangent function. The errors ε_t are often assumed iid $(0, \sigma^2)$. Many neural network modellers assume $\boldsymbol{\beta}_0 = \beta_{00}$, where β_{00} is called the 'bias'.

A theoretical argument for the use of ANN models is that they are universal approximators. Suppose that $y_t = H(\mathbf{z}_t)$, that is, there exists a functional relationship between y_t and $\mathbf{z}_t$. Then, under mild regularity conditions for H, there is a positive integer $q \leq q_0 < \infty$ such that for an arbitrary $\delta > 0$, $|H(\mathbf{z}_t) - \sum_{j=1}^{q} \beta_j G(\gamma_j'\mathbf{z}_t)| < \delta$. The importance of this result lies in the fact that q is finite, so that any unknown function H can be approximated arbitrarily accurately by a linear combination of squashing functions $G(\gamma_j'\mathbf{z}_t)$. This has been discussed in several papers, including Cybenko (1989), Funahashi (1989) and Hornik, Stinchcombe and White (1989). Neural network models are very generously parameterized and are only locally identified. The log-likelihood typically contains a large amount of local maxima, which makes parameter estimation difficult.

Testing linearity

All threshold models nest a linear model but they are not identified when the data are generated from this linear model. For this reason, testing linearity before fitting a threshold model is necessary in order to avoid the estimation of an unidentified model whose parameters cannot be estimated consistently. In this case, linearity testing has to precede any nonlinear estimation.

There exist general misspecification tests that are linearity tests if the specification to be tested is linear; see Bierens (1990) and Stinchcombe and White (1998). There also exist parametric tests that have been designed to be tests against an unspecified alternative but are not consistent against deviations from linearity. The popular Regression Error Specification Test (RESET) of Ramsey (1969) is such a test. Teräsvirta (1998) and van Dijk, Teräsvirta and Franses (2002) discuss tests against smooth transition regression models and Hansen (1999) surveys linearity testing against TAR models. Linearity testing in the Markov-switching framework is considered in Garcia (1998). Some recent econometrics textbooks discuss linearity tests against various threshold models.

As already mentioned, threshold models nest a linear model and are not identified if linearity holds. For example, the STR model (3) is not identified if $\gamma = 0$ in (4) or $\boldsymbol{\psi} = \mathbf{0}$. In the former case, $\boldsymbol{\psi}$ and $\mathbf{c}$ are not identified, and in the latter, the nuisance parameters are γ and $\mathbf{c}$. Consequently, the standard asymptotic theory is not applicable in testing linearity. This problem may be solved following Davies (1977). Let $\boldsymbol{\gamma}$ be the vector of nuisance parameters. For example, $\boldsymbol{\gamma} = (\gamma, \mathbf{c}')'$ in (3) when the null hypothesis is $\boldsymbol{\psi} = \mathbf{0}$. When $\boldsymbol{\gamma}$ is known testing linearity is straightforward. Let $S_T(\boldsymbol{\gamma})$ be the corresponding test statistic whose large values are critical and define $\Gamma = \{\gamma : \gamma \in \Gamma\}$, the set of admissible values of $\boldsymbol{\gamma}$. When $\boldsymbol{\gamma}$ is unknown, the statistic is not operational because it is a function of $\boldsymbol{\gamma}$. The problem is solved by defining another statistic $S_T = \sup_{\gamma \in \Gamma} S_T(\gamma)$ that is free of nuisance parameters $\boldsymbol{\gamma}$. The asymptotic distribution of S_T under the null hypothesis does not generally have an analytic form, but Davies (1977) gives an approximation to it that holds under certain conditions, including the assumption that $S(\gamma) = \text{plim}_{T \to \infty} S_T(\gamma)$ has a derivative. Other choices of test statistic include the average:

$$S_T = \text{ave}S_T(\gamma) = \int_\Gamma S_T(\gamma) dW(\gamma) \tag{6}$$

where $W(\boldsymbol{\gamma})$ is a weight function defined by the user such that $\int_\Gamma W(\gamma)\mathrm{d}\gamma = 1$, and the exponential

$$\exp S_T = \ln\left(\int_\Gamma \exp\{(1/2)S_T(\gamma)\} dW(\gamma)\right). \tag{7}$$

Andrews and Ploberger (1994) have recommended these tests and demonstrated their local asymptotic optimality properties. The statistics (6) and (7) are two special cases in the family of average exponential tests (for definitions and details, see Andrews and Ploberger, 1994). Hansen (1996) shows how to obtain asymptotic critical values for these statistics by simulation under rather general conditions. His method is computationally intensive but useful. It may be pointed out that it works for SR and STR models where s_t is observable. For MS models, the situation is more complicated, see Garcia (1998) for discussion.

A computationally simpler alternative is to circumvent the identification problem instead of directly solving it. It has been popular in testing linearity against smooth transition models, eq. (3). The idea is to replace the transition function (4) by its Taylor series approximation around the null hypothesis $\gamma = 0$. This transforms the testing problem into one of testing a linear hypothesis in a linear auxiliary regression; see Luukkonen, Saikkonen and Teräsvirta (1988) or Teräsvirta (1998).

Parameter estimation

Parameters of threshold models have to be estimated numerically. This is because the objective function to be optimized is not quadratic in parameters, so an analytical solution to the problem does not exist. The easiest models to estimate are the switching regression or threshold autoregressive models. Their parameters are estimated conditionally by ordinary least squares (OLS), given the switch parameters $c_1, \ldots, c_r$, and the combination of $c_1, \ldots, c_r$ yielding the smallest sum of squared residuals gives the estimates of these and the other parameters. For example, when $r = 1$ the OLS estimation is repeated for a set of c_1 values such that both regimes contain at least a certain minimum amount of observations, typically 10% or 15% of the total number. Under rather general conditions, including stationarity and ergodicity of the TAR process, the least squares estimators are $\sqrt{T}$-consistent and asymptotically normal. The threshold parameter estimators are super (T-) consistent.

Smooth transition models are estimated using standard maximum likelihood. The most efficient numerical method is the Newton–Raphson method that makes use of both the first (the score) and the second (the Hessian) partial derivatives of the log-likelihood function. It has many variants in which the Hessian is replaced by computationally simpler alternatives that either do not require second derivatives, such as the method of scoring or the so-called Berndt–Hall–Hall–Hausman (BHHH) algorithm, or avoid inverting the Hessian altogether. Examples of this include the steepest descent and variable metric methods. Of the latter, the Broyden–Fletcher–Goldfarb–Shanno (BFGS) algorithm can be found in a number of modern software packages.

Hidden Markov models cannot be estimated using standard optimization algorithms because they contain the latent variable s_t. Their parameters are typically estimated using the expectation–maximization (EM) algorithm (see Cappé, Moulines and Rydén, 2005, and Hamilton, 1994, ch. 22).

Estimation of ANN models using maximum likelihood is often numerically demanding because the likelihood function can contain a large number of local maxima, due to a large number of parameters. This problem has been discussed in Medeiros, Teräsvirta and Rech (2006) and White (2006). Because of this difficulty, the literature on ANN models contains a wide variety of estimation methods of various kinds; see for example Fine (1999) and White (2006).

Modelling

Typically, economic theory does not uniquely determine the functional form of a threshold model. This means that the model builder has to select an appropriate model for the problem at hand. In this case, applying a consistent modelling strategy is helpful. The modelling approach of Box and Jenkins (1970) for the ARIMA class of linear models is a case in point. When it comes to threshold models, modelling strategies consisting of stages of specification, estimation and evaluation have been worked out and applied for TAR or, more generally, SR classes of models as well as STR models. For the former, see Tsay (1989) (univariate models) and Tsay (1998) (multivariate models) and for the latter, Teräsvirta (1998) or Teräsvirta (2004). Medeiros, Teräsvirta and Rech (2006) suggest a similar procedure for ANN models. An essential first stage in all these strategies is testing linearity. If linearity is not rejected, the task of the model builder is considerably simplified.

Forecasting

The main purpose of univariate nonlinear models is forecasting. Multivariate models may also be useful for policy analysis. Forecasts are typically conditional means in which the conditioning set consists of a subset of the information available at the time of making the forecast. In nonlinear models such as threshold models, a typical situation is that making forecasts for more than one period ahead requires numerical methods. This is due to the fact that for a random variable X, generally $Eg(X) \neq g(EX)$. Equality holds if g is a linear function of X.

To illustrate, assume an information set $\mathscr{F}_T$ at time T. The optimal one-period mean forecast $f^y_{T,1} = E\{y_{T+1}|\mathscr{F}_T\}$, the conditional mean of y_{T+1}, given $\mathscr{F}_T$. For example, consider the simple bivariate model

$$y_t = g(x_{t-1}) + \varepsilon_t \tag{8}$$

where

$$x_t = \beta x_{t-1} + \eta_t \tag{9}$$

with $|\beta| < 1$, and $\{\eta_t\}$ is a sequence of independent, identically distributed random variables with zero mean. Function $g(\cdot)$ may define an SR, STR or ANN model. The forecast for y_{T+1} equals $f^y_{T,1} = g(x_T)$ as $E\{\varepsilon_{T+1}|\mathscr{F}_T\} = 0$. Thus, if one knows the function $g(\cdot)$, one-step forecasts can be obtained with no difficulty.

The optimum two-step forecast is

$$f^y_{T,2} = E\{y_{T+2}|\mathscr{F}_T\} = E\{g(x_{T+1})|\mathscr{F}_T\}.$$

As x_{T+1} is not usually known at time T, it has to be forecast from its autoregressive equation. This gives a one-step OLS forecast $f^x_{T,1} = \beta x_T$. The two-step forecast equals

$$f^y_{T,2} = E\{g(f^x_{T,1} + {}_{T+1})|\mathscr{F}_T\}. \tag{10}$$

The exact forecast equals

$$fe^y_{T,2} = \int_{-\infty}^{\infty} g(f^x_{T,1} + z)\mathrm{d}D(z)$$

where $D(z)$ is the cumulative distribution function of z. The integral has to be calculated numerically. It may, however, also be approximated by simulation or by bootstrapping the residuals of the estimated model; see Granger and Teräsvirta (1993) or Teräsvirta (2006a). This alternative becomes even more practical when the forecast horizon exceeds two periods. Yet another alternative is to ignore the error η_{T+1}, but the "naive" forecast $fn^y_{T,2} = g(f^x_{T,1})$ is biased. In practice, the function $g(\cdot)$ is not known and has to be specified and estimated from the data before forecasting.

One may also obtain the forecast directly as

$$fd^y_{T,2} = E(y_{T+2}|\mathscr{F}_T)$$

so that $y_{t+2} = g_2(x_t, y_t) + \varepsilon^*_t$, say, and the function $g_2(\cdot)$ has to be determined and estimated separately, rather than derived from the one-step representation (8).

A difficulty with this approach is that the errors will not necessarily be white noise. A separate forecast function is needed for each forecast horizon.

All forecasts from hidden Markov models can be obtained analytically by a sequence of linear operations. This is a direct consequence of the fact that the regimes in (1) when $I(c_{j-1} < s_t \leq c_j)$ is replaced by $I(s_t = j)$, where s_t is a latent discrete variable, are linear in parameters. This is discussed for example in Hamilton (1993) or Teräsvirta (2006a).

Experiences from large empirical studies in which macroeconomic variables are forecast with threshold models, are mixed. No model dominates the others, and in several cases nonlinear threshold models do not improve the accuracy of point forecasts compared to linear models. Recent studies of this type include Stock and Watson (1999), Marcellino (2004) and Teräsvirta, van Dijk and Medeiros (2005).

Further reading

Many statistics and econometrics monographs contain accounts of threshold models, among them Franses and van Dijk (2000), Granger and Teräsvirta (1993) and Guégan (1994). Tong (1990) focuses on TAR models. There are also useful book chapters and review articles such as Bauwens, Lubrano and Richard (2000) offering a Bayesian perspective, Brock and Potter (1993), Teräsvirta (2006a,b) and Tsay (2002). For hidden Markov models, see Cappé, Moulines and Rydén (2005) and Hamilton (1993, 1994, ch. 22). The latter reference concentrates on the autoregressive model (2). Several thorough treatments of ANN models exist; see for example Fine (1999) or Haykin (1999).

TIMO TERÄSVIRTA

See also **forecasting; nonlinear time series analysis.**

Bibliography

Andrews, D.W.K. and Ploberger, W. 1994. Optimal tests when a nuisance parameter is present only under the alternative. *Econometrica* 62, 1383–414.

Bauwens, L., Lubrano, M. and Richard, J.-F. 2000. *Bayesian Inference in Dynamic Econometric Models.* Oxford: Oxford University Press.

Bierens, H.J. 1990. A consistent conditional moment test of functional form. *Econometrica* 58, 1443–58.

Box, G.E.P. and Jenkins, G.M. 1970. *Time Series Analysis, Forecasting and Control.* San Francisco: Holden-Day.

Brock, W.A. and Potter, S.M. 1993. Nonlinear time series and macroeconometrics. In *Handbook of Statistics*, vol. 11, ed. G.S. Maddala, C.R. Rao and H.D. Vinod. Amsterdam: North-Holland.

Cappé, O., Moulines, E. and Rydén, T. 2005. *Inference in Hidden Markov Models.* New York: Springer.

Cybenko, G. 1989. Approximation by superposition of sigmoidal functions. *Mathematics of Control, Signals, and Systems* 2, 303–314.

Davies, R.B. 1977. Hypothesis testing when a nuisance parameter is present only under the alternative. *Biometrika* 64, 247–54.

Fine, T.L. 1999. *Feedforward Neural Network Methodology.* Berlin: Springer.

Franses, P.H. and van Dijk, D. 2000. *Non-Linear Time Series Models in Empirical Finance*. Cambridge: Cambridge University Press.

Funahashi, K. 1989. On the approximate realization of continuous mappings by neural networks. *Neural Networks* 2, 183–92.

Garcia, R. 1998. Asymptotic null distribution of the likelihood ratio test in Markov switching models. *International Economic Review* 39, 763–88.

Granger, C.W.J. and Teräsvirta, T. 1993. *Modelling Nonlinear Economic Relationships*. Oxford: Oxford University Press.

Guégan, D. 1994. *Séries chronologiques non linéaires à temps discret*. Paris: Economica.

Hamilton, J.D. 1993. Estimation, inference and forecasting of time series subject to changes in regime. In *Handbook of Statistics*, vol. 11, ed. G.S. Maddala, C.R. Rao and H.R. Vinod. Amsterdam: North-Holland.

Hamilton, J.D. 1994. *Time Series Analysis*. Princeton, NJ: Princeton University Press.

Hansen, B.E. 1996. Inference when a nuisance parameter is not identified under the null hypothesis. *Econometrica* 64, 413–30.

Hansen, B.E. 1999. Testing for linearity. *Journal of Economic Surveys* 13, 551–76.

Haykin, S. 1999. *Neural Networks. A Comprehensive Foundation*, 2nd edn. Upper Saddle River, NJ: Prentice-Hall.

Hornik, K., Stinchcombe, M. and White, H. 1989. Multi-layer feedforward networks are universal approximators. *Neural Networks* 2, 359–66.

Lindgren, G. 1978. Markov regime models for mixed distributions and switching regressions. *Scandinavian Journal of Statistics* 5, 81–91.

Luukkonen, R., Saikkonen, P. and Teräsvirta, T. 1988. Testing linearity against smooth transition autoregressive models. *Biometrika* 75, 491–99.

Marcellino, M. 2004. Forecasting EMU macroeconomic variables. *International Journal of Forecasting* 20, 359–72.

Medeiros, M.C., Teräsvirta, T. and Rech, G. 2006. Building neural network models for time series: a statistical approach. *Journal of Forecasting* 25, 49–75.

Ramsey, J.B. 1969. Tests for specification errors in classical least-squares regression analysis. *Journal of the Royal Statistical Society, Series B* 31, 350–71.

Stinchcombe, M.B. and White, H. 1998. Consistent specification testing with nuisance parameters present only under the alternative. *Econometric Theory* 14, 295–325.

Stock, J.H. and Watson, M.W. 1999. A comparison of linear and nonlinear univariate models for forecasting macroeconomic time series. In *Cointegration, Causality and Forecasting: A Festschrift in Honour of Clive W. J. Granger*, ed. R.F. Engle and H. White. Oxford: Oxford University Press.

Teräsvirta, T. 1998. Modeling economic relationships with smooth transition regressions. In *Handbook of Applied Economic Statistics*, ed. A. Ullah and D.E. Giles. New York: Dekker.

Teräsvirta, T. 2004. Smooth transition regression modeling. In *Applied Time Series Econometrics*, ed. H. Lütkepohl and M. Krätzig. Cambridge: Cambridge University Press.

Teräsvirta, T. 2006a. Forecasting economic variables with nonlinear models. In *Handbook of Economic Forecasting*, vol. 1, ed. G. Elliott, C.W.J. Granger and A. Timmermann. Amsterdam: North-Holland.

Teräsvirta, T. 2006b. Univariate nonlinear time series. In *Palgrave Handbook of Econometrics: Volume 1, Econometric Theory*, ed. T.C. Mills and K. Patterson. Basingstoke: Palgrave Macmillan.

Teräsvirta, T., van Dijk, D. and Medeiros, M.C. 2005. Smooth transition autoregressions, neural networks, and linear models in forecasting macroeconomic time series: a re-examination. *International Journal of Forecasting* 21, 755–74.

Tong, H. 1990. *Non-Linear Time Series: A Dynamical System Approach*. Oxford: Oxford University Press.

Tsay, R.S. 1989. Testing and modeling threshold autoregressive processes. *Journal of the American Statistical Association* 84, 231–40.

Tsay, R.S. 1998. Testing and modeling multivariate threshold models. *Journal of the American Statistical Association* 93, 1188–202.

Tsay, R.S. 2002. *Analysis of Financial Time Series*. New York: Wiley.

van Dijk, D., Teräsvirta, T. and Franses, P.H. 2002. Smooth transition autoregressive models – a survey of recent developments. *Econometric Reviews* 21, 1–47.

White, H. 2006. Approximate nonlinear forecasting methods. In *Handbook of Economic Forecasting*, ed. G. Elliott, C.W.J. Granger and A. Timmermann. Amsterdam: North-Holland.

time series analysis

Any series of observations ordered along a single dimension, such as time, may be thought of as a time series. The emphasis in time series analysis is on studying the dependence among observations at different points in time. What distinguishes time series analysis from general multivariate analysis is precisely the temporal order imposed on the observations. Many economic variables, such as GNP and its components, price indices, sales, and stock returns are observed over time. In addition to being interested in the contemporaneous relationships among such variables, we are often concerned with relationships between their current and past values, that is, relationships over time.

The study of time series of, for example, astronomical observations predates recorded history. Early writers on economic subjects occasionally made explicit reference to astronomy as the source of their ideas. For example, Cournot (1838) stressed that, as in astronomy, it is necessary to recognize the *secular* variations which are independent of the periodic variations. Similarly, Jevons (1884) remarked that his study of short-term fluctuations used the methods of astronomy and meteorology. During the 19th century interest in, and analysis of, social and economic time series evolved into a new field of study independent of developments in astronomy and meteorology (see Nerlove, Grether and Carvalho, 1979, pp. 1–21, for a historical survey).

Harmonic analysis is one of the earliest methods of analysing time series thought to exhibit some form of periodicity. In this type of analysis, the time series, or some simple transformation of it, is assumed to be the result of the superposition of sine and cosine waves of different frequencies. However, since summing a finite number of such strictly periodic functions always results in a perfectly periodic series, which is seldom observed in practice, one usually allows for an additive stochastic component, sometimes called 'noise'. Thus, an observer must confront the problem of searching for 'hidden periodicities' in the data, that is, the unknown frequencies and amplitudes of sinusoidal fluctuations hidden amidst noise. An early method for this purpose is *periodogram analysis*, suggested by Stokes (1879) and used by Schuster (1898) to analyse sunspot data and later by others, principally William Beveridge (1921; 1922), to analyse economic time series.

Spectral analysis is a modernized version of periodogram analysis modified to take account of the stochastic nature of the entire time series, not just the noise component. If it is assumed that economic time series are fully stochastic, it follows that the older periodogram technique is inappropriate and that considerable difficulties in the interpretation of the periodograms of economic series may be encountered.

At the time when harmonic analysis proved to be inadequate for the analysis of economic and social time series, another way of characterizing such series was suggested by the Russian statistician and economist, Eugen Slutsky (1927), and by the British statistician, G.U. Yule (1921; 1926; 1927). Slutsky and Yule showed that, if we begin with a series of purely random numbers and then take sums or differences, weighted or unweighted, of such numbers, the new series so produced has many of the apparent cyclic properties that were thought at the time to characterize economic and other time series. Such sums or differences of purely random numbers and sums or differences of the resulting series form the basis for the class of autoregressive moving-average (ARMA) processes which are used for modelling many kinds of time series. ARMA models are examples of time domain representations of time series. Although the latter may look very different from spectral representations of time series, there is a one-to-one mapping between time domain analysis and spectral analysis. Which approach is preferred in practice is a matter only of convenience. The choice is often determined by the transparency with which a given question can be answered. The remainder of this article explores these two complementary approaches to the analysis of economic time series.

1. Basic theory

1.1 Stationarity and ergodicity of time series processes

Consider a random variable x_t where $t \in N$, the set of integers; the infinite vector $\{x_t, t \in N\}$ is called a discrete time series. Let M denote a subset of T consecutive elements of N. The distribution of the finite dimensional vector $\{x_t, t \in M\}$ is a well-defined multivariate distribution function, $F_M(\cdot)$. The time series $\{x_t, t \in N\}$ is said to be *strictly stationary* if, for any finite subset M of N and any integer τ, the distribution function of $\{x_t, t \in M + \tau\}$ is the same as the distribution function of $\{x_t, t \in M\}$. In other words, the joint distribution function of the finite vector of observations on x_t is invariant with respect to the origin from which time is measured. All the unconditional moments of the distribution function, if they exist, are independent of the index t; in particular,

$$\begin{aligned} E(x_t) &= \mu \\ \gamma(\tau) &= E[x_t - \mu][x_{t+\tau} - \mu], \end{aligned} \tag{1}$$

where $\gamma(\tau)$ is the autocovariance function and depends only on the difference in indices, τ. Time-series processes for which (1) holds, but which are not necessarily strictly stationary according to the definition above, are said to be weakly *stationary*, *covariance stationary*, or *stationary to the second order*. Time-series processes for which $F_M(\cdot)$ is multivariate normal for any subset M of N are called *Gaussian processes*. For Gaussian processes covariance stationarity implies strict stationarity.

In practice, we usually observe only one realization of a finite subset of the time series of interest, corresponding to one of the many possible draws of length T from $F_M(\cdot)$. The question is whether the moments of x_t may be inferred from one such realization; for example, from the time averages of sums (or sums of products) of the observed values of a time series. If the process is what is known as *ergodic*, time averages of functions of the observations on the time series at T time points converge in mean square to the corresponding population expectations of x_t across alternative draws, as $T \to \infty$ (Priestley, 1981, pp. 340–3; Doob, 1953, p. 465). It is possible for a process to be stationary, yet not ergodic. Consider, for example, the process $x_t^{(i)} = \eta^{(i)} + \varepsilon_t$, where $x_t^{(i)}$ denotes the ith draw for observation x_t from the universe of all possible draws for x_t. Suppose that $\eta^{(i)} \sim N(0, \lambda^2)$ is the mean of the ith draw and that $\varepsilon_t \sim N(0, \sigma^2)$ is independent of $\eta^{(i)}$. This process is clearly stationary in that the probability limit of the ensemble average is zero, yet the time average $\sum_{t=1}^{T} x_t^{(i)}/T = \eta^{(i)} + \sum_{t=1}^{T} \varepsilon_t/T$ converges to $\eta^{(i)}$ rather than zero, thus violating ergodicity.

1.2 The Wold decomposition and general linear processes

Let $\{\varepsilon_t\}$ be one element of a time series of serially uncorrelated, identically distributed random variables with zero mean and variance σ^2. Then the infinite, one-sided moving average (MA) process

$$x_t = \sum_{j=0}^{\infty} b_j \varepsilon_{t-j}, \tag{2}$$

where $b_0 = 1$ and $\sum_{j=0}^{\infty} b_j^2 < \infty$, is also a well-defined stationary process with mean 0 and variance $\sigma^2 \Sigma_0^{\infty} b_j^2$. Processes of this form and, more generally, processes based on an infinite two-sided MA of the same form are called *linear processes*, are always ergodic, and play a key role in time series analysis (Hannan, 1970).

The importance of the process (2) is underscored by the Wold decomposition theorem (Wold, 1938), which states that any weakly stationary process may be decomposed into two mutually uncorrelated component processes, one an infinite one-sided MA of the form (2) and the other a so-called linearly deterministic process, future values of which can be predicted exactly by some linear function of past observations. The linearly deterministic component is non-ergodic.

2. Linear processes in time and frequency domains

2.1 Autocovariance and autocovariance generating functions

The autocovariance function of a stationary process, defined in (1) above, or its matrix generalization for vector processes, provides the basic representation of time dependence for weakly stationary processes. For the stationary process defined in (2), it is

$$\gamma(\tau) = \sigma^2 \sum_{j=0}^{\infty} b_j b_{j+\tau}. \tag{3}$$

Let z denote a complex scalar. Then the autocovariance generating transform is defined as

$$g(z) = \sum_{-\infty}^{\infty} \gamma(\tau) z^{\tau} \tag{4}$$

in whatever region of the complex plane the series on the right-hand side converges. If the series $\{x_t\}$ is covariance stationary, convergence will occur in an annulus about the unit circle. The autocovariance generating transform for the one-sided MA process defined in (2) is

$$g(z) = \sigma^2 B(z) B(z^{-1}) \tag{5}$$

where

$$B(z) = \sum_{k=0}^{\infty} b_k z^k.$$

If $B(z)$ has no zeros on the unit circle, the process defined in (2) is invertible and also has an infinite-order autoregressive (AR) representation as

$$A(L)x_t = \varepsilon_t, \tag{6}$$

where L is the lag operator such that $L^j x_t = x_{t-j}$ and $A(L) = a_0 + a_1 L + a_2 L^2 + \cdots$.

So-called ARMA processes have an autocovariance generating transform which is a rational function of z. If the ARMA process is both stationary and invertible, $g(z)$ may be written as

$$G(z) = \frac{P(z)P(z^{-1})}{Q(z)Q(z^{-1})} = \sigma^2 \frac{\prod_{k=1}^{m}(1-\beta_k z)(1-\beta_k z^{-1})}{\prod_{j=1}^{n}(1-\alpha_j z)(1-\alpha_j z^{-1})} \tag{7}$$

where $|\beta_k|, |\alpha_j| < 1 \ \forall j, k$. Then the corresponding ARMA model is

$$Q(L)x_t = P(L)\varepsilon_t, \tag{8}$$

where

$$Q(L) = \prod_{j=1}^{n}(1-\alpha_j L) \quad \text{and} \quad P(L) = \prod_{k=1}^{m}(1-\beta_k L).$$

2.2 Spectral density functions

If the value of z lies on the complex unit circle, it follows that $z = e^{-i\lambda}$, where $i = \sqrt{-1}$ and $-\pi \le \lambda \le \pi$. Substituting for z in the autocovariance generating transform (5) and dividing by 2π, we obtain the *spectral density function* of a linearly non-deterministic stationary process $\{x_t\}$in terms of the frequency λ:

$$\begin{aligned} f(\lambda) &= (1/2\pi)g(\mathrm{e}^{i\lambda}) = (\sigma^2/2\pi)B(\mathrm{e}^{i\lambda})B(\mathrm{e}^{-i\lambda}) \\ &= (1/2\pi)\sum_{-\infty}^{\infty}\gamma(\tau)\mathrm{e}^{-i\lambda\tau}, \qquad -\pi \leq \lambda < \pi. \end{aligned} \tag{9}$$

Thus, the spectral density function is the Fourier transform of the autocovariance function. It can be shown that for a process with absolutely summable autocovariances the spectral density function exists and can be used to compute all of the autocovariances, so the same time series can be characterized equivalently in terms of the autocovariance function in the time-domain or in terms of the spectral density function in the frequency domain.

The spectral density function for a linearly non-deterministic, stationary, real-valued time series is a real-valued, non-negative function, symmetric about the origin, defined in the interval $[-\pi, \pi]$:

$$f(\lambda) = (1/2\pi)\left[\gamma(0) + 2\sum_{\tau=1}^{\infty}\gamma(\tau)\cos\lambda\tau\right]. \tag{10}$$

Moreover,

$$E(x_t - \mu)^2 = \int_{-\pi}^{\pi} f(\lambda)\mathrm{d}\lambda, \tag{11}$$

so that the spectral density function is a frequency-band decomposition of the variance of $\{x_t\}$.

When the process generating $\{x_t\}$ is merely stationary, that is, when $\{x_t\}$ may have a linearly deterministic component, the spectral density function is

$$f(\lambda) = \int_{-\pi}^{\pi} \mathrm{e}^{i\lambda\tau}\mathrm{d}F(\lambda), \tag{12}$$

where $F(\lambda)$ is a distribution function (Doob, 1953, p. 488). Note that deterministic seasonal effects, for example, may cause a jump in the spectral distribution function.

The autocovariance function, its generating transform and the spectral distribution function all have natural generalizations to the multivariate case, in which $\{x_t\}$ can be thought of as a vector of time-series processes.

The estimation and analysis of spectral density and distribution functions play an important role in all forms of time-series analysis. More detailed treatments are Doob (1953), Fishman (1969), Koopmans (1974), Fuller (1976), Nerlove, Grether and Carvalho (1979, ch. 3) and Priestley (1981).

2.3 Unobserved components (UC) models

In the statistical literature dealing with the analysis of economic time series it is common practice to classify the types of movements that characterize a time series as trend, cyclical, seasonal, and irregular components. The idea that a time series may best be viewed as being composed of several unobserved components is by no means universal, but it plays a fundamental role in many applications, for example, the choice of methods for seasonal adjustment. Nerlove, Grether and Carvalho (1979, Ch. 1) review the history of the idea of unobserved components in economics from its origin early in the 19th century.

In the 1960s, Nerlove (1964; 1965; 1967) and Granger (1966) suggested that the typical spectral shape of many economic time series could be accounted for by the superposition of two or more independent components with specified properties. There are basically two approaches to the formulation of UC models: Theil and Wage (1964) and Nerlove and Wage (1964), Nerlove (1967) and Grether and Nerlove (1970) choose the form of components in such a way as to replicate the typical spectral shape of the series which represents their superposition. For example, let T_t represent the trend component, C_t the cyclical, S_t the seasonal, and I_t the irregular of a monthly time series; then the observed series can be represented as

$$y_t = T_t + C_t + S_t + I_t, \tag{13}$$

where

$$T_t = a_0 + a_1 t + a_2 t^2 + \cdots + a_p t^p,$$

$$C_t = \frac{1 + \beta_1 L + \beta_2 L^2}{(1 - \alpha_1 L)(1 - \alpha_2 L)} \varepsilon_{1t},$$

$$S_t = \frac{1 + \beta_3 L + \beta_4 L^2}{1 - \gamma L^{12}} \varepsilon_{2t},$$

$$I_t = \varepsilon_{3t},$$

and $\varepsilon_{1t}, \varepsilon_{2t}$, and ε_{3t} are i.i.d. normal variables with variances σ_{11}, σ_{22}, and σ_{33}, respectively. This approach has been carried forward by Harvey (1984), Harvey and Peters (1990) and Harvey and Todd (1984).

An alternative approach is to derive the components of the UC model from a well-fitting ARMA model (obtained after suitably transforming the data), given sufficient a priori identifying restrictions on the spectral properties of the components. See Box, Hillmer and Tiao (1978), Pierce (1978; 1979), Burman (1980), Hillmer and Tiao (1982), Hillmer, Bell and Tiao (1983), Bell and Hillmer (1984), Burridge and Wallis (1985), and Maravall (1981; 1984). The basis of this procedure is the fact that every stationary UC model, or the stationary part of every UC model, has an equivalent ARMA form, the so-called canonical form of the UC model (Nerlove and Wage, 1964; Nerlove, Grether and Carvalho, 1979, Ch. 4).

3. Specification, estimation, inference and prediction

3.1 Autocovariance and spectral density functions

Suppose we have a finite number of observations of a realization of the process generating the time series, say $x_1, \ldots, x_T$. For expository purposes it is assumed that all deterministic components of x_t have been removed. If μ is unknown, this may be accomplished by subtracting the sample mean of the time series observations from the data prior to the analysis. For a zero mean series x_t there are basically two ways of estimating $\gamma(\tau)$ defined in (1): the first is the biased estimator

$$c(\tau) = (1/T)\sum_{t=1}^{T-|\tau|} x_t x_{t+|\tau|}, \quad \tau = 0, \quad \pm 1, \ldots, \pm M, M \leq (T-1). \tag{14}$$

The second is the unbiased estimator

$$\tilde{c}(\tau) = [1/(T-|\tau|)]\sum_{t=1}^{T-|\tau|} x_t x_{t+|\tau|}, \quad \tau = 0, \quad \pm 1, \ldots, \pm M, \quad M \leq T-1. \tag{15}$$

Although $c(\tau)$ is biased in finite samples, it is asymptotically unbiased. The key difference between $c(\tau)$ and $\tilde{c}(\tau)$ is that $c(\tau)$ is a positive definite function of τ whereas $\tilde{c}(\tau)$ is not (Parzen, 1961, p. 981). The variance and covariances of the estimated autocovariances are derived, *inter alia*, by Hannan (1960), and Anderson (1971). As $T \to \infty$, both tend to zero, as the estimates are asymptotically uncorrelated and consistent. However,

$$E[c(\tau) - Ec(\tau)]^2/E[c(\tau)] \to \infty \quad \text{as} \quad \tau/T \to 1. \tag{16}$$

This property accounts for the failure of the estimated autocorrelation function

$$r(\tau) = c(\tau)/c(0) \tag{17}$$

to damp down as $\tau \to \infty$, as it should for a stationary, linearly non-deterministic process (Hannan, 1960, p. 43).

A 'natural' estimator of the spectral density function is obtained by replacing $\gamma(\tau)$ in (10) by $c(\tau)$ or $\tilde{c}(\tau)$. The resulting estimator is proportional, at each frequency, to a sample quantity called the periodogram:

$$I_T(\lambda) = (2/T)\left|\sum_{1}^{T} e^{i\lambda t} x_t\right|^2 \tag{18}$$

usually evaluated at the equi-spaced frequencies

$$\lambda = 2k\pi/T, \qquad k = 1, 2, \ldots, [T/2] \tag{19}$$

in the interval $[0, \pi]$. Although, for a stationary, nonlinearly deterministic process, the periodogram ordinates are asymptotically unbiased estimates of the spectral densities at the corresponding frequencies, they are not consistent estimates; moreover, the correlation between adjacent periodogram ordinates tends to zero with increasing sample size. The result is that the periodogram presents a jagged appearance which is increasingly difficult to interpret as more data become available.

In order to obtain consistent estimates of the spectral density function at specific frequencies, it is common practice to weight the periodogram ordinates over the frequency range or to form weighted averages of the autocovariances at different lags. There is a substantial literature on the subject. The weights are called a 'spectral window'. Essentially the idea is to reduce the variance of the estimate of an average spectral density around a particular frequency by averaging periodogram ordinates which are asymptotically unbiased and independently distributed estimates of the corresponding ordinates of the spectral density function. Related weights can also be applied to the estimated autocovariances which are substituted in (10); this weighting system is called a 'lag window'.

Naturally the sampling properties of the spectral estimates depend on the nature of the 'window' used to obtain consistency (see Priestley, 1981, pp. 432–94 for further discussion). Regardless of the choice of window, the 'bandwidth' used in constructing the window must decrease at a suitable rate as the sample size grows. In the spectral window approach, this means that the window width must decrease at a slower rate than the sample size. In the lag window approach, this means that the number of included autocovariances must increase at a slower rate than the sample size.

3.2 ARMA models

The autocovariance function and the spectral density function for a time series represent nonparametric approaches to describing the data. An alternative approach is to specify and estimate a parametric ARMA model for x_t This approach involves choosing the orders of the polynomials P and Q in (7) and (8) and perhaps also specifying that one or more coefficients are zero or placing other restrictions on P and Q. The problem then becomes one of estimating the parameters of the model.

Despite the poor statistical properties of the estimated autocovariance function and a related function called the partial autocorrelation function, these are sometimes used to specify the orders of the polynomials P and Q. An alternative approach is to select the model that minimizes the value of information-theoretic criteria of the form

$$\mathrm{IC}(i) = \log(\hat{\sigma}_i^2) + k_i c_T, \tag{20}$$

where k_i refers to the number of estimated parameters in the candidate models $i = 1, \ldots, M$, and $\hat{\sigma}_i^2$ to the corresponding maximum likelihood estimate of the residual variance. Such criteria incorporate a trade-off between the fit of a model and its degree of parsimony. That trade-off depends on the penalty term c_t (Akaike, 1970; 1974; Schwarz, 1978). There is no universally accepted choice for c_T. For $c_T = 2/T$ expression (20) reduces to the Akaike information criterion (AIC), for example, and for $c_T = \ln(T)/T$ to the Schwarz information criterion (SIC). The asymptotic properties of alternative criteria will depend on the objective of the user and the class of models considered.

Given the orders of the AR and MA components, a variety of maximum likelihood or approximate maximum likelihood methods are available to estimate the model parameters. Newbold (1974) shows that, if x_t is characterized by (8) with $\varepsilon_t \sim NID(0, \sigma^2)$, then the exact likelihood function for the parameters of $P(\cdot)$ and $Q(\cdot)$ is such that the maximum likelihood estimates of the parameters and the least-squares (LS) estimates (in general highly nonlinear) are asymptotically identical. Only in the case of a pure AR model are the estimates linear conditional on the initial observations. Several approximations have been discussed (Box and Jenkins, 1970; Granger and Newbold, 1977; Nerlove, Grether and Carvalho, 1979, pp. 121–5).

Exact maximum likelihood estimation of ARMA models has been discussed by, inter alia, Newbold (1974), Anderson (1977), Ansley (1979), and Harvey (1981). Following Schweppe (1965), Harvey suggests the use of the Kalman filter to obtain the value of the exact-likelihood function, which may be maximized by numerical methods. The Kalman filter approach is easily adapted to the estimation of UC models in the time domain.

An alternative to exact or approximate maximum-likelihood estimation in the time domain was suggested by Hannan (1969). Estimates may be obtained by maximizing an approximate likelihood function based on the asymptotic distribution of the periodogram ordinates defined in (18). These are asymptotically independently distributed (Brillinger, 1975, p. 95), and the random variables $2I_t(\lambda)/f(\lambda)$ have an asymptotic χ^2 distribution with two degrees of freedom (Koopmans, 1974, pp. 260–5). This means that the asymptotic distribution of the observations, $\{x_1, \ldots, x_T\}$ is proportional to

$$\prod_{j=0}^{[T/2]} [1/f(\lambda_j)]\exp[-I(\lambda_j)/f(\lambda_j)] \tag{21}$$

where $\lambda_j = 2j\pi/T, j = 0, \ldots, [T/2]$, are the equi-spaced frequencies in the interval $[0, \pi]$ at which the periodogram is evaluated (Nerlove, Grether and Carvalho, 1979, pp. 132–6). Since the true spectral density $f(\lambda)$ depends on the parameters characterizing the process, this asymptotic distribution may be interpreted as a likelihood function. Frequency domain methods, as these are called, may easily be applied in the case of UC models.

Whether approximate or exact maximum-likelihood estimation methods are employed, inference may be based on the usual criteria related to the likelihood function. Unfortunately, serious difficulties may be encountered in applying the asymptotic theory, since the small sample distribution of the maximum likelihood estimator may differ greatly from the limiting distribution in important cases (Sargan and Bhargava, 1983; Anderson and Takemura, 1986).

3.3 Prediction and extraction

The problem of prediction is essentially the estimation of an unknown future value of the time series itself; the problem of extraction, best viewed in the context of UC models described in section 2.3, is to estimate the value of one of the unobserved components at a particular point in time, not necessarily in the future. Problems of trend extraction and seasonal adjustment may be viewed in this way (Grether and Nerlove, 1970). How the prediction (or extraction) problem is approached depends on whether we are assumed to have an infinite past history and, if not, whether the parameters of the process generating the time series are assumed to be known.

In practice, of course, an infinite past history is never available, but a very long history is nearly equivalent if the process is stationary or can be transformed to stationarity. It is common, as well, to restrict attention to linear predictors, which involves no loss of generality if the processes considered are Gaussian and little loss if merely linear. To devise a theory of optimal prediction or extraction requires some criterion by which to measure the accuracy of a particular candidate. The most common choice is the minimum mean-square error (MMSE) criterion, which is also the conditional expectation of the unknown quantity. For a discussion of alternative loss functions see Granger (1969) and Christoffersen and Diebold (1996; 1997).

The theory of optimal prediction and extraction due to Kolmogorov (1941) and Wiener (1949) and elaborated by Whittle (1963) for discrete processes assumes a possibly infinite past history and known parameters. As a special case of the Wiener–Kolmogorov theory for non-deterministic, stationary processes, consider the linear process defined by (2). Since the ε_t are i.i.d. with zero mean and variance σ^2, it is apparent that the conditional expectation of x_{t+v}, given all innovations from the infinite past to t, is

$$\hat{x}_{t+v} = b_v\varepsilon_t + b_{v+1}\varepsilon_{t+1} + \ldots \tag{22}$$

Of course, even if the parameters $b_j, j = 0, 1, \ldots,$ are assumed to be known, the series $\{\varepsilon_t\}$ is not directly observable. The ε_t's are sometimes called the *innovations* of the process, since it is easy to show that $\varepsilon_{t+1} = x_{t+1} - \hat{x}_{t+1}$ is the one-step ahead prediction error. If the process is invertible, it has the autoregressive representation (6) and so can be expressed solely in terms of the, generally infinite-order, autoregression

$$\hat{x}_{t+v} = D(L)x_t, \tag{23}$$

where the generating transform of the coefficients of D is

$$D(z) = \frac{1}{B(z)}\left[\frac{B(z)}{z^v}\right]_+.$$

The operator $[\cdot]_+$ eliminates terms involving negative powers of z.

The problem of extraction is best viewed in the context of multiple time series; in general we wish to 'predict' one time series $\{y_t\}$ from another related series $\{x_t\}$. It is not necessary that the series $\{y_t\}$ actually be observed as long as its relationship to an observed series $\{x_t\}$ can be described (Nerlove, Grether and Carvalho, 1979, Ch. 5).

The Kalman filter approach to prediction and extraction (Kalman, 1960) is both more special and more general than the Wiener–Kolmogorov theory: attention is restricted to finite-dimensional parameter spaces and linear processes, but these processes need not be stationary. The parameters may vary with time, and we do not require an infinite past. This approach represents a powerful tool of practical time-series analysis and may be easily extended to multiple time series. A full discussion, however, requires a discussion of 'state-space representation' of time series processes and is beyond the scope of this entry (Harvey, 1989)

4. Multiple time series analysis

A general treatment of multiple time series analysis is contained in Hannan (1970). The two-variable case will serve to illustrate the matter. Two stationary time series $\{x_t\}$ and $\{y_t\}$ are said to be jointly stationary if their joint distribution function does not depend on the origin from which time is measured. Joint stationarity implies, but is not in general implied by, weak or covariance joint stationarity; that is, $cov(x_t, y_s)$ is a function of $s - t$ only. In this case the cross-covariance function is

$$\gamma_{yx}(\tau) = E[y_t - \mu_y][x_{t-\tau} - \mu_x], \tag{24}$$

where $\mu_x = Ex_t$ and $\mu_y = Ey_t$. Note that $\gamma_{yx}(\tau)$ and $\gamma_{xy}(\tau)$ are, in general, different. The cross-covariance generating function is defined as

$$g_{yx}(z) = \sum_{-\infty}^{\infty} \gamma_{yx}(\tau) z^{\tau} \tag{25}$$

in that region of the complex plane in which the right-hand side of (25) converges. For two jointly stationary series this occurs in an annulus containing the unit circle. In this case, the cross-spectral density function is defined as

$$f_{yx}(\lambda) = (1/2\pi) g_{yx}(e^{i\lambda}). \tag{26}$$

Since $\gamma_{yx}(\tau)$ and $\gamma_{xy}(\tau)$ are not equal, the cross-spectral density function is complex valued and can be decomposed into a real part (the co-spectral density) and a complex part (the quadrature spectral density):

$$f_{yx}(\lambda) = c_{yx}(\lambda) + iq_{yx}(\lambda). \tag{27}$$

In polar form, the cross-spectral density may be written as

$$f_{yx}(\lambda) = \alpha_{yx}(\lambda) \exp[i\phi_{yx}(\lambda)], \tag{28}$$

where $\alpha_{yx}(\lambda) = [c_{yx}^2(\lambda) + q_{yx}^2(\lambda)]^{1/2}$ is called the amplitude or gain, and where $\phi_{yx}(\lambda) = \arctan\{-q_{yx}(\lambda)/c_{yx}(\lambda)\}$ is called the phase. Another useful magnitude is the coherence between the two series, defined as

$$\rho_{yx}(\lambda) = \frac{|f_{yx}(\lambda)|^2}{f_{xx}(\lambda) f_{yy}(\lambda)}, \tag{29}$$

which measures the squared correlation between y and x at a frequency λ. Clearly, $\rho_{yx}(\lambda) = \rho_{xy}(\lambda)$. Estimation of cross-spectral density functions and related quantities is discussed in Priestley (1981, pp. 692–712).

Often it is convenient to impose additional parametric structure in modelling multiple time series. The workhorse multiple time series model in econometrics has been the covariance-stationary K-dimensional vector autoregressive model, which may be viewed as a natural generalization of the univariate AR model discussed earlier:

$$A(L)x_t = \varepsilon_t \tag{30}$$

where $A(L) = I_K - A_1L - \cdots - A_pL^p$. Here each variable in x_t is regressed on its own lags as well as lags of all other variables in x_t up to some pre-specified lag order p. This *vector autoregression* (VAR) can also be viewed as an approximation to a general linear process x_t, and may be estimated by LS.

Similarly, the formulation of ARMA and UC models discussed earlier may be extended to the multivariate case by interpreting the polynomials in the lag operator as matrix polynomials and by replacing the scalar random variables by vectors. Although these vector ARMA and UC models bear a superficial resemblance to the corresponding univariate ones, their structure is, in fact, much more complicated and gives rise to difficult identification problems. In the univariate case, we can formulate simple conditions under which a given covariance function identifies a unique ARMA or UC model, but in the multivariate case these conditions are no longer sufficient. Hannan (1970; 1971) gives a complete treatment. State-space methods have also been employed to study the structure of multivariate ARMA models (Hannan, 1976; and, especially, 1979).

5. Unit roots, co-integration and long memory

Standard tools for time series analysis have been developed for processes that are covariance stationary or have been suitably transformed to achieve covariance stationarity by removing (or explicitly modelling) deterministic trends, structural breaks, and seasonal effects. The presence of a *unit root* in the autoregressive lag order polynomial of an ARMA process also violates the assumption of stationarity. Processes with a unit root are also called integrated of order one (or $I(1)$ for short) because they become covariance-stationary only upon being differenced once. In general, $I(d)$ processes must be differenced d times to render the process covariance-stationary.

The presence of unit roots has important implications for estimation and inference. When the scalar process x_t is $I(1)$ the variance of x_t will be unbounded, model innovations will have permanent effects on the level of x_t, the autocorrelation function does not die out, and x_t will not revert to a long-run mean. Moreover, coefficients of $I(1)$ regressors will have nonstandard asymptotic distributions, invalidating standard tools of inference.

The simplest example of an autoregressive integrated moving-average (ARIMA) process is the random walk process: $x_t = x_{t-1} + \varepsilon_t$. The potential pitfalls of regression analysis with $I(1)$ data are best illustrated by the problem of regressing one independent random walk on another. In that case, it can be shown that R^2 and $\hat{\beta}$ will be random and that the usual t-statistic will diverge, giving rise to seemingly significant correlations between variables that are unrelated by construction.

This *spurious regression* problem was first discussed by Yule (1926), further illustrated by Granger and Newbold (1974), and formally analyzed by Phillips (1986) and Phillips and Durlauf (1986). Similar problems arise in deterministically detrending $I(1)$ series (Nelson and Kang, 1981; Durlauf and Phillips, 1988). Unbalanced regressions, that is, regressions in which the regressand is not of the same order of integration as the regressor, may also result in spurious inference. An exception to this rule is inference on coefficients of mean zero $I(0)$ variables in regressions that include a constant term (Sims, Stock and Watson, 1991).

The standard response to dealing with $I(1)$ data is to difference the data prior to the analysis. There is one important exception to this rule. There are situations in which several variables are individually $I(1)$, but share a common unit root component. In that case, a linear combination of these variables will be $I(0)$:

$$c'x_t = u_t \sim I(0), \quad c \neq 0 \tag{31}$$

where x_t denotes a K-dimensional vector of $I(1)$ variables and c is a $(K \times 1)$ parameter vector. In other words, these variables share a common stochastic trend. This phenomenon is known as *co-integration* (Granger, 1981; Engle and Granger, 1987) and c is known as the co-integrating vector. Clearly, c is not unique. It is common to normalize one element of c to unity. The LS estimator of c in (31) is consistent, but corrections for omitted dynamics are recommended (Stock and Watson, 1993; Phillips and Hansen, 1990). Co-integrating relationships have been used extensively in modelling long-run equilibrium relationships in economic data (Engle and Granger, 1991).

Variables that are co-integrated are linked by an *error correction* mechanism that prevents the integrated variables from drifting apart without bound. Specifically, by the Granger representation theorem of Engle and Granger (1987), under some regularity conditions, any K-dimensional vector of co-integrated variables x_t can be represented as a vector error correction (VEC) model of the form:

$$\Delta x_t = \sum_{i=1}^{p-1} \Gamma_i \Delta x_{t-i} - \Pi x_{t-p} \tag{32}$$

where $\Gamma_i, \; i = 1, \ldots, p-1$, and $\Pi \equiv BC$ are conformable coefficient matrices and Δ denotes the first-difference operator. Model (32) allows for up to r co-integrating relationships where r is the rank of Π. For $r = 0$, the error correction term in model (32) drops out and the model reduces to a difference-stationary VAR. For $r = K$, all variables are $I(0)$ and model (32) is equivalent to a stationary VAR in levels. Otherwise, there are $0<r<K$ common trends. If the $(r \times K)$ matrix of co-integrating vectors, C, is known, the model in (32) reduces to

$$\Delta x_t = \sum_{i=1}^{p-1} \Gamma_i \Delta x_{t-i} - B z_{t-p} \tag{32'}$$

where $z_{t-p} \equiv Cx_{t-p}$, and the model may be estimated by LS; if only the rank r is known, the VEC model in (32) is commonly estimated by full information maximum likelihood methods (Johansen, 1995).

Starting with Nelson and Plosser (1982), a large literature has dealt with the problem of statistically discriminating between $I(1)$ and $I(0)$ models for economic data. Notwithstanding these efforts, it has remained difficult to detect reliably the existence of a unit root (or of co-integration). The problem is that in small samples highly persistent, yet stationary processes are observationally equivalent to exact unit root processes. It may seem that not much could hinge on this distinction then, but it can be shown that $I(1)$ and $I(0)$ specifications that fit the data about equally well may have very different statistical properties and economic implications (Rudebusch, 1993).

For processes with roots near unity in many cases neither the traditional asymptotic theory for $I(0)$ processes nor the alternative asymptotic theory for exact $I(1)$ processes will provide a good small-sample approximation to the distribution of estimators and test statistics. An alternative approach is to model the dominant root, ρ, of the autoregressive lag order polynomial as *local-to-unity* in the sense that $\rho = 1 - c/T, \ c > 0$. This asymptotic thought experiment gives rise to an alternative asymptotic approximation that in many cases provides a better small-sample approximation than imposing the order of integration or relying on unit root pretests (Stock, 1991; Elliott, 1998).

Stationary ARMA processes are 'short memory' processes in that their autocorrelation function dies out quickly. For large τ, ARMA autocorrelations decay approximately geometrically, that is, $\rho(\tau) \approx r^{\tau}$, where r is a constant such that $|r| < 1$. In many applied contexts including volatility dynamics in asset returns, there is evidence that the autocorrelation function dies out much more slowly. This observation has motivated the development of the class of *fractionally integrated* ARMA (ARFIMA) models:

$$Q(L)(1-L)^{d}x_t = P(L)\varepsilon_t \tag{33}$$

where d is a real number, as opposed to an integer (Baillie, 1996). Stationarity and invertibility require $|d| < 0.5$, which can always be achieved by taking a suitable number of differences. The autocorrelation function of an ARFIMA process decays at a hyperbolic rate. For large τ, we have $\rho(\tau) \approx \tau^{2d-1}$, where $d < 1/2$ and $d \neq 0$. Such '*long memory*' models may be estimated by the two-step procedure of Geweke and Porter-Hudak (1983) or by maximum likelihood (Sowell, 1992; Baillie, Bollerslev and Mikkelsen, 1996). A detailed discussion including extensions to the notion of fractional co-integration is provided by Baillie (1996). Long memory may arise, for example, from infrequent stochastic regime changes (Diebold and Inoue, 2001) or from the aggregation of economic data (Granger, 1980; Chambers, 1998). Perhaps the most successful application of long-memory processes in economics has been work on modelling the volatility of asset prices and powers of asset returns, yielding new insights into the behaviour of markets and the pricing of financial risk.

6. Nonlinear time series models

The behaviour of many economic time series appears to change distinctly at irregular intervals, consistent with economic models that suggest the existence of floors and ceilings, buffer stocks and regime switches in the data. This observation has given rise to a large literature dealing with nonlinear time series models. Nonlinear time series models still have a Wold representation with linearly unpredictable innovations, but these innovations are nevertheless dependent over time. This has important implications for forecasting and for the dynamic properties of the model. For example, the effects of innovations in nonlinear models will depend on the path of the time series and the size of the innovation, and may be asymmetric.

6.1 Nonlinear dynamics in the conditional mean

The increasing importance of nonlinear time series models in econometrics is best illustrated by two examples: hidden Markov chain models and smooth transition regression models of the conditional mean.

The idea of hidden Markov chains first attracted attention in econometrics in the context of *regime switching models* (Hamilton, 1989). The original motivation was that many economic time series appear to follow a different process during recession phases of the business cycle than during economic expansions. This type of regime-switching behaviour may be modelled in terms of an unobserved discrete-valued state variable (for example, 1 for a recession and 0 for an expansion) that is driven by a Markov chain. The transition from one state to another is governed by a matrix of transition probabilities that may be estimated from past data. The essence of this method thus is that the future will in some sense be like the past. A simple example of this idea is the regime-switching AR(1) model:

$$x_t = a_{1st}x_{t-1} + \varepsilon_t, \quad \varepsilon_t \sim NID(0, \sigma^2) \tag{34}$$

where the regime s_t is the outcome of an unobserved two-state Markov chain with s_t independent of ε_τ for all t and τ. In this model, the time-varying slope parameter will take on different values depending on the state s. Once the model has been estimated by maximum likelihood methods, it is possible to infer how likely a given regime is to have generated the observed data at date t. An excellent review of the literature on hidden Markov models is provided by Cappé, Moulines and Ryden (2005); for a general treatment of state space representations of nonlinear models, also see Durbin and Koopman (2001).

The idea of *smooth transition regression models* is based on the observation that many economic variables are sluggish and will not move until some state variable exceeds a certain threshold. For example, price arbitrage in markets will only set in once the expected profit of a trade exceeds the transaction cost. This observation has led to the development of models with fixed thresholds that depend on some observable state variable. Smooth transition models allow for the possibility that this transition occurs not all of a sudden at a fixed threshold but gradually, as one would expect in time series data that have been aggregated across many market participants.

A simple example is the smooth-transition AR(1) model:

$$x_t = \Phi(z_{t-1}, \ldots, z_{t-d}, \Gamma)x_{t-1} + \varepsilon_t \qquad \varepsilon_t \sim NID(0, \sigma^2) \tag{35}$$

where $\Phi(.)$ denotes the transition function, z_t is a zero mean state variable denoting the current deviation of x_t from a (possibly time-varying) equilibrium level and Γ is the vector of transition parameters. Common choices for the transition function are the logistic or the exponential function. For example, we may specify $\Phi(.) = (\exp\{\gamma(z_{t-1})^2\})$ with $\gamma < 0$. If $z_{t-1} = 0$, $\Phi(.) = 1$ and the model in (35) reduces to a random walk model; otherwise, $\Phi(.) < 1$ and the model in (35) reduces to a stationary AR(1). The degree of mean reversion is increasing in the deviation from equilibrium. For further discussion see Granger and Teräsvirta (1993).

6.2. Nonlinear dynamics in the conditional variance

While the preceding examples focused on nonlinear dynamics in the conditional mean, nonlinearities may also arise in higher moments. The leading example is the conditional variance. Many economic and financial time series are characterized by volatility clustering. Often interest centres on predicting these *volatility dynamics* rather than the conditional mean. The basic idea of modelling and forecasting volatility was set out in Engle's (1982) path-breaking paper on autoregressive conditional heteroskedasticity (ARCH). Subsequently, Bollerslev (1986) introduced the class of generalized autoregressive conditionally heteroskedastic (GARCH). Consider a decomposition of x_t into the one-step ahead conditional mean, $\mu_{t|t-1} \equiv E(x_t|\Omega_{t-1})$, and conditional variance, $\sigma^2_{t|t-1} \equiv Var(x_t|\Omega_{t-1})$, where Ω_{t-1} denotes the information set at $t-1$:

$$x_t = \mu_{t|t-1} + \sigma_{t|t-1}\nu_t \qquad \nu_t \sim NID(0, 1) \tag{36}$$

The leading example of a GARCH model of the conditional variance is the GARCH(1,1) model, which is defined by the recursive relationship

$$\sigma^2_{t|t-1} = \omega + \alpha\varepsilon^2_{t-1} + \beta\sigma^2_{t-1|t-2} \tag{37}$$

where $\varepsilon_t \equiv \sigma_{t|t-1}\nu_t$, and the parameter restrictions $\omega > 0$, $\alpha \geq 0$, $\beta \geq 0$ ensure that the conditional variance remains positive for all realizations of ν_t. The standard estimation method is maximum likelihood. The basic GARCH(1,1) model may be extended to include higher-order lags, to allow the distribution of ν_t to have fat tails, to allow for asymmetries in the volatility dynamics, to permit the conditional variance to affect the conditional mean, and to allow volatility shocks to have permanent effects or volatility to have long memory. It may also be extended to the multivariate case.

It follows directly from the formulation of the GARCH(1,1) model that the optimal, in the MMSE sense, one-step-ahead forecast equals $\sigma^2_{t+1|t}$. Similar expressions for longer horizons may be obtained by recursive updating. There is a direct link from the arrival of news to volatility measures and from volatility forecasts to risk assessments. These and alternative volatility models and the uses of volatility forecasts are surveyed in Andersen et al. (2006). For a comparison of GARCH models with the related and complementary class of stochastic volatility models, see Andersen, Bollerslev and Diebold (2006) and Shephard (2005).

7. Applications

Time series analytic methods have many applications in economics. Here we consider five: (1) analysis of the cyclic properties of economic time series, (2) description of seasonality and seasonal adjustment, (3) forecasting, (4) dynamic econometric modelling, and (5) structural vector autoregressions.

7.1. Analysis of the cyclic properties of economic time series

Suppose that the time series $\{x_t\}$ is a linearly non-deterministic stationary series and that the series $\{y_t\}$ is formed from $\{x_t\}$ by the linear operator

$$y_t = \sum_{j=m}^{n} w_j x_{t-j}, \qquad \sum_{m}^{n} w_j^2 < \infty. \tag{38}$$

Such an operator is called a time-invariant linear filter. Analysis of the properties of such filters plays an important role in time series analysis since many methods of trend estimation or removal and seasonal adjustment may be represented or approximated by such filters. An interesting example that illustrates the potential pitfalls of using such filters is provided by Adelman (1965), who showed that the 20-year long swings in various economic series found by Kuznets (1961) may well have been the result of the trend filtering operations used in preliminary processing of the data. For a fuller treatment see Nerlove, Grether and Carvalho (1979, pp. 53–7).

Since the 1980s, there has been increased interest in the use of nonlinear filters for extracting the business cycle component of macroeconomic time series. Examples include the band-pass filter (Christiano and Fitzgerald, 2003) and the Hodrick–Prescott (HP) filter (Hodrick and Prescott, 1997; Ravn and Uhlig, 2002). The latter approach postulates that $y_t = \tau_t + c_t$, where τ_t denotes the trend component and c_t the deviation from trend or 'cyclical' component of the time series y_t. The trend component is chosen to minimize the loss function:

$$\sum_{t=1}^{T} c_t^2 + \lambda \sum_{t=1}^{T} [(\tau_{t+1} - \tau_t) - (\tau_t - \tau_{t-1})]^2 \tag{39}$$

where $c_t = y_t - \tau_t$ and λ is a pre-specified parameter that depends on the frequency of the observations. The trade-off in this optimization problem is between the degree to which the trend component fits the data and the smoothness of the trend.

7.2 Description of seasonality and seasonal adjustment

Many economic time series exhibit fluctuations which are periodic within a year or a fraction thereof. The proper treatment of such seasonality, whether stochastic or deterministic, is the subject of a large literature, summarized rather selectively in Nerlove, Grether and Carvalho (1979, ch. 1). More recent treatments can be found in Hylleberg (1992), Franses (1996) and Ghysels and Osborn (2001).

Seasonality may be modelled and its presence detected using spectral analysis (Nerlove, 1964) or using time domain methods. Deterministic seasonality, in the form of model parameters that vary deterministically with the season, offers no great conceptual problems but many practical ones. Stochastic seasonality is often modelled in the form of seasonal unit roots. In that case, seasonal differencing of the data removes the unit root component. Multiple time series may exhibit seasonal co-integration. Sometimes it is convenient to specify stochastic seasonality in the form of an UC model (Grether and Nerlove, 1970). Appropriate UC models may be determined directly or by fitting an ARIMA model and deriving a related UC model by imposing sufficient a priori restrictions (Hillmer and Tiao, 1982; Bell and Hillmer, 1984).

7.3 Forecasting

One of the simplest forecasting procedures for time series is exponential smoothing based on the relationship

$$\hat{x}_{t+1|t} = (1-\theta)x_t + \theta\hat{x}_{t|t-1} \tag{40}$$

where x_t, is the observed series and $\hat{x}_{j|k}$ is the forecast of the series at time j made on the basis of information available up to time k. Muth (1960) showed that (40) provides an MMSE forecast if the model generating the time series is $x_t - x_{t-1} = \varepsilon_t - \theta\varepsilon_{t-1}$. Holt (1957) and Winters (1960) generalized the exponential smoothing approach to models containing more complex trend and seasonal components. Further generalization and proofs of optimality are contained in Theil and Wage (1964) and Nerlove and Wage (1964).

Perhaps the most popular approach to forecasting time series is based on ARIMA models of time series processes (Box and Jenkins, 1970). The developments discussed in the preceding paragraph led to the development of UC models, which give rise to restricted ARIMA model forms (Nerlove, Grether and Carvalho, 1979). State-space representations of these models permit the application of the Kalman filter to both estimation and forecasting. Harvey (1984) presents a unified synthesis of the various methods.

More recently, the focus has shifted from traditional forecasting methods towards methods that exploit the increased availability of a large number of potential predictors. Consider the problem of forecasting $y_{t+h|t}$ based on its own current and past values as well as those of N additional variables, x_t. Of particular interest is the case in which the number of predictors, N, exceeds the number of time series observations, T. In that case, principal components analysis provides a convenient way of extracting a low-dimensional vector of common factors from the original data-set x_t (Stock and Watson, 2002a; 2002b). Forecasts that incorporate estimated common factors have proved successful in many cases in reducing forecast errors relative to traditional time series forecasting methods. Boivin and Ng (2005) provide a systematic comparison of alternative *factor model forecasts*. Another promising forecasting method is *Bayesian model averaging* across alternative forecasting models (Raftery,

Madigan and Hoeting, 1997). The latter method builds on the literature on forecast combinations (Bates and Granger, 1969).

7.4 Dynamic econometric modelling

There is a close connection between multivariate time-series models and the structural, reduced and final forms of dynamic econometric models; the standard simultaneous-equations model (SEM) is a specific and restricted case.

Suppose that a vector of observed variables y_t may be subdivided into two classes of variables, 'exogenous', $\{x_t\}$, and endogenous, $\{z_t\}$. A dynamic, multivariate simultaneous linear system may be written.

$$\begin{bmatrix} \Psi_{11}(L) & \Psi_{12}(L) \\ 0 & \Psi_{22}(L) \end{bmatrix} \begin{pmatrix} z_t \\ x_t \end{pmatrix} = \begin{bmatrix} \Theta_{11}(L) & 0 \\ 0 & \Theta_{22}(L) \end{bmatrix} \begin{pmatrix} \varepsilon_{1t} \\ \varepsilon_{2t} \end{pmatrix} \tag{41}$$

where $\Psi_{ij}(\cdot)$ and $\Theta_{ij}(\cdot), i,j = 1,2$ are matrix polynomials in the lag operator L. Such systems are known as vector ARMAX models and conditions for their identification are given by Hatanaka (1975). The reduced form of the system is obtained by expressing z_t as a function of lagged endogenous and current and lagged exogenous variables. The final form is then obtained by eliminating the lagged endogenous variables (see Zellner and Palm, 1974; Wallis, 1977).

7.5 Structural vector autoregressions

An important special case of the dynamic SEM is the structural vector autoregressive model in which all variables are presumed endogenous, the lag structure is unrestricted up to some order p, and identification of the structural from is achieved by imposing restrictions on the correlation structure of the structural innovations (Sims, 1980). The most common form of the structural VAR(p) model imposes restrictions on the contemporaneous interaction of structural innovations. Consider the structural form for a K-dimensional vector $\{x_t\}, t = 1, \ldots, T$:

$$B_0 x_t = \sum_{i=1}^{p} B_i x_{t-i} + \eta_t, \tag{42}$$

where $\eta_t \sim (0, \Sigma_\eta)$ denotes the $(K \times 1)$ vector of serially uncorrelated structural innovations (or shocks) and B_i, $i = 0, \ldots, p$, the $(K \times K)$ coefficient matrices. Without loss of generality, let $\Sigma_\eta = I$. The corresponding reduced form is

$$x_t = \sum_{i=1}^{p} B_0^{-1} B_i x_{t-i} + B_0^{-1} \eta_t = \sum_{i=1}^{p} A_i x_{t-i} + \varepsilon_t \tag{43}$$

where $\varepsilon_t \sim (0, \Sigma_\varepsilon)$. Since $\varepsilon_t = B_0^{-1} \eta_t$, it follows that $\Sigma_\varepsilon = B_0^{-1} B_0^{-1'}$. Given a consistent estimate of the reduced form parameters A_i, $i = 1, \ldots, p$, and Σ_ε, the elements of B_0^{-1}will be exactly identified after imposing $K(K-1)/2$ restrictions on the parameters of B_0^{-1} that reflect the presumed structure of the economy. Given estimates of B_0^{-1} and A_i, $i = 1, \ldots, p$, estimates of the remaining structural parameters may be recovered from $B_i = B_0 A_i$.

In practice, the number of restrictions that can be economically motivated may be smaller or larger than $K(K-1)/2$. Alternative estimation strategies that remain valid in the over-identified case include the generalized method of moments (Bernanke, 1986) and maximum likelihood (Sims, 1986). An instrumental variable interpretation of VAR estimation is discussed in Shapiro and Watson (1988). Semi-structural VAR models that are only partially identified have been proposed by Bernanke and Mihov (1998).

Alternative identification strategies may involve putting restrictions on the long-run behaviour of economic variables (Blanchard and Quah, 1989; King et al., 1991) or on the sign and/or shape of the impulse responses (Faust, 1998). Other possibilities include identification via heteroskedasticity (Rigobon, 2003) or the use of high-frequency data (Faust, Swanson and Wright, 2004).

The estimates of the structural VAR form may be used to compute the dynamic responses of the endogenous variables to a given structural shock, variance decompositions that measure the average contribution of each structural shock to the overall variability of the data, and historical decompositions of the path of x_t based on the contribution of each structural shock.

8. Conclusions

The literature on time series analysis has made considerable strides since the 1980s. The advances have been conceptual, theoretical and methodological. The increased availability of inexpensive personal computers in particular has revolutionized the implementation of time series techniques by shifting the emphasis from closed-form analytic solutions towards numerical and simulation methods. The ongoing improvements in information technology, broadly defined to include not only processing speed but also data collection and storage capabilities, are likely to transform the field even further. For example, the increased availability of large cross-sections of time series data, the introduction of ultra high-frequency data, the electronic collection of micro-level time series data (such as web-based data or scanner data), and the increased availability of data in real time all are creating new applications and spurring interest in the development of new methods of time series analysis. These developments already have brought together the fields of empirical finance and time series econometrics, resulting in the emergence of the new and fertile field of financial econometrics. As the use of time series methods becomes more widespread in applied fields, there will be increasing interest in the development of methods that can be adapted to the specific objectives of the end user. Another question of growing importance is how to deal with rapidly evolving economic environments in the form of structural breaks and other model instabilities. Finally, the improvement of structural time series models for macroeconomic policy analysis will remain a central task if time series analysis is to retain its importance for economic policymaking.

FRANCIS X. DIEBOLD, LUTZ KILIAN AND MARC NERLOVE

See also **cointegration; forecasting; long memory models; nonlinear time series analysis; seasonal adjustment; spectral analysis; spurious regressions; trend/cycle decomposition; vector autoregressions.**

Bibliography

Adelman, I. 1965. Long cycles – fact or artifact? *American Economic Review* 60, 443–63.

Akaike, H. 1970. Statistical predictor identification. *Annals of the Institute of Statistical Mathematics* 22, 203–17.

Akaike, H. 1974. A new look at the statistical model identification. *IEEE Transactions on Automatic Control* 19, 716–23.

Anderson, T. 1971. *The Statistical Analysis of Time Series*. New York: John Wiley.

Anderson, T. 1977. Estimation for autoregressive moving average models in the time and frequency domains. *Annals of Statistics* 5, 842–65.

Anderson, T. and Takemura, A. 1986. Why do noninvertible moving averages occur? *Journal of Time Series Analysis* 7, 235–54.

Andersen, T., Bollerslev, T. and Diebold, F. 2006. Parametric and nonparametric volatility measurement. In *Handbook of Financial Economics*, ed. L. Hansen and Y. Ait-Sahalia. Amsterdam: North-Holland.

Andersen, T., Bollerslev, T., Christoffersen, P. and Diebold, F. 2006. Volatility and correlation forecasting. In *Handbook of Economic Forecasting*, ed. G. Elliott, C. Granger and A. Zimmermann. Amsterdam: North-Holland.

Ansley, C. 1979. An algorithm for the exact likelihood of a mixed autoregressive-moving average process. *Biometrika* 66, 59–65.

Baillie, R. 1996. Long memory processes and fractional integration in econometrics. *Journal of Econometrics* 73, 5–59.

Baillie, R., Bollerslev, T. and Mikkelsen, H.-O. 1996. Fractionally integrated generalized autoregressive conditional heteroskedasticity. *Journal of Econometrics* 74, 3–30.

Bates, J. and Granger, C. 1969. The combination of forecasts. *Operational Research Quarterly* 20, 451–68.

Bell, W. and Hillmer, S. 1984. Issues involved with seasonal analysis of economic time series. *Journal of Business and Economic Statistics* 2, 291–349.

Bernanke, B. 1986. Alternative explanations of the money-income correlation. *Carnegie-Rochester Conference Series on Public Policy* 25, 49–100.

Bernanke, B. and Mihov, I. 1998. Measuring monetary policy. *Quarterly Journal of Economics* 113, 869–902.

Beveridge, W. 1921. Weather and harvest cycles. *Economic Journal* 31, 429–52.

Beveridge, W. 1922. Wheat prices and rainfall in western Europe. *Journal of the Royal Statistical Society* 85, 412–59.

Blanchard, O. and Quah, D. 1989. The dynamic effects of aggregate demand and supply disturbances. *American Economic Review* 79, 655–73.

Boivin, J. and Ng, S. 2005. Understanding and comparing factor-based forecasts. *International Journal of Central Banking* 1(3), 117–52.

Bollerslev, T. 1986. Generalized autoregressive conditional heteroskedasticity. *Journal of Econometrics* 31, 307–27.

Box, G., Hillmer, S. and Tiao, G. 1978. Analysis and modeling of seasonal time series. In *Seasonal Analysis of Economic Time Series*. ed. A. Zellner. Washington, DC: Bureau of the Census, Department of Commerce.

Box, G. and Jenkins, G. 1970. *Time Series Analysis: Forecasting and Control*. San Francisco: Holden-Day.

Brillinger, D. 1975. *Time Series: Data Analysis and Theory*. New York: Holt.

Burman, J. 1980. Seasonal adjustment by signal extraction. *Journal of the Royal Statistical Society. Series A* 143, 321–37.

Burridge, P. and Wallis, K. 1985. Calculating the variance of seasonally adjusted series. *Journal of the American Statistical Association* 80, 541–52.

Cappé, O., Moulines, E. and Ryden, T. 2005. *Inference in Hidden Markov Models*. New York: Springer-Verlag.

Chambers, M. 1998. Long memory and aggregation in macroeconomic time series. *International Economic Review* 39, 1053–72.

Christiano, L. and Fitzgerald, T. 2003. The band pass filter. *International Economic Review* 44, 435–65.

Christoffersen, P. and Diebold, F. 1996. Further results on forecasting and model selection under asymmetric loss. *Journal of Applied Econometrics* 11, 561–71.

Christoffersen, P. and Diebold, F. 1997. Optimal Prediction under asymmetric loss. *Econometric Theory* 13, 808–17.

Cournot, A. 1838. Researches into the Mathematical Principles of the Theory of Wealth, trans. N. Bacon. New York: Macmillan, 1927.

Diebold, F. and Inoue, A. 2001. Long memory and regime switching. *Journal of Econometrics* 105, 131–59.

Doob, J. 1953. *Stochastic Processes*. New York: John Wiley.

Durbin, J. and Koopman, S. 2001. *Time Series Analysis by State Space Methods*. Oxford: Oxford University Press.

Durlauf, S. and Phillips, P. 1988. Trends versus random walks in time series analysis. *Econometrica* 56, 1333–54.

Elliott, G. 1998. The robustness of co-integration methods when regressors almost have unit roots. *Econometrica* 66, 49–58.

Engle, R. 1982. Autoregressive conditional heteroskedasticity with estimates of the variance of UK inflation. *Econometrica* 50, 987–1008.

Engle, R. and Granger, C. 1987. Co-integration and error correction: Representation, estimation and testing. *Econometrica* 55, 251–76.

Engle, R. and Granger, C. 1991. *Long Run Economic Relations: Readings in Co-integration*. Oxford: Oxford University Press.

Faust, J. 1998. The robustness of identified VAR conclusions about money. *Carnegie-Rochester Conference Series on Public Policy* 49, 207–44.

Faust, J., Swanson, E. and Wright, J. 2004. Identifying VARs based on high frequency futures data. *Journal of Monetary Economics* 51, 1107–31.

Fishman, G. 1969. *Spectral Methods in Econometrics*. Cambridge: Harvard University Press.

Franses, P. 1996. *Periodicity and Stochastic Trends in Economic Time Series*. Oxford: Oxford University Press.

Fuller, W. 1976. *Introduction to Statistical Time Series*. New York: John Wiley.

Geweke, J. and Porter-Hudak, S. 1983. The estimation and application of long memory series models. *Journal of Time Series Analysis* 4, 221–38.

Ghysels, E. and Osborn, D. 2001. *The Econometric Analysis of Seasonal Time Series*. Cambridge: Cambridge University Press.

Granger, C. 1966. The typical spectral shape of an economic variable. *Econometrica* 34, 150–61.
Granger, C. 1969. Prediction with a generalized cost of error function. *Operations Research Quarterly* 20, 199–207.
Granger, C. 1980. Long memory relationships and the aggregation of dynamic models. *Journal of Econometrics* 14, 227–38.
Granger, C. 1981. Some properties of time series data and their use in econometric model specification. *Journal of Econometrics* 16, 121–30.
Granger, C. and Newbold, P. 1974. Spurious regressions in econometrics. *Journal of Econometrics* 2, 111–20.
Granger, C. and Newbold, P. 1977. *Forecasting Economic Time Series*. New York: Academic Press.
Granger, C. and Teräsvirta, T. 1993. *Modelling Nonlinear Economic Relationships*. Oxford: Oxford University Press.
Grether, D. and Nerlove, M. 1970. Some properties of 'optimal' seasonal adjustment. *Econometrica* 38, 682–703.
Hamilton, J. 1989. A new approach to the economic analysis of nonstationary time series and the business cycle. *Econometrica* 57, 357–84.
Hannan, E. 1960. *Time Series Analysis*. London: Methuen.
Hannan, E. 1969. The estimation of mixed moving average autoregressive systems. *Biometrika* 56, 223–5.
Hannan, E. 1970. *Multiple Time Series*. New York: John Wiley.
Hannan, E. 1971. The identification problem for multiple equation systems with moving average errors. *Econometrica* 39, 751–65.
Hannan, E. 1976. The identification and parameterization of ARMAX and state space forms. *Econometrica* 44, 713–23.
Hannan, E. 1979. The statistical theory of linear systems. In *Developments in Statistics*, ed. P. Krishnaiah. New York: Academic Press.
Harvey, A. 1981. *Time Series Models*. Oxford: Allan.
Harvey, A. 1984. A unified view of statistical forecasting procedures. *Journal of Forecasting* 3, 245–75.
Harvey, A. 1989. *Forecasting. Structural Time Series Models and the Kalman Filter*. Cambridge: Cambridge University Press.
Harvey, A. and Peters, S. 1990. Estimation procedures for structural time series models. *Journal of Forecasting* 9, 89–108.
Harvey, A. and Todd, P. 1984. Forecasting economic time series with structural and Box–Jenkins models: a case study (with discussion). *Journal of Business and Economic Statistics* 1, 299–315.
Hatanaka, M. 1975. On the global identification of the dynamic simultaneous equations model with stationary disturbances. *International Economic Review* 16, 545–54.
Hillmer, S., Bell, W. and Tiao, G. 1983. Modeling considerations in the seasonal analysis of economic time series. In *Applied Time Series Analysis of Economic Data*, ed. A. Zellner. Washington, DC: Bureau of the Census, Department of Commerce.
Hillmer, S. and Tiao, G. 1982. An ARIMA-model-based approach to seasonal adjustment. *Journal of the American Statistical Association* 77, 63–70.
Hodrick, R. and Prescott, E. 1997. Postwar US business cycles: an empirical investigation. *Journal of Money, Credit and Banking* 29, 1–16.
Holt, C. 1957. Forecasting seasonals and trends by exponentially weighted moving averages. ONR Research Memorandum No. 52, Carnegie Institute of Technology.
Hylleberg, S. 1992. *Modelling Seasonality*. Oxford: Oxford University Press.
Jevons, W. 1884. *Investigations in Currency and Finance*. London: Macmillan.

Johansen, S. 1995. *Likelihood-Based Inference in Cointegrated Vector Autoregressive Models*. Oxford: Oxford University Press.

Kalman, R. 1960. A new approach to linear filtering and prediction problems. *Transactions of the American Society of Mechanical Engineers – Journal of Basic Engineering* D 82, 35–45.

King, R., Plosser, C., Stock, J. and Watson, M. 1991. Stochastic trends and economic fluctuations. *American Economic Review* 81, 819–40.

Kolmogorov, A. 1941. Interpolation und Extrapolation von stationären zufälligen Folgen. *Bulletin of the Academy Science (Nauk), USSR, Mathematical Series* 5, 3–14.

Koopmans, L. 1974. *The Spectral Analysis of Time Series*. New York: Academic Press.

Kuznets, S. 1961. *Capital and the American Economy: Its Formation and Financing*. New York: Princeton University Press for the National Bureau of Economic Research.

Maravall, A. 1981. Desestacionalization y Politica Monetaria. Economic Studies No. 19. Madrid: Bank of Spain.

Maravall, A. 1984. Model-based treatment of a manic depressive series. Working paper, Bank of Spain.

Muth, J. 1960. Optimal properties of exponentially weighted forecasts. *Journal of the American Statistical Association* 55, 299–305.

Nelson, C. and Kang, H. 1981. Spurious periodicity in inappropriately detrended time series. *Econometrica* 49, 741–52.

Nelson, C. and Plosser, C. 1982. Trends and random walks in macroeconomic time series: Some evidence and implications. *Journal of Monetary Economics* 10, 139–62.

Nerlove, M. 1964. Spectral analysis of seasonal adjustment procedures. *Econometrica* 32, 241–86.

Nerlove, M. 1965. A comparison of a modified Hannan and the BLS seasonal adjustment filters. *Journal of the American Statistical Association* 60, 442–91.

Nerlove, M. 1967. Distributed lags and unobserved components in economic time series. In *Ten Economic Essays in the Tradition of Irving Fisher*, ed. W. Fellner et al. New York: John Wiley.

Nerlove, M., Grether, D. and Carvalho, J. 1979. *Analysis of Economic Time Series*. New York: Academic Press.

Nerlove, M. and Wage, S. 1964. On the optimality of adaptive forecasting. *Management Science* 10, 207–24.

Newbold, P. 1974. The exact likelihood function for a mixed autoregressive-moving average process. *Biometrika* 61, 423–6.

Parzen, E. 1961. An approach to time series analysis. *Annals of Mathematical Statistics* 32, 951–89.

Phillips, P. 1986. Understanding spurious regressions in econometrics. *Journal of Econometrics* 33, 311–40.

Phillips, P. and Durlauf, S. 1986. Multiple time series regression with integrated processes. *Review of Economic Studies* 53, 473–95.

Phillips, P. and Hansen, B. 1990. Statistical inference in instrumental variables regression with I(1) processes. *Review of Economic Studies* 57, 99–125.

Pierce, D. 1978. Seasonal adjustment when both deterministic and stochastic seasonality are present. In *Seasonal Analysis of Economic Time Series*, ed. A. Zellner. Washington, DC: Bureau of the Census, Department of Commerce.

Pierce, D. 1979. Signal extraction error in nonstationary time series. *Annals of Statistics* 7, 1303–20.

Priestley, M. 1981. *Spectral Analysis and Time Series*. New York: Academic Press.

Raftery, A., Madigan, D. and Hoeting, J. 1997. Bayesian model averaging for linear regression models. *Journal of the American Statistical Association* 92, 179–91.

Ravn, M. and Uhlig, H. 2002. On adjusting the HP-filter for the frequency of observations. *Review of Economics and Statistics* 84, 371–76.
Rigobon, R. 2003. Identification through heteroskedasticity. *Review of Economics and Statistics* 85, 777–92.
Rudebusch, G. 1993. The uncertain unit root in real GNP. *American Economic Review* 83, 264–72.
Sargan, J. and Bhargava, A. 1983. Maximum likelihood estimation of regression models with moving average errors when the root lies on the unit circle. *Econometrica* 51, 799–820.
Schuster, A. 1898. On the investigation of hidden periodicities with application to the supposed 26-day period of meteorological phenomena. *Terrestrial Magnetism and Atmospheric Electricity* [now *Journal of Geophysical Research*] 3, 13–41.
Schwarz, G. 1978. Estimating the dimension of a model. *Annals of Statistics* 6, 461–4.
Schweppe, F. 1965. Evaluation of likelihood functions for Gaussian signals. *IEEE Transactions on Information Theory* 11, 61–70.
Shapiro, M. and Watson, M. 1988. Sources of business cycle fluctuations. *NBER Macroeconomics Annual* 3, 111–56.
Shephard, N. 2005. *Stochastic Volatility: Selected Readings*. Oxford: Oxford University Press.
Sims, C. 1980. Macroeconomics and reality. *Econometrica* 48, 1–48.
Sims, C. 1986. Are forecasting models usable for policy analysis? *Quarterly Review, Federal Reserve Bank of Minneapolis* 10, 2–16.
Sims, C., Stock, J. and Watson, M. 1991. Inference in linear time series models with some unit roots. *Econometrica* 58, 113–44.
Slutsky, E. 1927. The summation of random causes as the source of cyclic processes. *Econometrica* 5, April 1937, 105–46.
Sowell, F. 1992. Maximum likelihood estimation of stationary univariate fractionally integrated time series models. *Journal of Econometrics* 53, 165–88.
Stock, J. 1991. Confidence intervals for the largest autoregressive root in US economic time series. *Journal of Monetary Economics* 28, 435–60.
Stock, J. and Watson, M. 1993. A simple estimator of cointegrating vectors in higher order integrated systems. *Econometrica* 61, 783–820.
Stock, J. and Watson, M. 2002a. Forecasting using principal components from a large number of predictors. *Journal of the American Statistical Association* 97, 1167–79.
Stock, J. and Watson, M. 2002b. Macroeconomic forecasting using diffusion indexes. *Journal of Business and Economic Statistics* 20, 147–62.
Stokes, G. 1879. Note on searching for hidden periodicities. *Proceedings of the Royal Society* 29, 122–5.
Theil, H. and Wage, S. 1964. Some observations on adaptive forecasting. *Management Science* 10, 198–206.
Wallis, K. 1977. Multiple time series analysis and the final form of econometric models. *Econometrica* 45, 1481–97.
Whittle, P. 1963. *Prediction and Regulation by Linear Least-squares Methods*. London: English Universities Press.
Wiener, N. 1949. *The Extrapolation, Interpolation and Smoothing of Stationary Time Series with Engineering Applications*. New York: John Wiley.
Winters, P. 1960. Forecasting sales by exponentially weighted moving averages. *Management Science* 6, 324–42.
Wold, H. 1938. *A Study in the Analysis of Stationary Time Series*. Stockholm: Almqvist and Wiksell.
Yule, G. 1921. On the time-correlation problem, with special reference to the variate-difference correlation method. *Journal of the Royal Statistical Society* 84, 497–526.

Yule, G. 1926. Why do we sometimes get nonsense correlations between time series? A study in sampling and the nature of time series. *Journal of the Royal Statistical Society* 89, 1–64.

Yule, G. 1927. On a method of investigating periodicities in disturbed series with special reference to Wolfer's sunspot numbers. *Philosophical Transactions of the Royal Society of London A* 226, 267–98.

Zellner, A. and Palm, F. 1974. Time series analysis and simultaneous equation econometric models. *Journal of Econometrics* 2, 17–54.

trend/cycle decomposition

Macroeconomists distinguish between the forces that cause long-term growth and those that cause temporary fluctuations such as recessions. The former include population growth, capital accumulation, and productivity change, and their effect on the economy is permanent. The latter are generally monetary shocks such as shifts in central bank policy that affect the real economy through price rigidities that cause output to deviate temporarily from its long-run path. This conceptual dichotomy motivates the decomposition of aggregate output, real GDP, into two components: the *trend* which accounts for long-term change, and the *cycle* which is a short-term deviation from trend. While economists no longer believe the 'business cycle' to be deterministically periodic, that terminology remains. Seasonal variation could be a third component, though it has been suppressed in 'seasonally adjusted' data such as GDP.

This suggests we may express the natural log of GDP (or any other 'trending' time series), denoting the observation at time t by 'y_t', as follows:

$$y_t = \tau_t + c_t.$$

Here τ_t denotes the value of the trend and c_t the cycle at time t, neither of which is observed directly. Since this single equation cannot be solved directly for the unknown trend and cycle, additional assumptions are required for 'identification', a procedure which allows estimates of them to be calculated from the GDP data. The fundamental identifying assumption is that the cycle component is temporary, that it dies out after a sufficiently long time. However, this assumption of 'stationarity' or 'ergodicity', which distinguishes it from trend, which is permanent, does not suffice by itself to achieve identification. More has to be said about the nature of the trend.

The simplest specification of trend is to make τ_t a linear function of time where the slope is the long-term growth rate. A second identifying assumption is that trend should account for as much of the variation in the data as possible, minimizing the amplitude of the implied cycle. This is achieved by least squares regression of y_t on time and the estimated trend is $\hat{\tau}_t = a + b\ time$ where a and b are estimates of intercept and slope respectively. The implied cycle component is then $\hat{c}_t \equiv y_t - \hat{\tau}_t$. Though successful in accounting for a large fraction of the change in GDP over long periods, this approach implies cycles of extraordinary length, well beyond the roughly seven years between recession dates identified by the National Bureau of Economic Research for the United States, and the pattern is contrary to economic intuition (for the United States the 1970s, a decade of poor economic performance, were well above the trend line while the 1990s, a decade of prosperity, were well below trend). A more flexible trend function is clearly called for, but quadratic or higher-order polynomials in time imply unstable paths when extrapolated into the future. Perron (1989) suggested a segmented trend function allowing for an occasional change level or slope to be captured by dummy variables.

A general approach to estimating a flexible and adaptive trend is filtering, where estimated trend is a weighted average of adjacent observations. Here it is the weighting scheme which identifies the components. For example, $\hat{\tau}_t = .25 \cdot y_{t-1} + .50 \cdot y_t + .25 \cdot y_{t+1}$ applies symmetric though unequal weights to the current observations and its immediate neighbours. No filter is perfect in the sense of revealing the actual trend, but a desirable filter is one that extracts as much of the trend as possible from the data. A criterion for choosing a filter would be that it produces cycles having characteristics that match our notions of the business cycle, for example that recessions occur on average about every seven years. A widely used filter that does this is the Hodrick and Prescott (1980), filter which penalizes deviations from trend and changes in trend through a loss function.

The distinction between trend and cycle implies that the forecast of GDP far in the future must be the trend, since the cycle will die away. The approach to trend/cycle decomposition proposed by Beveridge and Nelson (1981) turns this conclusion on its head by proposing that the trend at a date in time be defined as the forecast of the distant future (adjusted for average growth). Specifically, they estimate an autoregressive moving average (ARMA) time series model for the growth rate and compute the forecast of the level into the distant future, adjusting for average growth. The resulting measure of trend shows whether actual GDP is above or below its forecast growth path, the difference being the cycle. Since parameters of the ARMA model are identified, and computation of forecasts is straightforward, the Beveridge–Nelson decomposition is identified. It turns out that the trend component is a random walk with drift regardless of the specific ARMA model, and this accords with the intuition that only unexpected shocks can affect a long horizon forecast.

To obtain the general expressions for the components we rearrange the ARMA model as:

$$\phi(L)\Delta y_t = \theta(L)\varepsilon_t$$

$$\Delta y_t = \psi(L)\varepsilon_t$$

where the average growth rate has been suppressed, the statistical shock ε_t is serially random, Δ denoted first difference, and L is the lag operator, and $\psi(L) = \theta(L)/\phi(L)$. The growth rate of GDP can be thought of as a weighted history of all past shocks where the coefficient of ε_{t-k} is ψ_k plus the expected average growth rate μ. It may be shown that an algebraically equivalent expression is

$$y_t = \psi(1)\sum_{k=0}^{\infty}\varepsilon_{t-k} - \tilde{\psi}(L)\varepsilon_t \qquad \tilde{\psi}_k = \sum_{j=k+1}^{\infty}\psi_j.$$

Note that the first term is the sum of all past shocks each with weight equal to the total effect of all past shocks. The second term may be shown to be a stationary time series with mean zero. Thus the trend is always a random walk regardless of the ARMA model.

For example, growth in US GDP is roughly an AR(1) process with coefficient .25, so the effect of a shock on the trend is $\psi(1) = 1/(1-.25) = 1.33$. This illustrates the surprising implication that the trend component may be highly variable; indeed, the results obtained by Beveridge and Nelson imply that variation in observed GDP is largely the result of variation in the trend component and is therefore permanent.

Unobserved components models identify trend and cycle by specifying a separate and specific stochastic process for each. The trend is generally assumed to be a random walk with drift, allowing it to account for long-term growth while permitting it to be shifted by stochastic shocks. The cycle is assumed to be a process that is stationary in the sense of reverting to a mean over time. (The mean of the cycle is zero for symmetric variation around trend, but evidence exits for asymmetric cycles with a negative mean.) This approach was introduced to economics by Harvey (1985) and Clark (1987). An example would be the following:

$$\tau_t = \tau_{t-1} + \mu + \eta_t \qquad c_t = \phi \cdot c_{t-1} + \varepsilon_t.$$

The parameter μ is the long-term growth rate, the shock η is random and may be positive or negative, the parameter ϕ measures the persistence of the cycle, and shocks ε drive the cycle. The two shocks are often assumed to be uncorrelated, which reduces the number of parameters to be estimated by one but may also place an unwarranted restriction on the relation between the two components. More generally the cycle process may have a higher-order ARMA specification. Identification of the parameters depends on whether a specific model implies a sufficient number of estimable parameters in the corresponding ARMA reduced form representation of Δy_t (corresponding to identification of simultaneous equation models). Given an identified model and parameter estimates, the estimated trend and cycle may be computed using the Kalman filter.

A useful result is that the random walk trend in the unobserved components model is identified even if its parameters are not identified. Morley, Nelson and Zivot (2003) show that the Beveridge–Nelson trend is always the conditional expectation of the trend component given past data. What identifies the trend is the random walk specification for the trend along with the assumption that the cycle process does not persist indefinitely. Thus, the long-horizon forecast reflects only the trend, and such forecasts can always be computed from the reduced form ARMA model.

CHARLES R. NELSON

See also **data filters; state space models; time series analysis; unit roots.**

Bibliography

Beveridge, S. and Nelson, C.R. 1981. A new approach to decomposition of economic time series into permanent and transitory components with particular attention to measurement of the business cycle. *Journal of Monetary Economics* 7, 151–74.

Clark, P.K. 1987. The cyclical component of U.S. economic activity. *Quarterly Journal of Economics* 102, 797–814.

Harvey, A.C. 1985. Trends and cycles in macroeconomic time series. *Journal of Business and Economic Statistics* 3, 216–27.

Hodrick, R.J. and Prescott, E.C. 1980. Postwar US business cycles: an empirical investigation. Working paper, Carnegie-Mellon University.

Hodrick, R.J. and Prescott, E.C. 1997. Postwar U.S. business cycles: an empirical investigation. *Journal of Money Credit and Banking* 29, 1–16.

Morley, J.C., Nelson, C.R. and Zivot, E. 2003. Why are the Beveridge–Nelson and unobserved-components decompositions of GDP so different? *Review of Economics and Statistics* 85, 235–43.

Perron, P. 1989. The great crash, the oil price shock, and the unit root hypothesis. *Econometrica* 57, 1361–401.

unit roots

Economic and financial time series have frequently been successfully modelled by autoregressive moving-average (ARMA) schemes of the type

$$a(L)y_t = b(L)\varepsilon_t, \tag{1}$$

where ε_t is an orthogonal sequence (that is, $E(\varepsilon_t)=0$, $E(\varepsilon_t\varepsilon_s)=0$ for all $t\neq s$), L is the backshift operator for which $Ly_t=y_{t-1}$ and $a(L)$, $b(L)$ are finite-order lag polynomials

$$a(L) = \sum_{i=0}^{p} a_i L^i, \qquad b(L) = \sum_{j=0}^{q} b_j L^j,$$

whose leading coefficients are $a_0=b_0=1$. Parsimonious schemes (often with $p+q\leq 3$) are usually selected in practice either by informal 'model identification' processes such as those described in the text by Box and Jenkins (1976) or more formal order-selection criteria which penalize choices of large p and/or q. Model (1) is assumed to be irreducible, so that $a(L)$ and $b(L)$ have no common factors. The model (1) and the time series y_t are said to have an *autoregressive unit root* if $a(L)$ factors as $(1-L)a_1(L)$ and a *moving-average unit root* if $b(L)$ factors as $(1-L)b_1(L)$.

Since the early 1980s, much attention has been focused on models with autoregressive unit roots. In part, this interest is motivated by theoretical considerations such as the importance of martingale models of efficient markets in finance and the dynamic consumption behaviour of representative economic agents in macroeconomics; and, in part, the attention is driven by empirical applications, which have confirmed the importance of random walk phenomena in practical work in economics, in finance, in marketing and business, in social sciences like political studies and communications, and in certain natural sciences. In mathematics and theoretical probability and statistics, unit roots have also attracted attention because they offer new and important applications of functional limit laws and weak convergence to stochastic integrals. The unit root field has therefore drawn in participants from an excitingly wide range of disciplines.

If (1) has an autoregressive unit root, then we may write the model in difference form as

$$\Delta y_t = u_t = a_1(L)^{-1} b(L)\varepsilon_t, \tag{2}$$

where the polynomial $a_1(L)$ has all its zeros outside the unit circle. This formulation suggests more general nonparametric models where, for instance, u_t may be formulated in linear process (or Wold representation) form as

$$u_t = c(L)\varepsilon_t = \sum_{j=0}^{\infty} c_j \varepsilon_{t-j}, \qquad \text{with } \sum_{j=0}^{\infty} c_j^2 < \infty, \tag{3}$$

or as a general stationary process with spectrum $f_u(\lambda)$. If we solve (2) with an initial state y_0 at $t=0$, we have the important partial sum representation

$$y_t = \sum_{j=1}^{t} u_j + y_0 = S_t + y_0, \tag{4}$$

showing that S_t and hence y_t are 'accumulated' or 'integrated' processes proceeding from a certain initialization y_0. A time series y_t that satisfies (2) or (4) is therefore said to be integrated of order one (or a unit root process or an I(1) process) provided $f_u(0)>0$. The latter condition rules out the possibility of a moving-average unit root in the model for u_t that would cancel the effect of the autoregressive unit root (for example, if $b(L)=(1-L)b_1(L)$ then model (2) is $\Delta y_t=\Delta a_1(L)^{-1}b_1(L)\varepsilon_t$ or, after cancellation, just $y_t=a_1(L)^{-1}b_1(L)\varepsilon_t$, which is not I(1)). Note that this possibility is also explicitly ruled out in the ARMA case by the requirement that $a(L)$ and $b(L)$ have no common factors. Alternatively, we may require that $u_t \neq \Delta v_t$ for some weakly stationary time series v_t, as in Leeb and Pötscher (2001) who provide a systematic discussion of I(1) behaviour. The partial sum process S_t in (4) is often described as a *stochastic trend.*

The representation (4) is especially important because it shows that the effect of the random shocks u_j on y_t does not die out as the time distance between j and t grows large. The shocks u_j then have a *persistent* effect on y_t in this model, in contrast to stationary systems. Whether actual economic time series have this characteristic or not is, of course, an empirical issue. The question can be addressed through statistical tests for the presence of a unit root in the series, a subject which has grown to be of major importance since the mid-1980s and which will be discussed later in this article. From the perspective of economic modelling the issue of persistence is also important because, if macroeconomic variables like real GNP have a unit root, then shocks to real GNP have permanent effects, whereas in traditional business cycle theory the effect of shocks on real GNP is usually considered to be only temporary. In more recent real business cycle theory, variables like real GNP are modelled in such a way that over the long run their paths are determined by supply side shocks that can be ascribed to technological and demographic forces from outside the model. Such economic models are more compatible with the statistical model (4) or close approximations to it in which the roots are local to unity in a sense that is described later in this essay.

Permanent and transitory effects in (4) can be distinguished by decomposing the process u_t in (3) as follows

$$u_t = \{C(1) + (L-1)\tilde{C}(L)\}\varepsilon_t = C(1)\varepsilon_t + \tilde{\varepsilon}_{t-1} - \tilde{\varepsilon}_t, \tag{5}$$

where $\tilde{\varepsilon}_t = \tilde{C}(L)\varepsilon_t$, $\tilde{C}(L) = \sum_0^\infty \tilde{c}_j L^j$ and $\tilde{c}_j = \sum_{j+1}^\infty c_s$. The decomposition (5) is valid algebraically if

$$\sum_{j=0}^{\infty} j^{1/2}|c_j| < \infty, \tag{6}$$

as shown in Phillips and Solo (1992), where validity conditions are systematically explored. Equation (5) is sometimes called the Beveridge–Nelson (1981) or BN decomposition of u_t, although both specialized and more general versions of it were known and used beforehand. The properties of the decomposition were formally investigated and used for the development of laws of large numbers and central limit theory and invariance principles in the paper by Phillips and Solo (1992). When the decomposition is applied to (4) it yields the representation

$$y_t = C(1)\sum_1^t \varepsilon_j + \tilde{\varepsilon}_0 - \tilde{\varepsilon}_t + y_0 = C(1)\sum_1^t \varepsilon_j + \xi_t + y_0, \text{ say,} \tag{7}$$

where $\xi_t = \tilde{\varepsilon}_0 - \tilde{\varepsilon}_t$. The right side of (7) decomposes y_t into three components: the first is a martingale component, $Y_t = C(1)\sum_1^t \varepsilon_j$, where the effects of the shocks ε_j are permanent; the second is a stationary component, where the effects of shocks are transitory, viz. $\xi_t = \tilde{\varepsilon}_0 - \tilde{\varepsilon}_t$, since the process $\tilde{\varepsilon}_t$ is stationary with valid Wold representation $\tilde{\varepsilon}_t = \tilde{C}(L)\varepsilon_t$ under (6) when ε_t is stationary with variance σ^2; and the third being the initial condition y_0. The relative strength of the martingale component is measured by the magnitude of the (infinite dimensional) coefficient $C(1) = \sum_{j=0}^{\infty} c_j$, which plays a large role in the measurement of long-run effects in applications. Accordingly, the decomposition (7) is sometimes called the martingale decomposition (cf., Hall and Heyde, 1980) where it was used in various forms in the probability literature prior to its use in economics.

The leading martingale term $Y_t = C(1)\sum_{s=1}^t \varepsilon_s$ in (7) is a partial sum process or stochastic trend and, under weak conditions on ε_t (see Phillips and Solo, 1992, for details) this term satisfies a functional central limit theorem whereby the scaled process

$$n^{-1/2}Y_{[nr]} \Rightarrow B(r), \tag{8}$$

a Brownian motion with variance $\omega^2 = C(1)^2\sigma^2 = 2\pi f_u(0)$, a parameter which is called the long-run variance of u_t, and where $[\cdot]$ signifies the integer part of its argument. Correspondingly,

$$n^{-1/2}y_{[nr]} \Rightarrow B(r), \tag{9}$$

provided $y_0 = o_p(\sqrt{n})$ A related result of great significance is based on the limit

$$n^{-1}\sum_{t=1}^{[nr]} Y_{t-1}\varepsilon_t C(1) \Rightarrow \int_0^r BdB \tag{10}$$

of the sample covariance of Y_{t-1} and its forward increment, $C(1)\varepsilon_t$. The limit process $M(r) = \int_0^r BdB$ is represented here as an Ito (stochastic) integral and is a continuous time martingale. The result may be proved directly (Solo, 1984; Phillips, 1987a; Chan and Wei, 1988) or by means of martingale convergence methods (Ibragimov and

Phillips, 2004) which take advantage of the fact that $\sum_{t=1}^{k} Y_{t-1}\varepsilon_t$ is a martingale. The limit theory given (9) and (10) was extended in Phillips (1987b; 1988a) and Chan and Wei (1987) to cases where the model (2) has an autoregressive root in the vicinity of unity ($\rho = 1 + \frac{c}{n}$, for some fixed c) rather than precisely at unity, in which case the limiting process is a linear diffusion (or Ornstein–Uhlenbeck process) with parameter c. This limit theory has proved particularly useful in the analysis of asymptotic local power functions of unit root tests (Phillips, 1987b) and the construction of confidence intervals (Stock, 1991). Phillips and Magdalinos (2007) considered moderate deviations from unity of the form $\rho = 1 + \frac{c}{k}$, where $k \to \infty$ but $\frac{k}{n} \to 0$, so that the roots are local but further away from unity, showing that central limit laws rather than functional laws apply in this case (see also Giraitis and Phillips, 2006). This theory is applicable to mildly explosive processes (where $c>0$) and therefore assists in bridging the gap between the limit theory for the stationary, unit root and explosive cases.

Both (8) and (10) have important multivariate generalizations that play a critical role in the study of spurious regressions (Phillips, 1986) and cointegration limit theory (Phillips and Durlauf, 1986; Engle and Granger, 1987; Johansen, 1988; Phillips, 1988a; Park and Phillips, 1988; 1989). In particular, if $y_t = (y'_{at}, y'_{bt})'$, $u_t = (u'_{at}, u'_{bt})'$ and $\varepsilon_t = (\varepsilon'_{at}, \varepsilon'_{bt})'$ are vector processes and $E(\varepsilon_t\varepsilon'_t) = \Sigma$, then: (i) the decomposition (5) continues to hold under (6), where $|c_j|$ is interpreted as a matrix norm; (ii) the functional law (8) holds and the limit process is vector Brownian motion $B = (B'_a, B'_b)'$ with covariance matrix $\Omega = C(1)\,\Sigma C(1)'$; and (iii) sample covariances converge weakly to stochastic processes with drift, as in

$$n^{-1}\sum_{t=1}^{[nr]} y_{at-1}u'_{bt} \Rightarrow \int_0^r B_a dB'_b + \lambda_{ab} r, \tag{11}$$

where $\lambda_{ab} = \sum_{k=1}^{\infty} E(u_{a0}u'_{bk})$ is a one sided long-run covariance matrix. The limit process on the right side of (11) is a semimartingale (incorporating a deterministic drift function $\lambda_{ab}r$) rather than a martingale when $\lambda_{ab} \neq 0$.

The decomposition (7) plays an additional role in the study of cointegration (Engle and Granger, 1987). When the coefficient matrix $C(1)$ is singular and β spans the null space of $C(1)'$, then $\beta' C(1) = 0$ and (7) leads directly to the relationship

$$\beta' Y_t = 0, \quad \text{a.s.},$$

which may be interpreted as a long run equilibrium (cointegrating) relationship between the stochastic trends (Y_t) of y_t. Correspondingly, we have the empirical cointegrating relationship

$$\beta' y_t = v_t,$$

among the observed series y_t with a residual $v_t = \beta'(\xi_t + y_0)$ that is stationary. The columns of β span what is called the cointegration space.

The above discussion presumes that the initialization y_0 has no impact on the limit theory, which will be so if y_0 is small relative to the sample size, specifically, if $y_0 = o_p(\sqrt{n})$. However, if $y_0 = O_p(\sqrt{n})$, for example if $y_0 = y_{0\theta_n}$ is indexed to depend on past shocks u_{-j} (satisfying a process of the form (3)) to some point in the distant past θ_n which is measured in terms of the sample size n, then the results can differ substantially. Thus, if $\theta_n = [\kappa n]$, for some fixed parameter $\kappa > 0$, then $y_{0\theta_n} = \sum_1^{[\kappa n]} u_{-j}$, and $n^{-1/2} y_{0\theta_n} \Rightarrow B_0(\kappa)$, for some Brownian motion $B_0(\kappa)$ with covariance matrix Ω_{00} given by the long-run variance matrix of u_{-j}. Under such an initialization, (9) and (11) are replaced by

$$n^{-1/2} y_{[nr]} \Rightarrow B(r) + B_0(\kappa) := B(r, \kappa), \quad \text{say} \tag{12}$$

and

$$n^{-1} \sum_{t=1}^{[nr]} y_{at-1} u'_{bt} \Rightarrow \int_0^r B_a(s, \kappa)\, dB_b(s)' + \lambda_{ab} r,$$

so that initializations play a role in the limit theory. This role becomes dominant when κ becomes very large, as is apparent from (12). The effect of initial conditions on unit root limit theory was examined in simulations by Evans and Savin (1981; 1984), by continuous record asymptotics by Phillips (1987a), in the context of power analysis by Müller and Elliott (2003), for models with moderate deviations from unity by Andrews and Guggenberger (2006), and for cases of large κ by Phillips (2006).

Model (4) is of special interest to economists working in finance because its output, y_t, behaves as if it has no fixed mean and this is a characteristic of many financial time series. If the components u_j are independent and identically distributed (i.i.d.) then y_t is a random walk. More generally, if u_j is a martingale difference sequence (mds) (that is orthogonal to its own past history so that $E_{j-1}(u_j) = E(u_j | u_{j-1}, u_{j-2}, \ldots, u_1) = 0$) then y_t is a martingale. Martingales are the essential mathematical elements in the development of a theory of fair games and they now play a key role in the mathematical theory of finance, exchange rate determination and securities markets. Duffie (1988) provides a modern treatment of finance that makes extensive use of this theory.

In empirical finance much attention has recently been given to models where the conditional variance $E(u_j^2 | u_{j-1}, u_{j-2}, \ldots, u_1) = \sigma_j^2$ is permitted to be time varying. Such models have been found to fit financial data well and many different parametric schemes for σ_j^2 have been devised, of which the ARCH (autoregressive conditional heteroskedasticity) and GARCH (generalized ARCH) models are the most common in practical work. These models come within the general class of models like (1) with mds errors. Some models of this kind also allow for the possibility of a unit root in the determining mechanism of the conditional variance σ_j^2 and these are called integrated conditional heteroskedasticity models. The IGARCH (integrated GARCH) model of Engle and Bollerslev (1986) is an example, where for certain parameters $\omega \geq 0$, $\beta \geq 0$, and $\alpha > 0$, we have the specification

$$\sigma_j^2 = \omega + \beta\sigma_{j-1}^2 + \alpha u_{j-1}^2, \tag{13}$$

with $\alpha + \beta = 1$ and $u_j = \sigma_j z_j$, where the z_j are i.i.d. innovations with $E(z_j) = 0$ and $E(z_j^2) = 1$. Under these conditions, the specification (13) has the alternative form

$$\sigma_j^2 = \omega + \sigma_{j-1}^2 + \alpha\sigma_{j-1}^2(z_{j-1}^2 - 1), \tag{14}$$

from which it is apparent that σ_j^2 has an autoregressive unit root. Indeed, since

$$E(\sigma_j^2 | \sigma_{j-1}^2) = \omega + \sigma_{j-1}^2,$$

σ_j^2 is a martingale when $\omega = 0$. It is also apparent from (14) that shocks as manifested in the deviation $z_{j-1}^2 - 1$ are persistent in σ_j^2. Thus, σ_j^2 shares some of the characteristics of an I(1) integrated process. But in other ways, σ_j^2 is very different. For instance, when $\omega = 0$ then $\sigma_j^2 \to 0$ almost surely as $j \to \infty$ and, when $\omega > 0$, σ_j^2 is asymptotically equivalent to a strictly stationary and ergodic process. These and other features of models like (13) for conditional variance processes with a unit root are studied in Nelson (1990).

In macroeconomic theory also, models such as (2) play a central role in modern treatments. In a highly influential paper, R. Hall (1978) showed that under some general conditions consumption is well modelled as a martingale, so that consumption in the current period is the best predictor of future consumption, thereby providing a macroeconomic version of the efficient markets hypothesis. Much attention has been given to this idea in subsequent empirical work.

One generic class of economic model where unit roots play a special role is the 'present value model' of Campbell and Shiller (1988). This model is based on agents' forecasting behaviour and takes the form of a relationship between one variable Y_t and the discounted, present value of rational expectations of future realizations of another variable X_{t+i} $(i = 0, 1, 2, \ldots)$. More specifically, for some stationary sequence c_t (possibly a constant) we have

$$Y_t = \theta(1 - \delta)\sum_{i=0}^{\infty} \delta^i E_t(X_{t+i}) + c_t. \tag{15}$$

When X_t is a martingale, $E_t(X_{t+i}) = X_t$ and (15) becomes

$$Y_t = \theta X_t + c_t, \tag{16}$$

so that Y_t and X_t are cointegrated in the sense of Engle and Granger (1987). More generally, when X_t is I(1) we have

$$Y_t = \theta X_t + c_t, \tag{17}$$

where $c_t = c_t + \theta\sum_{k=1}^{\infty} \delta^k E_t(\Delta X_{t+k})$, so that Y_t and X_t are also cointegrated in this general case. Models of this type arise naturally in the study of the term structure of interest rates, stock prices and dividends and linear-quadratic intertemporal optimization problems.

An important feature of these models is that they result in parametric linear cointegrating relations such as (16) and (17). This linearity in the relationship between Y_t and X_t accords with the linear nature of the partial sum process that determines X_t itself, as seen in (4), and has been extensively studied since the mid-1980s. However, in more general models, economic variables may be determined in terms of certain nonlinear functions of fundamentals. When these fundamentals are unit root processes like (4), then the resulting model has the form of a nonlinear cointegrating relationship. Such models are relevant, for instance, in studying market interventions by monetary and fiscal authorities (Park and Phillips, 2000; Hu and Phillips, 2004) and some of the asymptotic theory for analysing parametric models of this type and for statistical inference in such models is given in Park and Phillips (1999; 2001), de Jong (2004), Berkes and Horváth (2006) and Pötscher (2004). More complex models of this type are nonparametric and different methods of inference are typically required with very different limit theories and typically slower convergence rates (Karlsen, Myklebust, and Tjøstheim, 2007; Wang and Phillips, 2006). Testing for the presence of such nonlinearities can therefore be important in empirical practice (Hong and Phillips, 2005; Kasparis, 2004).

Statistical tests for the presence of a unit root fall into the general categories of classical and Bayesian, corresponding to the mode of inference that is employed. Classical procedures have been intensively studied and now occupy a vast literature. Most empirical work to date has used classical methods but some attention has been given to Bayesian alternatives and direct model selection methods. These approaches will be outlined in what follows.

Although some tests are known to have certain limited (asymptotic) point optimality properties, there is no known procedure which uniformly dominates others, even asymptotically. Ploberger (2004) provides an analysis of the class of asymptotically admissable tests in problems that include the simplest unit root test, showing that the conventional likelihood ratio (LR) test (or Dickey–Fuller, 1979; 1981, t test) is not within this class, so that the LR test, while it may have certain point optimal properties, is either inadmissible or must be modified so that it belongs to the class. This fundamental difficulty, together with the nonstandard nature of the limit theory and the more complex nature of the asymptotic likelihood in unit root cases partly explains why there is such a proliferation of test procedures and simulation studies analysing performance characteristics in the literature.

Classical tests for a unit root may be classified into parametric, semiparametric and nonparametric categories. Parametric tests usually rely on augmented regressions of the type

$$\Delta y_t = a y_{t-1} + \sum_{i=1}^{k-1} \varphi_i \Delta y_{t-i} + e_t, \tag{18}$$

where the lagged variables are included to model the stationary error u_t in (2). Under the null hypothesis of a unit root, we have $a=0$ in (18) whereas when y_t is stationary we have $a<0$. Thus, a simple test for the presence of a unit root against a stationary alternative in (18) is based on a one-sided t-ratio test of $\mathscr{H}_0 : a = 0$ against $\mathscr{H}_1 : a<0$. This test is popularly known as the ADF (or augmented Dickey–Fuller) test (Said and Dickey, 1984) and follows the work of Dickey and Fuller (1979; 1981) for testing Gaussian random walks. It has been extensively used in empirical econometric work since the Nelson and Plosser (1982) study, where it was applied to 14 historical time series for the USA leading to the conclusion that unit roots could not be rejected for 13 of these series (all but the unemployment rate). In that study, the alternative hypothesis was that the series were stationary about a deterministic trend (that is, trend stationary) and therefore model (18) was further augmented to include a linear trend, viz.

$$\Delta y_t = \mu + \beta t + a y_{t-1} + \sum_{i=1}^{k-1} \varphi_i \Delta y_{t-i} + e_t. \tag{19}$$

When y_t is trend stationary we have $a<0$ and $\beta \neq 0$ in (19), so the null hypothesis of a difference stationary process is $a=0$ and $\beta=0$. This null hypothesis allows for the presence of a non-zero drift in the process when the parameter $\mu \neq 0$. In this case a joint test of the null hypothesis $\mathscr{H}_0 : a = 0, \beta=0$ can be mounted using a regression F-test. ADF tests of $a=0$ can also be mounted directly using the coefficient estimate from (18) or (19), rather than its t ratio (Xiao and Phillips, 1998).

What distinguishes both these and other unit root tests is that critical values for the tests are not the same as those for conventional regression F- and t-tests, even in large samples. Under the null, the limit theory for these tests is nonstandard and involves functionals of a Wiener process. Typically, the critical values for five or one per cent level tests are much further out than those of the standard normal or chi-squared distributions. Specific forms for the limits of the ADF t-test (ADF_t) and coefficient (ADF_a) test are

$$ADF_t \Rightarrow \frac{\int_0^1 W dW}{\left(\int_0^1 W^2\right)^{1/2}}, \quad ADF_a \Rightarrow \frac{\int_0^1 W dW}{\int_0^1 W^2}, \tag{20}$$

where W is a standard Wiener process or Brownian motion with variance unity. The limit distributions represented by the functionals (20) are known as unit root distributions. The limit theory was first explored for models with Gaussian errors, although not in the Wiener process form and not using functional limit laws, by Dickey (1976), Fuller (1976) and Dickey and Fuller (1979; 1981), who also provided tabulations. For this reason, the distributions are sometimes known as Dickey–Fuller distributions. Later work by Said and Dickey (1984) showed that, if the lag number k in (18) is allowed to increase as the sample size increases with a condition on the divergence rate that $k = o(n^{1/3})$, then the ADF test is asymptotically valid in models of the form (2) where u_t is not necessarily autoregressive.

Several other parametric procedures have been suggested, including Von Neumann ratio statistics (Sargan and Bhargava, 1983; Bhargava, 1986; Stock, 1994a), instrumental variable methods (Hall, 1989; Phillips and Hansen, 1990) and variable addition methods (Park, 1990). The latter also allow a null hypothesis of trend stationarity to be tested directly, rather than as an alternative to difference stationarity. Another approach that provides a test of a null of trend stationarity is based on the unobserved components representation

$$y_t = \mu + \beta t + r_t + u_t, \quad r_t = r_{t-1} + v_t, \tag{21}$$

which decomposes a time series y_t into a deterministic trend, an integrated process or random walk (r_t) and a stationary residual (u_t). The presence of the integrated process component in y_t can then be tested by testing whether the variance (σ_v^2) of the innovation v_t is zero. The null hypothesis is then $\mathscr{H}_0 : \sigma_v^2 = 0$, which corresponds to a null of trend stationarity. This hypothesis can be tested in a very simple way using the Lagrange multiplier (LM) principle, as shown in Kwiatkowski et al. (1992), leading to a commonly used test known as the KPSS test. If $\hat{e}_t$ denotes the residual from a regression of y_t on a deterministic trend (a simple linear trend in the case of (21) above) and $\hat{\omega}_e^2$ is a HAC (heteroskedastic and autocorrelation consistent) estimate constructed from $\hat{e}_t$, then the KPSS statistic has the simple form

$$LM = \frac{n^{-2}\sum_{t=1}^{n} S_t^2}{\hat{\omega}_e^2},$$

where S_t is the partial sum process of the residuals $\sum_{j=1}^{t}\hat{e}_j$. Under the null hypothesis of stationarity, this LM statistic converges to $\int_0^1 V_X^2$, where V_X is a generalized Brownian bridge process whose construction depends on the form (X) of the deterministic trend function. Power analysis indicates that test power depends importantly on the choice of bandwidth parameter in HAC estimation and some recent contributions to this subject are Sul, Phillips and Choi (2006) and Müller (2005) and Harris, Leybourne and McCabe (2007). Other general approaches to testing I(0) versus I(1) have been considered in Stock (1994a, 1999).

By combining r_t and u_t in (21) the components model may also be written as

$$y_t = \mu + \beta t + x_t, \quad \Delta x_t = a x_{t-1} + \eta_t. \tag{22}$$

In this format it is easy to construct an LM test of the null hypothesis that y_t has a stochastic trend component by testing whether $a = 0$ in (22). When $a = 0$, (22) reduces to

$$\Delta y_t = \beta + \eta_t, \quad \text{or} \quad y_t = \beta t + \sum_{1}^{t} \eta_i + y_0, \tag{23}$$

and so the parameter μ is irrelevant (or surplus) under the null. However, the parameter β retains the same meaning as the deterministic trend term coefficient under both the null and the alternative hypothesis. This approach has formed the basis of several tests for a unit root that have been developed (see Bhargava, 1986; Schmidt and Phillips, 1992) and the parameter economy of this model gives these tests some advantage in terms of power over procedures like the ADF in the neighbourhood of the null.

This power advantage may be further exploited by considering point optimal alternatives in the construction of the test and in the process of differencing (or detrending) that leads to (23), as pursued by Elliott, Rothenberg and Stock (1995). In particular, note that (23) involves detrending under the null hypothesis of a unit root, which amounts to first differencing, whereas if the root were local to unity, the appropriate procedure would be to use quasi differencing. However, since the value of the coefficient in the locality of unity is unknown (otherwise, there would be no need for a test), it can only be estimated or guessed. The procedure suggested by Elliott, Rothenberg and Stock (1995) is to use a value of the localizing coefficient in the quasi-differencing process for which asymptotic power is calculated by simulation to be around 50 per cent, a setting which depends on the precise model for estimation that is being used. This procedure, which is commonly known as generalized least squares (GLS) detrending (although the terminology is a misnomer because quasi-differencing not full GLS is used to accomplish trend elimination) is then asymptotically approximately point optimal in the sense that its power function touches the asymptotic power envelope at that value. Simulations show that this method has some advantage in finite samples, but it is rarely used in empirical work in practice, partly because of the inconvenience of using specialized tables for the critical values of the resulting test and partly because settings for the localizing coefficient are arbitrary and depend on the form of the empirical model.

Some unit root tests based on standard limit distribution theory have been developed. Phillips and Han (2008), for example, give an autoregressive coefficient estimator whose limit distribution is standard normal for all stationary, unit root and local to unity values of the autoregressive coefficient. This estimator may be used to construct tests and valid confidence intervals, but tests suffer power loss because the rate of convergence of the estimator is $\sqrt{n}$ uniformly over these parameter values. So and Shin (1999) and Phillips, Park and Chang (2004) showed that certain nonlinear instrumental variable estimators, such as the Cauchy estimator, also lead to t-tests for a unit root which have an asymptotic standard normal distribution. Again, these procedures suffer power loss from reduced convergence rates (in this case, $n^{1/4}$), but have the advantage of uniformity and low bias. Bias is a well known problem in autoregressive estimation and many procedures for addressing the problem have been considered. It seems that bias reduction is particularly advantageous in the case of unit root tests in panel data, where cross-section averaging exacerbates bias effects when the time dimension is small. Some simulation and indirect inference procedures for bias removal have been successfully used both in autoregressions (Andrews, 1993;

Gouriéroux, Renault and Touzi, 2000) and in panel dynamic models (Gouriéroux, Phillips and Yu, 2006).

Semiparametric unit root tests are among the most commonly used unit root tests in practical work and are appealing in terms of their generality and ease of use. Tests in this class employ nonparametric methods to model and estimate the contribution from the error process u_t in (2), allowing for both autocorrelation and heterogeneity. These tests and the use of functional limit theory methods in econometrics, leading to the limit formulae (20), were introduced in Phillips (1987a). Direct least squares regression on

$$\Delta y_t = a y_{t-1} + u_t \tag{24}$$

gives an estimate of the coefficient and its t-ratio in this equation. These two statistics are then corrected to deal with serial correlation in u_t by employing an estimate of the variance of u_t and its long-run variance. The latter estimate may be obtained by a variety of kernel-type HAC or other spectral estimates (such as autoregressive spectral estimates) using the residuals $\hat{u}_t$ of the OLS regression on (24). Automated methods of bandwidth selection (or order selection in the case of autoregressive spectral estimates) may be employed in computing these HAC estimates and these methods typically help to reduce size distortion in unit root testing (Lee and Phillips, 1994; Stock, 1994a; Ng and Perron, 1995; 2001). However, care needs to be exercised in the use of automated procedures in the context of stationarity tests such as the KPSS procedure to avoid test inconsistency (see Lee, 1996; Sul, Phillips and Choi, 2006).

This semiparametric approach leads to two test statistics, one based on the coefficient estimate, called the $Z(a)$ test, the other based on its t-ratio, called the $Z(t)$ test. The limit distributions of these statistics are the same as those given in (20) for the ADF coefficient and t-ratio tests, so the tests are asymptotically equivalent to the corresponding ADF tests. Moreover, the local power functions are also equivalent to those of the Dickey–Fuller and ADF tests, so that there is no loss in asymptotic power from the use of nonparametric methods to address autocorrelation and heterogeneity (Phillips, 1987b). Similar semiparametric corrections can be applied to the components models (21) and (22) leading to generally applicable LM tests of stationarity ($\sigma_v^2 = 0$) and stochastic trends ($a = 0$).

The Z tests were extended in Phillips and Perron (1988) and Ouliaris, Park and Phillips (1989) to models with drift, and by Perron (1989) and Park and Sung (1994) to models with structural breaks in the drift or deterministic component. An important example of the latter is the trend function

$$h_t = \sum_{j=0}^{p} f_j t^j + \sum_{j=0}^{p} f_{m,j} t_m^j, \quad \text{where} \quad t_m^j = \begin{cases} 0 & t \in \{1, \ldots, m\} \\ (t-m)^j & t \in \{m+1, \ldots, n\} \end{cases} \tag{25}$$

which allows for the presence of a break in the polynomial trend at the data point $t = m + 1$. Collecting the individual trend regressors in (25) into the vector x_t, there exists a continuous function $X(r) = (1, r, \ldots, r^p)'$ such that $D_n^{-1} x_{[nr]} \to X(r)$ as

$n \to \infty$ uniformly in $r \in [0, 1]$, where $D_n = \text{diag}(1, n, \ldots, n^p)$. If $\mu = \lim_{n\to\infty}(m/n) > 0$ is the limit of the fraction of the sample where the structural change occurs, then the limiting trend function $X_\mu(r)$ corresponding to (25) has a similar break at the point μ. All the unit root tests discussed above continue to apply as given for such broken trend functions with appropriate modifications to the limit theory to incorporate the limit function $X_\mu(r)$. Indeed, (25) may be extended further to allow for multiple break points in the sample and in the limit process. The tests may be interpreted as tests for the presence of a unit root in models where broken trends may be present in the data. The alternative hypothesis in this case is that the data are stationary about a broken deterministic trend of degree p.

In order to construct unit root tests that allow for breaking trends like (25) it is necessary to specify the break point m. (Correspondingly, the limit theory depends on $X_\mu(r)$ and therefore on μ.) In effect, the break point is exogenously determined. Perron (1989) considered linear trends with single break points intended to capture the 1929 stock market crash and the 1974 oil price shock in this way. An alternative perspective is that any break points that occur are endogenous to the data and unit root tests should take account of this fact. In this case, alternative unit root tests have been suggested (for example, Banerjee, Lumsdaine and Stock, 1992; Zivot and Andrews, 1992) that endogenize the break point by choosing the value of m that gives the least favourable view of the unit root hypothesis. Thus, if $ADF(m)$ denotes the ADF statistic given by the t-ratio for α in the ADF regression (19) with a broken trend function like (25), then the trend break ADF statistic is

$$ADF(\hat{m}) = \min_{\underline{m} \le m \le \bar{m}} ADF(m), \quad \text{where} \quad \underline{m} = [n\,\underline{\mu}], \quad \bar{m} = [n\,\bar{\mu}], \tag{26}$$

for some $0 < \underline{\mu} < \bar{\mu} < 1$. The limit theory for this trend break ADF statistic is given by

$$ADF(\hat{m}) \Rightarrow \inf_{\mu \in [\underline{\mu}, \bar{\mu}]} \left[\int_0^1 W_{X_\mu} dW\right] \left[\int_0^1 W_{X_\mu}^2\right]^{-1/2}, \tag{27}$$

where W_X is detrended standard Brownian motion defined by $W_X(r) = W(r) - [\int_0^1 WX][\int_0^1 XX]^{-1} X(r)$. The limit process $X_\mu(r)$ that appears in the functional W_{X_μ} is dependent on the trend break point μ over which the functional is minimized. Similar extensions to trend breaks are possible for other unit root tests and to multiple breaks (Bai, 1997; Bai and Perron, 1998; 2006; Kapetanios, 2005). Critical values of the limiting test statistic (27) are naturally further out in the tail than those of the exogenous trend break statistic, so it is harder to reject the null hypothesis of a unit root when the break point is considered to be endogenous.

Asymptotic and finite sample critical values for the endogenized trend break ADF unit root test are given in Zivot and Andrews (1992). Simulations studies indicate that the introduction of trend break functions leads to further reductions in the power of unit root tests and to substantial finite sample size distortion in the tests. Sample trajectories of a random walk are often similar to those of a process that is stationary about a broken trend for some particular breakpoint (and even more so when several break points are permitted in the trend). So continuing reductions in the power of unit root tests against competing models of this type is to be expected and discriminatory power between such different time series models is typically low. In fact, the limit Brownian motion process in (9) can itself be represented as an infinite linear random combination of deterministic functions of time, as discussed in Phillips (1998), so there are good theoretical reasons for anticipating this outcome. Carefully chosen trend stationary models can always be expected to provide reasonable representations of given random walk or unit root data, but such models are certain to fail in post-sample projections as the post-sample data drifts away from any given trend or broken trend line. Phillips (1998; 2001) explores the impact of these considerations in a systematic way.

From a practical standpoint, models with structural breaks attach unit weight and hence persistence to the effects of innovations at particular times in the sample period. In effect, break models simply dummy out the effects of certain observations by parameterizing them as persistent effects. To the extent that persistent shocks of this type occur intermittently throughout the entire history of a process, these models are therefore similar to models with a stochastic trend. However, if only one or a small number of such breaks occur then the process does have different characteristics from that of a stochastic trend. In such cases, it is often of interest to identify the break points endogenously and relate such points to institutional events or particular external shocks that are know to have occurred.

More general nonparametric tests for a unit root are also possible. These rely on frequency domain regressions on (24) over all frequency bands (Choi and Phillips, 1993). They may be regarded as fully nonparametric because they test in a general way for coherency between the series y_t and its first difference Δy_t. Other frequency domain procedures involve the estimation of a fractional differencing parameter and the use of tests and confidence intervals based on the estimate. The time series y_t is fractionally integrated with memory parameter d if $(1-L)^d y_t = u_t$ and u_t is a stationary process with spectrum $f_u(\lambda)$ that is continuous at the origin with $f_u(0)>0$, or a (possibly mildly heterogeneous) process of the form given in (3). Under some rather weak regularity conditions, it is possible to estimate d consistently by semiparametric methods irrespective of the value of d. Shimotsu and Phillips (2005) suggest an exact local Whittle estimator $\hat{d}$ that is consistent for all d and for which $\sqrt{n}(\hat{d} - d) \Rightarrow N(0, \frac{1}{4})$, extending earlier work by Robinson (1995) on local Whittle estimation in the stationary case where $|d|<1$. These methods are narrow band procedures focusing on frequencies close to the origin, so that long run behaviour is captured. The Shimotsu–Phillips estimator may be used to test the unit root hypothesis $\mathscr{H}_0 : d = 1$ against alternatives such as $\mathscr{H}_1 : d<1$. The limit theory may also be used to construct valid confidence intervals for d.

The $Z(a)$, $Z(t)$ and ADF tests are the most commonly used unit root tests in empirical research. Extensive simulations have been conducted to evaluate the performance of the tests. It is known that the $Z(a)$, $Z(t)$ and ADF tests all perform satisfactorily except when the error process u_t displays strong negative serial correlation. The $Z(a)$ test generally has greater power than the other two tests but also suffers from more serious size distortion. All of these tests can be used to test for the presence of cointegration by using the residuals from a cointegrating regression. Modification of the critical values used in these tests is then required, for which case the limit theory and tables were provided in Phillips and Ouliaris (1990) and updated in MacKinnon (1994).

While the Z tests and other semiparametric procedures are designed to cope with mildly heterogeneous processes, some further modifications are required when there is systematic time-varying heterogeneity in the error variances. One form of systematic variation that allows for jumps in the variance has the form $E(\varepsilon_t^2) = \sigma_t^2 = \sigma^2 g(\frac{t}{n})$, where the variance evolution function $g(\frac{t}{n})$ may be smooth except for simple jump discontinuities at a finite number of points. Such formulations introduce systematic time variation into the errors, so that we may write $\varepsilon_t = g(\frac{t}{n})\eta_t$, where ζ_t is a martingale difference sequence with variance $E\zeta_t^2 = \sigma^2$. These evolutionary changes then have persistent effects on partial sums of ε_t, thereby leading to alternate functional laws of the form

$$n^{-1/2} Y_{[nr]} \Rightarrow B_g(r) = \int_0^r g(s)\, dB(s),$$

in place of (8). Accordingly, the limit theory for unit root tests changes and some nonparametric modification of the usual tests is needed to ensure that existing asymptotic theory applies (Beare, 2006) or to make appropriate corrections in the limit theory (Cavaliere, 2004; Cavaliere and Taylor, 2007) so that there is less size distortion in the tests.

An extension of the theory that is relevant in the case of quarterly data is to the seasonal unit root model

$$(1 - L^4)y_t = u_t. \tag{28}$$

Here, the polynomial $1-L^4$ can be factored as $(1-L)(1+L)(1+L^2)$, so that the unit roots (or roots on the unit circle) in (28) occur at 1, -1, i, and $-i$, corresponding to the annual ($L=1$) frequency, the semi-annual ($L=-1$) frequency, and the quarter and three quarter annual ($L=i, -i$) frequency respectively. Quarterly differencing, as in (28), is used as a seasonal adjustment device, and it is of interest to test whether the data supports the implied hypothesis of the presence of unit roots at these seasonal frequencies. Other types of seasonal processes, say monthly data, can be analysed in the same way. Tests for seasonal unit roots within the particular context of (28) were studied by Hylleberg et al. (1990), who extended the parametric *ADF* test to the case of seasonal unit roots. In order to accommodate fourth differencing, the autoregressive model is written in the new form

$$\Delta_4 y_t = \alpha_1 y_{1t-1} + \alpha_2 y_{2t-1} + \alpha_3 y_{3t-2} + \alpha_4 y_{3t-1} + \sum\nolimits_{i=1}^{p} \varphi_i \Delta_4 y_{t-i} + \varepsilon_t, \tag{29}$$

where $\Delta_4 = 1 - L^4$, $y_{1t} = (1+L)(1+L^2)y_t$, $y_{2t} = -(1-L)(1+L^2)y_t$, and $y_{3t} = -(1-L^2)y_t$. The transformed data y_{1t}, y_{2t}, y_{3t} retain the unit root at the zero frequency (long run), the semi-annual frequency (two cycles per year), and the annual frequency (one cycle per year). When $\alpha_1 = \alpha_2 = \alpha_3 = \alpha_4 = 0$, there are unit roots at the zero and seasonal frequencies. To test the hypothesis of a unit root ($L = 1$) in this seasonal model, a t-ratio test of $\alpha_1 = 0$ is used. Similarly, the test for a semi-annual root ($L = -1$) is based on a t-ratio test of $\alpha_2 = 0$, and the test for an annual root on the t-ratios for $\alpha_3 = 0$ or $\alpha_4 = 0$. If each of the α's is different from zero, then the series has no unit roots at all and is stationary. Details of the implementation of this procedure are given in Hylleberg et al. (1990), the limit theory for the tests is developed in Chan and Wei (1988), and Ghysels and Osborne (2001) provide extensive discussion and applications.

Most empirical work on unit roots has relied on classical tests of the type described above. But Bayesian methods are also available and appear to offer certain advantages like an exact finite sample analysis (under specific distributional assumptions) and mass point posterior probabilities for break point analysis. In addressing the problem of trend determination, traditional Bayes methods may be employed such as the computation of Bayesian confidence sets and the use of posterior odds tests. In both cases prior distributions on the parameters of the model need to be defined and posteriors can be calculated either by analytical methods or by numerical integration. If (18) is rewritten as

$$y_t = \rho y_{t-1} + \sum_{1}^{k-1} \varphi_i \Delta y_{t-i} + e_t \tag{30}$$

then the posterior probability of the nonstationary set $\{\rho \geq 1\}$ is of special interest in assessing the evidence in support of the presence of a stochastic trend in the data. Posterior odds tests typically proceed with 'spike and slab' prior distributions (π) that assign an atom of mass such as $\pi(\rho = 1) = \theta$ to the unit-root null and a continuous distribution with mass $1-\theta$ to the stationary alternative, so that $\pi(-1 < \rho < 1) = 1 - \theta$. The posterior odds then show how the prior odds ratio $\theta/(1-\theta)$ in favour of the unit root is updated by the data.

The input of information via the prior distribution, whether deliberate or unwitting, is a major reason for potential divergence between Bayesian and classical statistical analyses. Methods of setting an objective correlative in Bayesian analysis through the use of model-based, impartial reference priors that accommodate nonstationarity are therefore of substantial interest. These were explored in Phillips (1991a), where many aspects of the subject are discussed. The subject is controversial, as the attendant commentary on that paper and the response (Phillips, 1991b) reveal. The simple example of a Gaussian autoregression with a uniform prior on the autoregressive coefficient ρ and with an error variance σ^2 that is known illustrates one central point of controversy between Bayesian and classical inference procedures. In this case, when the prior on ρ is uniform, the posterior for ρ is Gaussian and symmetric about the maximum likelihood estimate $\hat{\rho}$ (Sims and Uhlig, 1991), whereas the sampling distribution of $\hat{\rho}$ is biased downwards and skewed with a long left-hand tail. Hence, if the calculated value of $\hat{\rho}$ were found to be $\hat{\rho} = 1$, then Bayesian inference effectively assigns a 50 per cent posterior probability to stationarity $\{|\rho| < 1\}$, whereas classical methods, which take into account the substantial downward bias in the estimate $\hat{\rho}$, indicate that the true value of ρ is much more likely to be in the explosive region $\{\rho > 1\}$.

Another major point of difference is that the Bayesian posterior distribution is asymptotically Gaussian under very weak conditions, which include cases where there are unit roots ($\rho = 1$), whereas classical asymptotics for $\hat{\rho}$ are non-standard, as in (20). These differences are explored in Kim (1994), Phillips and Ploberger (1996) and Phillips (1996). The unit root case is one of very few instances where Bayesian and classical asymptotic theory differ. The reason for the difference in the unit root case is that Bayesian asymptotics rely on the local quadratic shape of the likelihood and condition on a given trajectory, whereas classical asymptotics rely on functional laws such as (9), which take into account the persistence in unit root data which manifest in the limiting trajectory.

Empirical illustrations of the use of Bayesian methods of trend determination for various macroeconomic and financial time series are given in DeJong and Whiteman (1991a; 1991b), Schotman and van Dijk (1991) and Phillips (1991a; 1992), the latter implementing an objective model-based approach. Phillips and Ploberger (1994; 1996) develop Bayes tests, including an asymptotic information criterion PIC (posterior information criterion) that extends the Schwarz (1978) criterion BIC (Bayesian information criterion) by allowing for potential nonstationarity in the data (see also Wei, 1992). This approach takes account of the fact that Bayesian time series analysis is conducted conditionally on the realized history of the process. The mathematical effect of such conditioning is to translate models such as (30) to a 'Bayes model' with time-varying and data-dependent coefficients, that is,

$$y_{t+1} = \hat{\rho}_t y_t + \sum_{1}^{k-1} \hat{\varphi}_{it} \Delta y_{t-i} + e_t, \tag{31}$$

where ($\hat{\rho}_t$, $\hat{\varphi}_{it}$; $i = 1, \ldots, k-1$) are the latest best estimates of the coefficients from the data available to point 't' in the trajectory. The 'Bayes model' (31) and its probability measure can be used to construct likelihood ratio tests of hypotheses such as the unit root null $\rho = 1$, which relate to the model selection criterion PIC. Empirical illustrations of this approach are given in Phillips (1994; 1995).

Nonstationarity is certainly one of the most dominant and enduring characteristics of macroeconomic and financial time series. It therefore seems appropriate that this feature of the data be seriously addressed both in econometric methodology and in empirical practice. However, until the 1980s this was not the case. Before 1980 it was standard empirical practice in econometrics to treat observed trends as simple deterministic functions of time. Nelson and Plosser (1982) challenged this practice and showed that observed trends can be better modelled if one allows for stochastic trends even when there is some deterministic drift. Since their work there has been a continuing reappraisal of trend behaviour in economic time series and substantial development in the econometric methods of nonstationary time series. But the general conclusion that stochastic trends are present as a component of many economic and financial time series has withstood extensive empirical study.

This article has touched only a part of this large research field and traced only the main ideas involved in unit root modelling and statistical testing. This overview also does not cover the large and growing field of panel unit root testing and panel stationarity tests. The reader may consult the following review articles devoted to various aspects of the field for additional coverage and sources: (*a*) on unit roots: Phillips (1988b), Diebold and Nerlove (1990), Dolado, Jenkinson and Sosvilla-Rivero (1990), Campbell and Perron (1991), Stock (1994b), Phillips and Xiao (1998), and Byrne and Perman (2006); (*b*) on panel unit roots: Phillips and Moon (1999), Baltagi and Kao (2000), Choi (2001), Hlouskova and Wagner (2006); and (*c*) special journal issues of the *Oxford Bulletin of Economics and Statistics* (1986; 1992), *the Journal of Economic Dynamics and Control* (1988), *Advances in Econometrics* (1990), *Econometric Reviews* (1992), and *Econometric Theory* (1994).

PETER C.B. PHILLIPS

See also **ARCH models; Bayesian time series analysis; cointegration.**

Bibliography

Andrews, D.W.K. 1993. Exactly median-unbiased estimation of first-order autoregressive/unit root models. *Econometrica* 61, 139–66.

Andrews, D.W.K. and Guggenberger, P. 2006. Asymptotics for stationary very nearly unit root processes. Mimeo, Yale University.

Bai, J. 1997. Estimating multiple break one at a time. *Econometric Theory* 13, 315–52.

Bai, L. and Perron, P. 1998. Estimating and testing linear models with multiple structural changes. *Econometrica* 66, 47–78.

Bai, L. and Perron, P. 2006. Multiple structural change models: a simulation analysis. In *Econometric Theory and Practice*, ed. Corbae, B.E. Hansen and S.N. Durlauf. New York: Cambridge University Press.

Baltagi, B.H. and Kao, C. 2000. Nonstationary panels, cointegration in panels and dynamic panels: a survey. *Advances in Econometrics* 15, 7–51.

Banerjee, A., Lumsdaine, R. and Stock, J. 1992. Recursive and sequential tests of the unit root and trend break hypotheses: theory and international evidence. *Journal of Business and Economic Statistics* 10, 271–87.

Beare, B.K. 2006. Unit root testing with unstable volatility. Mimeo, Yale University.

Berkes, I. and Horváth, L. 2006. Convergence of integral functionals of stochastic processes. *Econometric Theory* 22, 304–22.

Beveridge, S. and Nelson, C.R. 1981. A new approach to decomposition of economic time series into permanent and transitory components with particular attention to measurement of the 'business cycle'. *Journal of Monetary Economics* 7, 151–74.

Bhargava, A. 1986. On the theory of testing for unit roots in observed time series. *Review of Economic Studies* 52, 369–84.

Box, G.E.P. and Jenkins, G.M. 1976. *Time Series Analysis: Forecasting and Control*, rev. edn. San Francisco: Holden Day.
Byrne, J.P. and Perman, R. 2006. Unit roots and structural breaks: a survey of the literature. Mimeo, University of Strathclyde.
Campbell, J.Y. and Perron, P. 1991. Pitfalls and opportunities: what macroeconomists should know about unit roots (with discussion). In *NBER Macroeconomics Annual 1991*, ed. O.J. Blanchard and S. Fischer. Cambridge, MA: MIT Press.
Campbell, J.Y. and Shiller, R.J. 1988. Interpreting cointegrated models. *Journal of Economic Dynamics and Control* 12, 505–22.
Cavaliere, G. 2004. Unit root tests under time-varying variances. *Econometric Reviews* 23, 259–92.
Cavaliere, G. and Taylor, A.M.R. 2007. Testing for unit roots in time series models with non-stationary volatility. *Journal of Econometrics* 140(2), 919–47.
Chan, N.H. and Wei, C.Z. 1987. Asymptotic inference for nearly nonstationary AR(1) processes. *Annals of Statistics* 15, 1050–63.
Chan, N.H. and Wei, C.Z. 1988. Limiting distributions of least squares estimates of unstable autoregressive processes. *Annals of Statistics* 16, 367–401.
Choi, I. 2001. Unit roots for panel data. *Journal of International Money and Finance* 20, 249–72.
Choi, I. and Phillips, P.C.B. 1993. Testing for a unit root by frequency domain regression. *Journal of Econometrics* 59, 263–86.
Christiano, L.J. 1992. Searching for a break in GNP. *Journal of Business & Economic Statistics* 10, 237–50.
de Jong, R. 2004. Addendum to 'asymptotics for nonlinear transformations of integrated time series'. *Econometric Theory* 20, 623–35.
DeJong, D.N. and Whiteman, C.H. 1991a. Reconsidering trends and random walks in macroeconomic time series. *Journal of Monetary Economics* 28, 221–54.
DeJong, D.N. and Whiteman, C.H. 1991b. The temporal stability of dividends and stock prices: evidence from the likelihood function. *American Economic Review* 81, 600–17.
Dickey, D.A. 1976. Estimation and hypothesis testing in nonstationary time series. Ph.D. thesis, Iowa State University.
Dickey, D.A. and Fuller, W.A. 1979. Distribution of the estimators for autoregressive time series with a unit root. *Journal of the American Statistical Association* 74, 427–31.
Dickey, D.A. and Fuller, W.A. 1981. Likelihood ratio statistics for autoregressive time series with a unit root. *Econometrica* 49, 1057–72.
Diebold, F.X. and Nerlove, M. 1990. Unit roots in economic time series. *Advances in Econometrics* 8, 3–70.
Dolado, J.J., Jenkinson, T. and Sosvilla-Rivero, S. 1990. Cointegration and unit roots. *Journal of Economic Surveys* 4, 249–73.
Duffie, D. 1988. *Security Markets: Stochastic Models*. San Diego: Academic Press.
Elliott, G., Rothenberg, T.J. and Stock, J.H. 1995. Efficient tests of an autoregressive unit root. *Econometrica* 64, 813–36.
Engle, R.F. and Bollerslev, T. 1986. Modeling the persistence of conditional variances. *Econometric Reviews* 5, 1–50.
Engle, R.F. and Granger, C.W.J. 1987. Co-integration and error correction: representation, estimation, and testing. *Econometrica* 55, 251–76.
Evans, G.B.A. and Savin, N.E. 1981. Testing for unit roots: 1. *Econometrica* 49, 753–79.
Evans, G.B.A. and Savin, N.E. 1984. Testing for unit roots: 2. *Econometrica* 52, 1241–69.
Fuller, W.A. 1976. *Introduction of Statistical Time Series*. New York: Wiley.
Ghysels, E. and Osborn, D.R. 2001. *Econometric Analysis of Seasonal Time Series*. Cambridge: Cambridge University Press.

Giraitis, L. and Phillips, P.C.B. 2006. Uniform limit theory for stationary autoregression. *Journal of Time Series Analysis* 27, 51–60.

Gouriéroux, C., Renault, E. and Touzi, N. 2000. Calibration by simulation for small sample bias correction. In *Simulation-Based Inference in Econometrics: Methods and Applications*, ed. R.S. Mariano, T. Schuermann and M. Weeks. Cambridge: Cambridge University Press.

Gouriéroux, C., Phillips, P.C.B. and Yu, J. 2006. Indirect inference for dynamic panel models. *Discussion Paper No. 1550*, Cowles Foundation, Yale University.

Hall, R.E. 1978. Stochastic implications of the life cycle-permanent income hypothesis. *Journal of Political Economy* 86, 971–87.

Hall, A. 1989. Testing for a unit root in the presence of moving average errors. *Biometrika* 76, 49–56.

Hall, P. and Heyde, C.C. 1980. *Martingale Limit Theory and its Application*. New York: Academic Press.

Harris, D., Leybourne, S. and McCabe, B. 2007. Modified KPSS tests for near integration. *Econometric Theory* 23, 355–63.

Hlouskova, J. and Wagner, M. 2006. The performance of panel unit root and stationarity tests: results from a large scale simulation study. *Econometric Reviews* 25, 85–116.

Hong, S.H. and Phillips, P.C.B. 2005. Testing linearity in cointegrating relations with an application to purchasing power parity. Discussion Paper No. 1541, Cowles Foundation, Yale University.

Hu, L. and Phillips, P.C.B. 2004. Dynamics of the federal funds target rate: a nonstationary discrete choice approach. *Journal of Applied Econometrics* 19, 851–67.

Hylleberg, S., Engle, R.F., Granger, C.W.J. and Yoo, S. 1990. Seasonal integration and cointegration. *Journal of Econometrics* 44, 215–38.

Ibragimov, R. and Phillips, P.C.B. 2004. Regression asymptotics using martingale convergence methods. *Discussion Paper No. 1473*, Cowles Foundation, Yale University.

Johansen, S. 1988. Statistical analysis of cointegration vectors. *Journal of Economic Dynamics and Control* 12, 231–54.

Kapetanios, G. 2005. Unit root testing against the alternative hypothesis of up to m structural breaks. *Journal of Time Series Analysis* 26, 123–33.

Karlsen, H.A., Myklebust, T. and Tjøstheim, D. 2007. Nonparametric estimation in a nonlinear cointegration model. *Annals of Statistics* 35(1).

Kasparis, I. 2004. Detection of functional form misspecification in cointegrating relations. Mimeo, University of Nottingham.

Kim, J.Y. 1994. Bayesian asymptotic theory in a time series model with a possible nonstationary process. *Econometric Theory* 10, 764–73.

Kwiatkowski, D., Phillips, P.C.B., Schmidt, P. and Shin, Y. 1992. Testing the null of stationarity against the alternative of a unit root: how sure are we that economic time series have a unit root? *Journal of Econometrics* 54, 159–78.

Lee, C.C. and Phillips, P.C.B. 1994. An ARMA prewhitened long-run variance estimator. Mimeo, Yale University.

Lee, J.S. 1996. On the power of stationary tests using optimal bandwidth estimates. *Economics Letters* 51, 131–7.

Leeb, H. and Potscher, B. 2001. The variance of an integrated process need not diverge to infinity and related results on partial sums of stationary processes. *Econometric Theory* 17, 671–85.

MacKinnon, J.G. 1994. Approximate asymptotic distribution functions for unit-root and cointegration tests. *Journal of Business and Economic Statistics* 12, 167–76.

Müller, U. 2005. Size and power of tests for stationarity in highly autocorrelated time series. *Journal of Econometrics* 128, 195–213.

Müller, U. and Elliott, G. 2003. Tests for unit roots and the initial condition. *Econometrica* 71, 1269–86.

Nelson, D.B. 1990. Stationarity and persistence in the GARCH (1, 1) model. *Econometric Theory* 6, 318–34.

Nelson, C.R. and Plosser, C. 1982. Trends and random walks in macroeconomic time series: some evidence and implications. *Journal of Monetary Econometrics* 10, 139–62.

Ng, S. and Perron, P. 1995. Unit root tests in ARMA models with data dependent methods for the selection of the truncation lag. *Journal of the American Statistical Association* 90, 268–81.

Ng, S. and Perron, P. 2001. Lag length selection and the construction of unit root tests with good size and power. *Econometrica* 69, 1519–54.

Ouliaris, S., Park, J.Y. and Phillips, P.C.B. 1989. Testing for a unit root in the presence of a maintained trend. In *Advances in Econometrics and Modelling*, ed. B. Raj. Norwell, MA: Kluwer.

Park, J.Y. 1990. Testing for unit roots and cointegration by variable addition. *Advances in Econometrics* 8, 107–33.

Park, J.Y. and Phillips, P.C.B. 1988. Statistical inference in regressions with integrated processes: Part I. *Econometric Theory* 4, 468–97.

Park, J.Y. and Phillips, P.C.B. 1989. Statistical inference in regressions with integrated processes: Part II. *Econometric Theory* 5, 95–131.

Park, J.Y. and Phillips, P.C.B. 1999. Asymptotics for nonlinear transformations of integrated time series. *Econometric Theory* 15, 269–98.

Park, J.Y. and Phillips, P.C.B. 2000. Nonstationary binary choice. *Econometrica* 68, 1249–80.

Park, J.Y. and Phillips, P.C.B. 2001. Nonlinear regressions with integrated time series. *Econometrica* 69, 1452–98.

Park, J.Y. and Sung, J. 1994. Testing for unit roots in models with structural change. *Econometric Theory* 10, 917–36.

Perron, P. 1989. The great crash, the oil price shock and the unit root hypothesis. *Econometrica* 57, 1361–401.

Phillips, P.C.B. 1986. Understanding spurious regressions in econometrics. *Journal of Econometrics* 33, 311–40.

Phillips, P.C.B. 1987a. Time series regression with a unit root. *Econometrica* 55, 277–301.

Phillips, P.C.B. 1987b. Towards a unified asymptotic theory of autoregression. *Biometrika* 74, 535–47.

Phillips, P.C.B. 1988a. Regression theory for near-integrated time series. *Econometrica* 56, 1021–44.

Phillips, P.C.B. 1988b. Multiple regression with integrated processes. In *Statistical Inference from Stochastic Processes*, ed. N.U. Prabhu. *Contemporary Mathematics* 80, 79–106.

Phillips, P.C.B. 1991a. To criticize the critics: an objective Bayesian analysis of stochastic trends. *Journal of Applied Econometrics* 6, 333–64.

Phillips, P.C.B. 1991b. Bayesian routes and unit roots: de rebus prioribus semper est disputandum. *Journal of Applied Econometrics* 6, 435–74.

Phillips, P.C.B. 1992. The long-run Australian consumption function reexamined: an empirical exercise in Bayesian inference. In *Long Run Equilibrium and Macroeconomic Modelling*, ed. C. Hargreaves. Cheltenham: Edward Elgar.

Phillips, P.C.B. 1994. Model determination and macroeconomic activity. Fisher–Schultz Lecture to the European Meetings of the Econometric Society, Maastricht. Discussion Paper No. 1083, Cowles Foundation, Yale University.

Phillips, P.C.B. 1995. Bayesian model selection and prediction with empirical applications. *Journal of Econometrics* 69, 289–332.

Phillips, P.C.B. 1996. Econometric model determination. *Econometrica* 64, 763–812.

Phillips, P.C.B. 1998. New tools for understanding spurious regressions. *Econometrica* 66, 1299–326.

Phillips, P.C.B. 2001. New unit root asymptotics in the presence of deterministic trends. *Journal of Econometrics* 11, 323–53.

Phillips, P.C.B. 2006. When the tail wags the unit root limit distribution. Mimeo, Yale University.

Phillips, P.C.B. and Durlauf, S.N. 1986. Multiple time series regression with integrated processes. *Review of Economic Studies* 53, 473–96.

Phillips, P.C.B. and Han, C. 2007. Gaussian inference in AR(1) time series with or without a unit root. *Econometric Theory* 24(3).

Phillips, P.C.B. and Hansen, B.E. 1990. Statistical inference in instrumental variables regression with I(1) processes. *Review of Economic Studies* 57, 99–125.

Phillips, P.C.B. and Magdalinos, T. 2007. Limit theory for moderate deviations from a unit root. *Journal of Econometrics* 136, 115–30.

Phillips, P.C.B. and Moon, H.R. 1999. Linear regression limit theory for nonstationary panel data. *Econometrica* 67, 1057–111.

Phillips, P.C.B. and Moon, H.R. 2000. Nonstationary panel data analysis: an overview of some recent developments. *Econometric Reviews* 19, 263–86.

Phillips, P.C.B. and Ouliaris, S. 1990. Asymptotic properties of residual based tests for cointegration. *Econometrica* 58, 165–93.

Phillips, P.C.B., Park, J.Y. and Chang, Y. 2004. Nonlinear instrumental variable estimation of an autoregression. *Journal of Econometrics* 118, 219–46.

Phillips, P.C.B. and Perron, P. 1988. Testing for a unit root in time series regression. *Biometrika* 75, 335–46.

Phillips, P.C.B. and Ploberger, W. 1994. Posterior odds testing for a unit root with data-based model selection. *Econometric Theory* 10, 774–808.

Phillips, P.C.B. and Ploberger, W. 1996. An asymptotic theory of Bayesian inference for time series. *Econometrica* 64, 381–413.

Phillips, P.C.B. and Solo, V. 1992. Asymptotics for linear processes. *Annals of Statistics* 20, 971–1001.

Phillips, P.C.B. and Xiao, Z. 1998. A primer on unit root testing. *Journal of Economic Surveys* 12, 423–69.

Ploberger, W. 2004. A complete class of tests when the likelihood is locally asymptotically quadratic. *Journal of Econometrics* 118, 67–94.

Pötscher, B.M. 2004. Nonlinear functions and convergence to Brownian motion: beyond the continuous mapping theorem. *Econometric Theory* 20, 1–22.

Robinson, P.M. 1995. Gaussian semiparametric estimation of long range dependence. *Annals of Statistics* 23, 1630–61.

Said, S.E. and Dickey, D.A. 1984. Testing for unit roots in autoregressive moving average models of unknown order. *Biometrika* 71, 599–608.

Sargan, J.D. and Bhargava, A. 1983. Testing residuals from least squares regression for being generated by the Gaussian random walk. *Econometrica* 51, 153–74.

Schmidt, P. and Phillips, P.C.B. 1992. LM tests for a unit root in the presence of deterministic trends. *Oxford Bulletin of Economics and Statistics* 54, 257–87.

Shimotsu, K. and Phillips, P.C.B. 2005. Exact local whittle estimation of fractional integration. *Annals of Statistics* 33, 1890–933.

Schotman, P. and van Dijk, H.K. 1991. A Bayesian analysis of the unit root in real exchange rates. *Journal of Econometrics* 49, 195–238.

Sims, C.A. and Uhlig, H. 1991. Understanding unit rooters: a helicopter tour. *Econometrica* 59, 1591–9.

So, B.S. and Shin, D.W. 1999. Cauchy estimators for autoregressive processes with applications to unit root tests and confidence intervals. *Econometric Theory* 15, 165–76.

Solo, V. 1984. The order of differencing in ARIMA models. *Journal of the American Statistical Association* 79, 916–21.

Stock, J.H. 1991. Confidence intervals for the largest autoregressive root in US macroeconomic time series. *Journal of Monetary Economics* 28, 435–59.

Stock, J. 1994a. Deciding between I(1) and I(0). *Journal of Econometrics* 63, 105–31.

Stock, J.H. 1994b. Unit roots, structural breaks and trends. In *Handbook of Econometrics*, vol. 4, ed. R.F. Engle and D. McFadden. Amsterdam: North-Holland.

Stock, J.H. 1999. A class of tests for integration and cointegration. In *Cointegration, Causality and Forecasting: A Festschrift in Honour of Clive W. J. Granger*, ed. R.F. Engle and H. White. Oxford: Oxford University Press.

Stock, J.H. and Watson, M.W. 1988. Variable trends in economic time series. *Journal of Economic Perspectives* 2(3), 147–74.

Sul, D., Phillips, P.C.B. and Choi, C.-Y. 2006. Prewhitening bias in HAC estimation. *Oxford Bulletin of Economics and Statistics* 67, 517–46.

Wang, Q. and Phillips, P.C.B. 2006. Asymptotic theory for local time density estimation and nonparametric cointegrating regression. Discussion Paper no. 1594, Cowles Foundation, Yale University.

Wei, C.Z. 1992. On predictive least squares principles. *Annals of Statistics* 20, 1–42.

Xiao, Z. and Phillips, P.C.B. 1998. An ADF coefficient test for a unit root in ARMA models of unknown order with empirical applications to the U.S. economy. *Econometrics Journal* 1, 27–43.

Zivot, E. and Andrews, D.W.K. 1992. Further evidence on the great crash, the oil price shock and the unit root hypothesis. *Journal of Business and Economic Statistics* 10, 251–70.

variance decomposition

Variance decomposition is a classical statistical method in multivariate analysis for uncovering simplifying structures in a large set of variables (for example, Anderson, 2003). For example, *factor analysis* or *principal components* are tools that are in widespread use. Factor analytic methods have, for instance, been used extensively in economic forecasting (see for example, Forni et al. 2000; Stock and Watson, 2002). In macroeconomic analysis the term 'variance decomposition' or, more precisely, 'forecast error variance decomposition' is used more narrowly for a specific tool for interpreting the relations between variables described by vector autoregressive (VAR) models. These models were advocated by Sims (1980) and used since then by many economists and econometricians as alternatives to classical simultaneous equations models. Sims criticized the way the latter models were specified, and questioned in particular the exogeneity assumptions common in simultaneous equations modelling.

VAR models have the form

$$y_t = A_1 y_{t-1} + \cdots + A_p y_{t-p} + u_t, \tag{1}$$

where $y_t = (y_{1t}, \ldots, y_{Kt})'$ (the prime denotes the transpose) is a vector of K observed variables of interest, the A_i's are $(K \times K)$ parameter matrices, p is the lag order and u_t is a zero mean error process which is assumed to be white noise, that is, $E(u_t) = 0$, the covariance matrix, $E(u_t u_t') = \Sigma_u$, is time invariant and the u_t's are serially uncorrelated or independent. Here deterministic terms such as constants, seasonal dummies or polynomial trends are neglected because they are of no interest in the following. In the VAR model (1) all the variables are a priori endogenous. It is usually difficult to disentangle the relations between the variables directly from the coefficient matrices. Therefore it is useful to have special tools which help with the interpretation of VAR models. Forecast error variance decompositions are such tools. They are presented in the following.

An h steps ahead forecast or briefly h-step forecast at origin t can be obtained from (1) recursively for $h = 1, 2, \ldots$, as

$$y_{t+h|t} = A_1 y_{t+h-1|t} + \cdots + A_p y_{t+h-p|t}. \tag{2}$$

Here $y_{t+j|t} = y_{t+j}$ for $j \leq 0$. The forecast error turns out to be

$$y_{t+h} - y_{t+h|t} = u_{t+h} + \sum_{i=1}^{h-1} \Phi_i u_{t+h-i} \sim \left(0, \Sigma_h = \Sigma_u + \sum_{i=1}^{h-1} \Phi_i \Sigma_u \Phi_i'\right),$$

that is, the forecast errors have mean zero and covariance matrices Σ_h. Here the Φ_i's are the coefficient matrices of the power series expansion $(I_K - A_1 z - \cdots - A_p z^p)^{-1} = I_K + \sum_{i=1}^{\infty} \Phi_i z^i$. Note that the inverse exists in a neighbourhood of $z = 0$ even if the VAR process contains integrated and cointegrated variables. (For an introductory exposition of forecasting VARs, see Lütkepohl, 2005.)

If the residual vector u_t can be decomposed in instantaneously uncorrelated innovations with economically meaningful interpretation, say, $u_t = B\varepsilon_t$ with $\varepsilon_t \sim (0, I_K)$, then $\Sigma_u = BB'$ and the forecast error variance can be written as $\Sigma_h = \sum_{i=0}^{h-1} \Theta_i \Theta_i'$, where $\Theta_0 = B$ and $\Theta_i = \Phi_i B$, $i = 1, 2, \ldots$. Denoting the (n, m)th element of Θ_j by $\theta_{nm,\, j}$, the forecast error variance of the kth element of the forecast error vector is seen to be

$$\sigma_k^2(h) = \sum_{j=0}^{h-1} (\theta_{k1,j}^2 + \cdots + \theta_{kK,j}^2) = \sum_{j=1}^{K} (\theta_{kj,\,0}^2 + \cdots + \theta_{kj,\,h-1}^2).$$

The term $(\theta_{kj,0}^2 + \cdots + \theta_{kj,h-1}^2)$ may be interpreted as the contribution of the jth innovation to the h-step forecast error variance of variable k. Dividing the term by $\sigma_k^2(h)$ gives the percentage contribution of innovation j to the h-step forecast error variance of variable k. This quantity is denoted by $\omega_{kj,\,h}$ in the following. The $\omega_{kj,\,h}$, $j = 1, \ldots, K$, decompose the h-step ahead forecast error variance of variable k in the contributions of the ε_t innovations. They were proposed by Sims (1980) and are often reported and interpreted for various forecast horizons.

For such an interpretation to make sense it is important to have economically meaningful innovations. In other words, a suitable transformation matrix B for the reduced form residuals has to be found. Clearly, B has to satisfy $\Sigma_u = BB'$. These relations do not uniquely determine B, however. Thus, restrictions from subject matter theory are needed to obtain a unique B matrix and, hence, unique innovations ε_t. A number of different possible sets of restrictions and approaches for specifying restrictions have been proposed in the literature in the framework of *structural VAR models*. A popular example is the choice of a lower-triangular matrix B obtained by a Choleski decomposition of Σ_u (for example, Sims, 1980). Such a choice amounts to setting up a system in recursive form where shocks in the first variable have potentially instantaneous effects also on all the other variables, shocks to the second variable can also affect the third to last variable instantaneously, and so on. In recursive systems it may be possible to associate the innovations with variables, that is, the jth component of ε_t is primarily viewed as a shock to the jth observed variable. Generally, the innovations can also be associated with unobserved variables, factors or forces and they may be named accordingly. For example, Blanchard and Quah (1989) consider a bivariate model for output and the unemployment rate, and they investigate effects of supply and demand shocks. Generally, if economically meaningful innovations can be found, forecast error variance decompositions provide information about the relative importance of different shocks for the variables described by the VAR model.

Estimation of reduced form and structural form parameters of VAR processes is usually done by least squares, maximum likelihood or Bayesian methods. Estimates of the forecast error variance components, $\omega_{kj,\,h}$, are then obtained from the VAR parameter estimates. Suppose the VAR coefficients are contained in a vector $\boldsymbol{\alpha}$, then

$\omega_{kj,h}$ is a function of $\boldsymbol{\alpha}$, $\omega_{kj,h} = \omega_{kj,h}(\alpha)$. Denoting the estimator of $\boldsymbol{\alpha}$ by $\hat{\alpha}$, $\omega_{kj,h}$ may be estimated as $\hat{\omega}_{kj,h} = \omega_{kj,h}(\hat{\boldsymbol{\alpha}})$. If $\hat{\boldsymbol{\alpha}}$ is asymptotically normal, that is, $\sqrt{T}(\hat{\boldsymbol{\alpha}} - \boldsymbol{\alpha}) \xrightarrow{\delta} \mathcal{N}(0, \Sigma_{\hat{\alpha}})$, then, under general conditions, $\hat{\omega}_{kj,h}$ is also asymptotically normally distributed, $\sqrt{T}(\hat{\omega}_{kj,h} - \omega_{kj,h}) \xrightarrow{d} \mathcal{N}(0, \sigma^2_{kj,h} = \frac{\partial\omega_{kj,h}}{\partial\boldsymbol{\alpha}'}\Sigma_{\hat{\alpha}}\frac{\partial\omega_{kj,h}}{\partial\boldsymbol{\alpha}})$, provided the variance of the asymptotic distribution is non-zero. Here $\partial\omega_{kj,h}/\partial\boldsymbol{\alpha}$ denotes the vector of first-order partial derivatives of $\omega_{kj,h}$ with respect to the elements of $\boldsymbol{\alpha}$ (see Lütkepohl, 1990, for the specific form of the partial derivatives). Unfortunately, $\sigma^2_{kj,h}$ is zero even for cases of particular interest, for example, if $\omega_{kj,h} = 0$ and, hence, the jth innovation does not contribute to the h-step forecast error variance of variable k (see Lütkepohl, 2005, section 3.7.1, for a more detailed discussion). The problem can also not easily be solved by using bootstrap techniques (cf. Benkwitz, Lütkepohl and Neumann, 2000). Thus, standard statistical techniques such as setting up confidence intervals are problematic for the forecast error variance components. They can at best give rough indications of sampling uncertainty. The estimated $\omega_{kj,h}$'s are perhaps best viewed as descriptive statistics.

HELMUT LÜTKEPOHL

See also **impulse response function; structural vector autoregressions; vector autoregressions.**

Bibliography

Anderson, T. 2003. *An Introduction to Multivariate Statistical Analysis*, 3rd edn. New York: John Wiley.

Benkwitz, A., Lütkepohl, H. and Neumann, M. 2000. Problems related to bootstrapping impulse responses of autoregressive processes. *Econometric Reviews* 19, 69–103.

Blanchard, O. and Quah, D. 1989. The dynamic effects of aggregate demand and supply disturbances. *American Economic Review* 79, 655–73.

Forni, M., Hallin, M., Lippi, M. and Reichlin, L. 2000. The generalized dynamic factor model: identification and estimation. *Review of Economics and Statistics* 82, 540–52.

Lütkepohl, H. 1990. Asymptotic distributions of impulse response functions and forecast error variance decompositions of vector autoregressive models. *Review of Economics and Statistics* 72, 116–25.

Lütkepohl, H. 2005. *New Introduction to Multiple Time Series Analysis*. Berlin: Springer-Verlag.

Sims, C. 1980. Macroeconomics and reality. *Econometrica* 48, 1–48.

Stock, J. and Watson, M. 2002. Forecasting using principal components from a large number of predictors. *Journal of the American Statistical Association* 97, 1167–79.

varying coefficient models

One of the most interesting forms of nonlinear regression models is the varying coefficient model (VCM). Unlike the linear regression model, VCMs were introduced by Hastie and Tibshirani (1993) to allow the regression coefficients to vary systematically and smoothly in more than one dimension. It is worth noting the distinction between the VCM and the so-called random coefficients model, which assumes that the coefficients vary non-systematically (randomly). Versions of the VCM are encountered in the literature as functional coefficient models (see Cai, Fan and Yao, 2000) and smooth coefficient models (see Li et al., 2002).

VCMs are very useful tools in applied work in economics as they can be used to model parameter heterogeneity in a general way. For example, Durlauf, Kourtellos and Minkin (2001) estimate a version of the Solow model that allows the parameters for each country to vary as functions of initial income. This work is extended in Kourtellos (2005), who finds parameter dependence on initial literacy, initial life expectancy, expropriation risk and ethnolinguistic fractionalization. Li et al. (2002) use the above smooth coefficient model to estimate the production function of the non-metal mineral industry in China. Stengos and Zacharias (2006) use the same model to examine an intertemporal hedonic model of the personal computer market, where the coefficients of the hedonic regression were unknown functions of time. Hong and Lee (2003) forecast the nonlinearity in the conditional mean of exchange rate changes using a VCM that allows the autoregressive coefficients to vary with investment positions. Ahmad, Leelahanon and Li (2005) apply the VCM in the estimation of a production function in China's manufacturing industry to show that the marginal productivity of labour and capital depends on the firm's R&D values. Mamuneas, Savvides and Stengos (2006) study the effect of human capital on total factor productivity in an empirical growth framework. In what follows we present the basic structure of the standard VCM specification as it appears in the literature and then proceed to discuss certain aspects of estimation and some of its recent generalizations.

Basic specification

Consider the following VCM

$$y_i = \beta(z_i)'X_i + u_i \tag{1}$$

with $E(u_i|X_i) = 0$, where $X_i = (1, x_{i2}, \ldots, x_{ip})'$ is a p dimensional vector of slope regressors and $\beta(z_i)' = (\beta_1(z_{i1}), \beta_2(z_{i2}), \ldots, \beta_p(z_{ip}))$ is a p dimensional vector of varying coefficients, which take the form of unknown smooth functions of $z_{i1}, z_{i2}, \ldots, z_{ip}$, respectively. Notice that $\beta_1(z_i)$ is a varying intercept that measures the direct relationship between the tuning variable z_i and the dependent variable in a nonparametric way. We refer to the variables z_i's as tuning variables, and they can be one-dimensional or multidimensional. These functions map the tuning variables into a set of local regression coefficient estimates that imply that the effect of X_i on y_i will not be constant but rather it will vary smoothly with the tuning variables. These tuning variables could take the form of a scalar like time or it could be a vector of dimension q. A common situation in the literature arises when the z_j is the same for all j.

It is worth is noting that the VCM (1) is a very flexible and rich family of models. One of the reasons is that the general additive separable structure of (1) offers also a very useful compromise to the general high-dimensional nonparametric regression that is known to suffer from the curse of dimensionality. This allows for nonparametric estimation even when the conditioning regressor space is in high dimensional. Another is that it nests many well-known models as a special case. For instance, consider the following cases. If $\beta_j(z_{ij}) = \beta_j$, for all j then we are dealing with the usual linear model. If $\beta_j(z_{ij}) = \beta_j z_{ij}$ for some variable j, we simply have the interaction term $\beta_j x_{ij} z_{ij}$ entering the regression function. If $x_i = c$ (a constant) or if $z_{ij} = x_{ij}$ for all $j = 1, \ldots p$, then the model takes the generalized additive form where the additive components are unknown functions (see Hastie and Tibshirani, 1990; Linton and Nielsen 1995).

We now set out some estimation issues. A popular estimation approach is based on local polynomial regression, as illustrated by Fan (1992), Fan and Gijbels (1996), and Fan and Zhang (1999), which we present in the context of a random sample design. Given a random sample $\{(z_i, X_i, y_i)\}_{i=1}^{n}$, the estimation procedure solves a simple local least squares problem. To be precise, for each given point z_0 the functions $\beta_j(z)$, $j = 1 \ldots p$ are approximated by local linear polynomials $\beta_j(z) \approx c_{j0} + c_{j1}(z - z_0)$ for z in a neighborhood of z_0. This leads to the following weighted local least squares problem:

$$\sum_{i=1}^{n} \left[y_i - \sum_{j=1}^{p} \{c_{j0} + c_{j1}(z - z_0)\} X_{ij} \right]^2 K_h(z_i - z_0) \tag{2}$$

for a given kernel function K and bandwidth h, where $K_h(\cdot) = K(\cdot/h)/h$. While this method is simple, it is implicitly assumed that the functions $\beta_j(z)$ possess the same degrees of smoothness and hence can be approximated equally well in the same interval. Fan and Zhang (1999) allow for different degrees of smoothness for different coefficient functions by proposing a two-stage method. This is similar in spirit to what Huang and Shen (2004) do for global smoothers using regression splines but allowing each coefficient function to have different (global) smoothing parameters.

An attractive alternative to local polynomial estimation is a global smoothing method based on general series methods such as polynomial splines and trigonometric approximation (see Ahmad, Leelahanon and Li, 2005; Huang, Wu and Zhou, 2004; Huang and Shen, 2004; Xue and Yang, 2006a). All these papers emphasize the computational savings from having to solve only one minimization problem. Ahmad,

Leelahanon and Li stress the efficiency gains of the series approach over a kernel-based approach when one allows for conditional heteroskedasticity. We should note that the inference for the estimated coefficients will differ for different choices of approximation, and the asymptotic properties of such estimators are generally not easy to obtain.

Although the model was initially developed for i.i.d. data, it has been extended for time series data by Chen and Tsay (1993), Cai, Fan, and Yao (2000), Huang and Shen (2004), and Cai (2007) for strictly stationary processes with different mixing conditions. The coefficient functions typically now become functions of time and/or lagged values of the dependent variable. It is worth noting that estimation issues such as bandwidth selection are similar, as in the i.i.d. data case (see Cai, 2007). The varying coefficient model has also been employed to analyse longitudinal data (see Brumback and Rice (1998), Hoover et al. (1998), and Huang, Wu and Zhou (2004).

The partially linear varying coefficient model

An interesting special case of eq. (1), where the unknown coefficient functions depend on a common z_i, is the partially linear VCM. Here some of the coefficients are constants (independent of z_i). In that case, eq. (1) can be rewritten as

$$y_i = \alpha' W_i + \beta(z_i)' X_i + u_i \tag{3}$$

where W_i is the ith observation on a $(1 \times q)$ vector of additional regressors that enter the regression function linearly. The estimation of this model requires some special treatment as the partially linear structure may allow for efficiency gains, since the linear part can be estimated at a much faster rate, namely, $\sqrt{n}$.

The partially linear VCM has been studied by Zhang, Lee and Song (2002), Xia, Zhang and Tong (2004), Ahmad, Leelahanon and Li (2005), and Fan and Huang (2005). Zhang, Lee and Song (2002) suggest a two-step procedure where the coefficients of the linear part are estimated in the first step using polynomial fitting with an initial small bandwidth using cross validation (see Hoover et al., 1998). In other words, the approach is based on under-smoothing in the first stage. Then these estimates are averaged to yield the final first-step linear part estimates which are then used to redefine the dependent variable and return to the environment of eq. (1), where local smoothers can be applied as described above. Alternatively, Xia, Zhang and Tong (2004) separate the estimation of γ from that of $\beta(z_i)$ by noting that the former can be estimated globally, but the latter locally. This is what they call a 'semi-local least squares procedure', and they achieve a more efficient estimate of γ without under-smoothing using standard bandwidth selection methods. Once γ has been estimated, then again the linear part can be used to redefine the dependent variable and return to the environment of eq. (3).

More recently, Fan and Huang (2005) use a profile least squares estimation approach to provide a simple and useful method for (3). More precisely, they construct a Wald test and a profile likelihood ratio test for the parametric component that share similar sampling properties. More importantly, they show that the asymptotic distribution of the profile likelihood ratio test under the null is independent of nuisance parameters, and follows an asymptotic χ^2 distribution. They also propose a generalized likelihood ratio test statistic to test whether certain parametric functions fit the nonparametric varying coefficients. This hypothesis test includes testing for the significance of the slope variables X (zero coefficients) and the homogeneity of the model (constant coefficients). Other work on specification testing includes Li et al. (2002), Cai, Fan and Yao (2000), Cai (2007), Yang et al. (2006) that mainly rely on bootstrapping in their implementation.

Generalizations and extensions

A useful generalization of (1) is to allow the dependent variable to be related to the regression function nonlinearly $m(X_i, Z_i) = \beta(z_i)'X_i$ via some given link function $g(\cdot)$

$$y_i = g(\beta(z_i)'X_i) + u_i \tag{4}$$

This generalization is known as the generalized varying coefficient model and was originally proposed by Hastie and Tibshirani (1993). Cai, Fan, and Li (2000) study this model using local polynomial techniques and propose an efficient one-step local maximum likelihood estimator. Notice that if $g(\cdot)$ is the normal CDF then (4) generalizes the standard tool of the discrete choice literature, namely the probit model.

Another strand of the literature allowed for a multivariate tuning variable z_l, $l = 1, 2, \ldots, q$. Although Hastie and Tibshirani (1993) proposed a back-fitting algorithm to estimate the varying coefficient functions, they did not provide any asymptotic justification. The most notable advance in this context has been by Xue and Yang (2006a), who propose a generalization of the VCM as in (1) that allows the varying coefficients to have an additive coefficient structure on regression coefficients to avoid the curse of dimensionality

$$\beta_j(z) = \gamma_{j0} + \gamma_{j1}(z_1) + \cdots + \gamma_{jq}(z_q) \quad \text{for all } j.$$

Under mixing conditions, Xue and Yang (2006a) propose local polynomial marginal integration estimators, while Xue and Yang (2006b) study this model using polynomial splines.

Finally, Cai et al. (2006) have shifted the discussion to consider a structural VCM. They examine the case of endogenous slope regressors, and propose a two-stage IV procedure based on local linear estimation procedures in both stages. We believe that this line of research is fruitful for economic applications.

Conclusion

VCMs have increasingly been employed as useful tools that allow for a compromise between fully nonparametric and parametric models. This compromise allows for the desired flexibility to uncover hidden structures that underlie the response coefficients of standard regression models without running into the serious curse of the dimensionality issue. More importantly, the structure of the VCM that allows the regression coefficients to vary with a tuning variable is very appealing in many economic applications, for it has a natural interpretation of non-constant marginal effects.

ANDROS KOURTELLOS AND THANASIS STENGOS

Bibliography

Ahmad, I., Leelahanon, S. and Li, Q. 2005. Efficient estimation of a semiparametric partially varying linear model. *Annals of Statistics* 33, 258–83.

Brumback, B. and Rice, J. 1998. Smoothing spline models for the analysis of nested and crossed samples of curves. *Journal of the American Statistical Association* 93, 961–76.

Cai, Z. 2007. Trending time-varying coefficient time series models with serially correlated errors. *Journal of Econometrics* 136, 163–88.

Cai, Z., Fan, J. and Li, R. 2000. Efficient estimation and inferences for varying-coefficient models. *Journal of American Statistical Association* 95, 888–902.

Cai, Z., Fan, J. and Yao, Q. 2000. Functional coefficient regression models for nonlinear time series models. *Journal of American Statistical Association* 95, 941–56.

Cai, Z., Das, M., Xiong, H. and Wu, Z. 2006. Functional coefficient instrumental variables models. *Journal of Econometrics* 133, 207–41.

Chen, R. and Tsay, R. 1993. Functional coefficient autoregressive models. *Journal of the American Statistical Association* 88, 298–308.

Durlauf, S., Kourtellos, A. and Minkin, A. 2001. The local Solow growth model. *European Economic Review* 45, 928–40.

Fan, J. 1992. Design-adaptive nonparametric regression. *Journal of the American Statistical Association* 87, 998–1004.

Fan, J. and Gijbels, I. 1996. *Local Polynomial Modelling and Its Applications*. London: Chapman and Hall.

Fan, J. and Huang, T. 2005. Profile likelihood inferences on semiparametric varying-coefficient partially linear models. *Bernoulli* 11, 1031–57.

Fan, J. and Zhang, W. 1999. Statistical estimation in varying-coefficient models. *Annals of Statistics* 27, 1491–518.

Hastie, T. and Tibshirani, R. 1990. *Generalized Additive Models*. New York: Chapman and Hall.

Hastie, T. and Tibshirani, R. 1993. Varying coefficient models. *Journal of the Royal Statistical Society*, Series B 55, 757–796.

Hong, Y. and Lee, T.-H. 2003. Inference on predictability of foreign exchange rates via generalized spectrum and nonlinear time series models. *Review of Economics and Statistics* 85, 1048–62.

Hoover, D., Rice, C., Wu, C. and Yang, L. 1998. Nonparametric smoothing estimates of time-varying coefficient models with longitudinal data. *Biometrika* 85, 809–22.

Huang, J. and Shen, H. 2004. Functional coefficient regression models for nonlinear time series: a polynomial spline approach. *Scandinavian Journal of Statistics* 31, 515–34.

Huang, J., Wu, C. and Zhou, L. 2004. Polynomial spline estimation and inference for varying coefficient models with longitudinal data. *Statistica Sinica* 14, 763–88.

Kourtellos, A. 2005. Modeling parameter heterogeneity in cross-country growth regression models. Mimeo, Department of Economics, University of Cyprus.

Li, Q., Huang, C., Li, D. and Fu, T. 2002. Semiparametric smooth coefficient models. *Journal of Business and Economic Statistics* 20, 412–22.

Linton, O. and Nielsen, J. 1995. A kernel method of estimating structural nonparametric regression based on marginal integration. *Biometrika* 82, 93–100.

Mamuneas, T., Savvides, A. and Stengos, T. 2006. Economic development and the return to human capital: a smooth coefficient semiparametric approach. *Journal of Applied Econometrics* 21, 111–32.

Stengos, T. and Zacharias, E. 2006. Intertemporal pricing and price discrimination: a semiparametric hedonic analysis of the personal computer market. *Journal of Applied Econometrics* 21, 371–86.

Stone, C. 1977. Consistent nonparametric regression. *Annals of Statistics* 5, 595–620.

Xia, Y., Zhang, W. and Tong, H. 2004. Efficient estimation for semivarying-coefficient models. *Biometrika* 91, 661–81.

Xue, L. and Yang, L. 2006a. Estimation of semiparametric additive coefficient model. *Journal of Statistical Planning and Inference* 136, 2506–34.

Xue, L. and Yang, L. 2006b. Additive coefficient modeling via polynomial spline. *Statistica Sinica* 16, 1423–46.

Yang, L., Park, B., Xue, L. and Härdle, W. 2006. Estimation and testing for varying coefficients in additive models with marginal integration. *Journal of the American Statistical Association* 101, 1212–227.

Zhang, W., Lee, S.-Y. and Song, X. 2002. Local polynomial fitting in semivarying coefficient model. *Journal of Multivariate Analysis* 82, 166–88.

vector autoregressions

Vector autoregressions (VARs) are a class of dynamic multivariate models introduced by Sims (1980) to macroeconomics. These models arise mainly as a response to the 'incredible' identifying assumptions embedded in traditional large-scale econometric models of the Cowles Commission. The traditional approach uses predetermined or exogenous variables, coupled with many strong exclusion restrictions, to identify each structural equation. VARs, by contrast, explicitly recognize that all economic variables are interdependent and thus should be treated endogenously. The philosophy of VAR modelling begins with a multivariate time series model that has minimal restrictions, and gradually introduces identifying information, with emphasis always placed on the model's fit to data.

While the traditional econometric approach allows disturbances or shocks to structural equations to be correlated, the VAR methodology insists that structural shocks ought to be independent of one another. The independence assumption plays an essential role in achieving unambiguous economic interpretations about structural shocks such as technology and policy shocks; it can be tested using recently developed econometric tools (Leeper and Zha, 2003). The bulk of VAR work has focused on identifying structural shocks as a way to specify the contemporaneous relationships among economic variables. With most dynamic relationships unrestricted, the intent of such an identifying strategy is to construct models that have both economic interpretability and superior fit to data. Dynamic responses to a particular shock, called impulse responses, are often used as economic interpretations to the model. They summarize the properties of all systematic components of the system and have become a major tool in modern economic analysis.

Modelling policy shocks explicitly is important in addressing the practical importance of the Lucas critique. If policy switches regime, such a change may be viewed as a sequence of random shocks from the public's viewpoint (Sims, 1982). If this sequence displays a persistent pattern, the public will adjust its expectations formation accordingly and the Lucas critique may be consequential. For the practice of monetary policy, however, it is an empirical question how significant this adjustment is. Leeper and Zha (2003) construct an econometric measure from the sequence of policy shocks implied by regime switches to gauge whether the public's behaviour could be well approximated by a linear model. This measure is particularly useful if counterfactual exercises regarding the effects of policy changes are conducted with respect to the Lucas critique.

VARs have also been used for other tasks. Armed with a Bayesian prior, VARs have been known to produce out-of-sample forecasts of economic variables as well as, or even better than, those from commercial forecasting firms (Litterman, 1986; Geweke and Whiteman, 2006). Because of their ability to forecast, VARs have given researchers a convenient diagnostic tool to assess the feasibility or plausibility of real-time policy projections of other economic models (Sims, 1982). VARs have been increasingly used for policy analysis and as a benchmark for comparing different dynamic stochastic general equilibrium (DSGE) models. Restrictions on lagged coefficients have been gradually introduced to give more economic interpretations to individual equations. All these developments are positive and help narrow the gap between statistical and economic models.

This article discusses these and other aspects of VARs, summarizes some key theoretical results for the reader to consult without searching for different sources, and provides a perspective on where future research in this area will be headed.

General framework

Structural form

VARs are generally represented in a structural form of which the reduced form is simply a byproduct. The general form is

$$\mathbf{y}_t'\mathbf{A} = \sum_{l=1}^{p} \mathbf{y}_{t-l}'\mathbf{A}_l + \mathbf{z}_t'\mathbf{D} + \varepsilon_t', \tag{1}$$

where $\mathbf{y}_t$ is an $n \times 1$ column vector of endogenous variables, $\mathbf{A}$ and $\mathbf{A}_l$ are $n \times n$ parameter matrices, $\mathbf{z}_t$ is an $h \times 1$ column vector of exogenous variables, $\mathbf{D}$ is an $h \times n$ parameter matrix, p is the lag length, and ε_t is an $n \times 1$ column vector of structural shocks. The parameters of individual equations in (1) correspond to the columns of $\mathbf{A}$, $\mathbf{A}_l$, and $\mathbf{D}$. The structural shocks are assumed to be i.i.d. and independent of one another:

$$E(\varepsilon_t|\mathbf{y}_{t-s}, s>0) = \underset{n\times 1}{\mathbf{0}}, \quad E(\varepsilon_t\varepsilon_t'|\mathbf{y}_{t-s}, s>0) = \underset{n\times n}{\mathbf{I}},$$

where $\underset{n\times n}{\mathbf{0}}$ is the $n \times n$ matrix of zeros and $\underset{n\times n}{\mathbf{I}}$ is the $n \times n$ identity matrix. It follows that the reduced form of (1) is

$$\mathbf{y}_t' = \sum_{l=1}^{p} \mathbf{y}_{t-l}'\mathbf{B}_l + \mathbf{z}_t'\mathbf{C} + u_t', \tag{2}$$

where $\mathbf{B}_l = \mathbf{A}_l\mathbf{A}^{-1}$, $\mathbf{C} = \mathbf{D}\mathbf{A}^{-1}$, and $u_t' = \varepsilon_t'\mathbf{A}^{-1}$. The covariance matrix of u_t is $\Sigma = (\mathbf{A}\mathbf{A}')^{-1}$.

In contrast to the traditional econometric approach, the VAR approach puts emphasis almost exclusively on the dynamic properties of endogenous variables $\mathbf{y}_t$ rather than exogenous variables $\mathbf{z}_t$. In most VAR applications, $\mathbf{z}_t$ simply contains the constant terms.

Identification

One main objective in the VAR literature is to obtain economically meaningful impulse responses to structural shocks ε_t. To achieve this objective, it is necessary to impose at least $n(n-1)/2$ identifying restrictions, often on the contemporaneous coefficients represented by **A** in the structural system (1). In his original work, Sims (1980) makes the contemporaneous coefficient matrix **A** triangular for identification. The triangular system, often called the recursive identification, has a 'Wold chain causal' interpretation which is based on the timing of how shocks affect variables contemporaneously. It assumes that some shocks may influence only a subset of variables within the current period. This identification is still popular because it is straightforward to use and can yield some results that match widely held views. Christiano, Eichenbaum and Evans (1999) discuss extensively how recursive identification can be used in policy analysis.

There are fundamental economic applications that require identification under alternative assumptions rather than the recursive system. One familiar example is the determination of price and quantity as discussed in Sims (1986) and Gordon and Leeper (1994). Both variables are often determined *simultaneously* by the supply and demand equations in equilibrium; this simultaneity is inconsistent with recursive identification. Bernanke (1986) and Blanchard and Watson (1986) pioneered other applications of non-recursive identified VARs. Estimation of non-recursive VARs presents technical difficulties that are absent in recursive systems. These difficulties help explain the use of recursive VARs even if this maintained assumption is implausible. Recent developments in Bayesian econometrics, however, have made it feasible to estimate non-recursive VARs.

All of these works focus on the contemporaneous coefficient matrix. There are other ways to achieve identification. Blanchard and Quah (1993) and Gali (1992) propose using identifying restrictions directly on short-run and long-run impulse responses, which have been used in quantifying the effects of technology shocks and various nominal shocks, although the unreliable statistical properties of long-run restrictions are documented by Faust and Leeper (1997).

Many VAR applications rely on exact identification: the number of identifying restrictions equals $n(n-1)/2$. This counting condition is necessary but *not* sufficient for identification. To see this point, consider a three-variable VAR with the following restrictions

$$\mathbf{A} = \begin{bmatrix} * & * & 0 \\ 0 & * & * \\ * & 0 & * \end{bmatrix}$$

where $*$'s indicate unrestricted coefficients and 0's indicate exclusion restrictions. This VAR is *not* identified because in general there exist two distinct sets of structural parameters that deliver the same dynamics of $\mathbf{y}_t$. For larger and more complicated systems with both short-run and long-run restrictions, there has been, until recently, no practical guidance as to whether the model is identified. The paper by Rubio-Ramirez, Waggoner and Zha (2005) develops a theorem for a necessary and sufficient condition for a VAR to be exactly identified. This theorem applies to a wide range of identified VARs, including those used in the literature. The basic idea is to transform the original structural parameters to the $(np + h) \times n$ matrix F (which is a function of $\mathbf{A}, \mathbf{A}_1, \ldots, \mathbf{A}_p, \mathbf{D}$) so that linear restrictions can be applied to each column of F. The linear restrictions for the ith column of F can be summarized by the matrix Q_i of rank q_i, where q_i is the number of restrictions. According to their theorem, the VAR model is exactly identified if and only if $q_i = n - i$ for $1 \leq i \leq n$. This result gives the researcher a practical way to determine whether a VAR model is identified.

When the number of identifying restrictions is greater than $n(n-1)/2$, a VAR is over-identified. Allowing for over-identification is important since economic theory often implies more than $n(n-1)/2$ restrictions. Moreover, many economic applications call for restrictions on the model's parameters beyond the contemporaneous coefficients (Cushman and Zha, 1997). Restrictions on the lag structure, such as block recursions, offer an effective way to handle over-parameterization when the lag length is long (Zha, 1999). Classical or Bayesian econometric procedures can be used to test over-identifying restrictions. A review of theoretical results for Bayesian estimation and inference for both exactly identified and over-identified VARs is discussed below.

Impulse responses

Impulse responses are most commonly used in the VAR literature and are defined as $\partial \mathbf{y}_{t+s}/\partial \varepsilon'_t$ for $s \geq 0$. Let Φ_s be the $n \times n$ impulse response matrix at step s and the ith row of Φ_s be responses of the n endogenous variables to the ith one-standard-deviation structural shock. One can show that the impulse responses can be recursively updated as

$$\Phi_s = \Phi_{s-1}\mathbf{B}_1 + \cdots + \Phi_{s-p}\mathbf{B}_p \tag{3}$$

with the convention that $\Phi_0 = \mathbf{A}^{-1}$ and $\Phi_\upsilon = \underset{n\times n}{\mathbf{0}}$ for $\upsilon < 0$.

The concept of impulse response is economically appealing and is used in strands of literature other than VAR work. For example, impulse responses to technology shocks or monetary policy shocks in a DSGE have been often compared to those in a VAR model. In empirical monetary economics, impulse responses of various macroeconomic variables to policy shocks have been a focal point in the recent debate on the effectiveness of monetary policy. These shocks can be thought of as shifts (deviations) from the systematic part of monetary policy that are hard to predict from the viewpoint of the public.

It is sometimes argued that identified VARs are unreliable because certain conclusions are sensitive to the specific identifying assumptions. This argument is a sophism. All economic models, DSGE model and VARs alike, are founded on 'controversial' assumptions, and the results can be sensitive to these assumptions. What researchers should do is to select a class of models based on how well they fit to the data, analyse how reasonable the underlying assumptions are, and examine whether there are robust conclusions across models.

Christiano, Eichenbaum, and Evans (1999) and Rubio-Ramirez, Waggoner and Zha (2005) show some important robust results across different VAR models that have reasonable assumptions and fit to the data equally well. One prominent example is the robust conclusion that a large fraction of the variation in policy instruments, such as the short-term interest rate, can be attributed to the systematic response of policy to shocks originating from the private economy. Such a conclusion is expected of good monetary policy, but it also explains the subtle and difficult task of identifying monetary policy shocks separately from the other shocks affecting the economy.

Estimation and inference

Bayesian prior

When one estimates a VAR model for macroeconomic time series data, there is a trade-off between using short and long lags. A VAR with a short lag is prone to mis-specification, and a VAR with a long lag length is likely to suffer from the over-fitting problem. The Bayesian prior proposed by Sims and Zha (1998) is designed to eliminate the over-fitting problem without reducing the dimension of the model. It applies to not only reduced-form but also identified VARs.

To describe this prior simply, let $\mathbf{z}_t$ contain only a constant term and thus $\mathbf{D}$ is a $1 \times n$ vector of parameters. Rewrite the structural system (1) in the compact form of $\mathbf{y}_t'\mathbf{A} = \mathbf{x}_t'\mathbf{F} + \varepsilon_t'$, where

$$\underset{1\times k}{\mathbf{x}_t'} = [\mathbf{y}_{t-1}' \ \Lambda \ \mathbf{y}_{t-p}' \ \ \mathbf{z}_t'], \qquad \underset{n\times k}{\mathbf{F}'} = [\mathbf{A}' \ \Lambda \ \mathbf{A}_p' \ \ \mathbf{D}'],$$

and $k = np + h$. For $1 \le j \le n$, let $\mathbf{a}_i$ be the jth column of $\mathbf{A}$ and $\mathbf{f}_i$ be the jth column of $\mathbf{F}$. The first component of the prior is that $\mathbf{a}_j$ and $\mathbf{f}_i$ have Gaussian distribution

$$\mathbf{a}_j \sim N(\mathbf{0}, \mathbf{S}) \quad \text{and} \quad \mathbf{f}_j | \ \mathbf{a}_j \sim N(\mathbf{P}\mathbf{a}_j, \mathbf{H}), \tag{4}$$

where $\underset{n\times k}{\mathbf{P}'} = [\underset{n\times n}{\mathbf{I}} \ \ \underset{n\times n}{\mathbf{0}} \ \cdots \ \underset{n\times n}{\mathbf{0}} \ \ \underset{n\times 1}{\mathbf{0}}]$, which is consistent with the reduced-form random walk prior of Litterman (1986). The covariance matrices $\mathbf{S}$ and $\mathbf{H}$ are assumed to be diagonal matrices and are treated as hyperparameters. In principle, one could estimate these hyperparameters or integrate them out in a hierarchical framework. In practice, the values of these hyperparameters are specified before estimation. The ith diagonal element of $\mathbf{S}$ is λ_0/σ_i. The diagonal element of $\mathbf{H}$ that corresponds to the coefficient on lag l of variable i in equation j is $(\lambda_0\lambda_1\lambda_2^{\delta(i,j)})/(\sigma_i l^{\lambda_3})$, where $\delta(i,j)$ equals 1 if $i = j$ and 0 otherwise. The diagonal element of $\mathbf{H}$ corresponding to the constant term is the square of $\lambda_0\lambda_4$. The hyperparameter λ_0 controls the overall tightness of belief about the random walk feature, as well as tightness on the prior of $\mathbf{A}$ itself; λ_1 further controls the tightness of belief on random walk and the relative tightness on the prior of lagged coefficients; λ_2 controls the influence of variable i in equation j; λ_3 controls the rate at which the influence of lag decreases as its length increases; and λ_4 controls the relative tightness on the zero value of the constant term. The hyperparameters σ_i are scale factors to make the units uniform across variables, and are chosen at the sample standard deviations of residuals from univariate autoregressive models fitted to the individual time series in the sample (Litterman, 1986).

A VAR with many variables and a long lag is likely to produce relatively large coefficient estimates on distant lags and thus volatile sampling errors. The prior described here is designed to reduce the influence of distant lags and the unreasonable degree of explosiveness embedded in the system. It is essential for ensuring reasonable small-sample properties of the model, especially when there are relatively few degrees of freedom in a large VAR.

The aforementioned prior, however, does not take into account the features of unit roots and cointegration relationships embedded in many time series. For this reason, Sims and Zha (1998) add another component to their prior. This component uses Litterman's idea of dummy observations to express beliefs on unit roots and cointegration. Specifically, there are $n+1$ dummy observations added to the original system, which can be written as

$$\mathbf{Y}_d\mathbf{A} = \mathbf{X}_d\mathbf{F} + \mathbf{E}, \tag{5}$$

where $\mathbf{E}$ is a matrix of random shocks,

$$\underset{(n+1)\times n}{\mathbf{Y}_d} = \begin{bmatrix} \mu_5 y_1^0 & 0 & 0 \\ 0 & \ddots & 0 \\ 0 & 0 & \mu_5 y_n^0 \\ \mu_6 y_1^0 & \cdots & \mu_6 y_n^0 \end{bmatrix}, \qquad \underset{(n+1)\times 1}{\mathbf{c}_d} = \begin{bmatrix} 0 \\ \vdots \\ 0 \\ \mu_6 \end{bmatrix},$$

$$\underset{(n+1)\times(np+1)}{\mathbf{X}_d} = \begin{bmatrix} \mathbf{Y}_d & \cdots & \mathbf{Y}_d & \mathbf{c}_d \end{bmatrix},$$

and y_i^0 is the sample average of the p initial conditions for the ith variable of $\mathbf{y}_t$ and μ_5 and μ_6 are hyperparameters. The first $n+1$ dummy-observation equations in (5) express beliefs that all variables are stationary with means equal to y_i^0's or cointegration is present. The larger the values of μ_5 and μ_6, the stronger these beliefs. Since the values of λ's and μ's move in opposite directions to increase or loosen the tightness of the prior, the two symbols λ and μ are kept distinct. In applied work, the values of the hyperparameters for quarterly data are typically set to $\lambda_0 = 1$, $\lambda_1 = 0.2$, and $\lambda_2 = \lambda_3 = \lambda_4 = \mu_5 = \mu_6 = 1.0$. For monthly data, $\lambda_0 = 0.6$, $\lambda_1 = 0.1$, $\lambda_2 = 1.0$, $\lambda_4 = 0.1$, and $\mu_5 = \mu_6 = 5.0$, while the choice of the lag decay weight λ_3 is somewhat complicated and is elaborated in Robertson and Tallman (1999).

By taking into account the cointegration relationships among macroeconomic variables, this additional component of the prior helps improve out-of-sample forecasting, reduces the difference in forecasting accuracy between using the vintage and final data, and produces robust impulse responses to monetary policy shocks across VARs with different identification assumptions (Robertson and Tallman, 1999; 2001). Furthermore, Leeper, Sims and Zha (1996) demonstrate that with this prior it is feasible to estimate VAR models with as many as 18 variables – far more than the current DSGE models can handle. Because the prior proposed by Sims and Zha (1998) reflects *widely held* beliefs in the behaviour of macroeconomic time series, it has been often used as a base line prior in the Bayesian estimation and inference of VAR models.

Marginal data density

If a model is used as a candidate for the 'true' data-generating mechanism, it is imperative that the model's fit to the data is superior to those of alternative models. Recent developments in Bayesian econometrics have made it feasible to compare nested and non-nested models for their fits to the data (Geweke, 1999). With a proper Bayesian prior, one can numerically compute the marginal data density (MDD) defined as

$$\int_{\Theta} L(\mathbf{Y}_T|\phi)p(\phi)d\phi, \tag{6}$$

where ϕ is a collection of all the model's parameters, $\boldsymbol{\Theta}$ is the domain of ϕ, $\mathbf{Y}_T$ is all the data up to T, and $L(\mathbf{Y}_T|\phi)$ is the proper likelihood function. To determine the goodness of fit of a DSGE model, for example, one can compare its MDD with that of a VAR model (Smets and Wouters, 2003; Del Negro and Schorfheide, 2004).

As a VAR is often used as a benchmark for comparing different models, it is important that one compute its MDD efficiently and accurately. For an unrestricted reduced-form VAR as specified in (2), there is a standard closed-form expression for (6) so that no Markov chain Monte Carlo (MCMC) method is needed to obtain the MDD. For restricted (tightly parameterized) VARs implied by a growing number of economic applications, there is in general no closed-form solution to (6), and a numerical approximation to (6) is needed. Because of a high dimension in the VAR parameter space and possible simultaneity in an identified model, popular MCMC approaches such as importance sampling and modified harmonic mean methods require a long sequence of posterior draws to achieve numerical reliability in approximating (6), and thus are computationally very demanding.

Chib (1995) offers a procedure for accurate evaluations of the MDD that requires the existence of a Gibbs sampler by partitioning ϕ into a few blocks. One can sample alternately from the conditional posterior distribution of one block of parameters given other blocks. While sampling between blocks entails additional simulations, the Chib algorithm can be far more efficient than other methods because each conditional posterior probability density function (PDF) can be evaluated in closed form. The objects needed to complete this algorithm are the closed-form prior PDF and the conditional posterior PDF for each block.

Because the prior discussed so far includes the dummy observations component, there is a question as to whether this overall prior has a standard PDF. To answer this question, it can be shown from (4) and (5) that the overall prior PDF is

$$\mathbf{a}_j \sim N(\mathbf{0}, \mathbf{S}) \quad \text{and} \quad \mathbf{f}_j|\, \mathbf{a}_j \sim N(\mathbf{P}\mathbf{a}_j, \mathbf{H}), \tag{7}$$

where $\mathbf{S} = \mathbf{S}$ and $\mathbf{H} = (\mathbf{X}_d'\mathbf{X}_d + \mathbf{H}^{-1})^{-1}$. The result (7) follows from the two claims:

$$(\mathbf{X}_d'\mathbf{X}_d + \mathbf{H}^{-1})^{-1}(\mathbf{X}_d'\mathbf{Y} + \mathbf{H}^{-1}\mathbf{P}) = \mathbf{P};$$

$$\mathbf{Y}_d'\mathbf{Y}_d + \mathbf{P}'\mathbf{H}^{-1}\mathbf{P} = (\mathbf{Y}_d'\mathbf{X}_d + \mathbf{P}'\mathbf{H}^{-1})\mathbf{P}.$$

Given the prior (7), Waggoner and Zha (2003a) develop a Gibbs sampler for identified VARs with the linear restrictions studied in the VAR literature. These restrictions can be summarized as

$$\underset{n\times n}{\mathbf{Q}_j}\, \mathbf{a}_j = \underset{n\times 1}{\mathbf{0}}, \quad \underset{n\times k}{\mathbf{R}_j}\, \mathbf{f}_j = \underset{n\times 1}{\mathbf{0}}; \quad j = 1, \ldots, n. \tag{8}$$

If there are q_j restrictions on $\mathbf{a}_j$ and r_j restrictions on $\mathbf{f}_j$, the ranks of $\mathbf{Q}_j$ and $\mathbf{R}_j$ are q_j and r_j respectively. Let $\mathbf{U}_j$ ($\mathbf{R}_j$) be an $n \times q_j$ ($n \times r_j$) matrix whose columns form an orthonormal basis for the null space of $\mathbf{Q}_j$ ($\mathbf{R}_j$). The conditions in (8) are satisfied if and only if there exist a $q_j \times 1$ vector $\mathbf{b}_j$ and an $r_j \times 1$ vector $\mathbf{g}_j$ such that $\mathbf{a}_j = \mathbf{U}_j\mathbf{b}_j$ and $\mathbf{f}_j = \mathbf{V}_j\mathbf{g}_j$. The vectors $\mathbf{b}_j$ and $\mathbf{g}_j$ are the free parameters of $\mathbf{a}_j$ and $\mathbf{f}_j$ dictated by the conditions in (8). It follows from (7) that the prior distribution of $\mathbf{b}_j$ and $\mathbf{g}_j$ is jointly normal.

As for the conditional posterior PDFs, it can be shown that the posterior distribution of $\mathbf{g}_j$ conditional on $\mathbf{b}_j$ is normal and that the posterior distribution of $\mathbf{b}_j$ conditional on $\mathbf{b}_i$'s for $i \neq j$ has a closed-form PDF and can be simulated from it exactly. These results enable one to use the efficient method of Chib (1995). The MDD calculated this way is reliable and requires little computing time. For example, it takes less than one minute to obtain a very reliable estimate of the MDD for a large VAR with 13 lags and 10 variables. Such accuracy and speed make it feasible to compare a large number of identified VARs with different degrees of restriction.

Error bands

Because impulse responses are of central interest in interpreting dynamic multivariate models and helping guide the directions for new economic theory to be developed (Christiano, Eichenbaum and Evans, 2005), it is essential that measures of the statistical reliability of estimated impulse responses be presented as part of the process of evaluating models. The Bayesian methods reviewed so far in this essay make it feasible to construct the error bands around impulse responses. The error bands can contain any probability and are typically expressed in both .68 and .90 probability bands to characterize the shapes of the likelihood implied by the model.

The error bands of impulse responses reported in most VAR works are constructed as follows. One begins with the Gibbs sampler to draw $\mathbf{b}_j$ and $\mathbf{g}_j$ for $j = 1, \ldots, n$. For each posterior draw, the free parameters $\mathbf{b}_j$'s and $\mathbf{g}_j$'s are transformed to the original structural parameters $\mathbf{A}$, $\mathbf{A}_l$ ($l = 1, \ldots, p$), and $\mathbf{D}$; then the impulse responses are computed according to (3). The empirical distribution for each element of the impulse responses is formed and the equal-tail .68 and .90 probability intervals around each element are computed. The probability intervals have exact small-sample properties from a Bayesian point of view; and .90 or .95 probability intervals have been used in the empirical literature to approximate classical small-sample confidence intervals when the high dimensional parameter space and a large number of nuisance parameters make it difficult or impossible to obtain exact classical inferences.

One issue related to the error bands around impulse responses, whose importance is beginning to be recognized, is normalization. A normalization rule selects the sign of each draw of impulse responses from the posterior distribution. If there is no restriction imposed on the sign of each column of the contemporaneous coefficient matrix **A**, then the likelihood or the posterior function remains the same when the sign of a column of **A** is reversed. Without any sign restriction, the error bands for impulse responses would be symmetric around zero and thus the estimated responses would be determined to be imprecise.

The conventional normalization is to keep the diagonal of **A** always positive, based on the notion that a choice of normalization cannot have substantive effects on the results. But this notion is mistaken. If an identified VAR is non-recursive, normalization can generate ill-determined or unreasonably wide error bands around some impulse responses because some coefficients on the diagonal may be insignificantly different from zero.

Waggoner and Zha (2003b) show that normalized likelihoods can be different across normalization rules and that inappropriate normalization tends to produce a multi-modal likelihood. They propose a normalization rule designed to prevent the normalized likelihood from being spuriously multi-modal and thus avoid unreasonably wide error bands caused by the multi-modal likelihood. The algorithm for their normalization is straightforward to implement: for each posterior draw of $\mathbf{a}_j$, keep $\mathbf{a}_j$ if $\mathbf{e}_j'\mathbf{A}^{-1}\hat{\mathbf{a}}_j > 0$ and replace $\mathbf{a}_j$ with $-\mathbf{a}_j$ if $\mathbf{e}_j'\mathbf{A}^{-1}\hat{\mathbf{a}}_j < 0$, where $\mathbf{e}_j$ is the jth column of the $n \times n$ identity matrix. This algorithm works for not only short-run but also long-run restrictions (Evans and Marshall, 2002).

Another important issue related to error bands, not addressed until recently, is the characterization of the uncertainty around estimated impulse responses not only at one particular point but also around the shape of the responses as a whole. Let $\Phi_s(i, j)$ be the s-step impulse response of the jth variable to the ith structural shock. The associated error band is only pointwise. It is very unlikely in economic applications, however, that uncertainty about $\Phi_s(i, j)$ is independent across j or s. For example, the response of output to a policy shock is likely to be negatively correlated with the response of unemployment, and the response of inflation this period is likely to be positively correlated with the previous and next responses.

The procedure proposed by Sims and Zha (1999) takes into account these possible correlations across variables and across time. To use this procedure, one can simply stack all the relevant impulse responses into a column vector denoted by $\tilde{\mathbf{c}}$, where the tilde refers to a posterior draw. From a large number of posterior draws, the mean $\bar{\mathbf{c}}$ and covariance matrix $\mathbf{\Omega}$ of $\tilde{\mathbf{c}}$ are computed. For each posterior draw $\tilde{\mathbf{c}}$, the kth component $\tilde{\gamma}_k = (\tilde{\mathbf{c}} - \mathbf{c})'\mathbf{w}_k$ is calculated, where $\mathbf{w}_k$ is the eigenvector corresponding to the kth largest eigenvalue of $\mathbf{\Omega}$. From the empirical distribution of $\tilde{\gamma}_k$, one can tabulate different quantiles such as $\gamma_{k,.16}$ and $\gamma_{k,.84}$. Thus, the .68 probability error bands explained by the kth component of variation in the group of impulse responses can be computed as $\mathbf{c}_{.16} = \mathbf{c} + \gamma_{k,.16}\mathbf{w}_k$ and $\mathbf{c}_{.84} = \mathbf{c} + \gamma_{k,.84}\mathbf{w}_k$. For a particular economic application, if it turns out that only one to three eigenvalues dominate the covariance matrix of $\tilde{\mathbf{c}}$, these kinds of connecting-dots error bands can be useful in understanding the magnitudes and directions of uncertainty among a group of interrelated impulse responses. This method has proven to be particularly useful in economic applications that characterize the uncertainty around the entire paths, not just points one at a time (Cogley and Sargent, 2005; Nason and Rogers, 2006).

Markov-switching VARs

The class of VARs discussed thus far assumes that the parameters are constant over time. This assumption is made mainly for the technical constraint on estimation and inference, however. Many macroeconomic time series display patterns that seem impossible to capture by constant-parameter VARs. One prominent example is changes in volatility over time. In the VAR framework, volatility changes mean that the reduced-form covariance matrix Σ is not constant. In policy analysis, there is a serious debate on whether the coefficients in the policy rule have changed over time, or whether the variances of shocks in the private sector have changed over time, or both. Time-varying VARs are designed to answer these kinds of questions. Stock and Watson (2003) use the reduced-form VAR framework to show that fluctuations in US business cycles can be largely explained by changes in Σ. Sims and Zha (2006b) identify the behaviour of monetary policy from the rest of the VAR system and show that changes in the coefficients in monetary policy are, at most, modest and the variance changes in shocks originating from the private sector dominate aggregate fluctuations.

There have been a number of studies on time-varying VARs that allow the coefficients or the covariance matrix of residuals or both to change over time. These models typically let all the coefficients drift as a random walk or persistent process. To the extent that this kind of modelling tries to capture possible changes in the model's parameters, the model tends to over-fit because the dimension of time variation embedded in the data is much lower than the model's specification. Conceptually, there is a problem of distinguishing shocks to the residuals from shocks to the coefficients. The inability to distinguish among these shocks makes it difficult to interpret the effects of, say, monetary policy shocks.

The Markov-switching VAR introduced by Sims and Zha (2006a) is designed to overcome the over-fitting problems present in the other time-varying VARs and, at the same time, maintain clear interpretation of structural shocks. It builds on the Markov-switching model of Hamilton (1989), but emphasizes ways to restrict the degree of time variation allowed in the VAR. It has a capability to approximate parameter drifts arbitrarily well with the growing number of states, while restricting the transition matrix to be concentrated on the diagonal. This feature also allows discontinuous jumps from one state to another, which appears to matter for aggregate fluctuations.

To see how this method works, suppose that the parameter z_t drifts according to the process $z_t = \rho z_{t-1} + \nu_t$ where $\nu_t \sim N(0, \sigma^2)$. By discretizing this autoregressive process, one can let the probability of the transition from state j to i be proportional to

$$\Pr\left[z_t \in \left(\frac{\tau_i \sigma}{\sqrt{1-\rho^2}}, \frac{\tau_{i+1}\sigma}{\sqrt{1-\rho^2}}\right) \middle| z_{t-1} = \frac{\tau_j+\tau_{j+1}}{2}\frac{\rho\sigma}{\sqrt{1-\rho^2}}\right]$$
$$= \Psi\left(\frac{\tau_{i+1}}{\sqrt{1-\rho^2}} - \frac{\tau_j+\tau_{j+1}}{2}\frac{\rho}{\sqrt{1-\rho^2}}\right) - \Psi\left(\frac{\tau_i}{\sqrt{1-\rho^2}} - \frac{\tau_j+\tau_{j+1}}{2}\frac{\rho}{\sqrt{1-\rho^2}}\right),$$

where $\Psi(\ )$ is the standard normal cumulative probability function. The values of τ divide up the interval between -2 and 2 (two standard deviations). For nine states, for example, one has $\tau_1 = -2$, $\tau_2 = -1.5$, $\tau_3 = -1, \ldots, \tau_8 = 1.5$, and $\tau_9 = 2$. Careful restrictions on the degree of time variation, as well as on the constant parameters themselves, will put VARs a step closer to DSGE modelling. Recent work by Davig and Leeper (2005) shows an example of how to use a DSGE model to restrict a VAR on monetary and fiscal policy.

Conclusion

There is a tension between models that have clear economic interpretations but offer a poor fit to data and models that fit well but have few a priori assumptions and are therefore less interpretable (Ingram and Whiteman, 1994; Del Negro and Schorfheide, 2004). The original philosophy motivating VARs assumes that the economy is sufficiently complex and that simplified theoretical models, while useful in organizing thought about how the economy works, generally abstract from important aspects of the economy. VAR modelling begins with the minimal restrictions on dynamic time-series models, explores empirical regularities that have been ignored by simple models, and insists on the model's fit to data. The emphasis on fit has begun to bear fruit, as an increasing array of dynamic stochastic general equilibrium models have been tested and compared with VARs (Christiano, Eichenbaum and Evans, 2005; Smets and Wouters, 2003). Markov-switching VARs go a step further in bringing VARs even closer to the data and thus provide a new benchmark for model comparison.

At the same time, considerable progress has been made to narrow the gap between VARs and DSGE models. Some results from VARs have provided empirical support to the key assumption made by real business cycle (RBC) models that monetary policy shocks play insignificant roles in generating business fluctuations. Nason and Cogley (1994) and Cogley and Nason (1995) discuss similar results from both VAR and RBC approaches. Fernandez-Villaverde, Rubio-Ramirez and Sargent (2005) provide conditions and examples under which there exists the VAR representation of a DSGE model. Sims and Zha (2006a) display a close connection between an identified VAR and a DSGE model, and provide a measure for determining whether the 'invertibility problem' is a serious issue.

Undoubtedly there are payoffs in moving beyond the original VAR philosophy by imposing more restrictions on both contemporaneous relationships and lag structure while the restrictions are guided carefully by economic theory. Although moving in this direction is desirable, it is essential to maintain the spirit of VAR analysis as originally proposed by Sims (1980). This requires that heavily restricted VARs be subject to careful evaluation in terms of fit. Recent advances in Bayesian estimation and inference methods of restricted VARs make it feasible to compute the MDD accurately and efficiently and, therefore, to determine whether the restrictions have compromised the fit. These methods, however, still fall short of handling VARs with cross-equation restrictions implied by DSGE models. Thus, the challenge ahead of us is to develop new tools for VARs with possible cross-equation restrictions.

TAO ZHA

See also **Bayesian methods in macroeconometrics; structural vector autoregressions.**

Bibliography

Bernanke, B. 1986. Alternative exploration of the money-income correlation. *Carnegie-Rochester Conference Series on Public Policy* 25, 49–99.

Blanchard, O. and Quah, D. 1993. The dynamic effects of aggregate demand and supply disturbances. *American Economic Review* 83, 655–73.

Blanchard, O. and Watson, M. 1986. Are business cycles all alike? In *The American Business Cycle: Continuity and Change*, ed. R. Gordon. Chicago: University of Chicago Press.

Chib, S. 1995. Marginal likelihood from the Gibbs output. *Journal of the American Statistical Association* 90, 1313–21.

Christiano, L., Eichenbaum, M. and Evans, C. 1999. Monetary policy shocks: what have we learned and to what end? In *Handbook of Macroeconomics*, vol. 1A, ed. J. Taylor and M. Woodford. Amsterdam: North-Holland.

Christiano, L., Eichenbaum, M. and Evans, C. 2005. Nominal rigidities and the dynamics effects of a shock to monetary policy. *Journal of Political Economy* 113, 1–45.

Cogley, T. and Nason, J. 1995. Output dynamics in real business cycle models. *American Economic Review* 85, 492–511.

Cogley, T. and Sargent, T. 2005. Drifts and volatilities: monetary policies and outcomes in the post WWII U.S. *Review of Economic Dynamics* 8, 262–302.

Cushman, D. and Zha, T. 1997. Identifying monetary policy in a small open economy under flexible exchange rates. *Journal of Monetary Economics* 39, 433–48.

Davig, T. and Leeper, E. 2005. Fluctuating macro policies and the fiscal theory. Working Paper No. 11212. Cambridge, MA: NBER.

Del Negro, M. and Schorfheide, F. 2004. Priors from general equilibrium models for VARs. *International Economic Review* 45, 643–73.

Evans, C. and Marshall, D. 2002. Economic determinants of the nominal treasury yield curve. Working paper. Federal Reserve Bank of Chicago.

Faust, J. and Leeper, E. 1997. When do long-run identifying restrictions give reliable results? *Journal of Business and Economic Statistics* 15, 345–53.

Fernandez-Villaverde, J., Rubio-Ramirez, J. and Sargent, T. 2005. A, B, C's (and D's) for understanding VARs. Working Paper No. 2005-9. Federal Reserve Bank of Atlanta.

Gali, J. 1992. How well does the IS-LM model fit postwar U.S. data? *Quarterly Journal of Economics* 107, 709–38.

Geweke, J. 1999. Using simulation methods for Bayesian econometric models: inference, development, and communication. *Econometric Reviews* 18, 1–73.

Geweke, J. and Whiteman, C. 2006. Bayesian forecasting. In *The Handbook of Economic Forecasting*, ed. G. Elliott, C. Granger and A. Timmermann. Amsterdam: North-Holland.

Gordon, D. and Leeper, E. 1994. The dynamic impacts of monetary policy: an exercise in tentative identification. *Journal of Political Economy* 102, 1228–47.

Hamilton, J. 1989. A new approach to the economic analysis of nonstationary time series and the business cycle. *Econometrica* 57, 357–84.

Ingram, B. and Whiteman, C. 1994. Supplanting the Minnesota prior: forecasting macroeconomic time series using real business cycle model priors. *Journal of Monetary Economics* 34, 497–510.

Leeper, E., Sims, C. and Zha, T. 1996. What does monetary policy do? *Brookings Papers on Economic Activity* 2, 1–78.

Leeper, E. and Zha, T. 2003. Modest policy interventions. *Journal of Monetary Economics* 50, 1673–700.

Litterman, R. 1986. Forecasting with Bayesian vector autoregressions – five years of experience. *Journal of Business and Economic Statistics* 4, 25–38.

Nason, J. and Cogley, T. 1994. Testing the implications of long-run neutrality for monetary business cycle models. *Journal of Applied Econometrics* 9, S37–S70.

Nason, J. and Rogers, J. 2006. The present-value model of the current account has been rejected: round up the usual suspects. *Journal of International Economics* 68, 159–87.

Robertson, J. and Tallman, E. 1999. Vector autoregressions: forecasting and reality. *Federal Reserve Bank of Atlanta Economic Review* 84(1), 4–18.

Robertson, J. and Tallman, E. 2001. Improving federal-funds rate forecasts in VAR models used for policy analysis. *Journal of Business and Economic Statistics* 19, 324–30.

Rubio-Ramirez, J., Waggoner, D. and Zha, T. 2005. Markov-switching structural vector autoregressions: theory and applications. Working Paper No. 2005–27. Federal Reserve Bank of Atlanta.

Sims, C. 1980. Macroeconomics and reality. *Econometrica* 48, 1–47.

Sims, C. 1982. Policy analysis with econometric models. *Brookings Papers on Economic Activity* 1, 107–52.

Sims, C. 1986. Are forecasting models usable for policy analysis. *Federal Reserve Bank of Minneapolis Quarterly Review* 10(1), 2–16.

Sims, C. and Zha, T. 1998. Bayesian methods for dynamic multivariate models. *International Economic Review* 39, 949–68.

Sims, C. and Zha, T. 1999. Error bands for impulse responses. *Econometrica* 67, 1113–55.

Sims, C. and Zha, T. 2006a. Does monetary policy generate recessions? *Macroeconomic Dynamics* 10(2) 231–272.

Sims, C. and Zha, T. 2006b. Were there regime switches in US monetary policy? *American Economic Review* 96, 54–81.

Smets, F. and Wouters, R. 2003. An estimated dynamic stochastic general equilibrium model of the euro area. *Journal of the European Economic Association* 1, 1123–75.

Stock, J. and Watson, M. 2003. Has the business cycle changed? Evidence and explanations. Prepared for the Federal Reserve Bank of Kansas City Symposium 'Monetary Policy and Uncertainty: Adapting to a Changing Economy', Jackson Hole, Wyoming, 28–30 August.

Waggoner, D. and Zha, T. 2003a. Likelihood-preserving normalization in multiple equation models. *Journal of Econometrics* 114, 329–47.

Waggoner, D. and Zha, T. 2003b. A Gibbs simulator for structural vector autoregressions. *Journal of Economic Dynamics & Control* 28, 349–66.

Zha, T. 1999. Block recursion and structural vector autoregressions. *Journal of Econometrics* 90, 291–316.

wavelets

Wavelets provide a powerful tool of analysis for economics and finance, as well as for scientists in a wide variety of fields, such as signal processing, medical imaging, data compression and geology. One interpretation of wavelets is that they are a collection of functions that provide a basis for the representation of test functions that may be complicated with localized shocks, have abrupt changes in functional form, or are signals with time varying frequency and amplitude.

Another interpretation is that wavelets are a generalization of Fourier analysis in which stationarity of the time series is no longer critical and localization of a signal can be achieved. In this light, to borrow a great insight by Strang (see Strang and Nguyen, 1996), Fourier analysis is best at representing functions that are composed of linear combinations of stationary inputs, but wavelets are like musical notation in that each note is characterized by its frequency, its position in time, and its duration.

A further interpretation of wavelets is that of a filter bank, so that different classes of wavelet functions are generated by prescribing different banks of filters. Filter banks can achieve results that are not possible with a single filter (Strang and Nguyen, 1996). Yet another interpretation is that of a decomposition of a signal in terms of different time scales, an interpretation that is at the heart of much economic analysis as represented by the long-standing notions of the 'short, medium and long runs' and is fundamental to the concept of the 'business cycle'.

At this time there is a vast literature on wavelets in mathematics, statistics, and various branches of engineering, but relatively little in economics, although that situation is changing fast. An introduction to the economic literature that is highly recommended is Gencay, Selcut and Whitcher (2002). This is the most comprehensive and detailed coverage with numerous descriptions of economic applications and discussions of the statistical properties of the wavelet estimators. Bruce and Gao (1996) discuss the properties of wavelets and give instructions for calculating wavelets using S-Plus; Chui (1992), Percival and Walden (2000), and Strang and Nguyen (1996) develop the mathematics at a moderate level of difficulty and discuss the statistical properties of wavelets. Silverman and Vassilicos (1999) provide interesting examples of the applications of wavelets and further discussions of the statistical properties (see also Ramsey, 1999a). Two lower-level introductions for economists are Crowley (2005) and Schleicher (2002).

Two informative examples

Let **X** represent an N dimensional vector of observations on a time series or a function $f(\cdot)$ evaluated at the discrete points $t = 1, 2, \ldots, \mathrm{N}$. One may consider an orthonormal transformation from **X** to an N dimensional vector **W**, with elements, W_n, $n = 0, 1, \ldots, \mathrm{N}-1$, where **W** is generated from an $\mathrm{N} \times \mathrm{N}$ dimensional orthonormal matrix $\mathcal{W}$;

$\mathbf{W} = W\mathbf{X}$ and $I_N = W^T W$. Let $N = 2^J$ for some integer J. This assumption, while inessential, is analytically convenient and is useful for defining an efficient algorithm for evaluating the wavelet coefficients. The N^2 elements of the transformation matrix W are the filter elements to be defined below and the N elements of the vector **W** are the wavelet coefficients that are given by the inner product $W\mathbf{X}$.

The wavelet defined here is known as the discrete wavelet transform (DWT). By choosing an orthonormal transform it is immediate that the modulus of **W** is the same as that of **X**; that is, $\|\mathbf{X}^2\| = \|\mathbf{W}^2\|$, where $\|X^2\|$ is given by $\mathbf{X}^T\mathbf{X} = \sum x_i^2$, and where w_n^2 is the energy contributed by the nth wavelet coefficient.

The power of wavelet analysis lies in the choices made for the components of W. Two examples illustrate the choices that can be made and indicate the scope that a wavelet analysis offers. (I am indebted to Percival and Walden, 2000, for the examples to follow.) The first example is the oldest and simplest function used to generate a wavelet transformation. For the sake of simplicity of exposition, choose $N = 2^J = 2^4 = 16$. The notation, $W_{i,*}^T$ indicates the ith row of the matrix W, $i = 0, 1, \ldots, N-1$. The function used to generate the elements of W for the DWT is the Haar function.

$$\begin{aligned}
W_{0,*}^T &= \left\{-\frac{1}{\sqrt{2}}, \frac{1}{\sqrt{2}}, 0_1, \ldots, 0_{14}\right\}, \\
W_{8,*}^T &= \left\{-\frac{1}{2}, -\frac{1}{2}, \frac{1}{2}, \frac{1}{2}, 0_1, \ldots, 0_{12}\right\} \\
W_{12,*}^T &= \left\{-\frac{1}{\sqrt{8}_1}, \ldots, -\frac{1}{\sqrt{8}_4}, \frac{1}{\sqrt{8}_1}, \ldots, \frac{1}{\sqrt{8}_4}, 0_1, \ldots, 0_8\right\} \\
W_{14,*}^T &= \left\{-\frac{1}{4_1}, \ldots, -\frac{1}{4_8}, \frac{1}{4_1}, \ldots, \frac{1}{4_8}\right\}, \quad W_{15,*}^T = \left\{\frac{1}{4_1}, \ldots \frac{1}{4_{16}}\right\}
\end{aligned} \tag{1}$$

The unshifted filters are $W_{0,*}^T, W_{8,*}^T, W_{12,*}^T, W_{14,*}^T, W_{15,*}^T$. The remaining rows of W, are circularly shifted versions of the filter terms defined above. We have:

$$\begin{aligned}
W_{1,*}^T &= T^2 W_{0,*}^T = \left\{0, 0, -\frac{1}{\sqrt{2}}, \frac{1}{\sqrt{2}}, 0_1, \ldots, 0_{12}\right\} \\
W_{2,*}^T &= T^4 W_{0,*}^T = \left\{0, 0, 0, 0, -\frac{1}{\sqrt{2}}, \frac{1}{\sqrt{2}}, 0_1, \ldots, 0_{10}\right\} \\
W_{9,*}^T &= T^4 W_{8,*}^T = \left\{0, 0, 0, 0, -\frac{1}{2}, -\frac{1}{2}, \frac{1}{2}, \frac{1}{2}, 0_1, \ldots, 0_8\right\} \\
W_{13,*}^T &= T^8 W_{12,*}^T = \left\{0_1, \ldots, 0_8, -\frac{1}{\sqrt{8}_1}, \ldots, -\frac{1}{\sqrt{8}_4}, \frac{1}{\sqrt{8}_1}, \ldots, \frac{1}{\sqrt{8}_4}\right\}
\end{aligned} \tag{2}$$

The filter $W_{15,*}^T$ yields the 'scaling coefficient', and the remaining filters, $W_{i,*}^T$, yield the wavelet coefficients. From the above equations one sees that the Haar filters involve first differences between scale averages at non-overlapping intervals. At the lowest scale, 2^{j-1} for $j = 1$, the first differences are between adjacent observations, at the next scale level, 2^{j-1} for $j=2$, the differences are between adjacent pairs of observations, at the scale 2^{j-1} for $j=3$ the differences are between groups of four terms, and so on. At the highest scale, $J=4$, we have two filters of the data, a first difference between the first and last eight of the observations, and an average over the full set of observations. The latter is a 'father' wavelet transform and the remaining rows of W are the 'mother' wavelet transforms.

In order to gain further insight, consider a Daubchies wavelet designated as D(4) (see Daubchies, 1992), which is a member of a sequence of discrete wavelet filters in which the Haar is D(2). Define $Y_t \equiv aX_t + bX_{t-1}$ and form the backward second discrete difference, $Y_t^{(2)}$ by:

$$Y_t^{(2)} \equiv Y_t^{(1)} - Y_{t-1}^{(1)} = Y_t - 2Y_{t-1} + Y_{t-2}. \tag{3}$$

For a particular choice of 'a' and 'b', the nth D(4) wavelet can be written as:

$$\begin{aligned} W_n &= Y_{2n+1} - 2Y_{2n} + Y_{2n-1} \\ &= aX_{2n+1} + (b-2a)X_{2n} + (a-2b)X_{2n-1} + bX_{2n-2} \\ &= h_0 X_{2n+1} + h_1 X_{2n} + h_2 X_{2n-1} + h_3 X_{2n-2}. \end{aligned} \tag{4}$$

Imposing orthonormality {and some other conditions to ensure uniqueness}, enables one to derive the values for h_i, $i=0, 1, 2, 3$, *namely:*

$$h_0 = \tfrac{1-\sqrt{3}}{4\sqrt{2}}, \quad h_1 = \tfrac{-3+\sqrt{3}}{4\sqrt{2}}, \quad h_2 = \tfrac{3+\sqrt{3}}{4\sqrt{2}}, \quad h_3 = \tfrac{-1-\sqrt{3}}{4\sqrt{2}}.$$

Repeating the exercise above with $N=2^J=16$ observations, we have:

$$W_{0,*}^T = \{h_1,\ h_0,\ 0_1, \ldots, 0_{12},\ h_3,\ h_2\}, \quad W_{1,*}^T = \{h_3,\ h_2,\ h_1,\ h_0,\ 0_1, \ldots, 0_{12},\}$$

where $W_{1,*}^T = \mathrm{T}^2 W_{0,*}^T$, orthonormality requires that $\|W_{0,*}\|^2 = h_0^2 + h_1^2 + h_2^2 + h_3^2 = 1$ and $\langle W_{0,*}^T W_{1,*} \rangle = h_0 h_2 + h_1 h_3 = 0$. $W_{15,*}^T$ applied to the time series yields the mean.

A wavelet filter of length L must satisfy at a minimum for all nonzero integers n the following conditions.

$$\sum_{l=0}^{L-1} h_l = 0, \quad \sum_{l=0}^{L-1} h_l^2 = 1 \tag{5}$$

$$\sum_{l=0}^{L-1} h_l h_{l+2n} = \sum_{l=-\infty}^{\infty} h_l h_{l+2n} = 0.$$

For the Haar wavelet, D(2), and the Daubchies D(4) wavelet L is respectively 2 and 4.

Wavelets and multiresolution analysis (MRA)

One can approach the definition of wavelets from a related perspective that indicates the similarities to and differences from Fourier transforms. In both cases, one is considering a projection of a signal onto a set of basis functions for the space containing the test function. In the case of the Fourier transform the basis functions are rescalings of the fundamental frequency, for example, $\{e^{-in\,\omega_0}\}$ where ω_0 is the fundamental frequency and n provides the scaling. In this expression there is no resolution in the time domain, and the signal is assumed to be stationary. In contrast, the wavelet-generating functions are defined over very general spaces and each function is compact. One has the recursive relationship:

$$g_{s,k}(t) = \frac{1}{\sqrt{s}} g\left(\frac{t-k}{s}\right) \tag{6}$$

where s indicates the scale of the function, k is the location index; the term $\sqrt{s}$ ensures that the norm of $g(\cdot)$ is 1. The projection of the signal onto the scalable function $g(\cdot)$ depends on two parameters; s, which defines the time scale and implicitly designates a relevant range of frequencies, and k, which indicates the centre of location of the projection. The compactness of $g(\cdot)$ together with the time index k implies that the analysis of a time series is essentially local at each scale, whereas the Fourier analysis is essentially global. There is considerable latitude in the choice of the function $g(\cdot)$, or in the choice of the filters that generate the functions $g(\cdot)$. Desirable criteria include symmetry, smoothness and orthogonality. Whether one begins by specifying the properties of a basis function $g(\cdot)$ or one begins by specifying the properties of a filter $\{h_i\}$, the process generates two related classes of wavelet transforms, the 'father' wavelet that yields the 'scaling' coefficients and the 'mother' wavelets that yield the detail coefficients.

One can link the filter coefficients to the definition of the father and mother wavelets and link the father and mother transforms themselves by noting that, for a given sequence of low pass filter coefficients, $l(k)$, and the corresponding high pass filter coefficients, $h(k)$, one solves for $\Phi(t)$, father, and $\psi(t)$, mother, from:

$$\Phi(t) = \sqrt{2}\sum_{k=0}^{N} l(k)\Phi(2t-k) \tag{7}$$

$$\Psi(t) = \sqrt{2}\sum_{k=0}^{N} h(k)\Phi(2t-k).$$

For the Haar example above, the filter coefficients are: $l(k) = \{\frac{1}{\sqrt{2}}, \frac{1}{\sqrt{2}}\}$ and $h(k) = \{-\frac{1}{\sqrt{2}}, \frac{1}{\sqrt{2}}\}$.

From these equations, one derives the scaling, or 'smooth' coefficients and the wavelet or 'detail' coefficients of the function $f(\cdot)$ by the integrals,

$$s_{J,k} = \int f(t)\Phi_{J,\,k}dt \tag{8}$$

$$d_{j,k} = \int f(t)\Psi_{j,\,k}dt, \quad j = 1, 2, \ldots, J \tag{9}$$

where $\Phi_{J,k}$ and $\psi_{J,k}$ are the scaled and translated versions of Φ and ψ defined in equation (7). The function $f(t)$ can be synthesized by the equations:

$$f(t) = \sum_k s_{J,k}\Phi_{J,k}(t) + \sum_k d_{J,k}\Psi_{J,k}(t) + \cdots \sum_k d_{j,k}\Psi_{j,k}(t) \cdots + \sum_k d_{1,k}\Psi_{1,k}(t)$$

$$f(t) = S_J + D_J + D_{J-1} + \cdots D_j + \cdots D_1$$

$$S_J = \sum_k s_{J,k}\Phi_{J,k}, \quad D_j = \sum_k d_{j,k}\Psi_{j,k},$$

$$j = 1, 2, \cdots, J \quad f(t) \approx S_{j-1} = S_j + D_j. \tag{10}$$

An easy way to visualize the above scale and locational decomposition of the signal is as a series of maps of ever greater detail as elements of D_j are added; S_J provides a smooth outline, D_J adds the highest scale detail; and the D_j add ever more detail as j decreases. One can approximate the function $f(t)$ by truncating the expansion at some j, $1<j<J$. This is known as a multiresolution analysis (MRA), which can yield enormous data compression by representing the function $f(t)$ with relatively few coefficients.

There are many choices for the basis function, or alternatively choices for the filter banks that provide a great deal of richness for the wavelet approach. Two generalizations of note are wavelet packets and an exploratory technique known as waveform dictionaries. The former analyses signals by basis functions that differ by location and scale as for wavelets, but also by an oscillation index; wavelet packets are most useful in representing time series that have short term, localized oscillations (Bruce and Gao, 1996). Waveform dictionaries (Mallat and Zhang, 1993) provide a modification to wavelet analysis. The basic function providing the basis is a function $g(.)$ defined by:

$$g_\gamma(t) = \frac{1}{\sqrt{s}} g\left(\frac{t-u}{s}\right) e^{i\omega t}, \quad \gamma = (s, u, \omega). \tag{11}$$

The function $g_\gamma(t)$ has norm one, has scale s, and the time scale energy is centred at u and proportional to s. The Fourier transform of $g_\gamma(t)$ has its frequency energy centred at ω and is proportional to $1/s$. The dictionary of functions $g_\gamma(t)$ illustrate a very important principle of these transforms. Improved resolution in the time domain reduces resolution in the frequency domain and vice versa; this is a version of the Heisenberg uncertainty principle.

Applications in economics and finance

All analytical procedures can be assessed on the basis of their contribution in four categories: provide estimators in novel situations; improve efficiency or reduce bias;

enhance robustness to modelling errors; or provide new insights into the data-generating processes. Wavelets have provided benefits in all these categories.

One advantage of the waveform dictionary approach indicated in eq. (11) is that the researcher need not prejudge the presence of frequency components as well as the occurrence of short-term shocks. The process is exploratory and projection pursuit methods can be utilized to isolate local and global characteristics. Waveform dictionaries have been used as an exploratory tool in the analysis of financial and foreign exchange data (Ramsey and Zhang, 1996; 1997). In the analysis of daily stock-price data and tic-by-tic exchange rate data, there was no strong structural evidence for any frequency, but there was weak evidence for frequencies that appeared and disappeared or that waxed and waned in strength. Most of the power was summarized in terms of time-localized bursts of activity. The results in both papers indicated that, while for any given time period surprisingly few wavelet coefficients were needed to fit the data, the relevant coefficients varied randomly from period to period. Each burst was characterized by a rapid increase in amplitude and fast oscillation in frequency; in short, market adjustment processes seem to be characterized by a rapid increase in oscillation amplitude and frequency followed by a decay in frequency and amplitude; adjustment is neither smooth nor fast.

For a deep analysis of the scaling properties of volatilities and the relationship between risk and time scales, see Gencay, Selcut and Whitcher (2001; 2003). Another example in these references is the estimation of time varying Betas in the capital asset pricing model (CAPM). The analysis indicated that in the cases examined beta coefficients varied substantially over time, thereby modifying the structure of optimal investment strategies.

Wavelets have been instrumental in improving the robustness and efficiency of estimation in numerous examples (see Jensen, 1999; 2000) for efficiency gains and enhanced robustness of estimates for the fractional differencing parameter in long memory processes (see also Gencay, Selcut, and Whitcher, 2002). This reference is also useful for examples of estimation of covariance matrices and providing confidence intervals.

Wavelets have been successfully employed in situations not amenable to standard approaches – for example, forming estimators in testing for serial correlation of unknown form in panel models. As Hong and Kao (2004) state in their abstract: 'This paper proposes a new class of generally applicable wavelet-based tests for serial correlation of unknown form in the estimated residuals of a panel regression model, where error components can be one-way or two-way, individual and time effects can be fixed or random, and regressors may contain lagged dependent variables or deterministic/stochastic trending variables.'

Ramsey and Lampart (1998a; 1998b) discovered that the relationship between economic variables – for example, between money and income, or between consumption and income – can be decomposed into relationships at separate scales with interesting implications for economic theory. The insights gained include reconciling a number of anomalies in the consumer and monetary theory literature as well as introducing the notion that observed time series are an aggregation of behaviour at different time scales; for example, the time path of consumption totals represents the actions by consumers operating on a variety of time scales. In the same papers the authors also discovered that at certain time scales the relationship between economic variables may be subject to variations in the delay. These results have been confirmed by other researchers (for example, Gencay, Selcut, and Whitcher, 2002). For an alternative approach to testing for causality in the frequency domain using wavelets, see Dalkir (2004).

Yet another insight provided by wavelet analysis is the distinction between 'smoothing' and denoising. The former, traditional in econometrics, is based on the assumption that the signal is smooth relative to the noise, whereas the latter allows for the signal to be as irregular as the noise, but with greater amplitude. For smooth signals subject to noise, the obvious approach in order to minimize the effect of the noise is to average in some manner. However, if the signal is not smooth, averaging is not a suitable approach in that the averaging process distorts the signal itself. One can claim that denoising is often more relevant to economic and financial analysis than is smoothing (see Ramsey, 2004). These remarks are particularly relevant in the context of estimating relationships involving regime shifts, threshold models, and other non-differential changes in variable values. In an important series of papers Donoho, Johnstone and coauthors explored the use of wavelets and the concept of shrinkage whereby the size of the wavelet coefficient estimates is reduced to allow for the presence of noise (see Donoho et al., 1995). Further, shrinkage can be applied differentially across scales thereby refining the technique (see Ramsey, 2004, for more recent references and Gencay, Selcut, and Whitcher, 2002, for a thorough development of wavelet denoising).

Forecasting is an important topic; see Fryzlewicz, van Bellegem and von Sachs (2002) and Li and Hinich (2002), who demonstrate how the wavelet approach disentangles the variation in forecastability over time scales; that is, the ability to forecast varies across time scales. At the simplest level a given time series can be decomposed into trend, business cycle and seasonal components by wavelets and individually structured forecasting methods applied to each component separately before synthesizing the entire signal in order to produce forecasts for the whole series (see Ramsey, 2004, for a brief review of the literature on forecasting using wavelets).

JAMES B. RAMSEY

See also **seasonal adjustment; spectral analysis; time series analysis.**

Bibliography

Brillinger, D.R. 1994. Some uses of cumulants in wavelet analysis. *Journal of Nonparametric Statistics* 6, 93–114.

Bruce, A. and Gao, H. 1996. *Applied Wavelet Analysis with S-Plus*. New York: Springer.

Chui, C.K. 1992. *An Introduction to Wavelets*. New York: Academic Press.

Crowley, P. 2005. An intuitive guide to wavelets for economists. Discussion Paper No. 1/2005, Research Department, Bank of Finland.

Dalkir, M. 2004. A new approach to causality in the frequency domain. *Economics Bulletin* 3(44), 1–14.

Daubchies, I. 1992. *Ten Lectures on Wavelets*. Philadelphia: SIAM.

Donoho, D.L., Johnstone, I.M., Keryachariam, G. and Picard, D. 1995. Wavelet shrinkage: asymptopia? *Journal of the Royal Statistical Society* 57, 301–69.

Fryzlewicz, P., van Bellegem, S. and von Sachs, R. 2002. Forecasting non-stationary time series by wavelet process modeling. Technical Report No. 0208, Institut de Statistique, Université Catholique de Louvain.

Gencay, R., Selcut, F. and Whitcher, B. 2001. Scaling properties of foreign exchange volatility. *Physica A* 289, 249–66.

Gencay, R., Selcut, F. and Whitcher, B. 2002. *An Introduction to Wavelets and Other Filtering Methods in Finance and Economics*. New York: Academic Press.

Gencay, R., Selcut, F. and Whitcher, B. 2003. Systematic risk and time scales. *Quantitative Finance* 3, 108–16.

Hong, Y. and Kao, C. 2004. Wavelet testing for serial correlation of unknown form in panel models. *Econometrica* 72, 1519–63.

Jensen, M.J. 1999. Using wavelets to obtain a consistent ordinary least squares estimator of the long-memory parameter. *Journal of Forecasting* 18, 17–32.

Jensen, M.J. 2000. An alternative maximum likelihood estimator of long-memory processes using compactly supported wavelets. *Journal of Economic Dynamics and Control* 24, 361–87.

Li, T.H. and Hinich, M. 2002. A filter bank approach for modeling and forecasting seasonal patterns. *Technometrics* 44(1), 1–14.

Mallat, S. and Zhang, Z. 1993. Matching pursuit with time frequency dictionaries. *IEEE Transactions on Signal Processing* 41, 3397–415.

Percival, D.B. and Walden, A.T. 2000. *Wavelet Methods for Time Series Analysis*. Cambridge: Cambridge University Press.

Ramsey, J.B. 1999a. The contribution of wavelets to the analysis of economic and financial data. In *Philosophical Transactions of the Royal Society, Series A: Mathematical and Physical Sciences*. Repr. in *Wavelets*, ed. B.W. Silverman and J.C. Vassilicos. Oxford: Oxford University Press.

Ramsey, J.B. 1999b. Regression over time scale decompositions: a sampling analysis of distributional properties. *Economic Systems Research* 11(2), 163–83.

Ramsey, J.B. 2004. Wavelets in economics and finance: past and future. *Studies in Nonlinear Dynamics and Econometrics* 6(3), Article 1.

Ramsey, J.B. and Lampart, C. 1998a. The decomposition of economic relationships by time scale using wavelets: money and income. *Macroeconomic Dynamics* 2, 49–71.

Ramsey, J.B. and Lampart, C. 1998b. The decomposition of economic relationships by time scale using wavelets: expenditure and income. *Studies in Nonlinear Dynamics and Econometrics* 3(4), 23–42.

Ramsey, J.B. and Zhang, Z. 1996. The application of waveform dictionaries to stock market index data. In *Predictability of Complex Dynamical Systems*, ed. Y.A. Kravtsov and J. Kadtke. New York: Springer.

Ramsey, J.B. and Zhang, Z. 1997. The analysis of foreign exchange rates using waveform dictionaries. *Journal of Empirical Finance* 4, 341–72.

Schleicher, C. 2002. An introduction to wavelets for economists. Working Paper No. 2002-3, Bank of Canada.

Silverman, B.W. and Vassilicos, J.C. 1999. *Wavelets: The Key to Intermittent Information*? Oxford: Oxford University Press.

Strang, G. and Nguyen, T. 1996. *Wavelets and Filter Banks*. Wellesley, MA: Wellesley-Cambridge Press.

Index